Inorganic and Organometallic Polymers

ACS SYMPOSIUM SERIES 360

Inorganic and Organometallic Polymers

Macromolecules Containing Silicon, Phosphorus, and Other Inorganic Elements

Martel Zeldin, EDITOR
Indiana University-Purdue University at Indianapolis

Kenneth J. Wynne, EDITOR
Office of Naval Research

Harry R. Allcock, EDITOR
The Pennsylvania State University

Developed from a symposium sponsored
by the Divisions of Inorganic Chemistry,
of Polymer Chemistry, Inc., and
of Polymeric Materials: Science and Engineering
at the 193rd Meeting
of the American Chemical Society,
Denver, Colorado,
April 5-10, 1987

American Chemical Society, Washington, DC 1988

Library of Congress Cataloging-in-Publication Data

Inorganic and organometallic polymers/Martel Zeldin, editor, Kenneth J. Wynne, editor, Harry R. Allcock.

p. cm.—(ACS symposium series; 360)

"Developed from a symposium sponsored by the Divisions of Inorganic Chemistry, of Polymer Chemistry, Inc., and of Polymeric Materials: Science and Engineering at the 193rd Meeting of the American Chemical Society, Denver, Colorado, April 5–10, 1987."

Includes bibliographies and indexes.

ISBN 0-8412-1442-5

1. Inorganic polymers—Congresses.
2. Organometallic polymers—Congresses.

I. Zeldin, Martel, 1937- . II. Wynne, Kenneth J., 1940- . III. Allcock, H. R. IV. American Chemical Society. Division of Inorganic Chemistry. V. American Chemical Society. Division of Polymer Chemistry. VI. American Chemical Society. Division of Polymeric Materials: Science and Engineering. VII. Series.

QD196.I535 1988
546—dc19 87-30630
CIP

PRINTED IN THE UNITED STATES OF AMERICA

Second printing 1989

ACS Symposium Series

M. Joan Comstock, *Series Editor*

Foreword

The ACS SYMPOSIUM SERIES was founded in 1974 to provide a medium for publishing symposia quickly in book form. The format of the Series parallels that of the continuing ADVANCES IN CHEMISTRY SERIES except that, in order to save time, the papers are not typeset but are reproduced as they are submitted by the authors in camera-ready form. Papers are reviewed under the supervision of the Editors with the assistance of the Series Advisory Board and are selected to maintain the integrity of the symposia; however, verbatim reproductions of previously published papers are not accepted. Both reviews and reports of research are acceptable, because symposia may embrace both types of presentation.

Contents

Preface .... **xi**

1. **An Introduction to Inorganic and Organometallic Polymers** **1**
   Kenneth J. Wynne

POLYSILANES AND POLYCARBOSILANES

2. **Polysilane High Polymers: An Overview** **6**
   Robert West and Jim Maxka

3. **Polycarbosilanes: An Overview** **21**
   Dietmar Seyferth

4. **Soluble Polysilane Derivatives: Chemistry and Spectroscopy** **43**
   R. D. Miller, J. F. Rabolt, R. Sooriyakumaran, W. Fleming, G. N. Fickes, B. L. Farmer, and H. Kuzmany

5. **Poly(di-*n*-hexylsilane) in Room-Temperature Solution: Photophysics and Photochemistry** **61**
   J. Michl, J. W. Downing, T. Karatsu, K. A. Klingensmith, G. M. Wallraff, and R. D. Miller

6. **New Synthetic Routes to Polysilanes** **78**
   K. Matyjaszewski, Y. L. Chen, and H. K. Kim

7. **Polymerization of Group 14 Hydrides by Dehydrogenative Coupling** **89**
   John F. Harrod

8. **Polysilylene Preparations** **101**
   D. J. Worsfold

9. **Characterization of Copolydiorganosilanes with Varying Compositions** **112**
   Samuel P. Sawan, Yi-Guan Tsai, and Horng-Yih Huang

POLYSILAZANES AND POLYSILAZOXANES

10. **Synthetic Routes to Oligosilazanes and Polysilazanes: Polysilazane Precursors to Silicon Nitride** **124**
    Richard M. Laine, Yigal D. Blum, Doris Tse, and Robert Glaser

11. **Organosilicon Polymers as Precursors for Silicon-Containing Ceramics: Recent Developments** **143**
    Dietmar Seyferth, Gary H. Wiseman, Joanne M. Schwark, Yuan-Fu Yu, and Charles A. Poutasse

12. **NMR Characterization of a Polymethyldisilylazane: A Precursor to Si-C-N-O Ceramics** **156**
    Jonathan Lipowitz, James A. Rabe, and Thomas M. Carr

13. **Silicon-Nitrogen-Containing Rings and Polymers** **166**
    Z. Lasocki, B. Dejak, J. Kulpinski, E. Lesniak, S. Piechucki, and M. Witekowa

POLYSILOXANES

14. **Recent Advances in Organosiloxane Copolymers** 180
J. D. Summers, C. S. Elsbernd, P. M. Sormani, P. J. A. Brandt, C. A. Arnold, I. Yilgor, J. S. Riffle, S. Kilic, and J. E. McGrath

15. **Polysiloxanes Functionalized with 3-(1-Oxypyridinyl) Groups: Catalysts for Transacylation Reactions of Carboxylic and Phosphoric Acid Derivatives** 199
Martel Zeldin, Wilmer K. Fife, Cheng-xiang Tian, and Jian-min Xu

16. **Photochemical Behavior of Organosilicon Polymers Bearing Phenyldisilanyl Units** 209
Mitsuo Ishikawa and Kazuo Nate

17. **Routes to Molecular Metals with Widely Variable Counterions and Band-Filling: Electrochemistry of a Conductive Organic Polymer with an Inorganic Backbone** 224
Tobin J. Marks, John G. Gaudiello, Glen E. Kellogg, and Stephen M. Tetrick

18. **A New Approach to the Synthesis of Alkyl Silicates and Organosiloxanes** 238
George B. Goodwin and Malcolm E. Kenney

POLYPHOSPHAZENES

19. **Current Status of Polyphosphazene Chemistry** 250
Harry R. Allcock

20. **Phosphazene Polymers: Synthesis, Structure, and Properties** 268
Robert E. Singler, Michael S. Sennett, and Reginald A. Willingham

21. **Polyphosphazenes: Performance Polymers for Specialty Applications** 277
H. R. Penton

22. **Poly(alkyl/arylphosphazenes)** 283
Robert H. Neilson, R. Hani, G. M. Scheide, U. G. Wettermark, P. Wisian-Neilson, R. R. Ford, and A. K. Roy

23. **Hybrid Inorganic-Organic Polymers Derived from Organofunctional Phosphazenes** 290
Christopher W. Allen

24. **Polybis(pyrrolyl)phosphazene** 296
R. C. Haddon, S. V. Chichester, and T. N. Bowmer

25. **Skeletal Stabilization as the Basis for Synthesis of Novel Phosph(III)azane Oligomers and Polymers** 303
Elizabeth G. Bent, Joseph M. Barendt, R. Curtis Haltiwanger, and Arlan D. Norman

ORGANOELEMENT-OXO POLYMERS DERIVED FROM SOL-GEL PROCESSES

26. **Structure of Sol-Gel-Derived Inorganic Polymers: Silicates and Borates** 314
C. J. Brinker, B. C. Bunker, D. R. Tallant, K. J. Ward, and R. J. Kirkpatrick

27. **Organically Modified Silicates as Inorganic-Organic Polymers** 333
H. K. Schmidt

28. **Sol-Gel Preparation and Properties of Fibers and Coating Films** 345
S. Sakka, K. Kamiya, and Y. Yoko

29. **New Hybrid Materials Incorporating Poly(tetramethylene oxide) into Tetraethoxysilane-Based Sol-Gel Glasses: Structure-Property Behavior** 354
Hao-Hsin Huang, Raymond H. Glaser, and Garth L. Wilkes

BORON-CONTAINING POLYMERS

30. **Precursors to Nonoxide Macromolecules and Ceramics** 378
C. K. Narula, R. T. Paine, and R. Schaeffer

31. **Boron-Nitrogen Polymer Precursors** 385
S. Yvette Shaw, Donn A. DuBois, and Robert H. Neilson

32. **Boron Nitride and Its Precursors** 392
K. J. L. Paciorek, W. Krone-Schmidt, D. H. Harris, R. H. Kratzer, and Kenneth J. Wynne

OTHER METAL- AND METALLOID-CONTAINING POLYMERS

33. **Electron Transport in and Electrocatalysis with Polymeric Films of Metallotetraphenylporphyrins** 408
B. A. White, S. A. Raybuck, A. Bettelheim, K. Pressprich, and Royce W. Murray

34. **Electronically Conducting Films of Poly(trisbipyridine)-Metal Complexes** 420
C. Michael Elliott, J. G. Redepenning, S. J. Schmittle, and E. M. Balk

35. **Copper Chloride Complexes with Poly(2-vinylpyridine)** 430
A. M. Lyons, E. M. Pearce, M. J. Vasile, A. M. Mujsce, and J. V. Waszczak

36. **Cationic and Condensation Polymerization of Organometallic Monomers** 437
Kenneth E. Gonsalves and Marvin D. Rausch

37. **Soluble Metal Chelate Polymers of Coordination Numbers 6, 7, and 8** 463
Ronald D. Archer, Bing Wang, Valentino J. Tramontano, Annabel Y. Lee, and Ven O. Ochaya

38. **A New Class of Oligomeric Organotin Compounds** 469
Robert R. Holmes, Roberta O. Day, V. Chandrasekhar, Charles G. Schmid, K. C. Kumara Swamy, and Joan M. Holmes

39. **NMR Characterization of the Compositional and Configurational Sequencing of Tri-*n*-butyltin Polymers** 483
Jon M. Bellama and William F. Manders

INDEXES

Author Index 498

Affiliation Index 499

Subject Index 499

Preface

THE FIELD OF INORGANIC-ORGANIC MACROMOLECULES is entering a phase of rapid development and change. For the past 20 years this area has grown steadily, mainly through fundamental studies by a small number of academic, government, and industrial scientists. Today, the burgeoning interest in this field is driven by the search for new high-performance materials and by the recognition that substances that combine the properties of organic polymers with those of inorganic solids can provide solutions to a wide range of hitherto intractable engineering problems. Thus, great interest is directed toward the electrical, photochemical, mechanical, and biomedical properties of the new polymers as well as their unusual behavior at low and high temperatures.

Polysiloxanes (silicones) began as a scientific curiosity in the 1930s, but their current widespread use in industrial and consumer applications is well-known. The recent emergence of polyphosphazenes, polysilanes, and organoelement-oxo polymers derived from the sol-gel process appears to be following a similar pattern—led first by long-range, fundamental, academic research, and then developed into an expanding technology by work in industrial and government laboratories. It is a stimulating experience to be a part of, or to follow, the current growth and expansion of this diverse field.

It is a direct consequence of the needs, opportunities, challenges, and broad interdisciplinary nature of the subject that prompted the symposium on which this book is based to survey the state of the art and current perspectives in inorganic and organometallic polymers. The contributions in this book fall into two categories: topical reviews and specialist reports by symposium participants and invited contributors. The topical reviews provide a thorough survey of a particular subject area and may also contain recent results from the authors' laboratories. The specialist contributions are shorter chapters that describe particularly exciting recent research progress. Together, the topical reviews and the specialist contributions provide an in-depth look into past accomplishments and currently stimulating new efforts.

We gratefully acknowledge financial support for the symposium from the following organizations: Celanese Corporation, Dow Corning Corporation, ACS Division of Inorganic Chemistry, ACS Division of Polymer Chemistry, Inc., ACS Division of Polymeric Materials: Science and

Engineering, Eastman Kodak Laboratories, and Petroleum Research Fund of the American Chemical Society. We also acknowledge the generous financial support for the symposium and the preparation of the volume from the Office of Naval Research.

MARTEL ZELDIN
Indiana University-Purdue University at Indianapolis
Indianapolis, IN 46223

KENNETH J. WYNNE
Office of Naval Research
Arlington, VA 22217

HARRY R. ALLCOCK
The Pennsylvania State University
University Park, PA 16802

September 11, 1987

Chapter 1

An Introduction to Inorganic and Organometallic Polymers

Kenneth J. Wynne

Chemistry Division, Office of Naval Research, Arlington, VA 22217-5000

"Inorganic and organometallic" in the context of this Symposium is meant to describe macromolecules which usually contain inorganic elements in the chain and organic moieties as pendant groups. The subtitle delineates elements of primary focus: "Macromolecules Containing Silicon, Phosphorus and Other Inorganic Elements." The term "macromolecules" implies that the subject matter includes chain molecules that may be built up of repeat units, as well as more complex ring, branched, or crosslinked species (for example, see Organo-Oxo-Element Macromolecules Related to Sol-Gel Processes, and contributions by Murray (p. 408) or Seyferth (p. 143)).

Prior reviews concerning inorganic and organometallic macromolecules are contained in texts by Stone and Graham (1), Andrianov (2), Borisov (3), Allcock (4,5), and Voronkov (6), and volumes based on previous ACS symposia edited by Rheingold (7), and Carraher, Sheats and Pittman (8, 9). The present effort is topical in nature, and the order of presentation approximates that of the Symposium presentations.

The impetus for this symposium volume is the considerable progress which has been made in the last few years in inorganic and organometallic macromolecules. Totally new macromolecules have been brought into existence by the development of new synthetic methods or improvement of known synthetic routes. Thus, Neilson (p. 283) describes a new polymerization reaction that gives high molecular weight poly(diorganophosphazenes), $-(R_2PN)_n-$, in which organic pendant groups are bonded through P-C bonds. These polyphosphazene analogs of polysiloxanes were not previously accessible, and the development of structure-property relationships in this new subclass of macromolecules will no doubt yield important information and perhaps significant applications.

Another example of interesting new inorganic polymers is found in the work of Lasocki (p. 166), who reports the synthesis of polysilazoxanes, $-[(R_2SiO)_x(R_2SiNR)_y]_n-$, and finds surprisingly better thermal stability compared with their polysiloxane analogs. The design of functionalized polymers with a specific utilization is seen in new polysiloxanes used by Zeldin (p. 199) as phase transfer catalysts. Novel functional polyphosphazenes have been reported as well by Allcock (p. 250). The introduction of transition metal cyclopentadienyl, metal carbonyl and carborane moieties into polyphosphazene macromolecules is representative of truly novel chemistry achieved after careful model studies with corresponding molecular systems.

West (p. 6), Miller (p. 43), Zeigler (10), and Sawan (p. 112) outline the synthesis of a wide variety of soluble, processable polydiorganosilanes, a class of polymers which not long ago was thought to be intractable. Matyjaszewski (p. 78) has found significant improvements in the synthetic method for polydiorganosilane synthesis as well as new synthetic routes to unusual substituted polydiorganosilanes. Seyferth (p. 21, 143) reports synthetic routes to a number of new polycarbosilanes and polysilazanes which may be used as precursors to ceramic materials.

New catalytic polymerization routes to polysilanes (Harrod, (p. 89)) and polysilazanes (Laine, (p. 124)) have been discovered. Singler (p. 268) describes the use of BCl_3 in the more efficient synthesis of $-(PNCl_2)_n-$, the starting high polymer for most polyphosphazene polymers currently under investigation. These results in the area of polymerization catalysts are important, as the systematic development of efficient catalytic routes for inorganic and organometallic macromolecules will make such materials more generally accessible and utilizable. This research is also closely related to understanding of mechanisms of chain growth.

Efforts aimed at the elucidation of polymerization mechanisms include those of Singler (p. 268) in polyphosphazenes , Lipowitz (p. 156) in polysilazanes, and Zeigler and Worsfold in polysilanes. In contrast with carbon chemistry, the mechanisms of polymerization reactions leading to inorganic and organometallic macromolecules are often not well understood. Such studies are critical in elucidating pathways of chain growth, termination, and branching so that these features may be controlled. Oftentimes mechanistic studies lead to more efficient synthetic methods, for example by improving yields or shortening reaction times.

In the section on boron-containing polymers, Paine (p. 378), Neilson (p. 385), and Paciorek (p. 392) present pioneering work aimed at the preparation of linear chain macromolecules with B-N backbones. This work is significant because condensation reactions of B-N compounds tend to produce compounds with ring structures rather than chains. One obvious potential utilization of such novel macromolecules is as preceramic materials, much in the same way as polyacrylonitrile is used as a precursor for carbon. Orientation of the B-N polymer may be transferred in part to the ceramic solid state again in manner similar to carbon chemistry. However, the properties of BN polymers and ceramic materials differ greatly from their carbon analogs due to localized electronic states in the BN bond. The consequences of this contrasting electronic structure on materials properties will be interesting to see as this new research area develops.

A plurality of papers in this volume concern linear chain macromolecules. Fundamental to understanding the physical and mechanical behavior and chemical and physical stability of these macromolecules is a familiarity with phase transition behavior, an area well known in organic polymer chemistry (11). As with organic polymers, amorphous and semicrystalline inorganic and organometallic macromolecules are known. Crystallinity arises from main chain order or side chain crystallization as discussed by Singler (p. 268) for polyphosphazenes and West (p. 6) and Miller (p. 43) for polysilanes. The latter work demonstrates that crystallization behavior plays a critical role in controlling main chain conformation and optical transitions in poly(diorganosilanes). Thus, important structure/property relationships are emerging that are relevant to electronic and optical materials applications for these materials. In a different vein, side chain crystallization has resulted in the first liquid crystalline inorganic and organometallic macromolecules, viz., unusual poly(dialkoxyphosphazenes) described by Allcock (p. 250) and Singler (p. 268). In this case, the flexible nature of the P-N chain places stringent structural requirements on the nature of the pendant group.

Unique combinations of properties continue to be discovered in inorganic and organometallic macromolecules and serve to continue a high level of interest with regard to potential applications. Thus, Allcock describes his collaborative work with Shriver (p. 250) that led to ionically conducting polyphosphazene/salt complexes with the highest ambient temperature ionic conductivities known for polymer/salt electrolytes. Electronic conductivity is found via the partial oxidation of unusual phthalocyanine siloxanes (Marks, p. 224) which contain six-coordinate rather than the usual four-coordinate Si.

Part of this symposium was directed to the synthesis, properties and applications of inorganic and organometallic macromolecules with network structures. The section on organo-oxo macromolecules relevant to sol-gel processing addresses the interesting synthesis and challenging characterization efforts in this area. Brinker (p. 314) outlines the complex chemical and physical factors which affect network formation and structure resulting from the hydrolysis of a tetraalkoxysilane. The interesting properties of hybrid organic/inorganic network structures are described in the work of Schmidt (p. 333) and Wilkes (p. 354).

In conclusion, some trends can be gleaned from an examination of the content of the symposium as a whole. The growth in research efforts addressing the synthesis and properties of poly(diorganosilanes) will likely continue. The unique photophysical properties of this newly developed class of inorganic macromolecules (12) together with ready synthetic routes will be contributing forces here, and no doubt new vectors will arise. Another area of increased attention will be organo-oxo macromolecules derived from sol-gel processing methods, either as copolymers or blends. Complex dependencies of organo-oxo macromolecular composition and structure on starting materials and processing conditions (including kinetic effects) will lead to challenging and interesting science. Important mechanical, optical and structural applications coupled again with emerging synthetic approaches will be among the drivers for continued high activity in this area.

Finally, one additional comment concerning the nature of progress from interdisciplinary research is evident from the results reported in this Symposium Volume. Once again it is seen that most rapid progress is made when synthetic chemists collaborate with their colleagues in materials science or physics to determine properties of new inorganic and organometallic polymers.

Acknowledgment

The author thanks the Office of Naval Research for support of this contribution.

Literature Cited

1. Stone, F. G. A., Graham, W. A. G. Inorganic Polymers, Academic Press, New York 1962
2. Andrianov, K. A. Metalorganic Polymers, Interscience Publishers, New York 1965.
3. Borisov, S. N., Voronkov, M. G., Lukevits, E. Ya. Organosilicon Heteropolymers and Heterocompounds, Plenum Press, New York 1970.
4. Allcock, H. R. Phosphorus-Nitrogen Compounds, Academic Press, New York 1972.
5. Allcock, H. R. Heteroatom Ring Systems and Polymers, Academic Press, New York 1967.
6. Voronkov, M. G., Mileshkevich, V. P., Yuzhelevskii, Yu. A. The Siloxane Bond, Consultants Bureau, New York 1978.
7. Rheingold, A. L. Homoatomic Rings, Chains and Macromolecules of Main-Group Elements, Elsevier Scientific Publishing Co., New York 1977.
8. Carraher, C. E., Sheats, J. E., Pittman, C. U. Organometallic Polymers, Academic Press, New York 1978.
9. Carraher, C. E., Sheats, J. E., Pittman, C. U. Advances in Organometallic and Inorganic Polymers, Marcel Dekker, Inc. New York 1982
10. Zeigler, J. M., Harrah, L. A., Johnson, A. W., Polymer Preprints, 28 (1), 424 (1987).
11. Wunderlich, B. Macromolecular Physics/Crystal Melting, Academic Press, New York 1980.
12. Abkowitz, M., Knier, F. E., Stolka, M., Wigley, R. J., Yuh, H.-J., Solid State Commun., 62, 547 (1987).

RECEIVED October 27, 1987

POLYSILANES AND POLYCARBOSILANES

Chapter 2

Polysilane High Polymers: An Overview

Robert West and Jim Maxka

Department of Chemistry, University of Wisconsin, Madison, WI 53706

The history and development of polysilane chemistry is described. The polysilanes (polysilylenes) are linear polymers based on chains of silicon atoms, which show unique properties resulting from easy delocalization of sigma electrons in the silicon-silicon bonds. Polysilanes may be useful as precursors to silicon carbide ceramics, as photoresists in microelectronics, as photoinitiators for radical reactions, and as photoconductors.

The polysilanes are compounds containing chains, rings, or three-dimensional structures of silicon atoms joined by covalent bonds. Recently, polysilane high polymers have become the subject of intense research in numerous laboratories. These polymers show many unusual properties, reflecting the easy delocalization of sigma electrons in the silicon-silicon bonds. In fact, the polysilanes exhibit behavior unlike that for any other known class of materials.

In this chapter, an introduction and overview of polysilane chemistry will be presented, concentrating on the linear high polymers (polysilanes) and their technological applications. Polysilane polymers were reviewed in 1986,(1) and a more general review of polysilane chemistry appeared in 1982.(2)

Historical

Poly(diphenylsilylene) may have been prepared as early as the 1920's by F. S. Kipping, the grandfather of organosilicon chemistry; but the polymeric or oligomeric products were not characterized. The first certain preparation of a linear polysilane came in 1949, when Charles Burkhard of the General Electric Company Research Laboratories, published a classic paper describing the synthesis of poly(dimethylsilylene), $(Me_2Si)_n$.(3) The polymer was obtained by condensing dimethyldichlorosilane with sodium metal, in essence the same process used today for the synthesis of polysilanes. Burkhard described $(Me_2Si)_n$ quite clearly and accurately, as an insoluble, infusible, and generally quite intractable material. It is now clear that poly(dimethylsilylene) is atypical among polysilanes, but this

0097–6156/88/0360–0006$06.00/0

was not realized at the time. The discouraging properties of poly(dimethylsilylene) perhaps contributed to the neglect of this field over the following 25 years.(4)

In any event, between 1951 and 1975, no papers appeared on polysilane high polymers. However, linear permethylpolysilanes of the type $Me(SiMe_2)_nMe$ were prepared and studied, especially by Kumada and his students,(5) and cyclic polysilanes were being investigated in several laboratories.(6,7) Studies of the permethylcyclosilanes, $(Me_2Si)_n$ where n = 4 to 7, showed that these compounds exhibit remarkable delocalization of the ring sigma electrons, and so have electronic properties somewhat like those of aromatic hydrocarbons.(6)

Interest in polysilane polymers was finally reawakened by the work of Yajima and Hayashi, who found that poly(dimethylsilylene) could be used as a precursor to silicon carbide.(9) The discovery, or rediscovery, of soluble polysilanes at Wisconsin was quite accidental.(10) In one attempt to prepare cyclosilanes containing both phenyl and methyl groups, $PhMeSiCl_2$ and Me_2SiCl_2 were co-condensed with alkali metal. A polymer was obtained instead of the desired ring compound, and to our surprise it proved to be somewhat soluble and meltable. The introduction of phenyl groups along the chain breaks up the crystallinity of $(Me_2Si)_n$ polymer. This adventitious finding led to synthesis of the "polysilastyrene" family of Me_2Si-PhMeSi copolymers.(11) At almost the same time, soluble polysilanes were reported by Trujillo(12) at Sandia Laboratories and by Wesson and Williams(13) at Union Carbide Co.

Research in polysilane polymers grew slowly at first after this reawakening. But within the past few years, both the unusual scientific interest and the technological possibilities of the polysilanes have been recognized, and activity in this field has increased sharply. Commercial manufacture of both poly(dimethylsilylene) and "polysilastyrene" is now being carried out in Japan, so that these two polymers are readily available in quantity.

Synthesis of Polysilanes

Poly(silylene) polymers are usually made by the reaction of diorganodichlorosilanes with sodium metal, in an inert diluent at temperatures above 100°C.(11) Rapid stirring is ordinarily used so that the sodium is finely dispersed, speeding the rate of reaction. Either homopolymers or copolymers can be synthesized:

$$RR'SiCl_2 \xrightarrow[>100^\circ C]{Na,\ solvent} -\!(\underset{R'}{\overset{R}{Si}})_n\!-$$

$$R^1R^2SiCl_2 + R^3R^4SiCl_2 \xrightarrow[>100^\circ C]{Na,\ solvent} -\!(\underset{R^2}{\overset{R^1}{Si}})_n\!-\!(\underset{R^3}{\overset{R^3}{Si}})_m\!-$$

Considerable low molecular weight, oligomeric material is usually produced along with high polymer, so that the yield of true polymer is often less than 50%. In a typical workup, a small amount of an alcohol is added to react with any excess sodium, then water is added to the mixture to dissolve the sodium salts. The organic layer is separated, filtered or centrifuged if necessary, and then a large amount of an alcohol such as 2-propanol is added to precipitate the high polymer. Most of the oligomeric compounds remain in solution during the precipitation. A second precipitation can be carried out if more complete separation of polymer from oligomer is desired.

Variables in the condensation reaction include the temperature, nature of the solvent, order of addition (either chlorosilane to excess sodium or "inverse" addition, sodium to excess chlorosilane), and the rate of addition. A careful study of reaction variables by the Sandia group of Dr. John Zeigler(15) will be described in detail elsewhere in this volume.

Physical properties of the polysilanes depend greatly upon the nature of the organic groups bound to silicon. A few of the many polysilanes are listed in Table I. Typically the linear polysilanes are thermoplastics, soluble in organic solvents like toluene, ethers,

Table I. Properties of Some Polysilane Polymers

R^1	R^2	Yield(%)	$\bar{M}_w$ [a]	λ_{max}
A. Homopolymers, $(R^1R^2Si)_n$				
Me	n-Pr	32	640,000	306
Me	n-Bu	34	110,000	304
Me	n-Hex	11	520,000	306
Me	n-$C_{12}H_{25}$	9	480,000	309
Me	PhC_2H_4	35	290,000	303
Me	Cy-Hex	25	800,000	326
Me	Ph	55	190,000	335
Me	p-Tol	25	75,000	337
Me	p-Biphen	40	80,000	352
n-Bu	n-Bu	12	1,800,000	314
n-Hex	n-Hex	9	2,500,000	316

R^1	R^2	R^3	R^4	n/m	$\bar{M}_w$ [a]	λ_{max}
B. Copolymers, $(R^1R^2Si)_n(R^3R^4Si)_m$						
Me	Me	Me	n-Hex	1.52	170,000	303
Me	Me	Me	Ph	1.51	900,000	330
Me	Me	Ph	Ph	1.13	350,000	351
Me	Cy-Hex	Me	PhC_2H_4	1.49	150,000	310
Me	Cy-Hex	Me	p-Tol	1.78	92,000	338
Me	PhC_2H_4	Me	Ph	1.77	400,000	326

[a] High mol. wt. peak from GPC, usually bimodal. $\bar{M}_w$'s given are relative to polystyrene standards; actual $\bar{M}_w$ values are higher by 2-3x.

and chlorinated hydrocarbons, although insoluble in alcohols. The crystallinity varies greatly depending upon substitution. Polymers with a methyl group and a long-chain alkyl group, i.e. $(n\text{-Hexyl-SiMe})_n$, have glass transition temperatures well below 0° and are therefore elastomers.

Molecular weights of polysilane polymers depend upon the exact method of synthesis, as well as the purity of the dichlorosilane starting materials. Bimodal molecular weight distributions are commonly reported, as shown in Figure 1, but under some conditions monomodal distributions can also be obtained. Weight-average molecular weights, determined by light-scattering measurements, as high as 7 x 10(6) have been obtained for ($\underline{n}$-HexylSiMe)$_n$, corresponding to a degree of polymerization of 750,000.(16)

The polysilanes are inert to air and atmospheric moisture, and are not attacked by mild reagents such as dilute acids, etc. However in a solvent such as THF, solvolysis of the Si-Si bond by strong bases is fairly rapid. With alcohols or water and bases, hydrogen is produced:

$$\sim Si{-}Si \sim \; + \; 2\,ROH \xrightarrow[THF]{OR^-} \sim Si{-}OR \; + \; H_2 \; + \; RO{-}Si \sim$$

Strong oxidizing agents such as $\underline{m}$-chloroperbenzoic (MCPBA) acid also react with polysilanes, to insert oxygen atoms between the silicons:(17)

$$\sim{-}Si{-}Si{-}\sim \xrightarrow{MCPBA} \sim Si{-}O{-}Si \sim$$

Polysilanes containing Si-H groups are reactive in various ways, for instance in addition to olefins catalyzed by platinum. Reactivity of organic substituent groups has also been observed; an example is hydrogen halide addition to the C=C double bond in polysilanes containing cyclohexenylethyl groups:(18)

$$\text{(chlorocyclohexylethyl)}(Si)_n(Si)_m \xleftarrow[SnCl_4]{HCl} \text{(cyclohexenylethyl)}(Si)_n(Si)_m \xrightarrow[BBr_3]{HBr} \text{(bromocyclohexylethyl)}(Si)_n(Si)_m$$

(each Si bearing Me; the second Si also bearing R)

Photochemistry and Crosslinking

When irradiated with ultraviolet light, alkylpolysilanes undergo scission almost exclusively, but crosslinking as well as scission is observed for aryl-substituted polysilanes.(19,20) Although the mechanism of photolysis has not been elucidated in detail, exhaustive photolysis of polysilanes (RR'Si)$_n$ in the presence of a silylene trapping agent, triethylsilane, led to the silylene product Et_3Si-SiRR'-H as well as disilane, HSiRR'-SiRR'H.(21) These findings suggest that there are at least two primary steps, simple scission to silyl radicals (A) and elimination of silylenes, R_2Si (B). These two processes could also occur simultaneously(C).

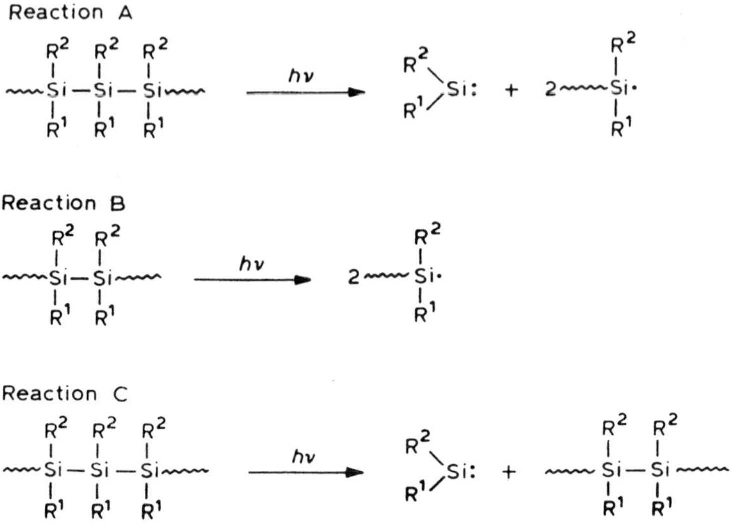

Reaction A

$$\sim\!\!\sim Si(R^2)(R^1)-Si(R^2)(R^1)-Si(R^2)(R^1)\sim\!\!\sim \xrightarrow{h\nu} R^2R^1Si: + 2\sim\!\!\sim Si(R^2)(R^1)\cdot$$

Reaction B

$$\sim\!\!\sim Si(R^2)(R^1)-Si(R^2)(R^1)\sim\!\!\sim \xrightarrow{h\nu} 2\sim\!\!\sim Si(R^2)(R^1)\cdot$$

Reaction C

$$\sim\!\!\sim Si(R^2)(R^1)-Si(R^2)(R^1)-Si(R^2)(R^1)\sim\!\!\sim \xrightarrow{h\nu} R^2R^1Si: + \sim\!\!\sim Si(R^2)(R^1)-Si(R^2)(R^1)\sim\!\!\sim$$

Scheme 1. Possible Reactions for the Photodegradation of Polysilane Polymers.

The silyl radicals formed in the initial scission appear to undergo further reactions, which may be complex. A possible secondary reaction is hydrogen transfer from an alpha carbon atom to give Si-H and a silicon carbon double bond:(21)

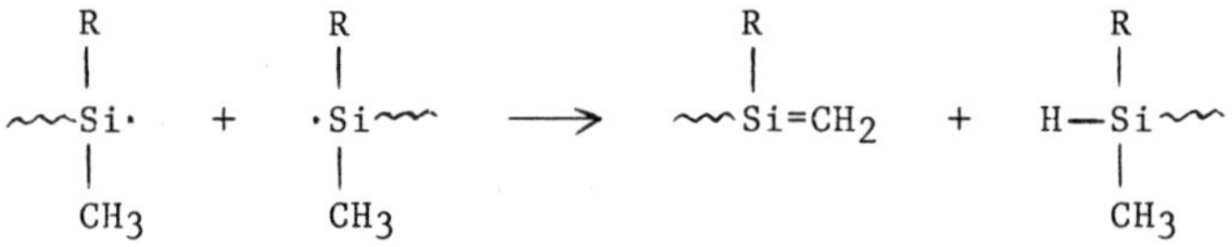

$$\sim\!\!\sim Si(R)(CH_3)\cdot + \cdot Si(R)(CH_3)\sim\!\!\sim \longrightarrow \sim\!\!\sim Si(R)=CH_2 + H-Si(R)(CH_3)\sim\!\!\sim$$

Photochemical crosslinking of alkenyl polysilanes takes place readily, presumably via radical addition to the unsaturated C=C linkages on adjacent chains. Crosslinking of any polysilane can be carried out photochemically if the polymer is mixed with a multiply-unsaturated compound such as phenyl trivinylsilane; addition of silyl groups to the carbon-carbon double bonds of the additive then provides the crosslinking.(20) Thermal crosslinking of polysilanes containing polyunsaturated crosslinking additives can also be carried out, with a free-radical initiator such as AIBN. Among the other crosslinking systems which have been devised, an interesting example is the crosslinking of a liquid mixture of an oligomeric polysilane containing Si-H bonds together with a polyunsaturated compound, brought about by the addition of H_2PtCl_6 as a hydrosilylation catalyst; this procedure provides "room temperature vulcanization" of polysilanes.(22)

Electronic Spectra

The polysilane polymers all show strong electronic absorption bands in the ultraviolet region, falling between 300 and 400 nm.(19) Absorption spectra for solutions of a few polymers are shown in Figure 2. The electronic transitions are of σ–σ^* type, reflecting extensive delocalization of σ-electrons in the catenated silicon atoms. The absorption maxima depend on the nature of the organic substituents. Simple, unhindered dialkylpolysilanes absorb near 300 nm, but the introduction of sterically-hindering groups shifts the maximum to longer wavelength. Aryl groups directly attached to silicon also produce bathochromic shifts; as an example, $(PhMeSi)_n$ has its absorption maximum at 340 nm. The longest wavelength maxima are found for the soluble diarylpolysilanes recently reported by Miller et al.,(24) which have λ_{max} 395 nm. Thus both electronic and steric effects influence the absorption spectra of poly-silanes.(25) Aryl substituents can influence the electronic spectrum by allowing mixing of sigma with pi states, as discussed in a recent paper from the NTT research group.(26)

Many polysilanes also show striking changes in their uv spectra with temperature.(27) An example is shown in Figure 3 for $(n\text{-pentyl}_2Si)_n$ in hexane solution. As the temperature is lowered the original absorption band at 313 nm decreases and a new band at 356 nm grows in. These changes are reversible, although micro-crystals apparently form if the solution is kept for a time at low temperatures, making the reversal quite slow.

The changes with steric hindrance and temperature must be due to conformational effects along the polymer chains. It is now generally agreed that the thermochromism results from an increase in the proportion of trans- to gauche conformations in the polymer chain as the temperature is decreased. Similarly, the introduction of sterically hindering substituents could increase the amount of trans junctions. Evidence in favor of this model is presented in several recent papers,(28) as well as the chapter by Michl in this volume.

Pure oligomeric polysilanes of moderate chain length would be very useful for determining conformational preferences and studying conformational changes, but none are yet available. The closest approximation is the cyclic oligomer, $(Me_2Si)_{16}$. The crystal structure for this compound, illustrated in Figure 4, shows that it has a conformation quite unlike that for organic 16-membered rings. Typical 16-ring carbocycles exhibit a "square" or diamond-lattice type structure with eight trans and eight gauche torsional angles. An example is 1,1,8,8-tetramethylcyclohexadecane, which has eight trans junctions between 175.6 and 180°, and eight gauche torsions between 50.3 and 59.3°, all normal values.(30) In contrast, the cyclosilane $(Me_2Si)_{16}$ shows no torsional angles in either the normal trans or normal gauche range.(30) Instead it has eight "gauche- eclipsed" junctions between 83.9 and 93.4°, and eight "trans- eclipsed" angles between 158.0 and 169.5°. These results suggest that the view of polysilane polymers as consisting of trans and gauche linkages may be oversimplified. In the polysilanes, intermediate torsional angles may be much more important than they are in carbon polymers.

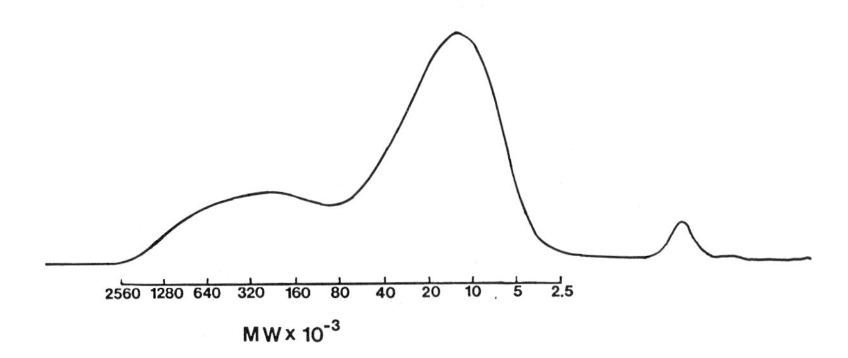

Figure 1. Gel permeation chromatograph for $(PhMeSi)_n(Me_2Si)_m$ copolymer, m=n, showing bimodal molecular weight distribution.

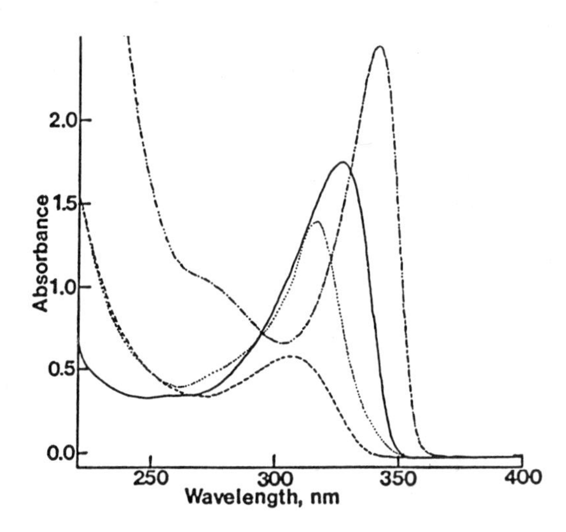

Figure 2. UV spectra of solutions of high molecular weight polysilanes in THF at 23°C; (---), $(\underline{n}\text{-DodecylMeSi})_n$; (···) $(\underline{n}\text{-Hexyl}_2\text{Si})_n$; (——), $(\text{CyclohexylMeSi})_n$ [in cyclohexane]; (-··-), $(PhMeSi)_n$.

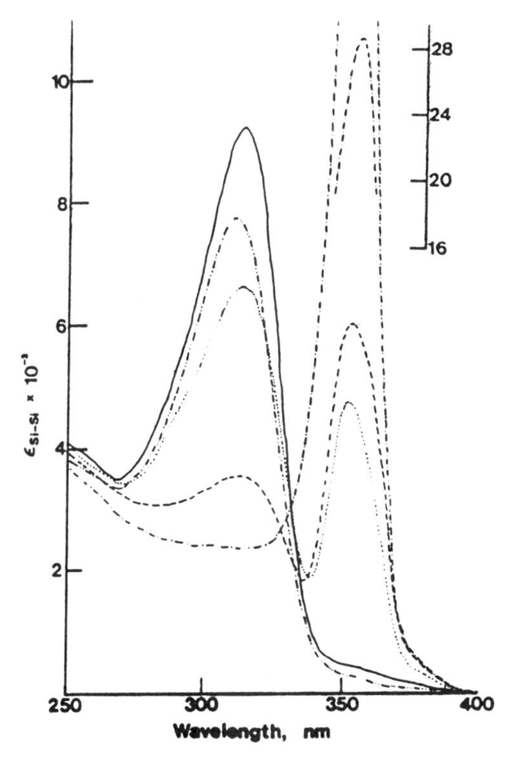

Figure 3. UV spectrum of a 0.0051% solution of ($\underline{n}$-$Pentyl_2Si)_n$ in hexane; (-···-), 74°C; (——), 23°C; (···), 0.0°C; (---), -7.2°C; (-·-), -52°C.

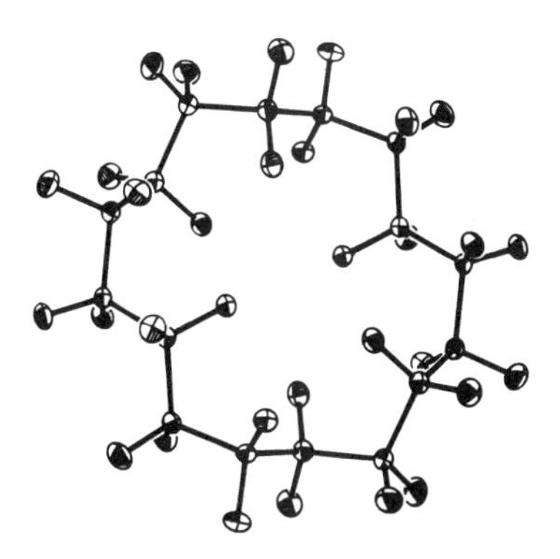

Figure 4. ORTEP diagram from x-ray crystal structure of $(Me_2Si)_{16}$. Hydrogens have been omitted for clarity.

Even more complicated changes take place in solid films of polysilane polymers as the temperature is changed. These are discussed in the chapter by Miller et al. in this volume.(31)

Nmr Spectra and Configuration

Proton, ^{13}C and ^{29}Si NMR spectra for polysilanes have been recorded.(32) The proton NMR provide little structural information, but integration of areas under the proton resonances is quite useful for determining the composition of copolymers.

The ^{29}Si NMR spectra are of particular interest because they reflect the configuration of the polymer chain.(33) Some ^{29}Si spectra of alkylpolysilanes are shown in Figure 5. Symmetrically-substituted polymers such as $(n\text{-hexyl}_2Si)_n$ have no chirality since there can be a plane of symmetry through each silicon atom. The ^{29}Si resonance is therefore a single narrow line. However for dialkylpolysilanes with two different alkyl groups on each silicon, $(RR'Si)_n$, each silicon atom is a chiral center and the resonance for a particular silicon will depend upon the relative stereochemistry of other nearby silicon atoms. For such polymers, a rather symmetrical cluster of peaks is observed (Figure 5). These results are consistent with atactic structures, having a statistical (Bernoullian) distribution of relative configurations.(32,33)

For arylpolysilanes the results are quite different.(34) The ^{29}Si NMR for $(PhSiMe)_n$ is shown in Figure 6; it consists of three broad lines with relative intensity 3:3:4, each line evidently containing a cluster of resonances. The patterns for other aryl-alkylpolysilanes differ, but in general two or three broad resonances are found; none of the aryl compounds studied so far has given a symmetrical pattern like those observed for the alkylpoly-silanes of Figure 5.

The results for arylsilanes are not fully understood, but the spectra may reflect partial tacticity in these polymers. Further work is needed; studies of model compounds with known relative configuration would be particularly helpful. Silicon-29 NMR of polysilane copolymers also shows great promise, especially for distinguishing block-like from fully random copolymers.

Technological Applications of Polysilanes

Possible ways in which polysilanes may be useful include, 1. As precursors to silicon carbide ceramics; 2. As photoinitiators in radical reactions; 3. As photoconductive materials, and 4. As photoresists in microelectronics. The last of these uses will be treated in the chapter by Miller,(31) and so will not be covered here.

Polysilanes as Precursors to Silicon Carbide. The original process for thermal generation of silicon carbide ceramic from polymeric precursors was pioneered by Yajima and Hayashi.(9) The starting materials are either poly(dimethylsilylene) or the cyclosilane $(Me_2Si)_6$. Thermolysis of these materials at 400-450°C leads to a complex series of changes in which insertion of CH_2 groups takes place into many of the Si-Si bonds, leaving hydrogen bound to silicon.

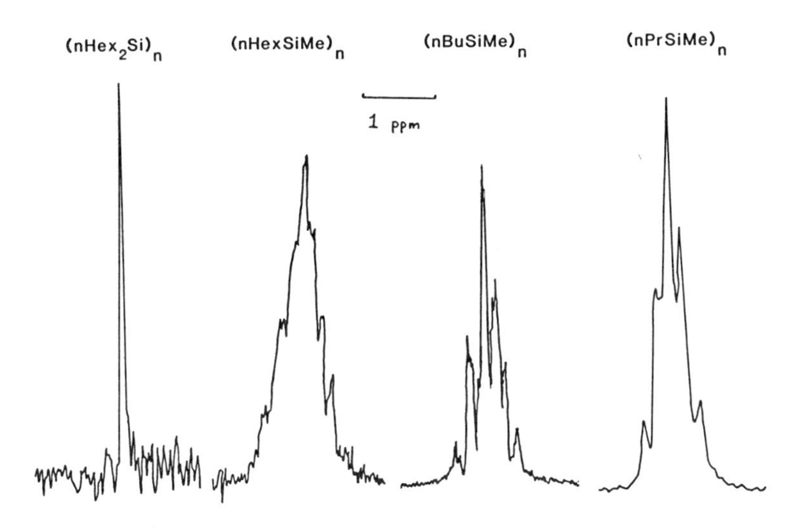

Figure 5. ^{29}Si NMR spectra for dialkylpolysilanes in benzene solution.

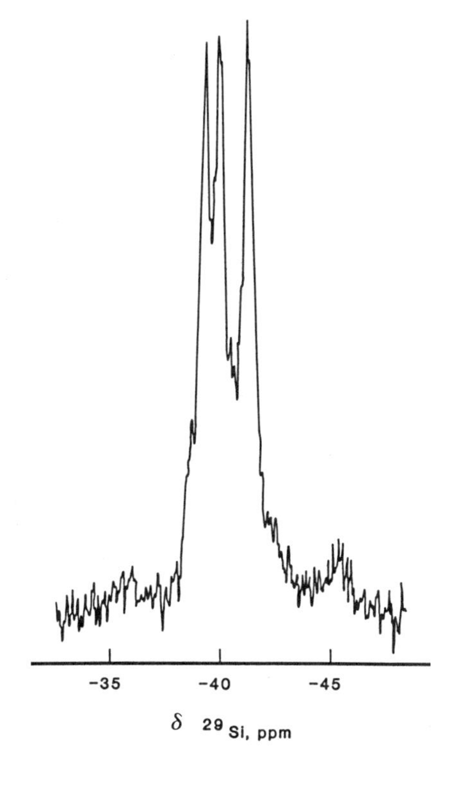

Figure 6. ^{29}Si NMR for $(PhSiMe)_n$.

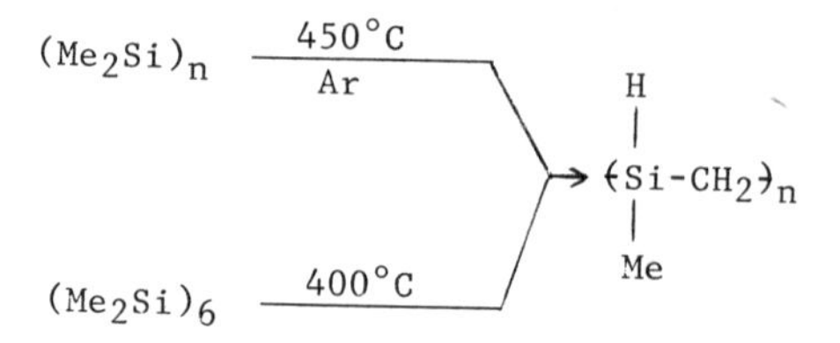

The product is a polymeric material with idealized structure shown, termed a polycarbosilane. The hexane-soluble, non-volatile portion of the polycarbosilane is fractionated by precipitation to give a material with a molecular weight of about 8000. The polymer fraction is then melt-spun into fibers, which are oxidized on the surface by heating in air. Surface oxidation makes the fibers rigid enough so they maintain their shape upon further heating. Finally, these fibers are pyrolyzed at higher temperatures, whereupon loss of methane and hydrogen takes place, producing silicon carbide:

$$-\!\!\left(\underset{CH_3}{\overset{H}{Si}}-CH_2\right)_{\!n}\xrightarrow[2)\ N_2,\ 1300^\circ C]{1)\ air,\ 350^\circ C}\beta\text{-}SiC$$

Amorphous silicon carbide is apparently formed near 800° C; as the temperature is raised to 1300° C, crystals of β-silicon carbide form in the fibers, which are actually a mixture of β-SiC with amorphous silicon carbide, carbon and silicon dioxide. With tensile strengths of 350 kg/mm^2, the fibers are among the strongest substances known. Silicon carbide fibers made in this way are now available in kilogram quantities from Nippon Carbon Company and are distributed in the United States by Dow Corning Corporation. This technology has the disadvantage that the thermolysis must be carried out in two steps, with fractionation of the intermediate polycarbosilane, and that air oxidation is necessary to strengthen the polycarbosilane fibers before final pyrolysis, introducing some oxygen into the final ceramic material.

Soluble polysilane polymers can also be used as precursors to silicon carbide. The first such application, using $(PhMeSi)_n$-$(Me_2Si)_m$ copolymers ("Polysilastyrene"), was to strengthen silicon nitride ceramics. The Si_3N_4 ceramic body was soaked in polysilane and refired, leading to the formation of silicon carbide whiskers in the pore spaces and a consequent increase in strength.(11)

As these experiments indicate, polysilanes can in some cases be converted to silicon carbide directly, without the necessity for formation of polycarbosilane, fractionation, or oxidation. For example, polysilastyrene copolymers can be formed into films or fibers and then crosslinked by irradiation with UV light. The crosslinked polysilane forms silicon carbide when heated to 1100° C in vacuum.(11) This method can be used in a "printing" mode, if a film of polysilane is cast onto a ceramic or metal substrate, then exposed to UV light through a mask, and finally heated. The unexposed polysilane is volatilized, leaving a silicon carbide coating only where the polysilane was exposed.

Recently, silicon carbide formed objects have been made by the Shin Nisso Kako Co. from polysilastyrene.(37) The polysilane is melted and mixed with submicron-sized silicon carbide powder, then injection molded in conventional equipment and fired. The resulting ceramic contains 20-30% void space, as microvoids. Although these reduce the ultimate strength slightly, they confer the advantage of inhibiting crack growth. Because of this unusual microstructure, the silicon carbide can be machined to exact dimensions. Quite large objects are now being manufactured, for example pipes more than 12 cm in diameter.

Polysilanes as Photoinitiators. Since polysilanes photolyze to give silyl radicals,(21) as explained in the section on photochemistry above, they can be used as photoinitiators for radical processes such as polymerization.[38] As can be seen from the information in Table II, this process is a fairly general one; a variety of

Table II. Photopolymerization with Polysilanes as Initiators

Polysilanes used	Monomers polymerized
$(PhC_2H_4SiMe)_n$	Styrene
$[(PhC_2H_4SiMe)_{0.8}(Me_2Si)_{1.0}]_n$	Ethyl acrylate
$[(PhC_2H_4SiMe)_{1.0}(PhMeSi)_{0.6}]_n$	Methylmethacrylate
$(PhMeSi)_n$	"Isooctyl" acrylate
$[(PhMeSi)(Me_2Si)]_m$	Acrylic acid
$[(CyHexSiMe)_n$	Phenoxyethyl acrylate
$[(CyHexSiMe)_{0.7}(Me_2Si)_{1.0}]_n$	1,6-hexanediol diacrylate
	Tripropyleneglycol diacrylate

polysilanes can be used to initiate chains in those monomers susceptible to radical polymerization.

The efficiency of photoinitiation by polysilanes is rather low, about 10^{-3} for $(PhMeSi)_n$ and styrene, but this is partly compensated by the very high extinction coefficients of the polysilanes. The net result is that the rate of polymerization is lower by about a factor of 10, when polysilanes are compared with conventional photoinitiators such as benzoin methyl ether.

An unusual property of polysilanes as photoinitiators is that they seem to be relatively insensitive to termination of the polymerization by oxygen. This lack of sensitivity to oxygen is further enhanced by the addition of amines as protective catalysts. Because protection of polymerizing systems from oxygen is expensive, this may be an important technical advantage. The reason for the oxygen-insensitivity of polysilanes is not known, but it is possible that a secondary species is formed in the photolysis which serves as an efficient oxygen scavenger.(38)

Although polysilanes have been used mostly as photoinitiators for polymerization, they may also find application as initiators for other radical reactions. Experiments to test this possibility are now being carried out.

Polysilanes as Photoconductors, etc. The delocalization in the Si-Si σ framework of polysilanes makes them possible electrical conductors. Although neutral polysilanes are insulators, they become semiconducting when doped with oxidizing agents such as AsF_5 or SbF_5.(1,11)

Conductivities up to 0.5 $ohm^{-1}cm^{-1}$ have been observed. The oxidative doping suggests that conduction is via a hole transport mechanism. However the oxidized polysilanes are highly moisture-sensitive, so it seems unlikely that they will be technologically useful.

A better possibility is offered by the finding that polysilane polymers are excellent photoconductors(39). Conduction operates through hole migration, with high mobilities being observed, 10^{14} cm^2/Vs at room temperature. Photoconductivity is found for dialkylpolysilanes as well as for $(PhSiMe)_n$, showing that the aromatic rings are not an essential part of the conduction process.(39,40,41) The photoconductivity of polysilanes has possible applications in electrophotography and in other communications technology.

Still another intriguing possibility is offered by the recent discovery that $(PhSiMe)_n$ shows nonlinear optical properties.(42) Thin films of this polymer generate third harmonic radiation upon irradiation at 1064 nm, believed to be due to a three-photon resonance. Thus polysilane polymers may eventually find use in laser technology.

Literature Cited

1. West R. J. Organomet. Chem. 1986, 300, 327-346.
2. West R. In Comprehensive Organometallic Chemistry; Stone, F.G.A.; Abel, E. W. Eds.; Pergamon Press: Oxford, 1983; Vol. 9, p 365-397.
3. Burkhard, C. A. J. Am. Chem. Soc. 1949, 71, 963.
4. Shortly after Burkhard's publication, a patent by Clark claimed the synthesis of soluble polysilanes, but since this work was never published it was somewhat overlooked. See H. A. Clark, U. S. Patent 2,563,005, Aug. 5, 1951, CA, 45 (1951) 60068 and U. S. Patent 2,606,879, Aug. 12, 1952, CA, 47 (1953) 347.
5. Kumada, M.; Tamao, K. Adv. Organomet. Chem. 1968, 6, 19.
6. West, R., Pure Appl. Chem., 1982, 54, 1041; West, R.; Carberry, E. Science, 1975, 189, 179.
7. Hengge, E. Topics Current Chem., 1974, 51, 1.
8. Brough, L. F.; Matsumura, K.; West, R. Angew. Chem. Int. Ed. Engl. 1979, 18, 955; Brough, L. F.; West, R. J. Am. Chem. Soc. 1981, 103, 3049.
9. Yajima, S.; Hayashi, J.; Omori, M. Chem. Lett. 1975, 931; Yamima, S.; Okamura, K.; Hayashi, J. ibid. 1975, 1209.
10. The history of this rediscovery has been colorfully told in a recent article: David, L. D. Chem. in Britain, 1987, 553.
11. Mazdyasni, K. S.; West, R.; David, L. D. J. Am. Ceram. Soc. 1978, 61, 504; West, R. David, L. D.; Djurovich, P. I.; Stearley, K. L.; Srinivasan, K.S.V.; Yu, H. J. Am. Chem. Soc. 1981, 103, 7352.
12. Trujillo, R. E.; J. Organomet. Chem. 1980, 198, C27.
13. Wesson, J. P.; Williams, T. C. J. Polym. Sci., Polym. Chem. Ed. 1979, 17, 2833.

14. Trefonas III, P.; West, R.; Miller, R. D.; Hofer, D. J. Polym. Sci., Polym. Lett. Ed. 1983, 21, 819; Trefonas III, P.; West, R. Inorg. Syn. in press.
15. Zeigler, J. M.; Harrah, L. A.; Johnson, A. W., chapter in this volume. See also Zeigler, J. M.; Harrah, L. A.; Johnson, A. W. Polym. Preprints, 1987, 28, 424.
16. Cotts, P. M.; Miller, R. D.; Trefonas III, P. T.; Maxka, J.; West. R.; Fickes, G. Macromolecules, in press.
17. Trefonas III, P.; West, R. J. Polym. Sci., Polym. Lett. Ed. 1985, 23, 469-473.
18. Stüger, H.; West, R. Macromolecules, 1985, 18, 2349-2352.
19. Trefonas III, P.; West, R.; Miller, R. D.; Hofer, D. J. Polym. Sci., Polym. Lett. Ed. 1983, 21, 823.
20. West, R; Zhang, X-H.; Djurovich, P. I.; Stüger, H. In Science of Ceramic Chemical Processing; Hench, L. L.; Ulrich, D. R., Eds.; Wiley, New York, N. Y., 1986; Chapter 36, p 337-344.
21. Trefonas III, P.; Miller, R.; West, R. J. Am. Chem. Soc. 1985, 107, 2737-2742. Photochemistry and photophysics of polysilanes is also discussed in the Chapter by Zeigler, Harrah and Johnson in this volume, ref. 15.
22. Wolff, A. R.; West, R. Applied Organomet. Chem. 1987, 1, 7-14.
23. Zhang, X-H.; West, R. Gaofenzi Tongxun (China), 1986, 4, 257-263.
24. Miller, R. D.; Sooriyakumaran, R. J. Polym. Sci., Polym. Lett. Ed. 1987, in press.
25. For short-chain oligomers the absorption maxima also depend on molecular weight, reaching a limiting value in silicon chains about 40-50 atoms in length. See ref. 17.
26. Takeda, K.; Teramae, H.; Matsumoto, N. J. Am. Chem. Soc. 1986, 108, 8186.
27. Trefonas III, P.; Damewood, Jr., J. R.; West, R.; Miller, R. D. Organometallics, 1985, 4, 1318-1319; Harrah, L. A.; Zeigler, J. M. J. Polym. Sci., Polym. Lett. Ed. 1985, 23, 209.
28. Klingensmith, K. A.; Downing, J. W.; Miller, R. D.; Michl, J. J. Am. Chem. Soc. 1986, 108, 7438; Bigelow, R. W.; McGrane, K. M. J. Polym. Sci., Polym. Phys. Ed. 1986, 24, 1233; Todesco, R. V.; Kamat, P. V. Macromolecules, 1986, 19, 196.
29. Michl, J.; Downing, J. W.; Karatsu, T.; Klingensmith, K. A.; Wallraff, G. M.; Miller, R. D., Chapter in this volume.
30. Shafiee, F.; Haller, K. J.; West, R. J. Am. Chem. Soc. 1986, 108, 5478-5482; Groth, P. Acta Chem. Scand. A, 1974, 29, 642.
31. Miller, R. D.; Rabolt, J. F.; Sooriyakumaran, R.; Fleming, W.; Fickes, G. N.; Farmer, B. L.; Kuzmany, H., Chapter in this volume.
32. Shilling, F. C.; Bovey, F. A.; Zeigler, J. M. Macromolecules, 1986, 19, 2309.
33. Wolff, A. R.; Maxka, J.; West, R. J. Polym. Sci., Polym. Chem. Ed., in press.
34. Wolff, A. R.; Nozue, I.; Maxka, J.; West, R. J. Polym. Sci., Polym. Chem. Ed., in press.
35. Mazdyasni, K. S.; West, R.; David, L. D. J. Am. Ceram. Soc. 1978, 61, 504.
36. West, R.; David, L. D.; Djurovich, P. I.; Yu, H.; Sinclair, R. Am. Cer. Soc. Bull. 1983, 62, 825-934; West, R. In Ultrastructure Processing of Ceramics, Glasses and Composites,

Hench, L. L.; Ulrich, D. R., Eds.; Wiley, New York, 1984; Chapter 19, Part 3.
37. Niihara, K.; Yamamoto, T.; Arima, J.; Takemoto, R.; Suganama K.; Watanabe, R.; Nishikawa, T.; Okumura, M. J. Polym. Sci., Polym. Chem. Ed., in press.
38. West, R.; Wolff, A. R.; Peterson, D. J. J. Radiat. Curing, 1986, 13, 35-40; Wolff, A. R.; West, R. Applied Organomet. Chem. 1987, 1, 7-14.
39. Kepler, R. G.; Zeigler, J. M.; Harrah, L. A.; Kurtz, S. R. Bull. Am. Phys. Soc. 1983, 28, 362; Kepler, R. G.; Zeigler, J. M.; Harrah, L. A. ibid. 1984, 29, 504; Kepler, R. G.; Zeigler, J.M.; Harrah, L. A.; Kurtz, S. R. Phys. Rev. B. 1987, 35, 2818.
40. Stolka, M.; Yuh, H.-J.; McGrane, K.; Pai, D. M. J. Polym. Sci., Polym. Chem. Ed., 1987, 25, 823.
41. Fujino, M. Chem. Phys. Lett. 1987, 136, 451.
42. Kazar, F.; Messier, J.; Rosilio, C. J. Appl. Phys. 1986, 60, 3040.

RECEIVED September 1, 1987

Chapter 3

Polycarbosilanes: An Overview

Dietmar Seyferth

Department of Chemistry, Massachusetts Institute of Technology, Cambridge, MA 02139

Polycarbosilanes, in their broadest definition, are organosilicon polymers whose backbone is composed of silicon atoms, appropriately substituted, and difunctional organic groups which bridge the silicon atoms, as shown in formula 1. The polycarbosilanes may be linear, or they can be cyclic or polycyclic.

$$[\mathrm{Si(R^1)(R^2)}-(\mathrm{C})]_n$$

1

Or they can be hybrids of two or all three of these. The organic "bridge," -(C)-, can be quite varied - a patent attorney's delight: CH_2, CH_2CH_2, higher $(CH_2)_n$ (≥ 3), CH=CH, $CH_2CH{=}CHCH_2$, C≡C, $CH_2C{\equiv}CCH_2$, arylene, xylylene, etc., etc. We shall restrict the scope of this discussion to the case where the connecting bridge is CH_2, i.e., the "polysilmethylenes," since a major emphasis will be on polymeric precursors for silicon carbide, i.e., on carbosilane-type preceramic polymers [1] whose pyrolysis gives a ceramic product as close as possible to the composition SiC. The polysilmethylenes may be viewed as the carbon analogs of the polysiloxanes:

$$[\mathrm{Si(CH_3)_2}-\mathrm{O}]_x \quad = \quad \text{polydimethylsiloxane}$$

$$[\mathrm{Si(CH_3)_2}-\mathrm{CH_2}]_x \quad = \quad \text{polydimethylsilmethylene}$$

0097–6156/88/0360–0021$06.25/0

Carbosilanes, defined this narrowly, as a class, monomers, cyclic and polycyclic oligomers and linear polymers, with emphasis on the cyclic and polycyclic systems, have been discussed in detail in an excellent recent book by Fritz and Matern [2].

The possibility of linear polysilmethylenes (although none were known at the time) was discussed by Rochow in his seminal book on organosilicon chemistry in 1949 [3]. At that time, two of the applicable preparative procedures had been reported, but had, as yet, only been applied to "small" molecules (eq. 1 and 2). Goodwin and Clark at Dow Corning Corporation prepared lower oligomers containing at least three silicon atoms by organomagnesium, organosodium or organoaluminum

$$(CH_3)_3SiCH_2MgCl + (CH_3)_3SiCl \xrightarrow{\text{ref.[4]}} (CH_3)_3SiCH_2Si(CH_3)_3 \quad (1)$$

$$(CH_3)_3SiCl + ClCH_2Si(CH_3)_2OEt + 2\ Na \xrightarrow{\text{ref.[5]}} (CH_3)_3SiCH_2Si(CH_3)_2OEt \quad (2)$$

routes [6] and Sommer, Mitch and Goldberg [7] using an organolithium route, prepared methylpolysilmethylenes, $CH_3[Si(CH_3)_2CH_2]_nH$, with n = 2, 3, 4, 5. These stepwise procedures, however, were not readily applicable to the synthesis of polysilmethylenes of higher molecular weight.

It was the action of reactive metals on (halomethyl)halosilanes (eq. 3) which provided a general route to such polymers. Thus, the action of sodium on $ClCH_2Si(CH_3)_2Cl$ gave $[(CH_3)_2SiCH_2]_n$ [8]. Removal

$$n\ XCH_2-\overset{R}{\underset{R'}{Si}}-X \xrightarrow{M} [RR'SiCH_2]_n + MX_2 \text{ (or 2 MX)} \quad (3)$$

of a volatile fraction, bp 150-250°C, left a residue of average molecular weight 850 (DP 11.8), with a viscosity of 300-400 cs at 25°C [8b].

In the sixties, such reactions were of greater interest as a source of 1,3-disilacyclobutanes (eq. 4) [9].

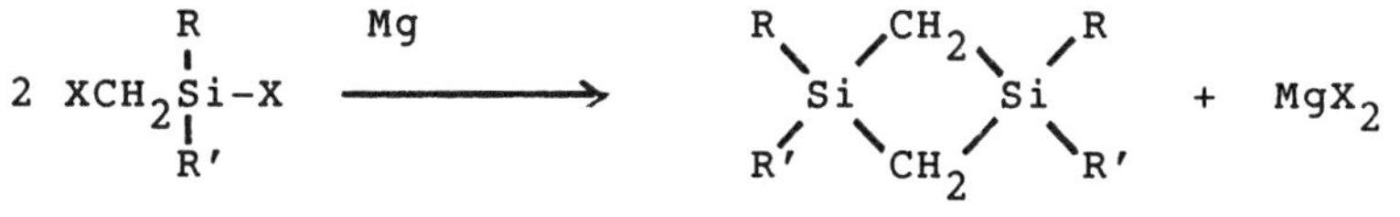

$$2\ XCH_2Si(R)(R')\text{-}X \xrightarrow{Mg} RR'Si(CH_2)_2SiRR' + MgX_2 \quad (4)$$

1,1,3,3-Tetramethyl-1,3-disilacyclobutane had been prepared earlier by Knoth and Lindsey [10], but a multistep synthesis was involved which was not generalizable to the synthesis of Si-functional 1,3-disilacyclobutanes. The reaction shown in eq. 4, provided it is carried out in the right way, represents a good, general route to 1,3-disilacyclobutanes. This reaction was reported first by Müller and his coworkers [11]. In this work, diethyl ether was used as reaction solvent and the product yield was only around 4%. Somewhat better yields were obtained by Russian workers [12], but it was the detailed studies of the (chloromethyl)chlorosilane/magnesium reaction by Kriner [13] which provided a good synthesis of 1,3-disilacyclobutanes. In most preparations of such compounds the yields generally are in the 30-60% range, although in some cases they can be greater. Higher cyclic oligomers, cyclo-$[(CH_3)_2SiCH_2]_3$ and cyclo-$[(CH_3)_2SiCH_2]_4$ in the case of $(CH_3)_2(CH_2Cl)SiCl$, also are formed.

In spite of reaction conditions chosen to optimize formation of 1,3-disilacyclobutanes ("inverse addition," i.e., addition of the magnesium to a solution of the $RR'(CH_2Cl)SiCl$, high dilution), some linear, higher molecular weight product is formed (although this is not always mentioned in the description of the experiments by other authors). Scheme 1 depicts these various products and Scheme 2 shows the sequential organomagnesium coupling reactions believed to be responsible for their formation [13a]. Hydrolytic workup linked the linear species via siloxane bonds. Formation of linear polysilmethylene was the major process which occurred when $(CH_3)_2Si(CH_2Cl)Cl$ was added to a suspension of magnesium in diethyl ether or THF. In one such reaction carried out in THF, the yield of cyclo-$[(CH_3)_2SiCH_2]_2$ was only 6.7%, that of $[(CH_3)_2SiCH_2]_3$ 1.4%, and the main product was a viscous, opaque residue, molecular weight about 930 [13b].

When starting (chloromethyl)chlorosilane contained more than one Si-Cl bond, lower yields of the 1,3-disilacyclobutane were obtained due to side reactions resulting from the availability of more Si-Cl functions (in the case of $CH_3Si(CH_2Cl)Cl_2$, 2 was produced in addition to cyclo-$[CH_3(Cl)SiCH_2]_2$) and to formation of higher yields of polysilmethylene [13, 14].

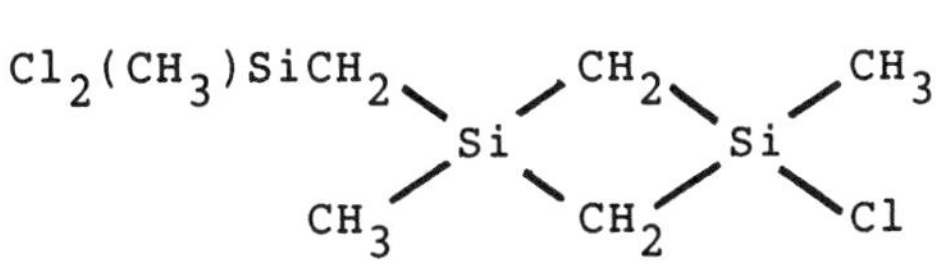

$$Cl_2(CH_3)SiCH_2(CH_3)Si(CH_2)_2Si(CH_3)Cl$$

2

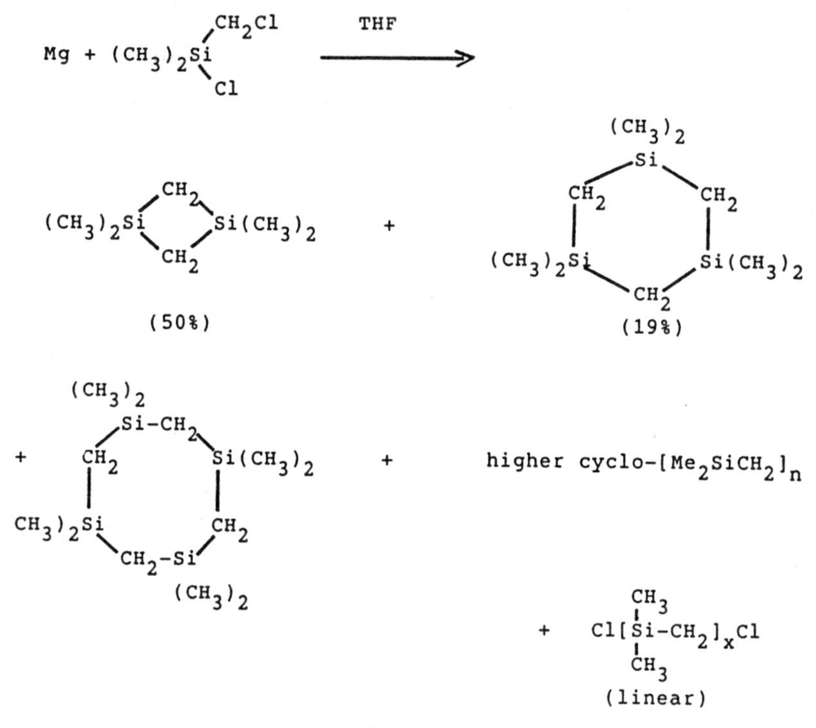
Mg + (CH3)2Si CH2Cl Cl
THF
(CH3)2Si CH2 CH2 Si(CH3)2
(50%)
+
(CH3)2 Si CH2 CH2 (CH3)2Si Si(CH3)2 CH2
(19%)
+ (CH3)2 Si-CH2 CH2 Si(CH3)2 CH3)2Si CH2 CH2-Si (CH3)2
+
higher cyclo-[Me2SiCH2]n
+ CH3 Cl[Si-CH2]xCl CH3
(linear)

Scheme I

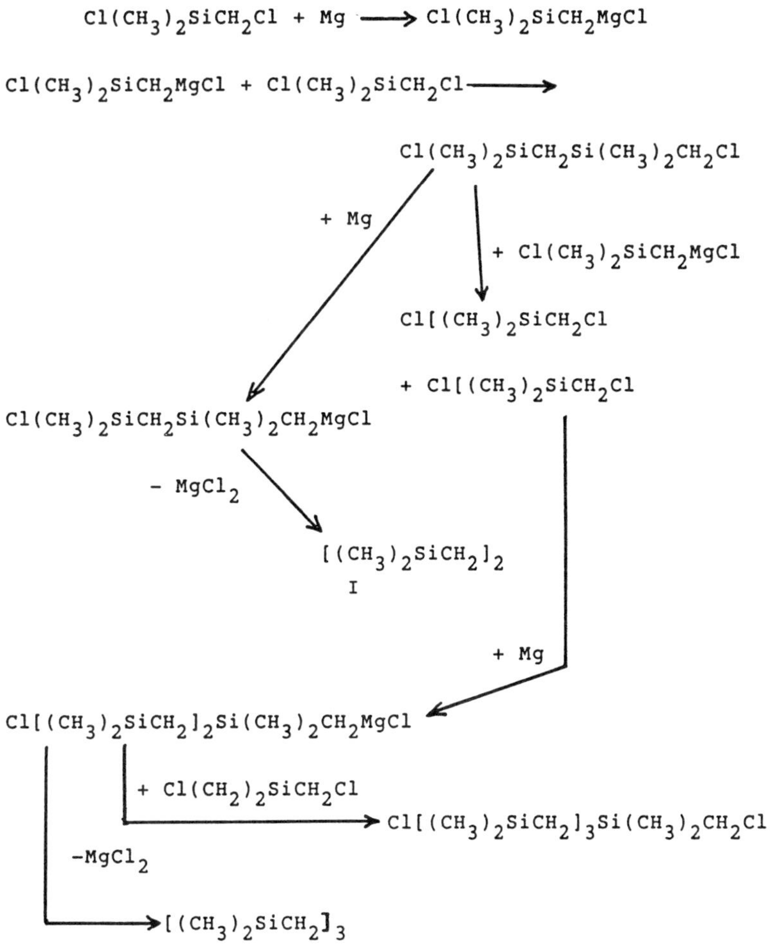

$$Cl(CH_3)_2SiCH_2Cl + Mg \longrightarrow Cl(CH_3)_2SiCH_2MgCl$$

$$Cl(CH_3)_2SiCH_2MgCl + Cl(CH_3)_2SiCH_2Cl \longrightarrow Cl(CH_3)_2SiCH_2Si(CH_3)_2CH_2Cl$$

$$Cl(CH_3)_2SiCH_2Si(CH_3)_2CH_2Cl \xrightarrow{+ Mg} Cl(CH_3)_2SiCH_2Si(CH_3)_2CH_2MgCl$$

$$Cl(CH_3)_2SiCH_2Si(CH_3)_2CH_2MgCl \xrightarrow{- MgCl_2} \underset{I}{[(CH_3)_2SiCH_2]_2}$$

$$Cl(CH_3)_2SiCH_2Si(CH_3)_2CH_2Cl \xrightarrow{+ Cl(CH_3)_2SiCH_2MgCl} Cl[(CH_3)_2SiCH_2Cl$$

$$+ Cl[(CH_3)_2SiCH_2Cl \xrightarrow{+ Mg} Cl[(CH_3)_2SiCH_2]_2Si(CH_3)_2CH_2MgCl$$

$$Cl[(CH_3)_2SiCH_2]_2Si(CH_3)_2CH_2MgCl \xrightarrow{+ Cl(CH_2)_2SiCH_2Cl} Cl[(CH_3)_2SiCH_2]_3Si(CH_3)_2CH_2Cl$$

$$Cl[(CH_3)_2SiCH_2]_2Si(CH_3)_2CH_2MgCl \xrightarrow{-MgCl_2} [(CH_3)_2SiCH_2]_3$$

Scheme II

According to the available reports discussed above, the RR'Si-$(CH_2Cl)Cl$/metal reactions (M = Na, Mg) do not appear to give high molecular weight polysilmethylenes and at this time, this reaction has not found useful application in preceramic polymer synthesis. However, the introduction of crosslinking processes into the basic metal-effected dehalogenative polymerization of $(CH_3)_2Si(CH_2Cl)Cl$ was shown by Schilling and his coworkers [15] to result in formation of polycarbosilanes whose pyrolysis gave higher ceramic yields. Thus, cocondensation of $(CH_3)_2Si(CH_2Cl)Cl$ with CH_3SiCl_3 by reaction with potassium in THF produced a branched polycarbosilane, $[((CH_3)_2SiCH_2)_x(CH_3Si)_y]_n$, which on pyrolysis to 1200°C gave a 31% yield of β-SiC. (This yield still is too low, and in further work, Schilling developed useful SiC polymer precursors which, however, are polysilanes, not polycarbosilanes [15].

Another procedure for the synthesis of 1,3-disilacyclobutanes is the pyrolysis of monosilacyclobutanes (eq. 5), [9, 14, 16], but this method has difficulties and disadvantages [14]. One of these is that polysilmethylene formation is a side-reaction when it is carried out in the gas-phase.

$$2\ RR'Si(CH_2)_3 \xrightarrow[\text{temp.}]{\text{high}} \text{cyclo-}[RR'SiCH_2]_2 + 2\ C_2H_4 \quad (5)$$

Such polymerization apparently is the sole process when the silacyclobutane thermolysis is carried out in the liquid phase (eq. 6).

$$RR'Si(CH_2)_3 \xrightarrow{\Delta} RR'Si{=}CH_2 + C_2H_4 \longrightarrow [RR'SiCH_2]_x \quad (6)$$

A more useful thermolytic polymerization which produces linear polysilmethylenes is that of 1,3-disilacyclobutanes carried out in the liquid phase. Such polymerization of 1,1,3,3,-tetramethyl-1,3-disilacyclobutane was reported first by Knoth [17] (eq. 7). This process was studied in some detail by Russian workers [18]. 1,1,3,3-Tetramethyl-1,3-disilacyclobutane is more thermally stable than 1,1-dimethyl-1-silacyclobutane.

$$(CH_3)_2Si(CH_2)_2Si(CH_3)_2 \xrightarrow{300°C} [(CH_3)_2SiCH_2]_n \quad (7)$$

Aryl and, more so, chlorine substituents on silicon enhance thermal stability of silacyclobutanes. The rate of the first-order thermal decomposition of silacyclobutanes varies inversely with the dielectric constant of the solvent used. Radical initiators have no effect on the thermal decomposition and a polar mechanism was suggested. Thermal polymerization of cyclo-$[Ph_2SiCH_2]_2$ has been reported to occur at 180-200°C. The product was a crystalline white powder which was insoluble in benzene and other common organic solvents [19].

Anionic polymerization of 1,3-disilacyclobutanes also is possible. Solid KOH and alkali metal silanolates were mentioned as being effective by Russian authors [18, 19. 20]. However, alkyllithiums, which can initiate polymerization of silacyclobutanes (eq. 8) [21], do not initiate polymerization of 1,3-disilacyclobutanes [18, 22]. The problem is one of steric hindrance.

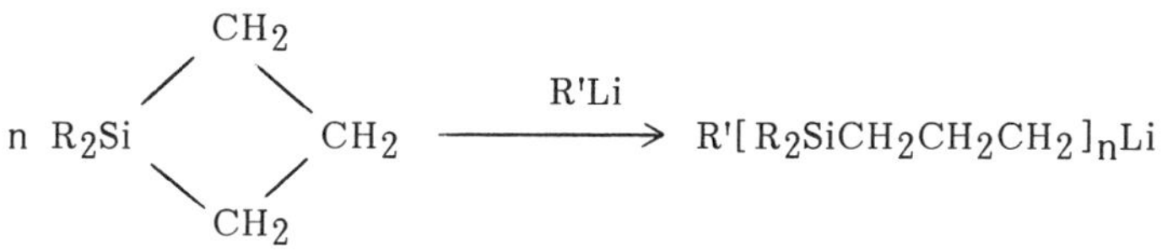

$$n\ R_2Si\langle(CH_2)_3\rangle \xrightarrow{R'Li} R'[R_2SiCH_2CH_2CH_2]_nLi \qquad (8)$$

The first intermediate, resulting from cyclo-$[(CH_3)_2SiCH_2]_2$ ring opening by RLi, is $R(CH_3)_2SiCH_2Si(CH_3)_2CH_2Li$, a bulky "neopentyl"-type reagent. Its attack at another cyclo-$[(CH_3)_2SiCH_2]_2$ molecule will not be very favorable and so the polymerization does not progress. In fact, it is possible to polymerize 3 selectively via monosilacyclobutane ring opening using an organolithium initiator [18].

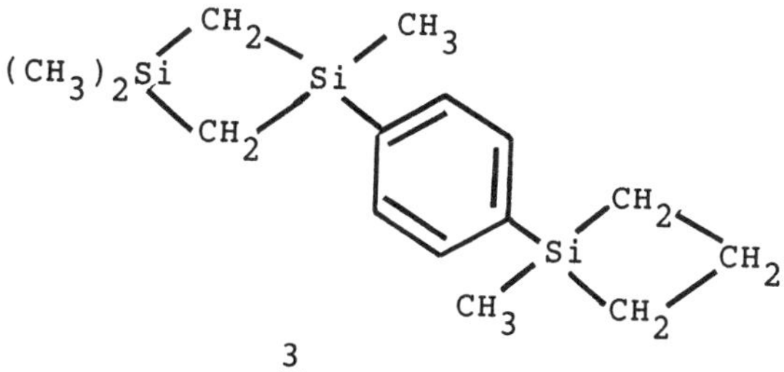

3

Transition metal catalysts are especially effective in initiating the ring-opening polymerization of 1,3-disilacyclobutanes to give polysilmethylenes, as reported by various workers [23-30]. Products ranging from low molecular weight telomers to high molecular weight polymers ($M_n \approx 10^5$) could be prepared, depending on experimental conditions. A large variety of transition metal catalysts was applicable: Pt/C, $H_2PtCl_6 \cdot 6H_2O$, $PtCl_2$, $IrCl_6^{2-}$, $RuCl_4^{2-}$, $AuCl_4^-$, $PdCl_2$, RuI_3, [olefin $PtCl_2]_2$, $(Et_2S)_2PtCl_2$, $(Bu_3P)_2Pt_2Cl_4$, $(Ph_3P)_2Pt(CH_3)_2$, $(Pr_3As)_2PtCl_2$, $PdBr_2$, $AuCl_3$, $CuCl_2$, CuCl, π-crotylnickel and -chromium compounds, and others. The telomeric products of lower molecular weight were obtained when the catalytic 1,3-disilacyclobutane ring opening was carried out in the presence of small amounts of trialkylsilanes, R_3SiH [23, 24, 26], or carbon tetrabromide [26].

The high molecular weight $[(CH_3)_2SiCH_2]_n$ polymers are very thermally stable (2.0% weight loss after heating at 450°C in vacuo for 30 min) [26]. However, pyrolysis under argon at 1000°C leaves little or no ceramic residue, so they are not useful preceramic polymers. Their thermal stability is poor in the presence of oxygen, polydimethylsiloxanes and formaldehyde and oxides of carbon being obtained as oxidation products. They are inert toward concentrated mineral acids and alkalis at room temeprature, but chlorinolysis to lower molecular weight products occurs on photoinduced chlorination at 20°C [26]. Attempted conversion of $[(CH_3)_2SiCH_2]_n$ to $[CH_3(Cl)SiCH_2]_n$ by treatment with the $(CH_3)_3$-$SiCl/AlCl_3$ silicon-methyl cleavage reagent did result in the desired methyl group cleavage, but the polymer of initial molecular weight 250,000 underwent chain scission, i.e., Si-CH_2 cleavage as well, to give $[CH_3(Cl)SiCH_2]_n$ products of average molecular weight 2300 [31].

Most of the polymerization studies were carried out with $[(CH_3)_2$-$SiCH_2]_n$. The platinum-catalyzed polymerization of cyclo-$[CH_3(EtO)$-$SiCH_2]_2$ and cyclo-$[CH_3(Cl)SiCH_2]_2$ gave only low yields of viscous liquid polymers [25]. Each of these monomers was copolymerized with cyclo-$[(CH_3)_2SiCH_2]_2$. Since the Si-H bond also is activated by transition metal catalysts, such catalyzed ring-opening polymerization of cyclo-$[CH_3(H)SiCH_2]_2$ very likely would present complications. Fairly low molecular weight $[CH_3(H)SiCH_2]_n$ oligomers were, however, accessible by another route: the $LiAlH_4$ reduction of the $[CH_3(Cl)SiCH_2]_n$ oligomers mentioned above [31]. In spite of their potentially useful reactive Si-H functionality, pyrolysis of these oligomers gave a ceramic yield of only 5%. Thus, a facile thermal cross-linking process did not occur.

Polydimethylsilylene, $[(CH_3)_2SiCH_2]_n$, has received detailed study by Mark and his coworkers at the University of Cincinnati [32]. This polymer was prepared by $H_2PtCl_6 \cdot 6H_2O$-catalyzed ring opening polymerization (at 25°C for 3 days) of 1,1,3,3-tetramethyl-1,3-disilacyclobutane. The polymeric product was fractionated by precipitation from benzene by addition of methanol and the fractions obtained were characterized using viscosity, osmotic pressure and dielectric constant measurements. Samples of the polymer which has been crosslinked by γ-radiation [32a] or a peroxide cure [32d] also were investigated (stress-strain isotherms and thermoelastic properties) in order to obtain useful information about the statistical properties of these chain molecules. The experimental and theoretical studies indicated that the most appropriate model for the $[(CH_3)_2SiCH_2]_n$ chain is one in which there is no strong preference for any conformation: most are of identical energy [32b]. Also studied was the stress-optical behavior of polydimethylsilylene [32c,d].

The polysilmethylenes, however, apparently are not useful polymers, either as such or as precursors for silicon carbide. A rubbery material could be prepared by treatment of $[(CH_3)_2SiCH_2]_n$ with dicumyl peroxide in the presence of fused silica and ground quartz, but no utility was claimed for this product [26]. The probable reason that polysilmethylenes are not useful preceramic polymers is that on pyrolysis they undergo chain scission. The reactive chain end thus generated, most likely a free radical center, then undergoes "back-biting," i.e., S_H2 attack further along the chain with resulting extrusion of a small, volatile, cyclic species (cyclo-$[(CH_3)_2SiCH_2]_n$ (n = 2,3,4...) in the case of polydimethylsilmethylene, as shown in Scheme 3. The result is that no residue is left behind as the polymer pyrolysis proceeds; all or most of the polymer is converted to volatiles. An apparent exception to this is the case of cyclo-$[H_2SiCH_2]_2$ which was reported to have been polymerized by catalytic activation with $H_2PtCl_6 \cdot 6H_2O$ at 75°C in vacuo to give a "viscous liquid or a clear cake" which was claimed to "have a linear polycarbosilane structure" [33]. Pyrolysis of this material to 900°C, it was claimed, gave an 85% yield of silicon carbide. It seems doubtful that a linear polymer has been formed.

Polysilmethylene chains also are formed in the thermal, copper-catalyzed reactions of methylene chloride and chloroform with elemental

```
   R      R      R      R      R      R
   |      |      |      |      |      |
[-Si-CH2-Si-CH2-Si-CH2-Si-CH2-Si-CH2-Si-CH2-]
   |      |      |      |      |      |
   R      R      R      R      R      R

                 |
                 |  Δ
                 ↓

   R      R      R      R      R                       R
   |      |      |      |      |    .                  |
[-Si-CH2-Si-CH2-Si-CH2-Si-CH2-Si-CH2       +        ·Si-CH2-]
   |      |      |      |      |                       |
   R      R      R      R      R                       R

               ↓
                              R
                              |
                        CH2-Si
                       /      |   \
   R      R      R    /       R    \
   |      |      |   /              \
[-Si-CH2-Si-CH2-Si                  CH2
   |      |      |  ↖               /
   R      R      R   \       R     /
                      \      |    /
                      ·CH2-Si          +    R
                              |             |
                              R            ·Si-CH2-]
                                            |
                                            R

                          ↘
                              R2
   R      R                   Si
   |      |    .           /      \
[-Si-CH2-Si-CH2    +     CH2       CH2
   |      |               |         |
   R      R             R2Si       SiR2
                           \      /
       |                     CH2
       ↓                 (volatile)

further SH2                    R
attack down                    |
the chain                  +  ·Si-CH2-]
                               |
                               R
```

Scheme III

silicon. Reported first by Patnode and Schiessler in 1945, the CH_2Cl_2 -Si/Cu reaction produced $Cl_3SiCH_2SiCl_3$, $Cl_3SiCH_2SiCl_2H$ and cyclo-$[Cl_2SiCH_2]_3$ as isolable products [34]. Also formed in this reaction was a viscous liquid product of approximate composition $[Cl_2SiCH_2]_n$. However, it was the work of Fritz and Wörsching [35] which led to the utilization of the CH_2Cl_2-Si/Cu reaction in the formation of silmethylenes of higher molecular weight. The production of such species was optimized when the reaction was carried out in a fluidized bed reactor. For ease of analysis, all Si-Cl bonds in the product mixture were reduced to Si-H; this permitted the application of HPLC in the separation of individual compounds in the product mixture. Those polysilmethylenes thus isolated and identified which contain four or more silicon atoms are shown in Figure 1. The $CHCl_3$-Si/Cu reaction was carried out under the same conditions and also was followed by Si-Cl to Si-H reduction and HPLC. The branched polycarbosilane products thus isolated that contain four or more silicon atoms are depicted in Figure 2. These mixtures (or the higher molecular weight portions of these mixtures) present interesting possibilities as preceramic materials.

The polycarbosilanes which are precursors to the Nicalon ceramic fibers are not strictly polysilmethylenes. However, their main repeat units are $[CH_3(H)SiCH_2]$ and $[(CH_3)_2SiCH_2]$, so they will be discussed here. The preparation of these polymeric precursors to silicon carbide as effected by S. Yajima and his coworkers was an important development [36]. The chemistry involved is fairly complex [36-39]. It is based on the thermal rearrangement of polydimethylsilylene (derived from sodium condensation of $(CH_3)_2SiCl_2$), initially, very likely to a polysilmethylene-type polymer as a result of a Kumada-type free radical rearrangement (which, in its simplest example converts, $(CH_3)_3SiSi(CH_3)_3$ to $(CH_3)_3SiCH_2Si(CH_3)_2H$ [40]). In the case of polydimethylsilylene, such a rearrangement would give as product $[CH_3(H)SiCH_2]_n$. However, it is clear that the thermal conversion effected at 450-470°C does not stop there. IR and NMR spectroscopic studies have shown that the Yajima polycarbosilane has a more complicated structure than $[CH_3(H)SiCH_2]_n$, and crosslinking and cyclization processes [such as that shown below (no mechanism implied)] have been suggested [41].

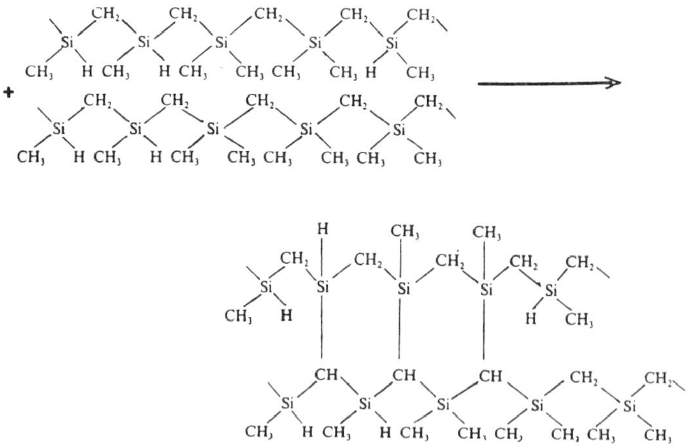

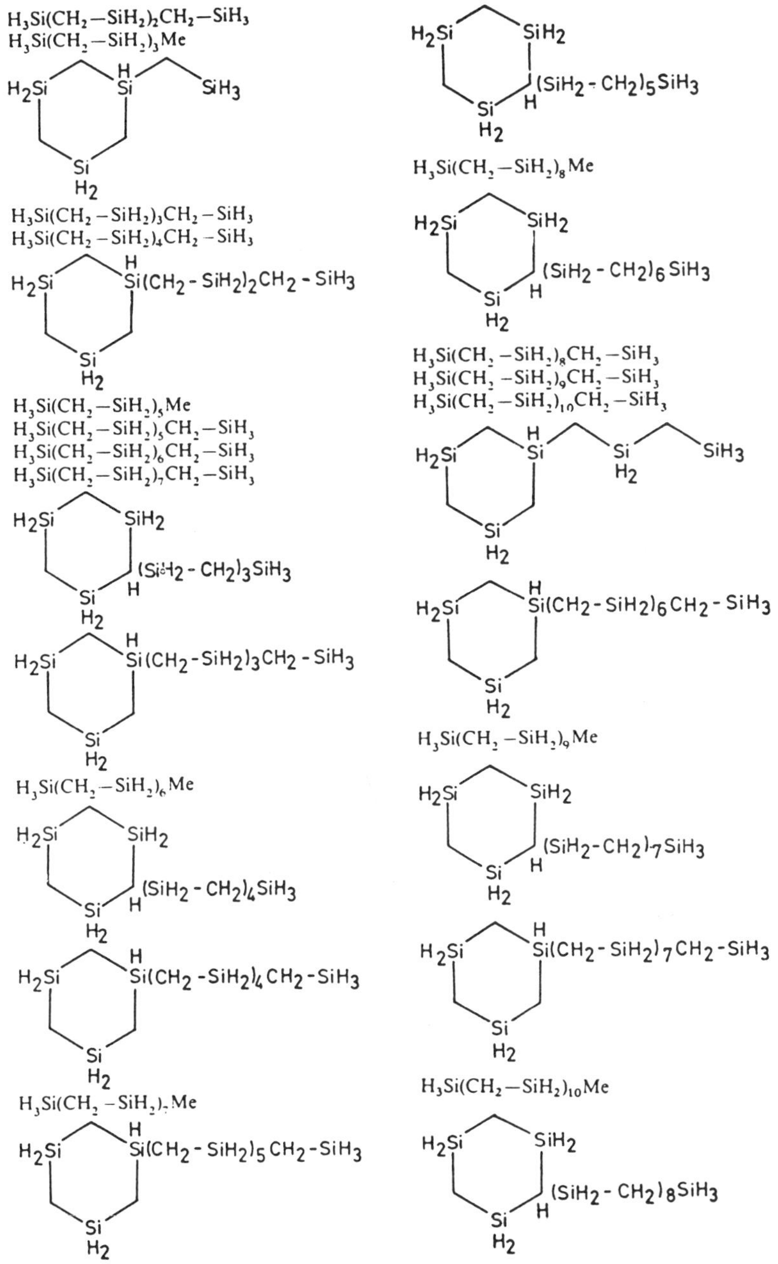

$H_3Si(CH_2{-}SiH_2)_2CH_2{-}SiH_3$

$H_3Si(CH_2{-}SiH_2)_3Me$

H_2Si / H Si / SiH_3 / Si H_2

$H_3Si(CH_2{-}SiH_2)_3CH_2{-}SiH_3$

$H_3Si(CH_2{-}SiH_2)_4CH_2{-}SiH_3$

H_2Si / H Si$(CH_2{-}SiH_2)_2CH_2{-}SiH_3$ / Si H_2

$H_3Si(CH_2{-}SiH_2)_5Me$

$H_3Si(CH_2{-}SiH_2)_5CH_2{-}SiH_3$

$H_3Si(CH_2{-}SiH_2)_6CH_2{-}SiH_3$

$H_3Si(CH_2{-}SiH_2)_7CH_2{-}SiH_3$

H_2Si / SiH_2 / $(SiH_2{-}CH_2)_3SiH_3$ H / Si H_2

H_2Si / H Si$(CH_2{-}SiH_2)_3CH_2{-}SiH_3$ / Si H_2

$H_3Si(CH_2{-}SiH_2)_6Me$

H_2Si / SiH_2 / $(SiH_2{-}CH_2)_4SiH_3$ H / Si H_2

H_2Si / H Si$(CH_2{-}SiH_2)_4CH_2{-}SiH_3$ / Si H_2

$H_3Si(CH_2{-}SiH_2)_7Me$

H_2Si / H Si$(CH_2{-}SiH_2)_5CH_2{-}SiH_3$ / Si H_2

H_2Si / SiH_2 / $(SiH_2{-}CH_2)_5SiH_3$ H / Si H_2

$H_3Si(CH_2{-}SiH_2)_8Me$

H_2Si / SiH_2 / $(SiH_2{-}CH_2)_6SiH_3$ H / Si H_2

$H_3Si(CH_2{-}SiH_2)_8CH_2{-}SiH_3$

$H_3Si(CH_2{-}SiH_2)_9CH_2{-}SiH_3$

$H_3Si(CH_2{-}SiH_2)_{10}CH_2{-}SiH_3$

H_2Si / H Si / Si H_2 / SiH_3 / Si H_2

H_2Si / H Si$(CH_2{-}SiH_2)_6CH_2{-}SiH_3$ / Si H_2

$H_3Si(CH_2{-}SiH_2)_9Me$

H_2Si / SiH_2 / $(SiH_2{-}CH_2)_7SiH_3$ H / Si H_2

H_2Si / H Si$(CH_2{-}SiH_2)_7CH_2{-}SiH_3$ / Si H_2

$H_3Si(CH_2{-}SiH_2)_{10}Me$

H_2Si / SiH_2 / $(SiH_2{-}CH_2)_8SiH_3$ H / Si H_2

Figure 1. Compounds containing four or more Si atoms from the reaction of CH_2Cl_2 with Si, after hydrogenation.

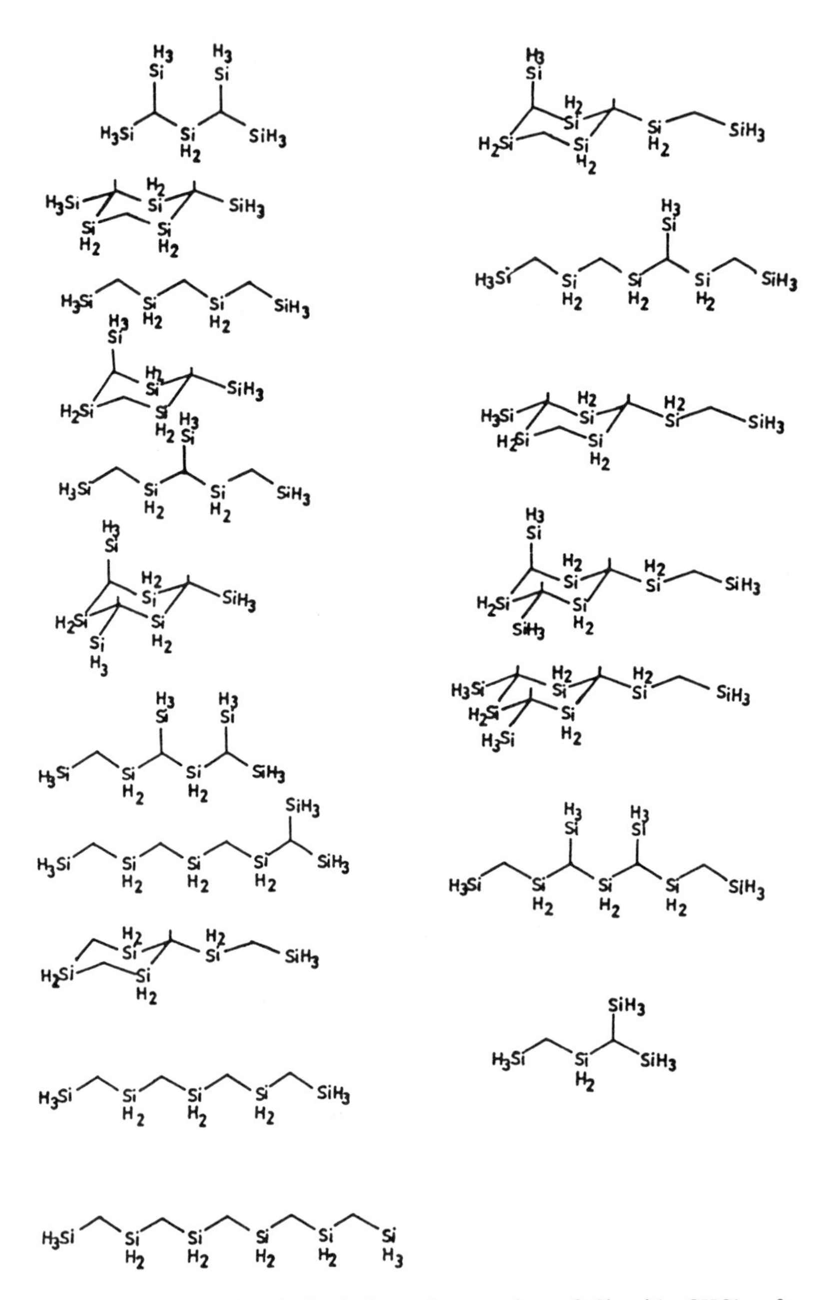

Figure 2. Compounds obtained from the reaction of Si with $CHCl_3$ after conversion of SiCl to SiH groups. Only compounds containing more than four Si atoms are shown.

Ultimately, this would result in a polymeric product containing cyclic and polycyclic units linked by linear bridges. A structure which contains such building blocks, suggested by Yajima [36], is shown below. Such a structure, on pyrolysis, should leave a higher yield of ceramic residue than the linear $[CH_3(H)SiCH_2]_n$. This is indeed the case. In one procedure, further heating of the yellowish-brown, glassy polycarbosilane which was obtained by heating the polydimethylsilylene at 450-470°C in an autoclave was heated at 280°C/1 mm Hg to remove volatiles. Pyrolysis of this material left a residue of silicon carbide and free carbon. The pyrolysis process has been investigated in some detail by Hasegawa and Okamura using spectroscopy, TGA-DTA and gas evolution studies [42]. The ceramic yields obtained on pyrolysis of such products usually are around 60%. These polycarbosilanes can be melt-spun into continuous fibers and the latter, after an oxidative cure step to render them infusible, can be pyrolyzed to give black ceramic fibers, the so-called Nicalon fibers which are commercially available [43].

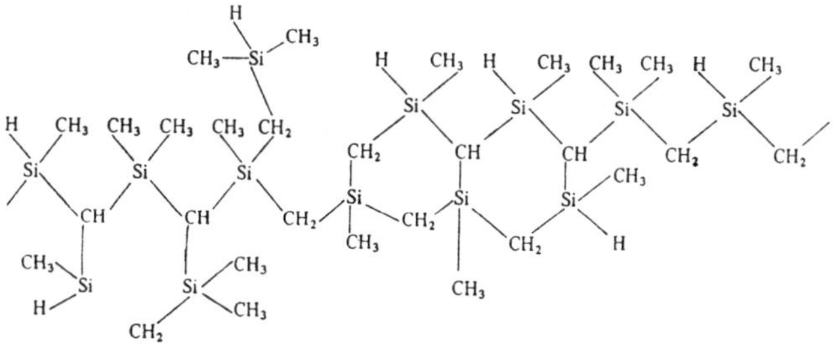

There are other variations of the basic Yajima polycarbosilane. In one case, the polydimethylsilylene is mixed with a few weight percent of a polyborodiphenylsiloxane prior to thermal rearrangement. This additive (which ends up in the polymeric product) serves to accelerate the rearrangement process so that it can be carried out at 350°C at amospheric pressure [42]. In another variation, the polycarbosilane obtained from the polyborodiphenylsiloxane-induced thermal rearrangement (M_n = 950) was heated in xylene at 130°C to 220°C with $Ti(OC_4H_9)_4$ to produce a more highly crosslinked (M_n = 1674) Ti-containing polycarbosilane [44]. Crosslinking occurs via Si-O-Ti-O-Si linkages and pyrolysis of this polymer in a stream of nitrogen to 1700°C gave a ceramic product that contained (by X-ray diffraction) TiC as well as SiC. Pyrolysis to 1300°C gave a ceramic yield of around 72%. This type of Ti-containing polycarbosilane is a commercial product of Ube Industries.

Another process for silicon carbide fibers, developed by Verbeek and Winter of Bayer AG [45], also is based on polymeric precursors which contain $[SiCH_2]$ units, although linear polysilmethylenes are not involved. The pyrolysis of tetramethylsilane at 700°C, with provision for recycling of unconverted $(CH_3)_4Si$ and lower boiling products, gave a polycarbosilane resin, yellow to red-brown in color, which was soluble in aromatic and in chlorinated hydrocarbons. Such resins could be melt-spun but required a cure-step to render them infusible before they were pyrolyzed to ceramic fibers. Oxidative and hydrolytic cures were mentioned. The pyrolysis of tetramethylsilane has been studied in great detail by G. Fritz and his coworkers and modern analytical, separation and spectroscopic techniques have enabled them to isolate and identify a large number of the products of this very complicated process [46] and to propose mechanisms for the thermolytic processes which result in their formation. (This work, which spans some 30 years, is summarized in ref. 1). During $(CH_3)_4Si$ pyrolysis, linear, cyclic and then polycyclic carbosilanes of ever increasing complexity are formed [46]. Some of the latter are shown in Figures 3 and 4. It will be appreciated that further condensation processes, occurring with further loss of the hydrogen and methyl substituents on carbon and silicon, will take place when products such as these are pyrolyzed. Their structures already give a hint of the final product, silicon carbide.

Yajima and his coworkers also studied tetramethylsilane thermolysis and used the resulting polycarbosilane in a fiber spinning investigation [42]. In their experiments, in which the recirculating pyrolysis procedure of Fritz [47] was used, the yield of usable polycarbosilane (bp > 200°C/1 mm Hg) of sufficiently high molecular weight (M_n = 774) was 6.8%. Its pyrolysis in a nitrogen stream to ca. 1300°C gave a ceramic yield of about 73%.*

In conclusion, the lesson learned from the research carried out to date on the subject of polycarbosilanes is that the general rule that linear, noncrosslinked polymers are not suitable preceramic polymers applies here as well. Crosslinked network-type polymers are needed. Such structures can be generated in more than one way, but in the case of the polycarbosilanes they have, to date, been obtained mainly by thermolytic routes: thermal treatment (with or without other chemical additives) in the case of the Yajima polycarbosilanes and the thermolysis of tetramethylsilane in the case of the Bayer process-derived polycarbosilane.

Although we have not discussed other types of polycarbosilanes, some may be mentioned to show what is available in terms of other specific examples of general formula $\underline{1}$. Already mentioned during the course of our discussion have been polymers of type $[R_2SiCH_2CH_2CH_2]_n$, obtained by ring-opening polymerization of monosilacyclobutanes. Such polymerization may be carried out thermally [48] or catalytically [20, 21, 24] (for reviews see ref. 9 and 18). Ring-opening polymerization of 1,1-disubstituted silacyclopentanes can be induced with an aluminum halide catalyst at temperatures ranging from 20°-80°C (eq. 9) [49]. (Silacyclohexanes did not polymerize under these conditions).

$$R_2Si\langle(CH_2-CH_2)(CH_2-CH_2)\rangle \xrightarrow{[AlX_3]} [R_2SiCH_2CH_2CH_2CH_2]_n \qquad (9)$$

Also of interest are the thermal ring opening polymerizations of 1,1-dimethyl-1-silabenzocyclobutene (eq. 10) [50] and of 1,1,2,2-tetramethyl-1,2-disilabenzocyclobutene (eq. 11) [51]. The polymer produced in the latter reaction is a polycarbodisilane.

---

* Excessively high pyrolysis temperatures should be avoided: D. Seyferth and J. J. Pudvin, CHEMTECH $\underline{11}$ (1981) 230.

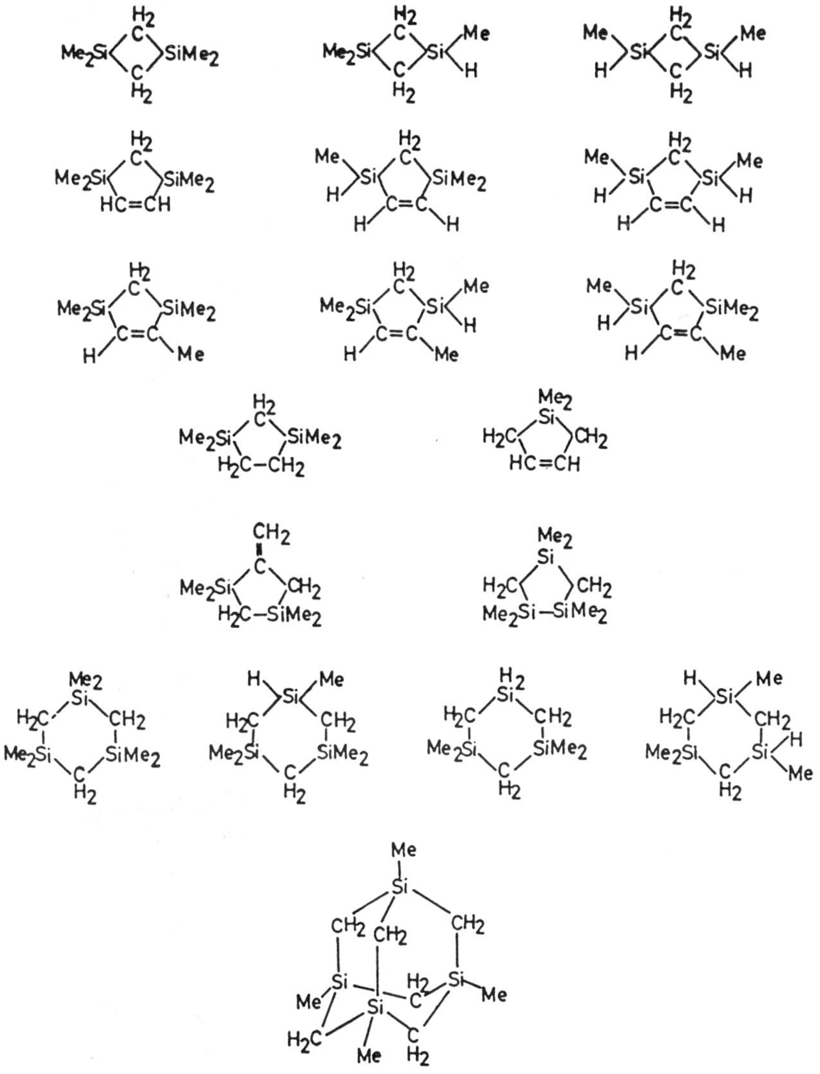

Figure 3. Compounds produced by the pyrolysis of $SiMe_4$. *Continued on next page.*

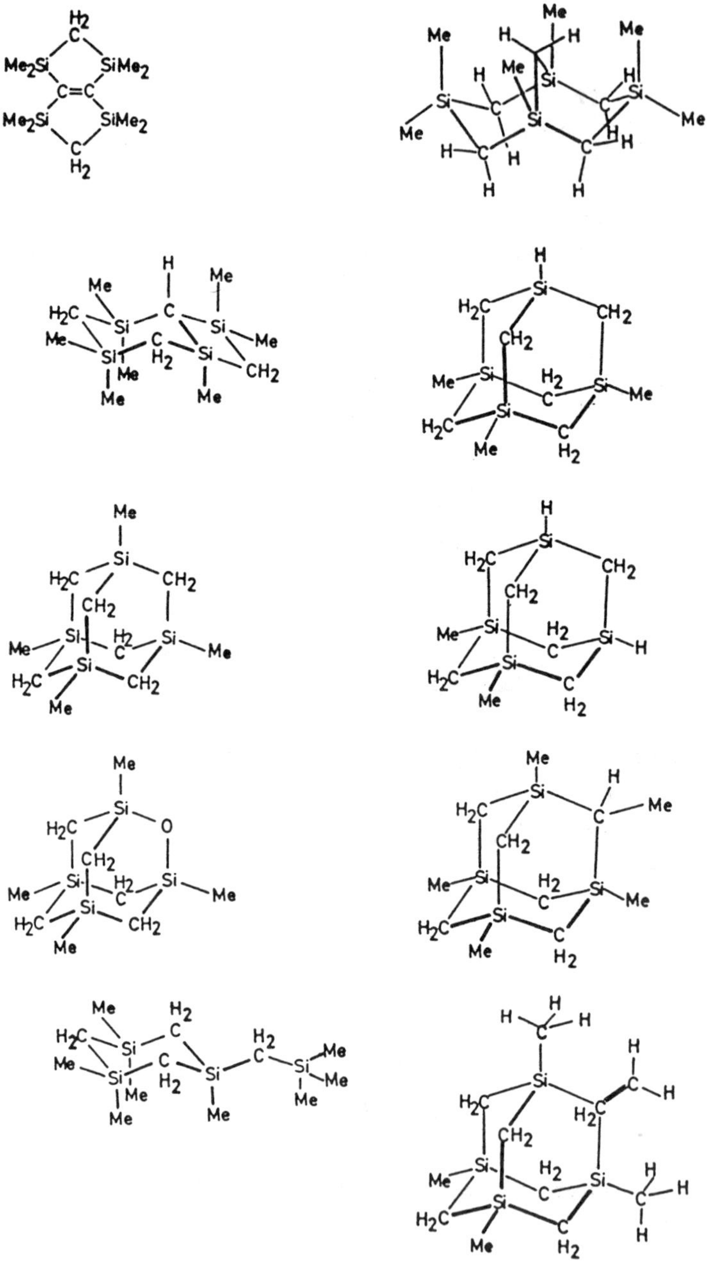

Figure 3. *Continued.* Compounds produced by the pyrolysis of $SiMe_4$.

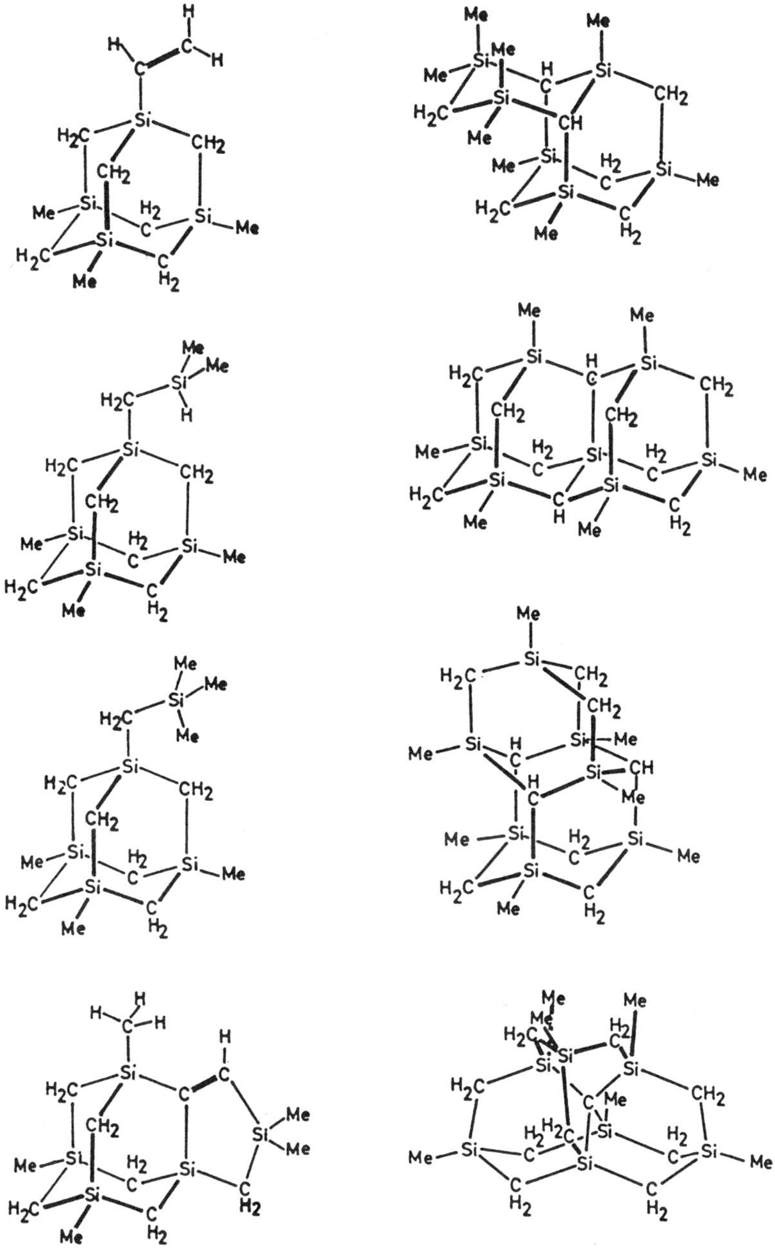

Figure 3. *Continued.* Compounds produced by the pyrolysis of $SiMe_4$.

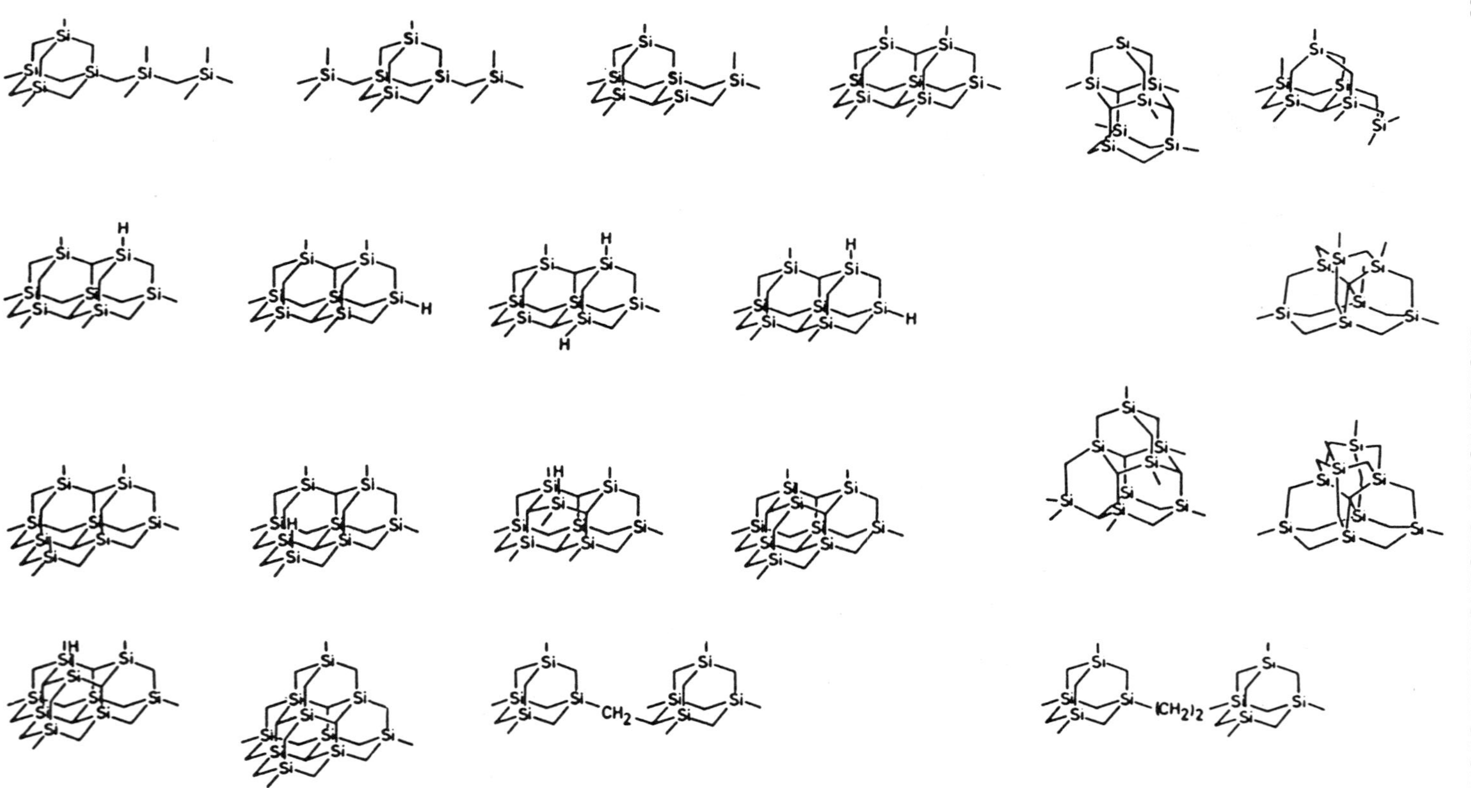

Figure 4. Carbosilanes containing five to eight Si atoms, isolated by column chromatography and HPLC.

$$\text{C}_6\text{H}_4(\text{CH}_2)(\text{Si}(\text{CH}_3)_2) \xrightarrow{\Delta} [\text{CH}_2\text{C}_6\text{H}_4\text{Si}(\text{CH}_3)_2]_n \quad (10)$$

$$\text{C}_6\text{H}_4(\text{Si}(\text{CH}_3)_2)_2 \xrightarrow{\text{room temp.}} [(\text{CH}_3)_2\text{Si}\text{C}_6\text{H}_4\text{Si}(\text{CH}_3)_2]_n \quad (11)$$

These ring opening reactions proceed via Si-C (or in the latter case, Si-Si) bond scission. Another type of ring-opening polymerization involves C=C bond scission (eq. 12) [52].

$$\text{(CH}_3)_2\text{SiC}_4\text{H}_6 \xrightarrow[\text{iBu}_3\text{Al, room temp.}]{\text{WCl}_6,\ \text{Na}_2\text{O}_2,} [-\text{CH=CHCH}_2\text{Si(CH}_3)_2\text{CH}_2-]_n \quad (12)$$

The hydrosilylation reaction has served in the polymerization of vinylsilicon hydrides (eq. 13) [53]. For the most part, β-addition predominated.

$$\text{CH}_2\text{=CHSiR}_2\text{H} \xrightarrow{\text{H}_2\text{PtCl}_6\cdot 6\text{H}_2\text{O}} [(\text{CH}_2\text{CH}_2\text{SiR}_2)_x(\text{CH(CH}_3)\text{-SiR}_2)_y]_n \quad (x>>y) \quad (13)$$

Polysilaacetylides have been prepared by the reaction of BrMgC≡CMgBr [54], $(\text{MgC}\equiv\text{C})_n$ and LiC≡CLi [55] with dichlorosilanes (eq. 14). In general, high molecular weight products were not obtained.

$$\text{RR}'\text{SiCl}_2 + (\text{MgC}\equiv\text{C})_n \longrightarrow [\text{SiRR}'\text{-C}\equiv\text{C}]_n \quad (14)$$

Silaarylene and silaarylene-siloxane polymers have been reviewed [56]. Recent work of interest on such systems is that of Koide and Lenz on poly(silaarylene-siloxanes) [57] and of Ishikawa et al. [58] on poly(p-(disilanylene)phenylenes), synthesized as shown in eq. 15.

$$\mathrm{ClSi(R)(CH_3)-C_6H_4-Si(R)(CH_3)Cl \xrightarrow{Na} [-Si(R)(CH_3)-Si(R)(CH_3)-C_6H_4-]_n} \quad (15)$$

(R = Et and Ph)

Ishikawa's polymers readily underwent photochemical degradation at specific frequencies, an observation which suggested their application as positive UV resists.

Acknowledgment

I acknowledge with thanks the generous support of our research in the polycarbosilane area by the Air Force Office of Scientific Research.

Literature Cited

1. For a review of preceramic polymers, "Ceramics via Polymer Pyrolysis," see: K. J. Wynne and R. W. Rice, Ann. Rev. Mater. Sci., 14 (1984) 297-334.
2. G. Fritz and E. Matern, "Carbosilanes. Syntheses and Reactions," Springer-Verlag, Berlin, 1986.
3. E. G. Rochow, "An Introduction to the Chemistry of the Silicones," Wiley, New York; 1st edition, 1949, Chapter 3.
4. (a) L. H. Sommer, G. M. Goldberg, J. Gold and F. C. Whitmore, J. Am. Chem. Soc., 69 (1947) 980; (b) B. Bluestein, J. Am. Chem. Soc., 70 (1948) 3068.
5. J. T. Goodwin, Jr., W. E. Baldwin and R. R. McGregor, J. Am. Chem. Soc., 69 (1947) 2247.
6. (a) J. T. Goodwin, Jr., U.S. patent 2,507,512 (1950); Chem. Abstr., 45 (1951) 3410; (b) H. A. Clark, U.S. patent 2,507,517 (1950); Chem.Abstr., 45 (1951) 2259; (c) H. A. Clark, U.S. patent 2,507,521 (1950); Chem. Abstr., 45 (1951) 2265; (d) J. T. Goodwin, Jr., U.S. patent 2,544,079 (1951); Chem. Abstr., 45 (1951) 6654; (e) J. T. Goodwin, Jr., Brit. patent 684,102 (1952).
7. L. H. Sommer, F. A. Mitch and G. M. Goldberg, J. Am. Chem. Soc., 71 (1949) 2746.
8. (a) J. T. Goodwin, Jr., U.S. patent 2,483,972 (Oct. 4, 1949); Chem. Abstr., 44 (1950) 2011e; Brit. patent 624,550 (June 10, 1949); Chem. Abstr., 44 (1950) 6632; (b) J. T. Goodwin, Jr., U.S. patent 2,607,791 (1952); Chem. Abstr., 48 (1954) 13732.
9. For a review, which includes the synthesis, properties and reactions of 1,3-disilacyclobutanes see: R. Damrauer, Organometal. Chem. Rev., A, 8 (1972) 67-133.
10. W. H. Knoth and R. V. Lindsey, Jr., J. Org. Chem., 23 (1958) 1392.
11. R. Müller, R. Köhne and H. Beyer, Chem. Ber., 95 (1962) 3030.
12. N. S. Nametkin, V. M. Vdovin, V. I. Zav'yalov and P. L. Grinberg, Izv. Akad. Nauk SSSR, Ser. Khim. (1965) 929.
13. (a) W. A. Kriner, J. Org. Chem., 29 (1964) 1601. (b) W. A. Kriner, U.S. patent U.S. 3,178,392 (1965).
14. N. Auner and J. Grobe, J. Organomet. Chem., 188 (1980) 151.
15. (a) C. L. Schilling, Jr., J. P. Wesson and T. C. Williams, Amer. Ceram. Soc. Bull., 62 (1983) 912. (b) C. L. Schilling, Jr., J. P. Wesson and T. C. Williams, J. Polymer Sci., PolymerSymposia, No.

70 (1983) 121. (c) C. L. Schilling, Jr., Brit. Polymer J., 18 (1986) 355.

16. (a) N. S. Nametkin, V. M. Vdovin, L. E. Gusel'nikov and V. I. Zav'yalov, Izv. Akad. Nauk SSSR, Ser. Khim., (1966) 584 [563]. (b) N.S. Nametkin, L. E. Gusel'nikov, V. M. Vdovin, P. L. Grinberg, V. I. Zav'yalov and V. D. Oppengeim, Dokl. Akad. Nauk S.S.S.R., 171 (1966) 630[1116]. (c) N. S. Nametkin, R. L. Ushakova, L. E. Gusel'nikov, E. D. Babich and V. M. Vdovin, Izv. Akad. Nauk S.S.S.R., Ser. Khim., (1970) 1676 [1589]. (d) L. E. Gusel'nikov and M. C. Flowers, Chem. Commun., (1967) 864. (e) M. C. Flowers and L. E. Gusel'nikov, J. Chem. Soc. B, (1968) 419. (f) C. M. Golino, R. D. Bush and L. H. Sommer, J. Am. Chem. Soc., 97 (1975) 7371. (g) N. S. Nametkin, V. M. Vdovin and V. I. Zav'yalov, Izv. Akad. Nauk S.S.S.R., Ser. Khim., (1965) 1448. (h) N. Auner and J. Grobe, J. Organomet. Chem., 197 (1980) 13. (i) N. Auner and J. Grobe, J. Organomet. Chem., 197 (1980) 147. (j) N. Auner and J. Grobe, Z. anorg. allg. Chem., 485 (1982) 53.
17. W. H. Knoth, U.S. 2,850,514 (1958); Chem. Abstr., 53 (1959) 4166a.
18. Review of ring opening reactions of silacyclobutanes: N. S. Nametkin and V. M. Vdovin, Izv. Akad. Nauk SSSR, Ser. Khim. (1974) 1153.
19. N. S. Nametkin, V. M. Vdovin and V. I. Zav'yalov, Dokl. Akad. Nauk SSSR, 162 (1965) 824.
20. N. S. Nametkin, V. M. Vdovin and V. I. Zav'yalov, Izv. Akad. Nauk SSSR Ser. Khim. (1964) 203.
21. N. S. Nametkin, V. M. Vdovin, V. A. Poletaev and V. I. Zav'yalov, Dokl. Akad. Nauk. SSSR, 175 (1967) 1068.
22. D. Seyferth and J. Mercer, unpublished work, 1976.
23. N. S. Nametkin, V. M. Vdovin and P. L. Grinberg, Izv. Akad. Nauk SSSR, Ser. Khim. (1964) 1123.
24. D. R. Weyenberg and L. E. Nelson, J. Org. Chem. 30 (1965) 2618.
25. W. A. Kriner, J. Polymer Sci., Part A-1, 4 (1966) 444.
26. W. R. Bamford, J. C. Lovie and J. A. C. Watt, J. Chem. Soc. C (1966) 1137.
27. G. Levin and J. B. Carmichael, J. Polymer Sci., Part A-1, 6 (1968) 1.
28. N. S. Nametkin, N. V. Ushakov and V. M. Vdovin, Vysokomolekul. Soedin., 1 (1971) 29.
29. V. S. Poletaev, V. M. Vdovin and N. S. Nametkin, Dokl. Akad. Nauk SSSR, 203 (1972) 1324.
30. V. A. Poletaev, V. M. Vdovin and N. S. Nametkin, Dokl. Akad. Nauk SSSR, 208 (1973) 1112.
31. E. Bacque', J.-P. Pillot, M. Birot and J. Dunogues, Paper presented at the NATO Advanced Workshop on "The Design, Activation and Transformation of Organometallics into Common and Exotic Materials," Cap d'Agde, France, Sept. 1-5, (1986).
32. (a) J. H. Ko and J. E. Mark, Macromolecules, 8 (1975) 869. (b) J. E. Mark and J. H. Ko, Macromolecules, 8 (1975) 874. (c) M. A. Llorente, J. E. Mark and E. Saiz, J. Polymer Sci., Polymer Phys. Ed., 21 (1983) 1173. (d) V. Galiatsatos, and J. E. Mark, Polymer Preprints, 28 (1987) 258.
33. T. L. Smith, Jr., U.S. patent 4,631,179 (Dec. 23, 1986).
34. (a) W. I. Patnode and R. W. Schiessler, U.S. 2,381,000 and 2,381,002 (1945); Chem. Abstr., (1945) 4888. (b) See also: A. V.

Topchiev, N. S. Nametkin and V. I. Setkin, Chem. Techn., 4 (1952) 508; Dokl. Akad. Nauk SSSR, 82 (1952) 927.

35. G. Fritz and A. Worsching, Z. anorg. allg. Chem., 512 (1984) 103.
36. S. Yajima, Am. Ceram. Soc. Bull., 62 (1983) 893.
37. S. Yajima, M. Omori, J. Hayashi, K. Okamura, T. Matsuzawa and C. Liaw, Chem. Lett. (1978) 551.
38. For early Yajima papers see: (b) S. Yajima, J. Hayashi and M. Omori, Chem. Lett. (1975) 931. (c) S. Yajima, K. Okamura and J. Hayashi, Chem. Lett. (1975) 1209. (d) S. Yajima, M. Omori, J. Hayashi, K. Okamura, T. Matsazawa and C. Liaw, Chem. Lett. (1976) 551. (e) S. Yajima, J. Hayashi, M. Omori and K. Okamura, Nature, 260 (1976) 683. (f) S. Yajima, K. Okamura, J. Hayashi and M. Omori, J. Am. Ceram. Soc., 59 (1976) 324.
39. S. Yajima, Y. Hasegawa, J. Hayashi and M. Iimura, J. Mater. Sci., 13 (1978) 2569.
40. (a) K. Shiina and M. Kumada, J. Org. Chem., 23 (1958) 139. (b) H. Sakurai, R. Koh, A. Hosomi and M. Kumada, Bull. Chem. Soc. Japan, 39 (1966) 2050. (c) H. Sakurai, A. Hosomi and M. Kumada, Chem. Commun. (1986) 930.
41. Y. Hasegawa and K. Okamura, J. Mater. Sci., 21 (1986) 321.
42. Y. Hasegawa and K. Okamura, J. Mater. Sci., 18 (1983) 3633.
43. Nippon Carbon Co. in Japan, via Dow Corning Corp. in the United States.
44. S. Yajima, T. Iwai, T. Yamamura, K. Okamura and Y. Hasegawa, J. Mater. Sci., 16 (1981) 1349.
45. W. Verbeek and G. Winter, Ger. Offenlegungsschrift 2,236,078 (1974).
46. G. Fritz and K.-P. Worns, Z. anorg. allg. Chem., 512 (1984) 103.
47. G. Fritz, J. Grobe and D. Kummer, Advan. Inorg. Chem. Radiochem., 7 (1965) 349.
48. N. S. Nametkin, V. M. Vdovin and V. I. Zav'yalov, Izv. Akad. Nauk SSSR Ser. Khim. (1965) 1448.
49. (a) V. M. Vdovin, K. S. Pushchevaya, N. A. Belikova, R. Sultanov, A. F. Plate and A. D. Petrov, Dokl. Akad. Nauk SSSR, 136 (1961) 96. (b) N. S. Nametkin, V. M. Vdovin, K. S. Pushchevaya and V. I. Zav'yalov, Izv. Akad. Nauk SSSR, Ser. Khim. (1965) 1453.
50. (a) N. S. Nametkin, V. M. Vdovin, E. Sh. Finkel'shtein, M. S. Yatsenko and N. V. Ushakov, Vysokomolekul. Soedin. B11 (1969) 207. (b) J. C. Salamone and W. L. Fitch, J. Polymer Sci., Part A 9 (1971) 1741.
51. K. Shiina, J. Organomet. Chem., 310 (1986) C57.
52. H. Lammens, G. Sartori, J. Siffert and N. Sprecher, J. Polymer Sci. B, Polymer Lett., 9 (1971) 341.
53. (a) J. W. Curry, J. Am. Chem. Soc., 78 (1956) 1686. (b) J. W. Curry, J. Org. Chem., 26 (1961) 1308.
54. L. K. Luneva, A. M. Sladkov and V. V. Korshak, Vysokomolekul. Soedin. A9, No. 4 (1967) 910.
55. D. Seyferth, Third International Conference on Ultrastructure Processing of Ceramics, Glasses and Composites, San Diego, Feb. 23-27, 1987.
56. W. R. Dunnavant, Inorg. Macromol. Rev. 1 (1971) 165.
57. N. Koide and R. W. Lenz, J. Polymer Sci., Polymer Symposia, No. 70 (1983) 91.
58. (a) M. Ishikawa, N. Hongzhi, K. Matsusaki, K. Nate, T. Inoue and H. Yokono, J. Polymer Sci., Polymer Lett. Ed., 22 (1984) 669. (b) M. Ishikawa, Polymer Preprints, 28 (1987) 426.

RECEIVED October 23, 1987

Chapter 4

Soluble Polysilane Derivatives: Chemistry and Spectroscopy

R. D. Miller[1], J. F. Rabolt[1], R. Sooriyakumaran[1], W. Fleming[1], G. N. Fickes[2], B. L. Farmer[3], and H. Kuzmany[4]

[1]IBM Almaden Research Center, San Jose, CA 95120–6099
[2]Department of Chemistry, University of Nevada, Reno, NV 89557
[3]Mechanical and Materials Engineering, Washington State University, Pullman, WA 99164
[4]Institut of Festkorperphysik, University of Vienna, Vienna, Austria

Polysilanes represent an interesting class of radiation sensitive polymers for which new applications have been discovered. Even though the polymer backbone is composed only of silicon-silicon single bonds, all high molecular weight polysilanes absorb strongly in the UV. The position of this absorption depends not only on the nature of the substituents, but also on the conformation of the backbone. In this regard, materials where the polymer backbone is locked into a planar zigzag conformation absorb at much longer wavelengths than comparable materials where the backbone is either disordered or nonplanar (e.g., helical). A planar zigzag backbone is often observed for symmetrically substituted derivatives in the solid state when there are also significant intermolecular interactions such as side chain crystallization. These materials are strongly thermochromic. We will discuss the thermochromism exhibited by a number of polysilanes in the solid state. Polysilane derivatives are also quite sensitive to light and ionizing radiation. The predominant process is chain scission leading to the production of lower molecular weight fragments. Certain applications which are dependent on this radiation sensitivity are described.

Polysilane derivatives, i.e., polymers containing only silicon in the backbone, are attracting considerable attention as a new class of radiation sensitive polymers with interesting spectroscopic properties (1). The first diaryl substituted polymers were probably first synthesized over 60 years ago by Kipping (2). In 1949, Burkhard (3) reported the preparation of the simplest substituted derivative poly(dimethylsilane). These early materials attracted relatively little scientific interest because of their intractable nature until the pioneering work of Yajima, who showed that poly(dimethylsilane) could be converted through a series of pyrolytic steps into β-SiC fibers with very high tensile strengths (4-7). Around the same time, a number of workers reported the preparation of a soluble homopolymer (8) and some soluble copolymers (9-10). The preparation of soluble derivatives restimulated interest in substituted silane polymers and, as a result, a number of new applications for high molecular weight polysilane derivatives have recently appeared. In this regard, they have been investigated as oxygen insensitive initiators for vinyl polymerization (11)

0097–6156/88/0360–0043$06.00/0

and as a new class of polymeric photoconductors exhibiting high hole mobilities (12-14). The discovery that polysilane derivatives were radiation sensitive, coupled with their stability in oxygen plasmas, has also led to a number of microlithographic applications (15-19).

Synthesis

High molecular weight, linear, substituted silane polymers are usually prepared by a Wurtz type coupling of the corresponding dichlorosilanes initiated by sodium dispersed in an aromatic or hydrocarbon solvent (1,15,23). Recently, lower molecular weight materials have been prepared from monosubstituted silanes using a variety of titanium containing catalysts (20,21). We have prepared a large number of soluble polysilanes by the modified Wurtz coupling (23). The mechanism of polymer formation is not known with certainty, but it appears to be surface initiated, since the use of soluble aromatic radical anions does not lead to the production of high polymer. Recently, some evidence has been presented to support the involvement of radicals in the polymerization process (22).

$$R^1R^2SiCl_2 + Na \xrightarrow{\Delta} \left(-\underset{R^2}{\overset{R^1}{\mathrm{Si}}}- \right)_n + 2NaCl$$

Our initial studies (23) were performed in toluene, and Table I shows the results from the polymerization of a number of representative monomers. The data reported in Table I are for "direct" addition of the monomer to the sodium dispersion. Inverse addition often leads to higher molecular weights, although the overall polymer yields are usually lower (15,23). The results in Table I show that, under these reaction conditions, a bimodal molecular molecular weight distribution is normally obtained. Furthermore, it is obvious that the crude polymer yields drop precipitously as the steric hindrance in the monomer increases.

In an effort to improve the yield of high polymer, a solvent study was conducted using a typical sterically hindered monomer dichloro-di-n-hexylsilane, and the results are shown in Table II. The use of as little as 5% by volume of diglyme in toluene led to a 6-7 fold increase in the yield of polymer. As the proportion of diglyme cosolvent decreased, the fraction of high molecular weight material in the isolated polymer increased. This polymerization can be also conducted around room temperature by using ultrasound activation both during the generation of the dispersion (24) and during the monomer addition. Interestingly, the characteristic blue color of the reaction mixture was not observed during the addition of monomer to the dispersion in pure xylene, but appeared soon after some diglyme cosolvent was added, with an attending large increase in the viscosity of the reaction mixture. The use of diglyme to improve the yield of high polymer is not limited to dichloro-di-n-hexylsilane and we have achieved significant improvements with a number of other dialkyl substituted monomers. Our preliminary studies indicate that there is usually little advantage in the use of polyether cosolvents for the polymerization of aryl substituted monomers unless the aryl ring contains a polar group such as an alkoxy substituent. Initially, it was felt that the role of the polyether solvent was cation complexation, thus promoting anionic propagation (15). However, the use of varying amounts of the specific sodium ion complexing agent, 15-crown-5, in the polymerization of dichloro-di-n-hexylsilane failed to significantly improve the polymerization yield and led to low molecular weight materials (entries 6 and 7). Furthermore, the addition of 16% by volume of the nonpolar solvent heptane to the toluene also resulted in a good yield of high molecular weight material. These results imply that specific cation complexation probably does not play a significant role in the polymerization process and suggest instead that there is a significant bulk solvent effect for this particular example.

Table I. Some Representative Polysilanes Produced in Toluene Using Sodium in a Direct Addition Mode

Polymer	Yield (%)	$\overline{M}_n \times 10^{-3}$	$\overline{M}_w \times 10^{-3}$	$\overline{M}_w/\overline{M}_n$	R
$(PhMeSi)_n$	60	107 5.6	193 9.9	1.81 1.69	0.72
$(\beta$-Phenethyl $MeSi)_n$	24	83.2 2.5	240 3.1	2.9 1.2	4.6
(n-PrMe)Si_n	32	297 7.4	644 13.3	2.17 1.79	0.27
(n-Hexyl $MeSi)_n$	12	281 14.6	524 20.5	1.86 1.40	2.4
(n-Dodecyl $MeSi)_n$	8	172	483	2.81	—
(cyc-Hexyl $MeSi)_n$	10	300 3.2	804 4.5	2.67 1.40	8.7
$[(C_6H_{13})_2Si]_n$	6	1120 1.10	1982 1.2	1.77 1.13	3.12
$[(C_{10}H_{21})_2Si]_n$	3	521 1.18	1693 1.42	3.2 1.21	5.3

Table II. Solvent Effects on the Polymerization of Dichloro-di-*n*-Hexylsilane

Entry	Additive in Toluene	Yield (%)	$\overline{M}_n \times 10^{-3}$	$\overline{M}_w \times 10^{-3}$	$\overline{M}_w/\overline{M}_n$	R $\left(\frac{\text{High Molecular wt}}{\text{Low Molecular wt}}\right)$
1	Toluene only	6	1982 1.2	1120 1.1	1.77 1.13	3.12
2	30% Diglyme	37	1073 31.7	561 13.5	1.9 2.3	1.42
3	10% Diglyme	36	1358 26.6	712 13.0	1.9 2.1	2.61
4	5% Diglyme	34	1008 22.3	539 12.6	1.87 1.79	3.41
5	25% Diglyme 75% Xylene (Ultrasound)	22	37.6	14.0	2.7	—
6	15-Crown-5 0.5 mmol/mmol Na	18	6.4	4.5	1.42	—
7	15-Crown-5 1.1 mmol/mmol Na	13	6.4	4.6	1.38	—
8	16% Heptane	27	1386 1.12	679 1.09	2.04 1.03	9.61

Spectral Properties and Polymer Morphology

One of the most intriguing properties of catenated silane derivatives is their unusual electronic spectra (25). The observation that low molecular weight silicon catenates absorb strongly in the UV was first reported by Gilman (26,27) and has been extensively studied (25). In this regard, the high molecular weight, substituted silane polymers all absorb strongly in the UV-visible region. This is an unusual feature for a polymer containing only sigma bonds in the backbone framework and differs markedly from comparable carbon backbone analogs. Earlier theoretical studies successfully treated this phenomenon using a linear combination of bond orbitals model (LCBO) (25). These predictions have been confirmed more recently using more sophisticated theoretical approaches (28-32).

Our initial spectroscopic studies on unsymmetrically substituted, atatic polysilanes derivatives resulted in a number of preliminary conclusions (33). We found that both the λ_{max} and the ϵ/Si-Si bond depended somewhat on the molecular weight, increasing rapidly at first with increased catenation and finally approaching limiting values for high molecular weight materials (33). These limiting values seem to be reached around a degree of polymerization of 40-50. Subsequent substituent studies at ambient temperatures suggested that the alkyl derivatives absorb from 305-325 nm with those polymers containing sterically demanding substituents being red shifted to longer wavelengths. The effect of the steric bulk of substituents on the absorption spectra of alkyl substituted polymers has also been reported by other workers (19). Those polymers containing aryl substituents directly bonded to the silicon backbone are usually red shifted to longer wavelengths ($\lambda_{max} \sim$ 335-350 nm). This shift, which has been previously documented for short silicon catenates (34), was suggested to result from interactions between the σ and σ^* levels of the silicon backbone and the π and π^* states of the aromatic moiety (25). It has recently been predicted from theoretical studies, that the extent of this perturbation should also be dependent on the geometric orientation of the aromatic substituent relative to the silicon backbone (31).

Recently, variable temperature studies on certain polysilane derivatives both in solution and in the solid state indicate that the nature of the substituent effects is more complex than first indicated. This is shown dramatically in Figure 1. The absorption spectrum of poly(di-n-hexylsilane)(PDHS) in solution is not unusual and shows a broad featureless long wavelength maximum around 316 nm. The spectral properties of films of PDHS were, however, quite unexpected (35). Figure 1 shows that a spectrum of PDHS run immediately after baking at 100°C to remove solvent resembles that taken in solution except for the anticipated broadening. However, upon standing at room temperature, the peak at 316 nm is gradually replaced by a sharper band around 372 nm. This process is completely thermally reversible upon heating above 50°C and recooling. DSC analysis of the sample revealed a strong endotherm (ΔH = 4.0 Kcal/mol) around 42°C. The corresponding cooling exotherm was observed around 20°C, the exact position being somewhat dependent on the cooling rate. This thermal behavior is reminiscent of that observed for a number of isotactic polyethylene derivatives substituted with long chain alkyl substituents (36-39). For the polyolefins, the thermal transition has been attributed to the crystallization of the hydrocarbon side chains into a regular matrix. However, the significant changes in the electronic spectrum of PDHS at the transition temperature suggest that the nature of the backbone chromophore is also being altered during the process.

IR and Raman spectroscopic studies on films and powders of PDHS indicate that the hexyl side chains are crystallizing into a hydrocarbon type matrix (40). This is indicated by the presence of a number of sharp characteristic alkane bands which become dramatically broadened above the transition temperature. Similar changes are observed for n-hexane below and above the melting point. CPMAS ^{29}Si NMR studies on PDHS also show that the rotational freedom of the side chains increases markedly above the transition temperature (41,42). All of the spectral evidence supports the suggestion that the side chains of PDHS are crystallizing below 42°C and that side chain melting is to a large extent associated with the observed thermal endotherm.

While it was somewhat unexpected that conformational changes in a sigma bonded backbone would produce such a dramatic effect on the absorption spectrum, this is not totally without precedent in silicon catenates. In this regard, Bock et al. (43) have suggested that conformational equilibration in short chain silicon catenates is responsible for the complexity of their photoelectron spectra. Recent theoretical studies (32) on larger silicon catenates have not only confirmed the dependence of the electronic properties on the backbone conformation, but have also suggested that the all trans, planar zigzag conformation should absorb at significantly longer wavelengths than the gauche conformations. This data strongly suggests that the red shift observed for PDHS upon cooling below 42°C is associated with a backbone conformational change initiated by side chain crystallization and, furthermore, that the backbone below the transition temperature is trans or planar zigzag. This hypothesis was confirmed by wide-angle X-ray diffraction and Raman studies on stretch-aligned films of PDHS (44,45). At room temperature, WAXD analysis yielded over 40 distinct reflections from 5-45°, 20 of which were indexed (44). The layer line spacing of 4.07Å and the appearance of a strong meridional spot on the third layer line in the inclined film confirms the planar zigzag nature of the polymer backbone. At 60°C, a single sharp reflection remains on the equator (13.5Å) corresponding to the silicon-silicon interchain distance. The maintenance of this strong reflection for long periods of time above the transition temperature suggests that all order is not lost, even though the side chain melting transition is associated with a conformational disordering of the silicon backbone resulting in a 60 nm blue shift in the absorption spectrum (44).

This thermochromic behavior is not limited to PDHS itself but extends to other symmetrical derivatives with longer n-alkyl chains (38,42). The thermochromism of the higher homologs is, however, frequently more complex than that observed for PDHS. Although all of the higher homologs examined (up to C_{14}) show side chain crystallization as evidenced by their IR and Raman spectra and the observation of a reversible endotherm (ΔH = 4.5-8.0 Kcal/mol) by DSC (Tm ~ 43-55°C), the nature of the spectral shifts displayed by these materials is variable. For example, although the heptyl and octyl derivatives behave almost identically to PDHS, the decyl derivative is more complex (see Table III). In the latter case, there exist at least two phases, each observable by DSC analysis. The low temperature phase, characterized by a thermal transition around 31°C, has a UV absorption maximum around 350 nm. Another phase, with a transition near 45°C, absorbs at much longer wavelengths, around 380 nm. The previous analysis would suggest that the phase which absorbs at the longer wavelength is more likely to contain the silicon backbone in a planar zigzag conformation. The tetradecyl derivative apparently forms a single phase (Tm ~ 54°C) and absorbs at 347 nm. The nature of the backbone conformation(s) for those materials absorbing around 350 nm has not yet been determined but it is most probably not planar zigzag. Apparently, the longer alkyl chains can still crystallize without enforcing a planar zigzag conformation of the backbone. A less likely alternative than a nonplanar backbone for those materials which absorb around 350 nm would be a planar zigzag backbone conformation where the silicon-silicon valence angles are significantly distorted from those values observed for PDHS by the size of the substituents and/or the mode in which they crystallize. Due to the nature of the orbital interactions in sigma bonded systems, the HOMO-LUMO excitation energy will be quite sensitive to changes in these valence angles even more so than to changes in the silicon-silicon bond lengths (31). On the basis of the above information, we submit that significant intermolecular interactions in the solid state (e.g., side chain crystallization) are a necessary but not always sufficient condition to force a planar zigzag conformation of the silicon backbone and in their absence the backbone either disorders or adopts alternative nonplanar conformations.

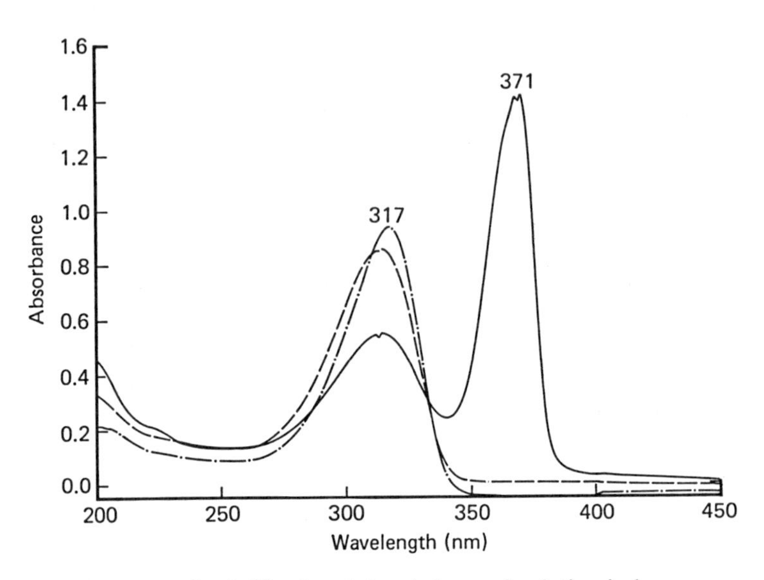

Figure 1. UV spectra of poly(di-n-hexylsilane); (– • –) solution in hexane; (– – –) film immediately after baking at 100°C; (——) film upon standing 3h at 21°C.

Table III. Transition Temperatures and Absorption Maxima of Some Di-*n*-alkyl Polysilane Films

Polymer	λ shift (nm)	T_m(°C)
poly(di-n-hexyl silane)	317 → 374	42
poly(di-n-heptyl silane)	317 → 374	45*
poly(di-n-octyl silane)	317 → 374	43*
poly(di-n-decyl silane)	320 → 383	45
	(320 → 350)	(28)
poly(di-n-tetradecyl silane)	322 → 347	54

*measured by IR

If the alkyl substituent in a dialkyl substituted polysilane is either too short or is branched, it cannot pack properly to allow side chain crystallization, and hence, one would expect very different, properties from PDHS and its higher homologs. In order to study this possibility, we synthesized and characterized the di-n-butyl (PDBS), di-n-pentyl (PDPS) and di-5-methylhexyl polymers (PDMHS). A film of the branched chain 5-methylhexyl polymer absorbed at 315 nm and the position of this maximum was relatively insensitive to temperature. Likewise, no thermal transition was observed in the region where the side chain melting transition was seen for PDHS and its higher homologs.

The thermal and optical characteristics of PDBS and PDPS are very similar and the latter will be used for illustrative purposes. The λ_{max} of PDNPS both in solution and as a solid film occurs around 314 nm. The absorption maximum of the film is unusually narrow relative to other polysilane samples with a full width at half height of 22 nm (see Figure 2). Cooling to $-60°C$ has little effect except for the appearance of a small, broad shoulder at 340 nm. Heating, however, results in a dramatic change which is shown in Figure 2. Above 75°C, the absorption broadens markedly ($w_{1/2} = 58$ nm) but the λ_{max} remains essentially unchanged. DSC analysis reveals the presence of a weak reversible endotherm ($\Delta H = 0.45$ Kcal/mol) at 74°C. It is significant that the nature of this transition is very different from that observed in PDHS (42°C, $\Delta H = 5$ Kcal/mol). In the latter case, the transition has been attributed to a combination of side chain melting and the subsequent disordering of the planar zigzag backbone. The other spectroscopic data for PDPS (46) is also very different from that described for PDHS. Figure 3 shows a comparison of the Raman spectra for the two materials. Particularly obvious is the absence of the strong characteristic band at 689 cm^{-1} for PDHS which had previously been assigned as a silicon-carbon vibration in the rigidly locked planar zigzag conformation. The CPMAS ^{29}Si NMR spectra of PDHS and PDPS are also markedly different (Figure 4). In particular, it should be noted that the silicon signal for PDPS, which is relatively narrow, is significantly upfield from that of PDHS below the transition temperature. It is interesting that at 87°C, which is above the respective phase transition temperatures of each polymer, the silicon signal of each sample becomes a narrow, time averaged resonance around -23.5 ppm relative to TMS. All of the comparative spectral data suggests that PDPS exists in a regular structure in the solid state below 74°C which is, however, different from the planar zigzag form of PDHS.

We have successfully stretch-aligned samples of PDPS, and the film diffraction pattern is compared with that of PDHS in Figure 5. From an analysis of this data, it is obvious that the backbone of PDPS is not planar zigzag but is, in fact, helical (44). The diffraction data suggests that PDPS exists as a 7/3 helix in the solid state where each silicon-silicon bond is advanced by approximately 30° from the planar zigzag conformation (Figure 6). A stable helical conformation for polysilane derivatives containing bulky substituents in the absence of intermolecular interactions is also consistent with recent calculations (47). The data shows for the first time that significant deviation from the trans backbone conformation, even in a regular structure, results in a large blue shift (~60 nm) in the UV absorption spectrum. On the basis of the results described, it now seems that the absorption spectra normally observed for alkyl polysilane derivatives ($\lambda_{max} \sim 305$-325 nm) may arise either from conformational disorder in the backbone or from systematic deviations from the planar zigzag conformation even in regular structures and do not represent the intrinsic limiting absorption for an alkyl substituted polysilane chain.

In a continuing study of substituent effects on the spectra of polysilane derivatives, we have succeeded in the preparation of the first soluble poly(diarylsilane) homopolymers. Materials of this type have traditionally proved to be insoluble and intractable. Very recently, West and co-workers have reported the preparation of some soluble copolymers which contain diphenylsilylene units (48,49).

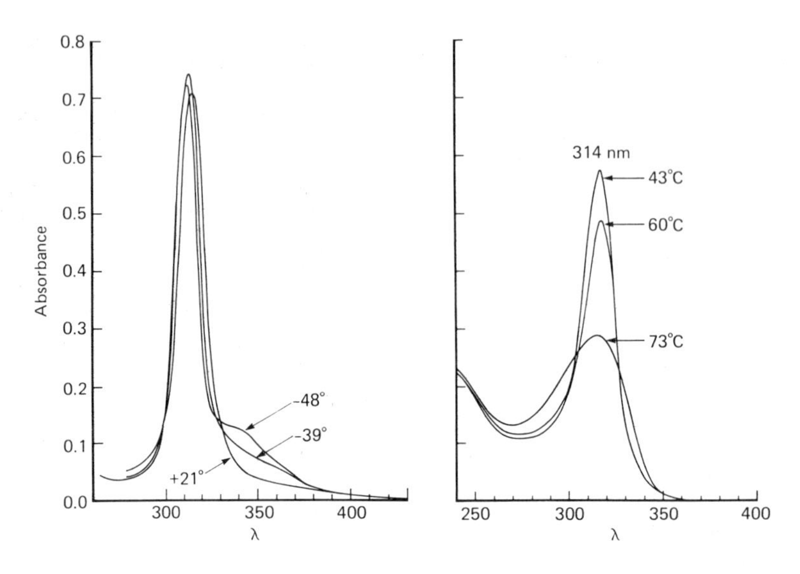

Figure 2. Variable temperature UV spectra of a poly(di-n-pentylsilane) (PDPS) film.

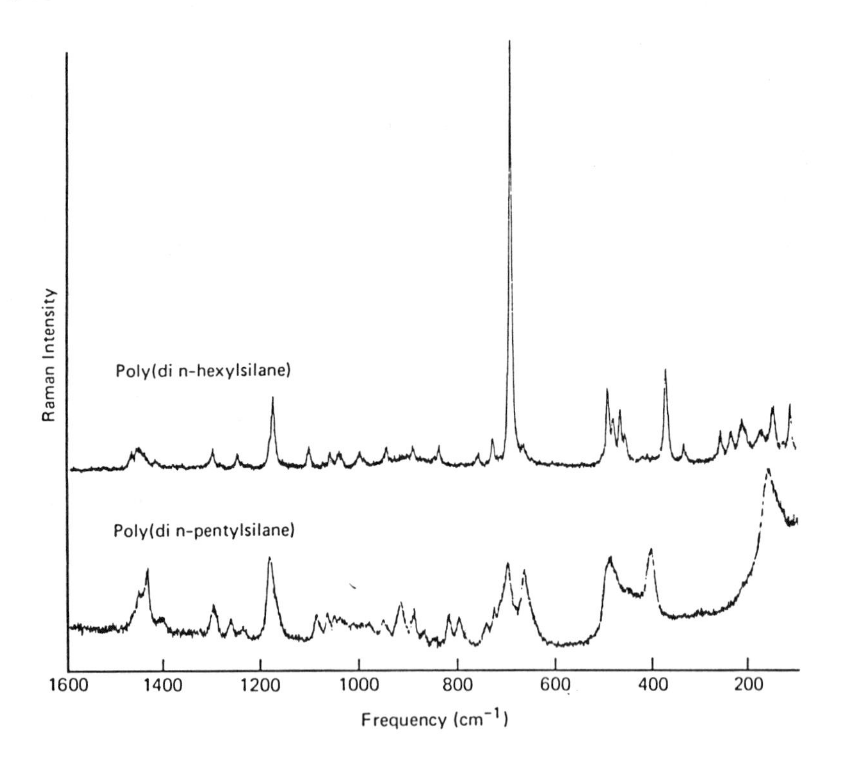

Figure 3. Raman spectra of PDHS (upper) and PDPS (lower) at 21°C.

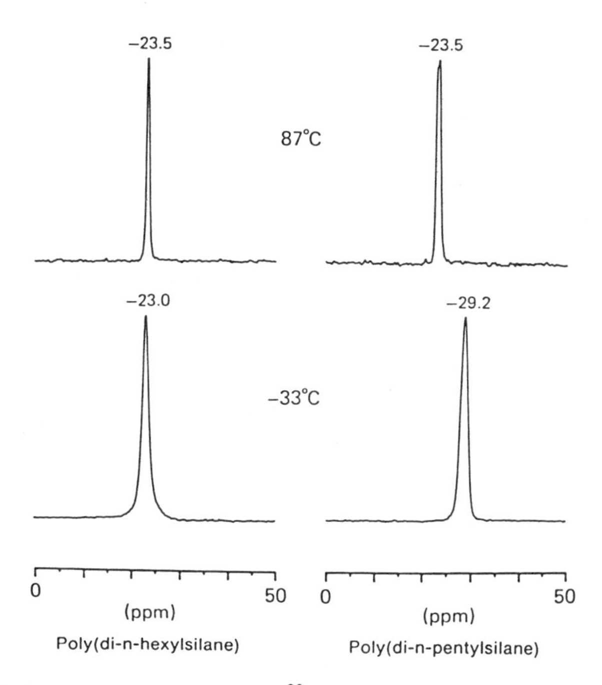

Figure 4. Variable temperature CPMAS ^{29}Si NMR spectra of PDHS (left) and PDPS (right).

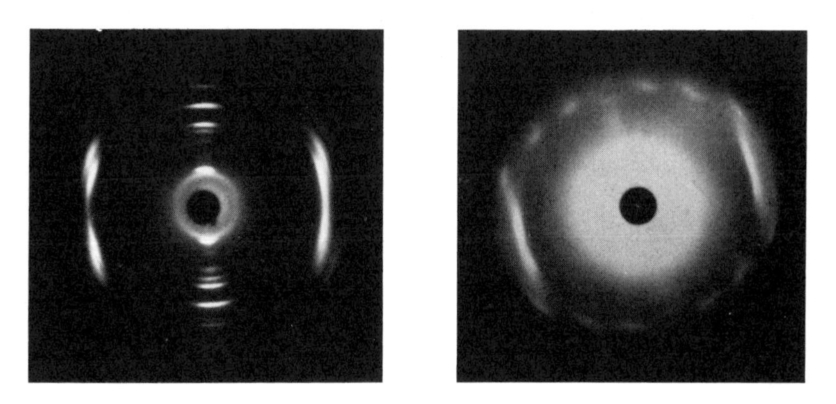

Figure 5. Room-temperature, wide-angle X-ray diffraction (WAXD) pictures of stretch-oriented samples of PDHS (left) and PDPS (right).

The soluble substituted diaryl homopolymers were prepared by the condensation of the respective substituted silyl dichlorides with sodium (50). The yields of these polymerizations were predictably low (~10%) as expected based on previous studies on sterically hindered monomers.

The most extraordinary feature of the soluble poly(diarylsilanes) is their remarkable absorption spectra. In this regard, they absorb around 400 nm, making them by far the most red shifted of any polysilane derivatives studied thus far. Some representative spectral data is shown in Table IV. Included in the table for comparison are data for poly(methylphenylsilane) and a random copolymer containing diphenylsilylene units. The long wavelength absorption band of the diaryl polysilanes in solution is exceptionally narrow (11-16 nm) for a polymeric material containing a backbone chromophore. Surprisingly, there is very little difference in the spectra recorded either in dilute solution or as films except for some broadening in the solid sample.

The origin of the large spectral red shifts observed for these materials is under active investigation. At this point, however, they seem much too large to be attributed to typical electronic substituent effects alone. In this regard, poly(phenylmethylsilane) is red shifted from typical atatic dialkyl derivatives by ~25-30 nm. This shift has been attributed to the electronic interaction of the substituent π and π^* orbitals with the backbone σ and σ^* levels. Even if the addition of a second aromatic group contributed similarly to the spectrum, which is highly unlikely based on the study of model compounds (34,51,52), the materials would be expected to absorb around 360-365 nm. In this regard, the random 1:1 copolymer prepared from the condensation of diphenyl silyl dichloride with hexylmethylsilyl dichloride absorbs around 350 nm. It therefore seems likely that the explanation of the unexpectedly large red shifts observed for the soluble diaryl polysilane homopolymers requires something other than a simple electronic substituent effect.

One intriguing, albeit tentative, explanation is that these spectral shifts also are conformational in origin. It is interesting that the magnitude of the observed shifts (50-60 nm) relative to poly(methylphenylsilane) is very similar to that observed for the poly(di-n-alkylsilanes) when they undergo a change from a disordered backbone to a regular planar zigzag arrangement (35,40). Although the close correspondence between the magnitude of the spectral shifts observed in each system may be fortuitous, it does raise the possibility that the diaryl derivatives adopt an extended planar zigzag conformation even in solution without the intervention of significant intermolecular interactions. This argument does not necessarily require that these polymers must exist exclusively as rigid rods, but the position of the long wavelength transition would argue that all-trans runs of significant length be present in solution. While the suggestion that the origin of the red shifts observed for the soluble diaryl polysilanes is conformational is intriguing, the final answer must await further structural and spectroscopic studies both in solution and in the solid state.

Photochemistry

Since both the λ_{max} and the ϵ_{SiSi} depend on the molecular weight up to a degree of polymerization of 40-50 (33), any process which induces significant chain scission should result in bleaching of the initial absorbance. Conversely, crosslinking processes which maintain or increase the molecular weight of the polymer should lead to very little bleaching. All of the polysilane derivatives which we have examined thus far undergo significant bleaching upon irradiation (15) both in solution and as films (see Figure 7 for a representative example) albeit the rate varies somewhat with structure. In general, the alkyl substituted materials bleach more rapidly than the aryl derivatives in the solid state and among the former those with bulky substituents usually bleach most rapidly. GPC examination of the polymer solutions after irradiation show that the molecular weight is continuously

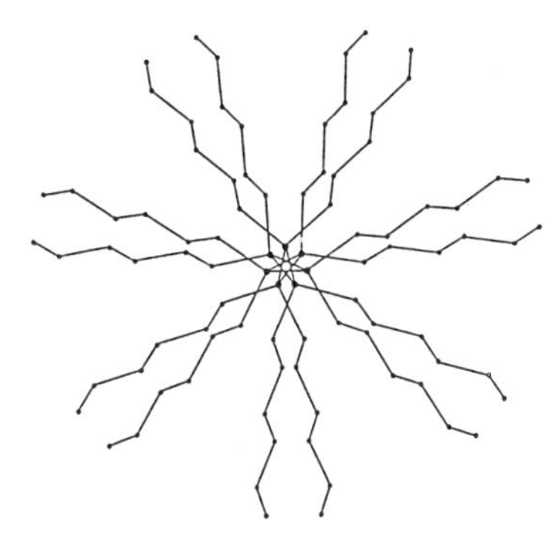

Figure 6. Projection of the 7/3 helical conformation of PDPS onto a plane normal to the helical axis. Silicon atoms are represented by the larger filled circles, carbon atoms by the smaller.

Table IV. UV–Visible Spectroscopic Data for Some Soluble Diaryl Polysilanes in Hexane

Polymer	λ_{max}	$\epsilon \times 10^{-3}$/SiSi
[structure]	390	10.2
[structure]	395	26.6
[structure]	390	16.2
[structure]	376	3.4
[structure]	397	23.3
[structure] (n-C_6H_{13})	400	21.3
[structure]	394	18.6
$(PhMeSi)_n$	340	9.0
$-[(Ph)_2Si]_x-$co$-[$Me Hexyl Si$]_y-$	350	5.5

reduced. Alkyl derivatives such as poly(methyldodecylsilane) cleanly scission and the molecular weight decreases regularly upon irradiation. The GPC traces of such materials show no high molecular weight tail. On the other hand, aromatic polysilane derivatives such as poly(phenylmethylsilane) are not cleanly converted to lower molecular weight materials upon irradiation and the GPC analysis of these irradiated materials clearly show the presence of a higher molecular tail suggesting that concurrent crosslinking is occurring (33). Materials such as PDHS which absorb at ~370 nm after standing for some time to allow side chain crystallization to orient the backbone into a planar zigzag configuration can also be bleached upon irradiation at 365 nm with the formation of lower molecular weight fragments. The soluble poly(diarylsilanes) are bleached very readily upon irradiation at 404 nm in solution, but much more slowly in the solid state. One possible explanation for this latter observation is that the absence of readily available α-hydrogens for disproportionation of the silyl chain radicals produced by photoscission promotes rapid recombination resulting in chain repair (vide infra).

The quantum yields for both chain scission (Φ_s) and crosslinking (Φ_x) can be estimated from plots of $1/\overline{M}n$ and $1/\overline{M}_w$ versus the absorbed dose in photochemical and other radiation induced processes (51,52). In general, the photochemical quantum yields for scission are quite high for the polysilane derivatives in solution and vary from ~0.5-1.0. The alkyl substituted derivatives show little tendency to crosslink during irradiation ($\Phi_x = 0\text{-}0.08$). On the other hand, aromatic materials such as poly(methylphenylsilane) and related derivatives are more prone to crosslinking ($\Phi_x \sim 0.12\text{-}0.18$), although chain scission is still the predominate process ($\Phi_s/\Phi_x \sim 5\text{-}7$). One particularly clear trend is that the efficiency for both scission and crosslinking is significantly reduced in the solid state relative to solution by as much as a factor of 50-100.

The presence of air is not necessary for photobleaching, although in vacuum, the efficiency depends somewhat on the nature of the substituents. This suggested that the polysilanes might have some potential as positive functioning e-beam resists (vide infra). To conveniently assess their sensitivity toward ionizing radiation, samples of poly(p-t-butylphenyl methylsilane) poly(di-n-butylsilane) and poly(di-n-pentylsilane) were placed in Pyrex tubes, sealed under vacuum and subjected to γ-radiolysis from a ^{60}Co source. After irradiation, the samples were opened and analyzed by GPC. In each case, polymer scission was the predominate pathway and the G(s) values varied from ~0.14 to 0.4. Under the same conditions, the G(s) value of PMMA was calculated to be 1.2. The alkyl polysilane derivatives were ~3 times more sensitive than the aromatic polymer. The observed crosslinking was relatively minor in each case and the ratios of G(s)/G(x) were all greater than 20.

Previous photochemical studies on acyclic silane oligomers have shown that both silyl radicals and substituted silylenes are intermediates in the irradiation (55,56). It has recently been suggested that monomeric silylene species are also major gas phase intermediates in the laser induced photoablation of a number of alkyl substituted copolymers (19). Accordingly, we have exhaustively irradiated a number of high molecular weight polysilane derivatives at 254 nm in the presence of trapping reagents such as triethylsilane and various alcohols (57). With the former reagent, the major products were the triethylsilane adducts of the substituted silylene fragment extruded from the polymer chain. Also produced in reasonable yield were disilanes carrying the substitution pattern of the polymer chain. The disilane fragments apparently accumulate in the photolysis mixture, because they are only weakly absorbing at 254 nm. These results suggest that silylenes and silyl radicals are both generated upon exhaustive photolysis of the linear substituted silane polymers. The silanes produced by abstraction of a hydrogen atom from the trapping reagent are subjected to further chain degradation until they reach the stage where they no longer efficiently absorb the light. The results from the photolysis in alcoholic solvents were also consistent with silylenes and silyl radicals as intermediates (57). These studies lead us to postulate the mechanism shown below for the photodegration of polysilane high polymers in solution. It is not known whether the silylenes are produced concurrently with or subsequently to the formation of the silyl chain radicals.

$$\sim\!\!\sim Si(R_2)(R_1)-Si(R_2)(R_1)-Si(R_2)(R_1)\sim\!\!\sim \xrightarrow{h\nu} R_1R_2Si\!: + 2 \sim\!\!\sim Si(R_1)(R_2)\cdot$$

$$R_1R_2Si\!: + (Et)_3SiH \longrightarrow (Et)_3Si-Si(R_2)(R_1)-H$$

$$\sim\!\!\sim Si(R_1)(R_2)\cdot + (Et)_3SiH \longrightarrow \sim\!\!\sim Si(R_1)(R_2)-H + (Et)_3Si\cdot$$

$$\sim\!\!\sim Si(R_1)(R_2)-H \xrightarrow[(Et)_3SiH]{h\nu} \xrightarrow{h\nu} H-Si(R_1)(R_2)-Si(R_1)(R_2)-H + (Et)_3Si-Si(R_1)(R_2)-H$$

Polysilanes for Microlithography

The trend in microlithography is toward denser circuitry containing features with smaller lateral dimensions. Due to unavoidable chip topography, the lateral dimensions are shrinking more rapidly than the vertical with an attending increase in the feature aspect ratio (height/width). This creates numerous problems for classical single layer resist processes and has led lithographers to explore multilayer alternatives in spite of their increased processing complexity (58). A basic description of a typical single and multilayer process is shown in Figure 8. The classical single layer, wet development process requires exposure of the resist, followed by development of the pattern using appropriate solvent developer. Since wet development processes are usually isotropic, for small features there is often loss in linewidth control and sloping nonvertical profiles may be produced. This can create problems at later stages in the process. On the right side of Figure 8 is pictured an idealized multilayer process. The wafer topography is first covered with a thick, inert planarizing polymer layer followed by a thin layer of resist. Since the resist layer can be much thinner than is acceptable in a single layer application, it can be imaged with high resolution. This pattern can be either wet developed or alternatively developed during imaging (ablative exposure) down to the planarizing

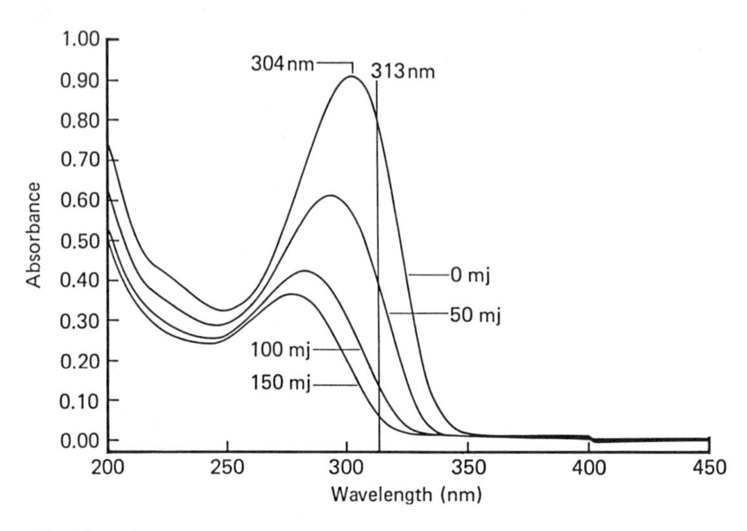

Figure 7. The photochemical bleaching of poly(hexylmethylsilane) at 313 nm.

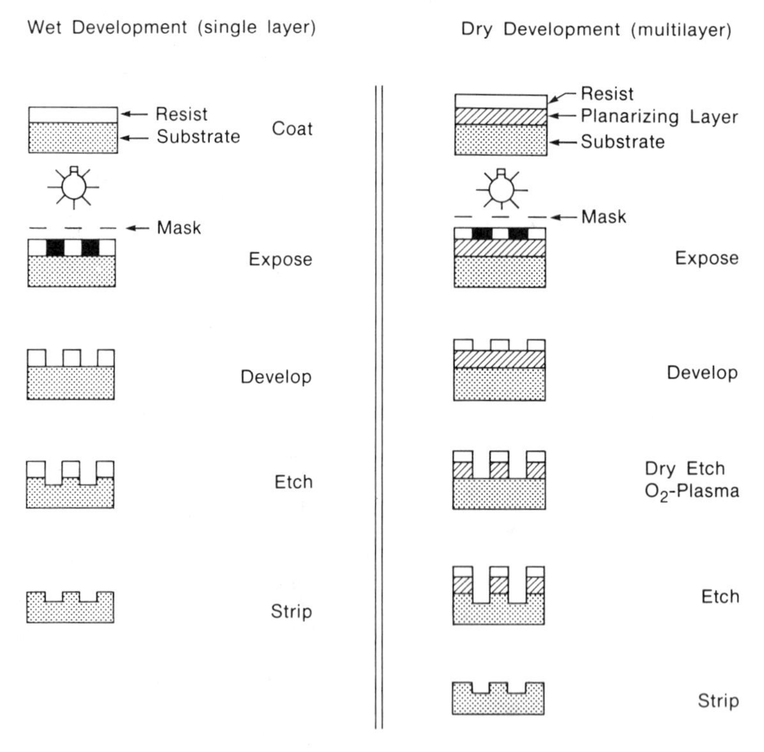

Figure 8. A schematic for a typical single layer (left) and a multilayer lithographic process.

layer. The problems associated with wet development in this case are ameliorated relative to the single layer because the resist layer is so thin (500Å to 0.2μ relative to $\sim 1\mu$ for a single layer). The image is then transferred through the planarizing polymer layer using a reactive oxygen plasma. This dry etch process can be made highly anisotropic so the linewidth control and feature profile is excellent.

One potential problem with the multilayer process is that the photoresist remaining after imaging must be resistant to the oxygen plasma in order to effectively mask the underlying polymer. Since oxygen plasmas are very aggressive, the best organic etch barriers are materials which form refractory oxides upon plasma treatment. This has stimulated the investigation of organometallic polymers, particularly those containing silicon (59). The polysilanes are ideal for such purposes, since they are thermally stable, soluble for coating, imagable by light and ionizing radiation and stable to oxygen etching due to the formation of a thin layer of inert SiO_2.

To date, we have exercised these materials in basically three types of multilayer lithographic applications (1) as short wavelength contrast enhancing layers, (2) as imagable O_2-RIE resistant materials in bilayer processes and (3) as radiation sensitive materials for multilayer, e-beam processes.

Contrast enhancement lithography is a clever procedure which uses a bleachable contrast enhancing layer to restore the distorted aerial image of the mask which has been blurred by diffraction effects into a sharp image at the underlying photoresist surface. The process is too complicated to explain in detail here and the interested reader is referred to the cited literature (60,61). Suffice it to say that the large extinction coefficients of most polysilane derivatives coupled with their ready bleachability make them ideally suited for such purposes and we have demonstrated this application at 313 nm (16,18).

The use of polysilanes as imagable, oxygen plasma resistant materials in bilayer applications has been explored in great detail. Using this technique and employing wet development techniques to develop the polysilane, we have generated submicron images with vertical wall profiles as shown in Figure 9 (17,18). In addition, a variation of this procedure where the polysilane is photoablatively removed during imaging has been used to create submicron images with vertical wall profiles in an "all dry" process (i.e., one which does not require wet development of the initial image (18,62).

Finally, the sensitivity of the polysilanes to ionizing radiation has allowed us to utilize these materials in a maskless e-beam imaging process. High resolution patterns such as shown in Figure 10 have been created by imaging a suitable polysilane derivative in a bilayer configuration. The e-beam image was wet developed and transferred by oxygen reactive ion etching.

In summary, the polysilanes comprise a new class of scientifically interesting, radiation sensitive materials for which many applications have become recently apparent. There is every reason to believe that future investigations will continue to be scientifically rewarding and result in new applications.

Acknowledgments

The authors wish to thank C. Cole and E. Hadziioannou of IBM for the GPC and thermal analyses, respectively. Similarly, we acknowledge the contributions of D. Le Vergne and D. Hofer (IBM) in the analyses of the results from the γ-radiolysis studies. R. D. Miller also gratefully acknowledges the partial financial support of the Office of Naval Research.

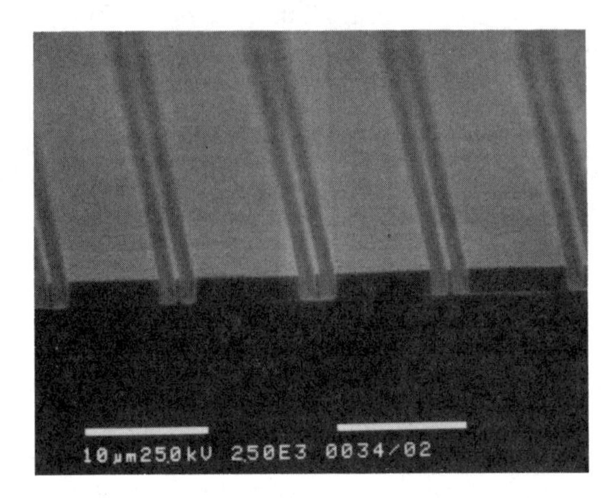

Figure 9. 0.75-µm features produced in a bilayer composed of 0.2-µm poly(methylcyclohexylsilane) coated over 2.0 µm of a hardbaked Novolac–naphthoquinone–2-diazide photoresist; mid-UV projection lithography, 100 mJ/cm^2, O_2-RIE image transfer. (Reproduced with permission from Ref. 17. Copyright 1984 Society of Photo-Optical Instrumentation Engineers.)

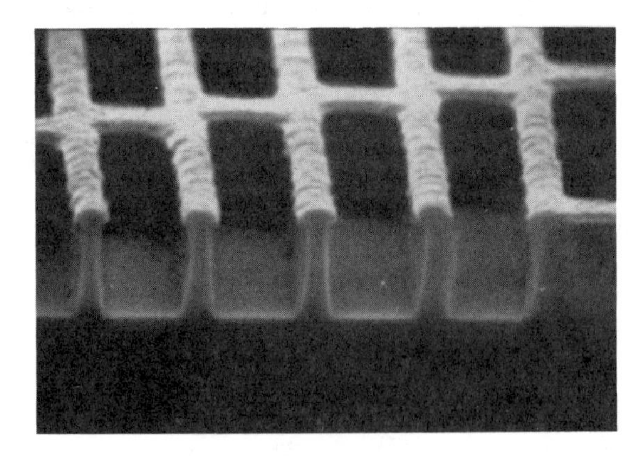

Figure 10. Submicron features generated by e-beam imaging of 0.14 µm of an aliphatic polysilane over 2.0 µm of hardbaked Novolac-naphthoquinone-2-diazide photoresist; 20 µC, O_2-RIE image transfer.

Literature Cited

1. West, R. J. Organomet. Chem. 1986, 300, 327, and references cited therein.
2. Kipping, F. S. J. Chem. Soc. 1924, 125, 2291.
3. Burkhard, C. A. J. Am. Chem. Soc. 1949, 71, 963.
4. Yajima, S.; Hayashi, J.; Omori, M. Chem. Lett. 1975, 931.
5. Yajima, S.; Okamura, K.; Hayashi, J.; Chem. Lett. 1975, 720.
6. Hasegawa, Y.; Iimura, M.; Yajima, S. J. Mater. Sci. 1980, 15, 1209.
7. West, R. In Ultrastructure Processing of Ceramics, Glasses, and Composites Hench, L. and Ulrich, D. C., Eds.; John Wiley and Sons, Inc.: New York, 1984.
8. Trujillo, R. E. J. Organomet. Chem. 1980, 198, C27.
9. Wesson, J. P.; Williams, T. C. J. Polym. Sci., Polym. Chem. Ed., 1980, 180, 959.
10. West, R.; David, L. D.; Djurovich, P. I.; Stearley, K. L.; Srinivasan, K. S. V.; Yu, H. G. J. Am. Chem. Soc. 1981, 103, 7352.
11. West, R.; Wolff, A. R.; Peterson, D. J. J. Rad. Curing 1986, 13, 35.
12. Kepler, R. G.; Zeigler, J. M.; Harrah, L. A.; Kurtz, S. R. Bull. APS 1983, 28, 362.
13. Kepler, R. G.; Zeigler J. M.; Harrah, L. A. Bull. APS 1984, 29, 504.
14. Stolka, M.; Yuh, H.-J.; McGrane, K.; Pai, D. M. J. Polym. Sci. Part A: Polym. Chem. 1987, 25, 823.
15. Miller, R. D.; McKean, D. R.; Hofer, D.; Willson, C. G.; West, R.; Trefonas, III, P. T. In Materials for Microlithography Thompson, L. F.; Willson, C. G.; Féchet, J. M. J., Eds.; ACS Symposium Series, No. 266; American Chemical Society: Washington D.C., 1984; Chapter 3.
16. Hofer, D. C.; Miller, R. D.; Willson, C. G.; Neureuther, A. R. Proc. of SPIE 1984, 469, 108.
17. Hofer, D. C.; Miller, R. D.; Willson, C. G. Proc. of SPIE 1984 469, 16.
18. Miller, R. D.; Hofer, D.; Fickes, G. N.; Willson, C. G.; Marinero, E.; Trefonas, III, P.; West, R. Polym. Eng. Sci. 1986, 26, 1129.
19. Zeigler, J. M.; Harrah, L. A.; Johnson, A. W. Proc. of SPIE 1985, 539, 166.
20. Aitken, C. T.; Harrod, J. F.; Samuel, E. J. Am. Chem. Soc. 1986, 108, 4059.
21. Aitken, C. T.; Harrod, J. F.; Samuel, E. Can. J. Chem. 1986, 64, 1677.
22. Zeigler, J. M.; Harrah, L. A.; Johnson, A. W. Polym. Preprints 1987, 28, 424.
23. Trefonas, III, P.; Djurovich, P. I.; Zhang, X.-H.; West, R.; Miller, R. D.; Hofer, D. J. Polym. Sci., Polym. Lett. Ed. 1983, 21, 819.
24. Luche, J. L.; Dupuy, C. P. Tet. Lett. 1984, 25, 753.
25. Pitt, C. G. In Homoatomic Rings, Chains and Macromolecules of Main Group Elements Rheingold, A. L., Ed.; Elsevier: Amsterdam, 1977; p. 203, and references cited therein.
26. Gilman, H.; Atwell, W. H.; Schwebke, G. L. Chem. Ind. (London) 1964, 1063.
27. Gilman, H.; Atwell, W. H.; Schwebke, G. L. J. Organometal. Chem. 1964, 371.
28. Bigelow, R. W. Organometallics 1964, 5, 1502.
29. Bigelow, R. W. Chem. Phys. Lett. 1986, 126, 63.
30. Takeda, K.; Matsumoto, N.; Fukuchi, M. Phys. Rev. B 1984, 30, 5871.
31. Takeda, K.; Teramar, K. A.; Matsumoto, N. J. Am. Chem. Soc. 1986, 108, 8199.
32. Klingensmith, K. A.; Downing, J. W.; Miller, R. D.; Michl, J. J. Am. Chem. Soc. 1986, 108, 7438.
33. Trefonas, III, P.; West, R.; Miller, R. D.; Hofer, D. J. Polym. Sci., Polym. Lett. Ed. 1983, 21, 823.
34. Pitt, C. G.; Carey, R. N.; Toren, Jr., E. C. J. Am. Chem. Soc. 1972, 94, 3806.

35. Miller, R. D.; Hofer, D.; Rabolt, J.; Fickes, G. N. J. Am. Chem. Soc. 1985, 107, 2172.
36. Reding, F. P. J. Polym. Sci. 1956, 21, 547.
37. Clark, K. J.; Jones, A. T.; Sandeford, D. J. H. Chem. Ind. (London) 1962, 2010.
38. Jones, A. T. Makromol. Chem. 1964, 71, 1.
39. Trafara, G.; Koch, R.; Blum, K.; Hummel, D. Makromol. Chem. 1976, 177, 1089.
40. Rabolt, J. F.; Hofer, D.; Miller, R. D.; Fickes, G. N. Macromolecules 1986, 19, 611.
41. Gobbi, G. C.; Fleming, W. W.; Sooriyakumaran, R.; Miller, R. D. J. Am. Chem. Soc. 1986, 108, 5624.
42. Schilling, F. C.; Bovey, F. A.; Lovinger, A. J.; Zeigler, J. M. Macromolecules 1986, 19, 2660.
43. Bock, H.; Ensslin, W.; Fehér, F.; Freund, R. J. Am. Chem. Soc. 1976, 98, 668.
44. Kuzmany, H.; Rabolt, J. F.; Farmer, B. L.; Miller, R. D. J. Chem. Phys. 1986, 85, 7413.
45. Lovinger, A. J.; Schilling, F. C.; Bovey, F. A; Zeigler, J. M. Macromolecules 1986, 19, 2657.
46. Miller, R. D.; Farmer, B. L.; Fleming, W.; Sooriyakumaran, R.; Rabolt, J. J. Am. Chem. Soc. 1987, 109, 2509.
47. Farmer, B. L.; Rabolt, J. F.; Miller, R. D. Macromolecules 1987.
48. Zhang, X. H.; West, R. J. Polym. Sci., Polym. Lett. Ed. 1985, 23, 479.
49. Zhang, X. H.; West, R. J. Polym. Sci., Polym. Chem. Ed. 1984, 22, 159.
50. Miller, R. D.; Sooriyakumaran, R. J. Polym. Sci., Polym. Lett. Ed. 1987 (in press).
51. Sakurai, H. J. Organomet. Chem. 1980, 200, 261.
52. Kumada, M.; Tamao, K. Adv. Organomet. Chem. 1968, 6, 81.
53. Kilb, R. W. J. Phys. Chem. 1959, 63, 1838.
54. Schnabel, W.; Kiwi, J. In Aspects of Degradation and Stabilization of Polymers Jellinek, H. H. G., Ed.; Elsevier: New York, 1978; Chapter 4.
55. Ishikawa, M.: Takaoka, T.; Kumada, M. J. Organomet. Chem. 1972, 42, 333.
56. Ishikawa, M.; Kumada, M. Adv. Organomet. Chem. 1981, 19, 51, and references cited therein.
57. Trefonas, III, P.; West, R.; Miller, R. D. J. Am. Chem. Soc. 1985, 107, 2737.
58. Lin, B. J. In Introduction to Microlithography Thompson, L. F.; Willson, C. G.; Bowden, M. G., Eds.; ACS Symposium Series, No. 219, American Chemical Society: Washington D.C., 1983; Chapter 5.
59. Reichmanis, E.; Smolinsky, G.; Wilkins, Jr., C. W. Solid State Technol. 1985, 28(8), 130, and references cited therein.
60. Griffing, B. F.; West, P. R. Polym. Eng. Sci. 1983, 23, 947.
61. West, P. R.; Griffing, B. F. Proc. of SPIE 1984, 33, 394.
62. Marinero, E. E.; Miller, R. D. Appl. Phys. Lett. 1987 (in press).

RECEIVED October 9, 1987

Chapter 5

Poly(di-*n*-hexylsilane) in Room-Temperature Solution

Photophysics and Photochemistry

J. Michl, J. W. Downing, T. Karatsu, K. A. Klingensmith, G. M. Wallraff, and R. D. Miller [1]

Center for Structure and Reactivity, Department of Chemistry, University of Texas at Austin, Austin, TX 78712–1167

Recent studies of substituted silane high polymers which contain only silicon in the backbone have created considerable scientific interest. Although polysilane derivatives were probably first prepared over sixty years ago by Kipping (1) these materials drew little attention because of their insoluble and intractable nature. The recent synthesis of a large number of soluble derivatives has ushered in a new era of investigation (2). The increased activity has resulted in a number of new applications for these materials. They have been utilized as thermal precursors to β-SiC (3-5), as oxygen insensitive initiators for vinyl polymerization (6), as a new class of polymeric charge conductors (7-9) and recently as materials for a variety of microlithographic applications (10-13).

Much of the scientific interest in polysilanes has centered on their very unusual electronic structure and their radiation sensitivity. The oligomeric silanes and substituted silane high polymers all absorb strongly in the ultraviolet region, which is an unusual feature for a sigma bonded framework. Early on, this was interpreted as evidence for extensive sigma delocalization and the interaction was modeled by a linear combination of bond orbitals (LBCO) (14). These studies suggested that in some ways the formally saturated polysilanes more closely resemble π-conjugated unsaturated polymers such as polyacetylene than their sigma-bonded carbon analogs.

The recent interest in substituted silane polymers has resulted in a number of theoretical (15-19) and spectroscopic (19-21) studies. Most of the theoretical studies have assumed an all-trans planar zig-zag backbone conformation for computational simplicity. However, early PES studies of a number of short chain silicon catenates strongly suggested that the electronic properties may also depend on the conformation of the silicon backbone (22). This was recently confirmed by spectroscopic studies of poly(di-n-hexylsilane) in the solid state (23-26). Complementary studies in solution have suggested that conformational changes in the polysilane backbone may also be responsible for the unusual thermochromic behavior of many derivatives (27,28). In order to avoid the additional complexities associated with this thermochromism and possible aggregation effects at low temperatures, we have limited this report to polymer solutions at room temperature.

The high radiation sensitivity of substituted silane polymers is an interesting phenomenon upon which a number of current applications are based. Detailed

[1]Current address: IBM Almaden Research Center, San Jose, CA 95120–6099

0097–6156/88/0360–0061$06.00/0

understanding requires a fundamental knowledge of the electronic structure and the photochemical and photophysical processes of polysilane derivatives. Previous photochemical studies on short chain and cyclic silane oligomers suggested that substituted silyl radicals and silylenes were produced as intermediates (29). Recent investigation by exhaustive irradiation (254 nm) of a number of polysilane high polymers in the presence of suitable trapping reagents have indicated the formation of similar intermediates (30). In addition, mass spectroscopic evidence for the formation of gaseous monomeric silylene fragments in the in vacuo photovolatilization of some substituted silane copolymers has also been presented (13). We have recently examined the photoablation of a number of polysilanes induced by excimer laser exposure (308 nm) and conclude that monomeric silylenes do not represent the bulk of the ablated gaseous material (31). Clearly, the details of the mechanism of photochemical degradation in solution or in the solid are not yet known. The present investigations of the photophysics and photochemistry of poly(di-n-hexylsilane) in room temperature solutions represents a step towards the resolution of this intriguing problem.

Experimental Section

Absorption spectra were obtained on a Cary 17D spectrophotometer. Flow linear dichroism was measured on a JASCO J-500C spectropolarimeter in a Couette cell using flow gradients up to 2000 s^{-1}. Fluorescence spectra were recorded on an SLM Aminco SPF-500C spectrophotometer. Polarized fluorescence spectra were obtained on a home-built spectrofluorimeter using a 1-kW xenon arc lamp, a 0.5-m double monochromator (SPEX 1302) for excitation, two Glan prism polarizers (Karl Lambrecht MGTYE 15), two wedge depolarizers, a 0.75-m single monochromator (SPEX 1702) for collection, a cooled S-20 response Centronics photomultiplier, and SPEX PC photon-counting electronics interfaced to an IBM PC. All emission spectra were measured using a wide-bore flow cell to prevent the photodegradation of the sample from affecting the results. Fluorescence lifetimes were measured using single-photon counting electronics and a mode-locked argon-ion pumped dye laser. Electron spin resonance spectra were measured on an IBM-Brucker ER/200 spectrometer. The 3-methylpentane solvent (Phillips) was purified by filtration through a $AgNO_3-Al_2O_3$ column and dried by refluxing over sodium. Cyclohexane was washed with concentrated H_2SO_4, water, sodium bicarbonate solution, and distilled from LAH. Tetrahydrofuran was pre-dried with KOH and distilled from LAH. The preparation of poly(di-n-hexylsilane) (1) has been described (32). The weight-average molecular weight was 1.1×10^6 as measured by GPC using polystyrene standards and the polydispersity was 2.4. Samples for the ESR studies were dissolved in pentane, freeze-thaw degassed, and sealed under vacuum.

Results and Discussion

Absorption and Emission Spectra. The excitation-emission spectrum of 1 (bottom half of Fig. 1) shows that the relatively narrow emission band is nearly independent of the excitation wavelength and that the excitation spectrum is not only nearly independent of the wavelength at which the emission is monitored, but is also very similar to the absorption spectrum, both being somewhat broader than the emission band. This leaves no doubt that the observed emission is due to the polysilane, and its shape, location and the mirror image relation to the absorption permit its assignment as fluorescence.

This result is further supported by the very short 130 ± 10 ps lifetime combined with the remarkably high quantum yield of 0.43, both observed upon excitation at 335 nm, near the long-wavelength edge of the absorption band. These values combine to a radiative lifetime of 300 ps, which corresponds (33a) to an oscillator strength of 1.8. Similarly short emission lifetimes have been observed for other poly(di-n-alkylsilanes): di-n-pentyl, 200 ps, di-n-decyl, 150 ps. The average oscillator

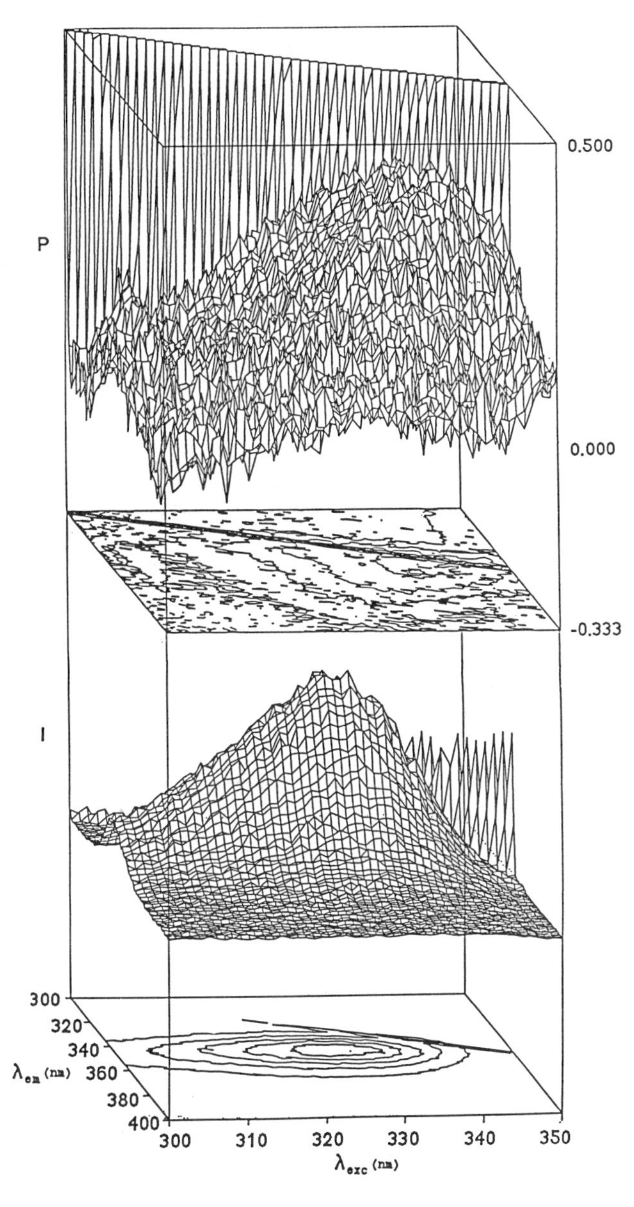

Figure 1. Fluorescence emission–excitation spectrum (bottom) and the degree of polarization P (top) for 1 in 3-methylpentane at 25 °C. (Reproduced from Ref. 19. Copyright 1986 American Chemical Society.)

strength per Si-Si bond obtained by integration (33b) of the absorption spectrum of 1 is 0.092.

Closer inspection of the results (Fig. 2) reveals some interesting anomalies, which make it clear that the situation is not quite as simple as outlined so far. Although the long-wavelength edge of the fluorescence band is indeed independent of the wavelength of excitation, its short-wavelength edge recedes to the red as the latter increases. Moreover, the fluorescence decay curves measured at short excitation wavelengths do not follow a single exponential, and, as the excitation wavelength is decreased, the fluorescence quantum yield drops precipitously to about half of the value quoted above. We have found this drop to be a general property of all polysilanes tested, including those with aromatic substituents. This makes reports of quantum yields measured at a single wavelength of limited value.

All of these anomalies can be rationalized by postulating that the random-coil (34) sigma-conjugated polysilane backbone is similar to the pi-conjugated backbones of polymers such as polyacetylene (35) in that it is effectively separated into a whole series of mutually and only weakly interacting chromophores capable of carrying localized electronic excitation. These chromophores would most likely correspond to conformationally well defined chain segments with a distribution of lengths and thus a distribution of excitation energies. The absorption spectrum would reflect a superposition of weighted contributions from a variety of segment lengths. This model has been proposed before to account for the thermochromic behavior of alkylated polysilanes (27,28). Long-wavelength absorption would then excite only the longest segments and would be followed by a competition between fluorescence and other processes such as intersystem crossing and photochemical reactions, leading to ordinary exponential decay. Short-wavelength absorption, however, would excite a wider distribution of segments of various excitation energies. If the shorter segments are connected to the ground state by transitions of higher energy, smaller oscillator strengths, and thus a longer radiative lifetime, their fluorescence will compete less successfully with processes such as intersystem crossing, photochemical reaction, and, particularly, singlet energy transfer to longer chain segments, which have lower excitation energy. This accounts qualitatively for the lower quantum yields and nonexponential decay of the fluorescence observed at shorter excitation wavelengths, as well as for the changes in the spectral shapes noted in Fig. 2. The relatively minor nature of these changes suggests, however, that the energy transfer is quite rapid on the time scale of the excited state lifetime.

A rough idea of the number of Si-Si bonds contained in a typical low-energy emitting chromophore can be obtained from a comparison of its oscillator strength of 1.8 with the average oscillator strength per Si-Si bond, 0.092. If the oscillator strength per Si-Si bond in the emitting long segment is similar to the average value, the chromophore consists of about twenty Si-Si bonds. We shall argue below that the oscillator strength per Si-Si bond in the longest emitting segments is larger than average. For this reason, we view the estimate of about twenty Si-Si bonds in the longest emitting chromophore as a likely upper limit and would not be surprised if the correct number were as few as a dozen.

Polarization of the Emission. We have sought support for the weakly interacting chain segment model from measurements of room temperature fluorescence polarization (19) on dilute solutions of 1 in 3-methylpentane. An independent preliminary report of similar measurements on a dilute glassy solution at 77K and on a neat polymer has also appeared (21). In the latter case, the analysis is complicated by inter-chain energy transfer.

First, we note that the σ-σ^* nature of the lowest excited state of long-chain polysilanes predicted by the semiempirical calculations described in detail below implies a transition moment direction lying approximately along the chain direction rather than perpendicular to it. Evidence for such an orientation of the transition moment has been obtained in measurements on solid polysilane samples (19,36). This

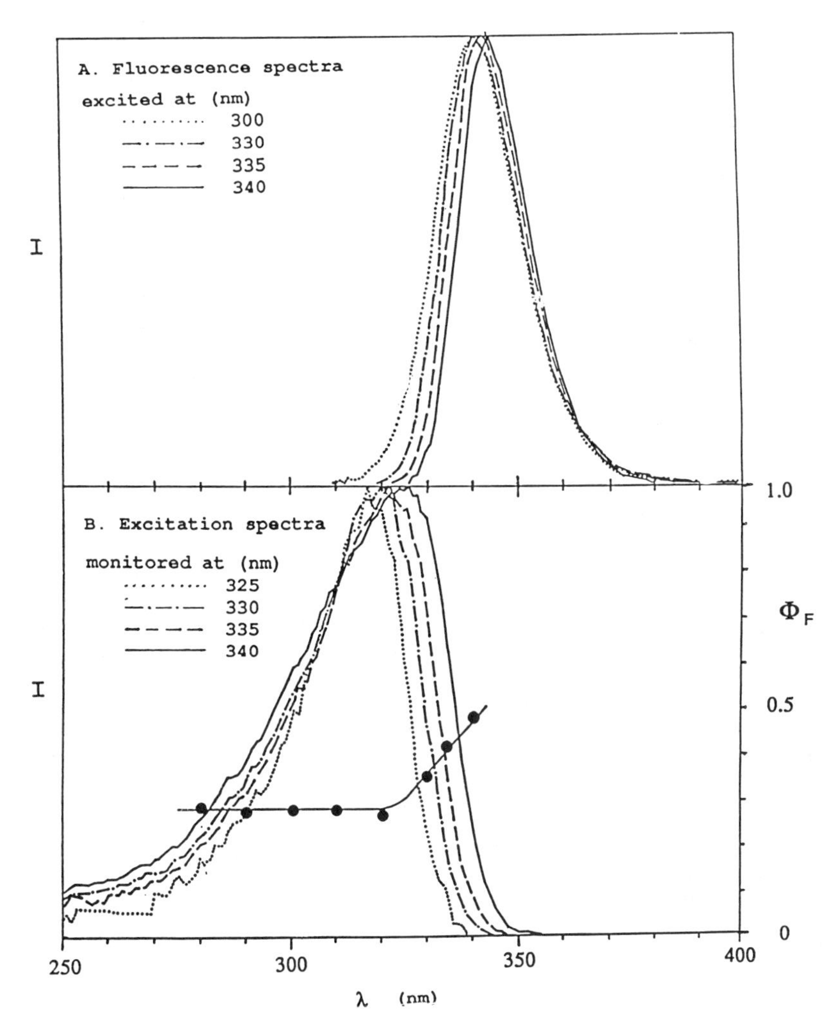

Figure 2. Room-temperature fluorescence and excitation spectra of 1 in cyclohexane at various choices of excitation and emission wavelengths, and the fluorescence quantum yield Φ_F at various wavelengths of excitation.

is subject to the objection that in the solid state the polymer chain conformation and properties could be quite different.

We now report that in the region of the absorption band the flow linear dichroism of a solution of 1 is positive (Fig. 3). Assuming that the nature of the flow orientation is of the usual kind, *i.e.*, that the polymer chains in a random coil conformation which dominates in solution (34) tend to align with the flow direction, this observation provides additional support for the absolute assignment of the transition moment direction along the chain direction, even in solution. Similar conclusions based on polarization studies on a stretched film of poly(di-n-hexyl silane) have recently been reported (36).

If the weakly interacting chain segment model is correct, the degree of fluorescence polarization P should be strongly positive for long-wavelength excitation, when only the longest segments absorb, since energy transfer will then be minimized and the same transition moment will be utilized in the absorption and emission steps. The theoretical value of P in the limit of no energy transfer (and no experimental artifacts) is 0.5 (33c). As the excitation wavelength is decreased, energy transfer to other segments located nearby should compete with emission that utilizes the transition moment responsible for original absorption. Since the chain forms a random coil, the orientation of the segments from which fluorescence will eventually occur must then differ from that of the segment that originally absorbed. As the transition moment of a segment is parallel to its chain direction, the absorbing and the emitting moments will now no longer be parallel, and the value of P will drop. In the limit of total directional randomization, P will be equal to zero.

The top half of Fig. 1 displays the observed values of P as a function of the wavelengths of excitation and observation and confirms fully the above expectations. For short wavelengths of excitation, P is close to zero, and at the longest wavelengths of excitation, it is close to 0.4. This agreement lends strong support to the weakly interacting chain segment model, but of course does not strictly prove that it is valid.

The Nature of the Chromophoric Chain Segments. ***Computations.*** In the following, we accept the description of the electronic states of the random-coil polysilane chain in solution in terms of a series of weakly interacting chromophores associated with conformationally well defined chain segments, the longest of which have lengths of the order of 10-20 Si-Si bonds. The next obvious question concerns the nature of these segments and the reasons for the limited communication between them, given that sigma conjugation cannot be fully turned off by any chain twist.

We have attempted (19) to answer this question using one of the standard tools of quantum chemistry that is applicable to large molecules, the INDO/S method (37). Although it is clearly only approximate, the method has a good track record in describing correctly the nature of the excited states of organic molecules, their excitation energies and transition moment directions. We have used this technique successfully on previous occasions for compounds of various types, including organosilicon compounds (38). Other authors have recently used the related CNDO and CNDO/2 (S + DES CI) methods for the calculation of the electronic spectra of short all-trans oligosilanes, up to octasilane (15-18). While it is not certain that the answer provided by the INDO/S model will be correct, the probability is high, and the results will surely be much more reliable than those of previous calculations by the Hückel method (14). Obviously, a direct experimental verification of the result would be desirable.

In view of the presence of the long alkyl side chains, the usual gauche and trans conformations of the backbone need not necessarily be the only ones that ought to be considered, and indeed there is evidence that in the solid state other conformations may sometimes be preferred (39,40). However, molecular mechanics (41) calculations for di-n-alkyloligosilanes (41-44) suggest strongly that in room-temperature solution the gauche and the trans conformations of the silicon chain dominate, and it is very likely that the chromophoric chain segments are constructed from them.

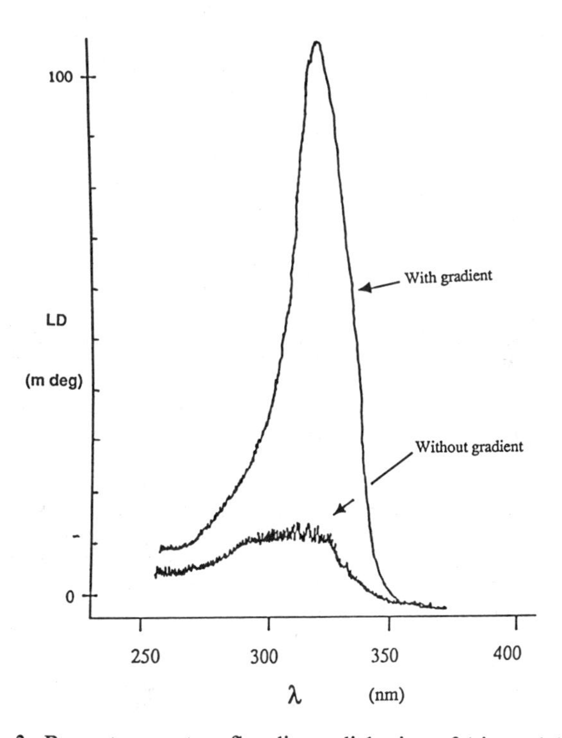

Figure 3. Room-temperature flow linear dichroism of 1 in cyclohexane.

In order to extend our calculations to long chains, we have first replaced the n-hexyl side chains with methyl groups and then by hydrogens. While the former may not be entirely accurate with respect to the molecular mechanics computation of the backbone geometry, the electronic effects of the two alkyl groups are surely very similar. The use of the hydrogen substituent is more likely to introduce serious problems. We therefore compared the INDO/S results for the parent polysilanes with those for the permethylated polysilanes for the first few members of the series and were relieved to note that the nature of the frontier orbitals, of the electronic wavefunctions, and of the transition moments were virtually identical in the two cases. The excitation energies were systematically higher for the parent series, but the trends were the same. Low-energy excitations were of the σ-σ^* type, in agreement with the experimental polarization results, and both the σ and the σ^* orbitals were localized nearly exclusively in the bonds between silicon atoms.

This result encouraged us to perform the main series of calculations (19) on the parent perhydrogenated polysilanes, using geometries optimized by molecular mechanics. The calculated transition energies are shown in Fig. 4 as a function of the number of Si atoms in the chain both for a series of all-trans and a series of all-gauche conformers. Both series converge rapidly to limiting values, and beyond about ten silicon atoms, the transition energies remain nearly constant as the chain is further extended. This limiting behavior was anticipated both on general theoretical principles and on the basis of previous calculations by more approximate methods (14). It also agrees with the fact that even in very high molecular weight solid polysilanes containing extended all-trans segments, the observed first transition energy does not drop below 27,000 cm^{-1}. This is true regardless of the nature of the alkyl substituents (23-25). At the moment, there is no experimental evidence for or against the INDO/S result that the limiting absorption wavelength in an all-trans chain is already reached at ten Si atoms, and in an all-gauche chain even earlier (Fig. 4). However, the INDO/S result is compatible with the above estimate of about 10-20 Si-Si bonds in the emitting low-energy chromophores.

The oscillator strengths per Si-Si bond calculated both in the dipole length and in the dipole velocity formulation are distinctly larger for the all-trans than for the all-gauche chain. At least in part, this is due to simple geometric reasons, as the local dipoles add much less efficiently in the all-gauche chain. This result has led us to comment above that the oscillator strength per Si-Si bond in the lowest-energy emitting chromophores, which we assign as all-trans segments, can be assumed to be higher than the average over all conformations present in a very long random coil chain in room temperature solution.

A lower excitation energy for the all-trans than the all-gauche conformation was suggested previously (23,25,27,28), without a theoretical justification. The difference we compute is very large and suggests strongly that the chromophoric chain segments in the polysilane chain are all-trans sequences separated from each other by one or more gauche or other nonplanar links. In order to test this hypothesis we have performed a series of calculations for a twenty-silicon chain in a variety of conformations. We have found that the lowest-energy excitation always localizes preferentially on the longest all-trans segment present, even if it is quite short. This is illustrated in Fig. 5, which displays the transition densities of the lowest excitation in several of the conformers. They are shown in the basis of the hybrid orbitals of the Si-Si bonds, into which we have transformed the results for easy visualization (very close to sp^3). The contributions from the hybrid Si orbitals on the Si-H bonds and from the 1s orbitals on hydrogen are very small and are therefore not shown. It is striking to note the degree to which the transition density for the lowest singlet excitation in a twenty-silicon atom chain consisting of an eleven-silicon all-trans segment separated by a gauche link from a nine-silicon all-trans segment is nearly exclusively localized in the longer all-trans segment. The transition density for the next higher singlet excitation is similarly localized on the shorter all-trans segment. It is perhaps even more striking to observe the degree to which the

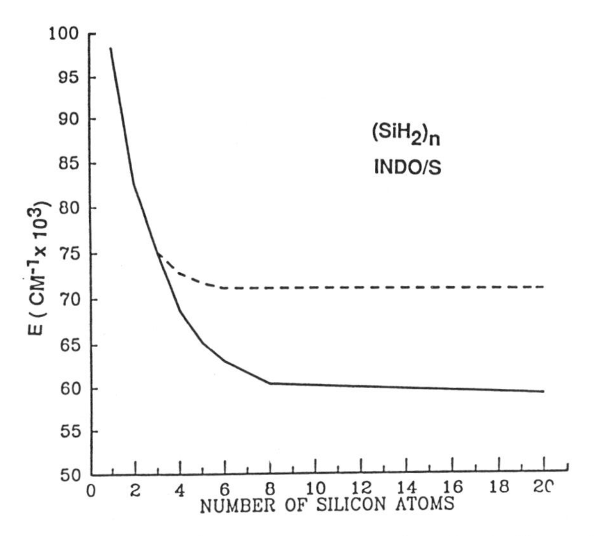

Figure 4. The INDO/S first singlet excitation energies of polysilanes, $(SiH_2)_n$: full line, all trans; broken line, all gauche.

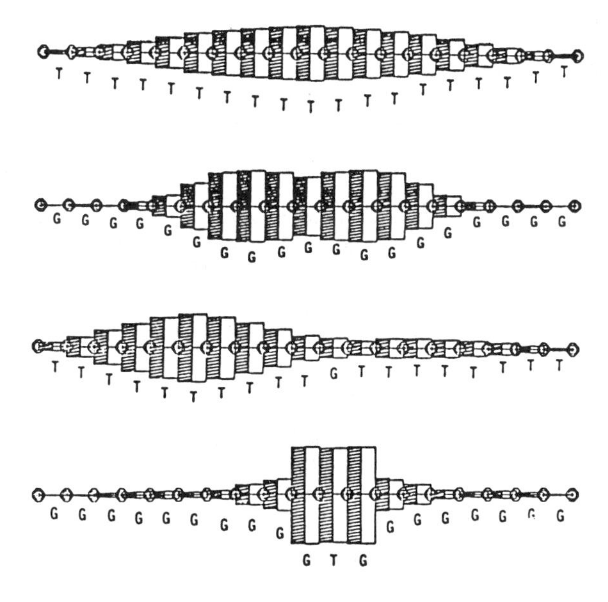

Figure 5. The INDO/S transition densities for four conformers of $Si_{20}H_{42}$ (G = gauche, T = trans) in terms of the hybrid orbital basis set (contributions from Si–H bonds are negligible). Coefficient sign (bar color) and magnitude (bar length) are shown for each Si atom (circle). (Reproduced from Ref. 19. Copyright 1986 American Chemical Society.)

transition density localizes in the immediate neighborhood of a single trans linkage in an otherwise all-gauche chain of silicons.

It should be noted that no geometrical relaxation of the excited state geometry was considered in our calculations. Such relaxations may well occur to some degree and would then enhance the localization of the electronic excitation. However, the relatively narrow spectral bandwidth suggests that the geometrical changes associated with excitation into the first excited singlet state are only minor.

Although the INDO/S description of the lowest-energy singlet excitations involves a moderate amount of configuration mixing, it is a fairly good approximation to describe them as one-electron transitions from the highest occupied to the lowest unoccupied molecular orbital (HOMO to LUMO). This permits a simple interpretation of the localization of the transition density in terms of a localization of the two molecular orbitals, since in the single-configuration approximation the transition density is given by the product of the HOMO and the LUMO. These two orbitals are shown in Fig. 6 for the same selection of conformers of the twenty-silicon chain just discussed and it is clearly apparent how the localization of the HOMO dictates the localization of the transition density.

The localization of the HOMO is also important for another reason. Since it describes the distribution of a hole in a radical cation, it relates to the hindrance that a positive charge will encounter as it propagates along the chain. There is indeed experimental evidence (9) that the "hole" states of the polysilane chain are localized and that they move by a hopping mechanism.

We have already mentioned the limiting behavior of the lowest excitation energy of an all-trans chain, which is calculated to converge rapidly. The just described picture of electronic excitation in a randomly coiled chain, with all-trans segments randomly distributed along the chain and separated from each other by one or more gauche links, also calls for a limiting behavior, and the two need to be carefully distinguished. While the distribution of segment lengths in the infinite random coil is not known, it is clear that upon gradual decrease in the chain length a point will be reached where the shorter segments will begin to be favored disproportionately. Since the shorter segments absorb at shorter wavelengths, this should lead to a gradual hypsochromic shift of the observed absorption maximum, which consists of a weighted superposition of contributions from segments of all lengths.

There is excellent experimental support for this picture (45): as the length of a randomly coiled (34) polysilane chain in room-temperature increases, the absorption maximum shifts to longer wavelengths, and the maximum of the absorption peak converges to a limiting value of about 31,000 cm^{-1} in a chain of 40-50 Si atoms. This, then, is the length of a chain required to support the segment length distribution characteristic of an infinitely long chain. The exact value of the limiting excitation energy depends on the nature of the alkyl substituent, which dictates the Si-Si-Si valence angles, the torsion angles in the conformers, and the segment length distribution function.

The observed (45) increase in the oscillator strength per Si-Si bond to a limiting value as the chain length grows is also qualitatively compatible with this picture if one accepts the above proposal that the Si-Si bonds in all-trans segments make a higher than average contribution to the total oscillator strength.

We realize that a more advanced method of calculation might place a larger weight on the role played by the orbitals of the two substituents carried by the Si atoms, and thus produce a larger difference between the parent polysilanes and their permethylated derivatives (cf. Ref. 18), but an inspection of the results suggests that this will not affect our conclusions. Another possible concern has to do with the absence of d orbitals in the INDO/S basis set. In a large basis set calculation, they would undoubtedly contribute to some degree to the description of both the σ and the σ^* Si-Si orbitals, which are symmetrical with respect to a local SiSiSi plane, and which are the only ones that enter the INDO/S description of the lower excited states. They

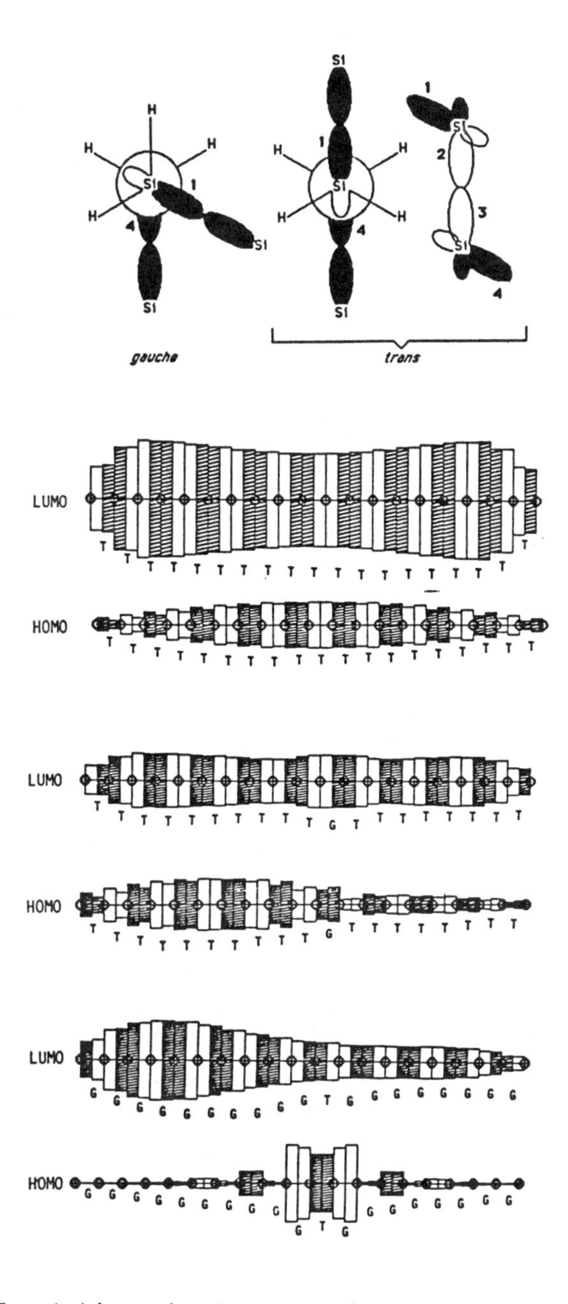

Figure 6. Top: 1–4 interactions between hybrid orbitals on neighboring Si atoms. Bottom: The frontier orbitals of three conformers of $Si_{20}H_{42}$ in the notation of Figure 5.

would also improve the description of both the σ and σ^* Si-H or Si-C orbitals, which yield symmetric and antisymmetric ("π") combinations with respect to the local SiSiSi symmetry plane, and which are unimportant for our purposes according to the INDO/S results. However, since the introduction of d orbitals does not introduce new types of low lying states, but only improves the description of states already present, it can hardly be expected to introduce qualitatively new features unless it changes the order of the low-lying excited states. This is not anticipated, since the INDO/S order, with the Si-Si σ-σ^* states lowest, agrees with the observed direction of the transition moment.

A Simple Rationalization of the Lower Excitation Energy of All-Trans Chains. While the higher HOMO energy and lower excitation energy of the all-trans conformational arrangement, which is ultimately responsible for the localization of the positive holes and of the electronic excitation on the all-trans segments, can be accepted simply as a computational result, it is interesting to ask whether it can be understood in simple terms. Such understanding should also help us estimate whether this result will survive in more complex computational models.

Inspection of Fig. 6 indeed provides a simple intuitive understanding of the HOMO-LUMO energy difference and therefore of the excitation energy and, ultimately, of the localization phenomenon as a function of the chain conformation. Consider the effect of a twist around the central Si-Si bond of a four-silicon chain segment on the resonance integrals between the members of the hybrid basis set, using the orbital numbering given in Fig. 6. Such twisting clearly does not affect the magnitudes of the primary interaction integrals between adjacent orbitals, β_{12} (geminal) and β_{23} (vicinal). These depend on the valence angles at the silicon atoms and are therefore sensitive to factors such as the bulkiness of the substituents they carry. They can only depend on the twisting motion to the degree to which it is accompanied by changes in the valence angles, and these are expected to be relatively small, justifying our concentration on the pure twisting coordinate as the main factor differentiating the trans and gauche conformers.

The next-nearest-neighbor-orbital resonance integrals, β_{13}, also remain unaffected by the pure twisting motion. We conclude that a pure twisting motion can therefore represent at best only a relatively small perturbation of the electronic structure of the polysilane chain, suitable for treatment by first-order perturbation theory. The perturbation is represented by changes in the resonance integrals between more distant hybrid orbitals, among which β_{14} clearly is the most important.

In the trans conformation, β_{14} represents an anti-periplanar interaction (46) and is positive, being dominated by the out-of-phase π-type overlap between the p components of the two hybrid orbitals (Fig. 6). In the cis conformation, not shown in the figure, β_{14} represents a syn-periplanar interaction and is negative, since the π-type overlap is now in-phase and reinforces the contribution derived from the σ-type part of the overlap. Part-way along the twisting coordinate, β_{14} must pass through zero, and at the gauche geometry, it is relatively small and negative.

Consider now the effect of a stepwise introduction of resonance integrals describing bonding between hybrid orbitals in a polysilane chain. The presence of the dominant interactions β_{12} and β_{23} produces the nodal structures for HOMO and LUMO that are shown in Fig. 6. The HOMO is bonding across all β_{23} interactions and can therefore be described as a bonding Si-Si σ orbital, but it is the least stable of all such orbitals since all of its β_{12} interactions are destabilizing. In contrast, LUMO is antibonding across all β_{23} interactions and therefore is an antibonding Si-Si* orbital, but is the most stable of all such orbitals since all of its β_{12} interactions are stabilizing. The picture is not modified substantially by the introduction of β_{13}.

Next, β_{14} is introduced and viewed as a weak perturbation. Given the just described nodal properties of HOMO and LUMO, all of the β_{14} integrals in the chain will act in the same direction, which is then easy to predict using first-order perturbation theory. In the HOMO, any two coefficients that are in a 1-4 relation have identical signs; in the LUMO, they have opposite signs. In the trans conformation, where β_{14} is positive, this implies a destabilization for the HOMO and stabilization for the LUMO, hence a decrease in the HOMO-LUMO gap. In the cis, and to a lesser degree, in the gauche conformation, where β_{14} is negative, it implies a stabilization for the HOMO and a destabilization for the LUMO, hence an increase in the HOMO-LUMO gap. This then accounts in a very simple manner for the difference in the lowest singlet excitation energies of the all-trans and the all-gauche chain segments. The relation to other anti-syn periplanar effects (46) is obvious.

The Primary Photochemical Process. The initial bond cleavage might occur in the lowest excited singlet or the lowest triplet state. If the reaction path consists merely of an increase of one of the Si-Si distances, yielding a pair of radicals, the triplet state is a far stronger candidate on theoretical grounds since the triplet σ-σ^* state between two saturated atoms is repulsive, while the singlet σ-σ^* state is bound (47). Since more than two Si atoms are involved and since it has not been demonstrated that the initial photochemical step is a simple cleavage of one sigma bond, the case for triplet involvement is not unequivocal. However, the initial formation of at least some radicals is probable since irradiation of polysilanes in the presence of polymerizable monomers induces radical photopolymerization (6).

The proposal that the chain scission proceeds via the triplet state was first prompted (13,20) by the observation of an extremely weak ($\phi < 10^{-4}$) emission band peaking near 430 nm upon excitation of neat alkylated polysilanes with a nitrogen laser at 77K. Unlike the relatively narrow fluorescence band, this very broad band shows indications of structure. The assignment of this emission as the phosphorescence of the intact polysilane (13,20) was supported by the observation of delayed fluorescence and of a long emission lifetime on the order of 1 ms. The poorly developed structure was assigned to vibrational progressions. Together with the large width of the band, this led to the suggestion that the triplet excitation is quite localized and presumably better able to produce bond scission, whereas the lowest singlet excitation is delocalized over many bonds.

While suggestive, this argument can hardly be considered compelling, and we have therefore searched for more unequivocal evidence that the direct photochemical reaction of polysilanes proceeds through the triplet state. We have found that the efficient triplet quencher, piperylene, (E_T = 20,700 cm^{-1}) (48) does not affect the rate of the photochemical reaction of 1 (nor does it quench its fluorescence). If the proposed (13,20) triplet state identification in 1 (E_T = 26,700 cm^{-1}) is correct, and if this triplet is indeed involved in the chain scission step, as suggested (13,20), the photodegradation should have been quenched observably in neat piperylene unless the room-temperature lifetime of the triplet is exceedingly short ($<$ 100 ps), as opposed to the 77K lifetime on the order of 1 ms. Such a large acceleration factor ($> 10^7$) is possible if the triplet reaction is thermally activated. Some indirect support for the triplet mechanism is provided by our recent results on poly(methylphenylsilane), whose photochemical degradation in room-temperature solution shows clear evidence of quenching with piperylene. Still, at this point, although the triplet mechanism appears likely, we believe that it would be premature to rule out alternatives in the case of 1, and we feel that its reaction need not necessarily proceed in the state responsible for the observed weak emission.

Secondary Photochemical Processes. While the nature of the primary photochemical step may be described as still uncertain, the nature of the subsequent secondary steps is best characterized as obscure. A previous trapping study during exhaustive irradiation (30) demonstrated that silylenes are formed somewhere along the line and implicated silyl radicals as well since the formation of Si-H bonds was observed, presumably by hydrogen atom abstraction.

On the assumption that the initial photochemical step produces a pair of radicals rather than two closed-shell species, we have used ESR spectroscopy to search for the initial radicals or products of their subsequent transformations. We have irradiated (254 nm) a number of dialkylpolysilanes in pentane solution at room temperature and have indeed observed strong overlapping ESR signals from at least two and probably a larger number of persistent radicals, which appear to be silicon-based judging by their g values, but have not thus far been able to assign their structures in an unequivocal fashion. However, none of the radicals exhibits a splitting by four β protons, which would be expected for the primary radical products. A persistent radical that appears early in the irradiation shows a apparent ~14G splitting by a single proton, and a ^{29}Si hyperfine splitting of ~75G. Another, which dominates the spectrum after a long irradiation period, shows no proton splitting at all and ^{29}Si satellites suggest a ^{29}Si coupling constant of ~56G. Such low ^{29}Si hyperfine coupling constants have been previously observed (49) in polysilylated silyl radicals. It is possible that restricted motion of the radicals centers caused by their polymeric nature may be responsible for some of the spectral features observed.

Structures that come to mind for assignment to the observed spectra are those that would be expected from an initial disproportionation of two primary silyl radicals to a silene and a hydrosilane, followed by an addition of another silyl radical to the silene, producing a β silylated silyl radical. Repetition of the process would eventually lead to a highly sterically encumbered and undoubtedly persistent silicon-based radical carrying only silicons in its β positions:

```
          |    |            |    |             |    |         \ /   |    |
         CH2  •Si—          CH  •Si—           CH —Si—         Si\  C  —Si—
      |   |    |         |  ||   |          |   |    |       /   |    |
    —Si—Si•  ——→       —Si—Si   ——→       —Si—Si• ——→ ——→  —Si—Si•   |
      |   |              |                  |   |            |    |    |
                                                             \Si/ C  —Si—
                                                             / \  |    |
```

The disproportionation of trimethylsilyl radicals to dimethylsilene and trimethylsilane is a well established process, kinetically competitive with the recombination to hexamethyldisilane (50), so that the proposed reaction sequence is well precedented. The formation of silenes upon irradiation of polysilanes and their well-known oxygen-scavenging ability have been invoked previously to account for the fact that the photoinitiation of radical polymerization in the presence of polysilanes is relatively insensitive to oxygen quenching (6). Also the observation (13) that carbonyl compounds and oligomers of silanones are formed upon irradiation of neat polysilanes in air suggests the possible intermediacy of silenes. It should be noted however, that this is not necessarily the only source of the silanone oligomers. For example, silylenes reacting with O_2 could conceivably generate these products as could a number of other reasonable processes.

In order to document the radical disproportionation reaction, we have used FT-IR spectroscopy to characterize the irradiation products. Upon irradiation of 1 in pentane, the formation of the characteristic peak near 2100 cm^{-1} due to Si-H stretching vibrations was readily apparent. The IR spectrum obtained in perdeuterated pentane was identical, suggesting that radical processes other than abstraction from the solvent are involved. Furthermore the ESR spectrum obtained in this solvent is identical to that already described. This raises the question whether the initially formed silyl radicals really abstract hydrogen from carbon with the formation of carbon-based radicals as suggested (13), particularly in light of the endothermicity of such a process.

When 1, fully deuterated in all α positions, was irradiated in pentane, the IR spectrum showed both the expected strong Si-D stretching peak and also a strong Si-H peak. The ESR spectra were also different and, in particular, the 14G splitting was absent. This is consistent with the proposal that silyl radicals disproportionate in the anticipated fashion, but also suggests that more complex processes play a role as well. The result can be accounted for without invoking C-Si bond cleavage if it is postulated that silyl radicals abstract hydrogen atoms from silenes with the formation of vinylsilyl radicals.

$$\begin{array}{c} | \\ CH_2 \\ | \\ CH \\ \| \\ -Si \\ | \end{array} + \begin{array}{c} | \\ \bullet Si- \\ | \end{array} \longrightarrow \begin{array}{c} | \\ CH \\ \| \\ CH \\ | \\ -Si\bullet \\ | \end{array} + \begin{array}{c} | \\ H-Si- \\ | \end{array}$$

While the abstraction of a hydrogen atom from an unactivated CH_2 group by a silyl radical is considered too endothermic to be of much importance, the $Si=C$ bond should activate a CH_2 group strongly by permitting a trade of an $Si=C$ bond for a 20 kcal/mol stronger (51) $C=C$ bond. This would also be consistent with the known propensity of silenes towards ene reactions (52). However, at this point we cannot exclude the possibility that C-Si bonds are cleaved as well, producing carbon-based radicals which then lose hydrogens that were originally in position β with respect to silicon.

These results indicate that the reactions triggered by the irradiation of polysilanes are rather complex and that the present understanding of these processes is clearly still quite limited.

Conclusions

The conclusions that can be drawn as the present stage of our study of poly(di-n-hexylsilane) in a room-temperature solution can now be summarized briefly.

The lowest singlet excitation is of the σ-σ^* type, is localized in all-trans chain segments, and travels rapidly along the polymer backbone from shorter to longer segments. The energy transfer to longer chains competes with fluorescence and chain scission, which is possibly triggered by an intersystem crossing event. The length of the low-energy emitting segments has been estimated at 10-20 Si-Si units. The reason for the localization of singlet excitation in all-trans chain segments is readily understood in terms of simple MO theory. The reactions that follow the initial chain rupture are complex and are presently not understood, but it is certain that complex silicon-based radicals are formed. Most likely, they abstract hydrogen atoms from each other as well as from the resulting silenes, but they do not abstract hydrogens from the solvent.

Acknowledgment

We thank Professor F. de Schryver, Katholiecke Universiteit, Leuven, Belgium, and Professor B. Nordén, Chalmers University of Technology, Gothenburg, Sweden, in whose respective laboratories the fluorescence lifetimes and the flow dichorism were measured. We are grateful to the National Science Foundation for financial support. R. D. Miller also gratelfully acknowledges the partial financial support of the Office of Naval Research.

Literature Cited

1. Kipping, F. S. J. Chem. Soc. 1924, 125, 2291.
2. West, R. J. Organomet. Chem. 1986, 300, 329.
3. Yajima, S.; Hayashi, J.; Omori, M. Chem. Lett. 1975, 931.
4. Hasegawa, Y.; Iimura, M.; Yajima, S. J. Mat. Sci. 1980, 15, 720.
5. West, R. in Ultrastructure Processing of Ceramics, Glasses, and Composites Hench, L.; Ulrich, D. L., eds.; John Wiley and Sons, Inc.: New York, 1984.
6. West, R.; Wolff, A. R.; Peterson, D. J. J. Radiat. Curing 1986, 13, 35.
7. Kepler, R. G.; Zeigler, J. M.; Harrah, L. A.; Kurtz, S. R. Bull. APS 1983, 28, 362.
8. Kepler, R. G.; Zeigler, J. M.; Harrah, L. A.; Kurtz, S. R. Phys. Rev. B 1987, 35, 2818.
9. Stolka, M.; Yuh, H.-J.; McGrane, K.; Pai, D. M. J. Polym. Sci.: Part A: Polym. Chem. 1987, 25, 823.
10. Hofer, D. C.; Miller, R. D.; Willson, C. G. Proc. of SPIE, Advances in Resist Technology 1984, 469, 108.
11. Hofer, D. C.; Miller, R. D.; Willson, C. G. Neureuther, A. R. Proc. of SPIE, Advances in Resist Technology 1984, 469, 108.
12. Miller, R. D.; Hofer, D.; Fickes, G. N.; Willson, C. G. Marinero, E.; Trefonas, P. III; West, R. Polym. Eng. Sci. 1986, 26, 1129.
13. Zeigler, J. M; Harrah, L. A.; Johnson, A. W. Proc. of SPIE, Advances in Resist Technology and Processing II 1985, 539, 166.
14. Pitt, G. G. In Homoatomic Rings, Chains and Macromolecules of the Main Group Elements, Rheingold, A. L., ed.; Elsevier: New York, 1977; and references cited therein.
15. Bigelow, R. W. Organometallics 1986, 5, 1502.
16. Bigelow, R. W. Chem. Phys. Lett. 1986, 126, 63.
17. Takeda, K.; Matsumoto, N.; Fukuchi, M. Phys. Rev. B. 1984, 30, 5871.
18. Takeda, K.; Teramae, A.; Matsumoto, N.; J. Am. Chem. Soc. 1986, 108, 8186.
19. Klingensmith, K. A.; Downing, J. W.; Miller, R. D.; Michl, J. J. Am. Chem. Soc. 1986, 108, 7438.
20. Harrah, L. A.; Zeigler, J. M. Polym. Preprints 1986, 27, 356.
21. Johnson, G. E.; McGrane, K. M. Polym. Preprints 1986, 27, 352.
22. Bock, A.; Ensslen, W.; Fehér, F.; Freund, R. J. Am. Chem. Soc. 1976, 98, 668.
23. Miller, R. D.; Hofer, D.; Rabolt, J.; Fickes, G. N. J. Am. Chem. Soc. 1985, 107, 2172.
24. Rabolt, J. F.; Hofer, D.; Miller, R. D.; Fickes, G. N. Macromolecules 1986, 19, 611.
25. Kuzmany, H.; Rabolt, J. F.; Farmer, B. L.; Miller, R. D. J. Chem. Phys. 1986, 85, 7413.
26. Lovinger, A. J.; Schilling, F. C.; Bovey, F. A.; Zeigler, J. M. Macromolecules 1986, 19, 2657.
27. Trefonas, P., III; Damewood, J. R.; West, R.; Miller, R. D. Organometallics 1985, 4, 1318.
28. Harrah, L. A.; Zeigler, J. M. J. Polym. Sci., Polym. Lett. Ed. 1985, 23, 209.
29. Ishikawa, M.; Takaoka, T.; Kumada, M. Adv. Organomet. Chem. 1981, 19, 51; and references cited therein.
30. Trefonas, P. III; West, R.; Miller, R. D. J. Am. Chem. Soc. 1985, 107, 2737.
31. Magnera, T. F.; Balaji, V.; Miller, R. D.; Michl, J. submitted for publication.
32. Miller, R. D.; Hofer, D.; McKean, D. R.; Willson, C. G.; West, R.; Trefonas, P. T. III In Materials for Microlithography Thompson, L. F.; Willson, C. G.; Fréchet, J. M. J., Eds.; ACS Symposium Series 266; American Chemical Society, Washington, D.C., 1984; p 293.
33. Michl, J.; Thulstrup, E. W. Spectroscopy with Polarized Light VCH Publishers, Deerfield Beach, FL, 1986; (a) p 43, $f = 3/2\pi_o n^3 \tilde{v}_o^2$, (b) p 28, $f = 4.319 \times 10^{-9} n \int E(\nu) d\nu$, (c) p 319.

34. Cotts, P. M.; Miller, R. D.; Trefonas III, P. T.; West, R.; Fickes, G. N. Macromolecules 1987, 20, 1046.
35. Kuzmany, H.; Mehring, M.; Roth, S. Electronics Properties of Polymers and Related Compounds, Springer Series in Solid State Sciences Vol. 63; Springer Verlag, New York, 1985; Kuzmany, H. Phys. Stat. Sol. 1980, 97, 521; Brédas, J. L. in Handbook of Conducting Polymers Skotheim, T. A.; ed.; Marcel Dekker, New York, 1986; Vol. II.
36. Harrah, L. A.; Zeigler, J. M. Macromolecules 1987, 20, 601.
37. Zerner, M.; Ridley, J. Theor. Chim Acta 1973, 32, 111.
38. Miller, R. D.; McKean, D. R. ; Thompkins, T. L. ; Clecak, N.; Willson, C. G.; Michl, J.; Downing, J. In Polymers in Electronics, Davidson, T.; ed.; ACS Symposium Series, American Chemical Society, Washington, D.C., 1984; p 25; Michl, J.; Radziszewski, J. G.; Downing, J. W.; Kopecky, J.; Kaszynski, P. Pure Appl. Chem., in press; Willson, G.; Miller. R.; McKean, D. R.; Michl, J.; Downing, J. J. Polymer Eng. Sci. 1983, 23, 1004; Raabe, G.; Vancik, H.; West, R.; Michl, J. J. Am. Chem. Soc. 1986, 108, 671.
39. Miller, R. D.; Farmer, B. L.; Fleming, W.; Sooriyakumaran, R.; Rabolt, J. F. J. Am. Chem. Soc. 1987, 105, 2509.
40. Farmer, B. L.; Rabolt, J. F.; Miller, R. D. Macromolecules 1987, 20, 1169.
41. Hummel, J. P.; Stackhouse, J.; Mislow, K. Tetrahedron 1977, 33, 1925.
42. Damewood, J. R., Jr.; West, R. Macromolecules 1985, 18, 159.
43. Welsh, W. J.; DeBolt, L.; Mark, J. E. Macromolecules 1986, 19, 2978.
44. Damewood, J. R., Jr. Macromolecules 1985, 18, 1793.
45. Trefonas, P. III; West, R.; Miller, R. D.; Hofer, D. J. Polym. Sci. Polym. Lett. Ed. 1983, 21, 823.
46. E.g., Hoffman, R.; Imamura, A.; Hehre, W. J. J. Am. Chem. Soc. 1968, 90, 1499; David, S.; Eisenstein, O.; Hehre, W. J.; Salem, L.; Hoffman, R. J. Am. Chem. Soc. 1973, 95, 3807.
47. Michl, J. Top. Curr. Chem. 1974, 46, 1.
48. Murov, S. L. In Handbook of Photochemistry, Marcel Dekker, New York, 1973; p 5.
49. Sakurai, H. In Free Radicals Kochi, J.; ed.; John Wiley and Sons, New York, 1973; Vol. 2; p 741.
50. Doyle, D. J.; Tokach, S. K.; Gordon, M. S.; Koob, R. D. J. Phys. Chem. 1982, 863626; cf. ref. 50, Table XVI; p 451.
51. Raabe, G.; Michl, J. Chem. Rev. 1985, 85, 419.
52. Dubac, J.; Laporterie, A. Chem. Rev. 1987, 87, 319.

RECEIVED September 22, 1987

Chapter 6

New Synthetic Routes to Polysilanes

K. Matyjaszewski, Y. L. Chen, and H. K. Kim

Department of Chemistry, Carnegie Mellon University, Pittsburgh, PA 15213

Three new synthetic routes for the synthesis of polysilanes have been applied: low temperature coupling in the presence of ultrasound, modification of poly(alkylarylsilylene) with triflic acid, and ring-opening polymerization. Low temperature reductive coupling of dichlorosilanes with sodium in the presence of ultrasound leads to monomodal polymers with low polydispersity ($\overline{M}_w/\overline{M}_n$=1.2) in contrast to bimodal polymers prepared by the thermal coupling. Modification of polysilanes with triflic acid gives access to new polysilanes with pendant alkoxy or amino substituents. Ring-opening polymerization have been successful only for the strained cyclic polysilanes with small substituents.

A rapidly increasing number of publications on polysilanes documents current interest in these polymers (1). Polysilanes are potentially applicable in microlithography as high resolution UV-resists (2), imageable etch barriers (3), or contrast enhancement layers (4). They have been successfully used as precursors to Si-C fibers (5) and ceramic reinforcing agents (6). Polysilanes have also initiated polymerization of vinyl monomers (7). Doping of polysilanes have increased their conductivity to the level of semiconductors (8). Very recently polysilanes were used as photoconductors (9) and non-linear optical materials (10).

At present, the only synthetic method leading to high molecular weight polysilanes is based on the Wurtz condensation reaction between dichlorodisubstituted silanes and alkali metals (low molecular weight oligomers were also prepared by dehydrogenation of secondary organosilanes). This reaction often gives a low yield of the desired linear high molecular weight polymer. Low molecular weight linear and cyclic oligomers in addition to the insoluble gel are the major side products. Target high molecular weight polysilane often has a very broad molecular weight distribution which limits its application in high resolution microlithography. Some improvements in the properties of the polymer and the yields were recently reported(11,12). Addition of

0097–6156/88/0360–0078$06.00/0

different solvents as well as reverse addition of alkali metal dispersions were reported. Nevertheless the desired product was rarely formed in yields higher than 50% and bimodality of molecular weight distribution remained. The high molecular weight polymer could be separated by fractional precipitation but it still has a high polydispersity.

We have been recently studying new pathways leading to polysilanes with low polydispersity and controlled structures. Our research is focused on three areas. The first one is low temperature reductive coupling in the presence of ultrasound. This leads to monomodal polymers with molecular weights in the range from $\overline{M}_n$=50,000 to $\overline{M}_n$=300,000 and polydispersities as low as $\overline{M}_w/\overline{M}_n$=1.20 (in addition to usually formed cyclic oligomers).

The second approach to linear polysilanes is based on the modification of polysilanes prepared by the reductive coupling method. The severe conditions of this reaction allow only alkyl or aryl substituents at the silicon atom in the starting dichlorosilane. Therefore only alkyl or aryl substituted polysilanes are known. We have successfully prepared new polysilanes with pendant alkoxy and amino side groups. This approach allows fine tuning of the properties of polysilanes.

The third synthetic pathway leading to polysilanes employs the ring-opening polymerization of strained cyclic polysilanes. The target is the preparation of polysilanes with active end groups which will be capable of reacting with other monomers leading to block and graft copolymers.

Experimental

Phenylmethyldichlorosilane (Petrarch) was distilled prior to use and dried over CaH_2. Toluene was distilled from CaH_2 and dried over CaH_2. The known amounts of sodium were placed in a flask filled with toluene and purged with dry argon. This flask was placed in the ultrasonic bath (75-1970 Ultramet II Sonic Cleaner, Buehler Ltd.) until stable dispersion of sodium was formed. In some experiments an immersion-type ultrasonic probe was used (W-140, Heat Systems-Ultrasonics, Inc.). A toluene solution of dichlorosilane was added to the reaction flask in a controlled manner under inert gas. The reaction was quenched after the required time by using equimolar mixtures of water and ethanol. The organic phase was later added to a large excess of isopropanol leading to the precipitation of the polymer. The polymer was dried and the yield determined gravimetrically. Molecular weights and polydispersities were determined by gpc using polystyrene standards. The compositions of the polymers were measured by NMR. The filtrate remaining after the evaporation of the isopropanol was analysed by gc/ms, gpc and hplc.

All other reactions were carried out under dry inert gas (nitrogen or argon). Reagents were distilled before use. Octaphenylcyclotetrasilane (mp=322 ^{0}C) was prepared from diphenyldichlorosilane and Li in 27% yield.

NMR spectra were recorded on 300 MHz (GE) or 80 MHz (IBM) spectrometers. UV spectra were recorded using a 9420 IBM spectrometer. GPC analysis was performed using Waters systems 10^6, 10^5, 10^4 10^3 (or linear ultrastyragel) and 500Å Waters columns. Molecular weights are based on the narrow polystyrene standards (actual molecular weights (VPO) are approximately two times higher).

Results and Discussion

Low Temperature Coupling in the Presence of Ultrasound

The coupling of dichlorosilanes with sodium was previously studied in boiling toluene. Under these conditions (above the melting point of sodium) the surface of sodium was continuously reactivated by rapid stirring. This reaction has not yet been described at lower temperatures. There is a little information on the mechanism of the coupling reaction. Formation of high polymer at relatively low conversion suggests the chain growth mechanism and not the step growth mechanism which is usually observed for the condensation process. Thus, undoubtedly some active species should be present at the chain ends. Silyl radicals, radical anions and silylenes were proposed as the hypothetical active species (1,12). The effect of addition of the triethylsilane (trap of silylene) is similar to the effect of the inert hexane (11). Addition of diglyme which is used as an agent stabilizing anions by solvation of counterions considerably reduced the degree of polymerization although it increased the final yield of the polymer. Thus, silyl radicals are the most probable chain carriers.

The first step in the polymerization is the electron transfer from sodium to dichlorosilane and the formation of the corresponding radical anion. The latter upon elimination of the chloride anion is transformed to the silyl radical. To fit the chain growth mechanism, the reactivities of the macromolecular radicals must be higher than the reactivities of the monomeric radicals. The latter after electron transfer and elimination of chloride anion could be transformed to the reactive silylenes. Thus, in principle, two or more mechanisms of chain growth are possible:

$$Cl_2SiR^1R^2 + Na \rightarrow [Cl_2SiR^1R^2]^{\cdot -} \rightarrow [ClSiR^1R^2]^{\cdot} + Na^+ \dashrightarrow \dashrightarrow \text{"propagation 1"}$$

$$\downarrow Na \qquad (1)$$

$$[ClSiR^1R^2]^- + Na^+ \rightarrow SiR^1R^2: + NaCl$$

$$\downarrow \qquad\qquad \downarrow$$

$$\text{"propagation 2"} \qquad \text{"propagation 3"}$$

This can result in polymodality of the molecular weight distribution when the exchange between the chain carriers is slow and they lead to the independent growth. Polymodality of the MWD is observed indeed. This phenomenon has been explained by differences in diffusion of the short and long polymer chains, but multiplicity of the chain carriers seems to us to be a more probable explanation.

The origin of the reaction which limits the chain growth is not known. Probability of cyclization, one of the side reactions, should decrease strongly with the polymerization degree in agreement with the

Jacobson-Stockmayer theory (13). Usually transfer and termination reactions in polymerization have energies of activation higher than propagation. Thus, we attempted coupling reactions at temperatures lower than usually applied. In order to assure the access of dichlorosilane to the surface of the sodium, the dispersion was continuously regenerated by ultrasonication.

We have observed a dependence of the yield, polymerization degree, and polydispersity of polysilanes on temperature and also on the power of ultrasonication. In the ultrasonication bath the simplest test of the efficiency of cavitation is the stability of the formed dispersion. It must be remembered that the ultrasonic energy received in the reaction flask placed in the bath depends on the position of the flask in the bath (it is not the same in each bath), on the level of liquid in the bath, on temperature, on the amount of solvent, etc. When an immersion probe is used the cavitation depends on the level of the meniscus in the flask as well. The power is usually adjusted close to 50% of the output level but it varies with the reaction volume, flask shape, and other rection conditions. The immersion-type probe is especially convenient at lower temperatures.

GPC traces of poly(phenylmethylsilylenes) prepared in the ultrasonication bath are shown in Fig. 1. In contrast to thermal condensation, monomodal high molecular weight polymer is formed. Oligomeric cycles (mostly cyclic pentamer), formed usually in high yield (cf. Table 1), can be very easily separated from the reaction mixture by precipitation with isopropanol. The molecular weight of polysilanes decreases and polydispersity increases with temperature.

The coupling reaction is usually rapid and completed in a short time after the addition of dichlorosilane. Ultrasonication accelerates the recovery of the sodium surface and enables large momentary excess of sodium to chlorosilanes. We have, however, observed a change in the molecular weight and polydispersity at longer reaction times. This is illustrated in Fig. 2.

Because the coupling reaction is usually carried out in excess sodium ($[Na]_0/[Si\text{-}Cl]_0=1.2$), degradation of the polymer is possible. We had prepared polysilanes and tested this possibility by their reaction with and without sodium in the presence of ultrasonication. Monomodal and bimodal polysilane was used. Reaction without sodium led to random degradation whereas ultrasonication with sodium favored degradation of the low molecular weight polymer. Thus, the monomodality of polysilanes obtained by ultrasonication can have its origin in the selective and more rapid degradation of the low molecular weight polymer, which might have a different microstructure than the high polymer, and may also favor the formation of the high polymer by reaction conditions different than those in a typical thermal process. If two independent mechanisms of chain growth are assumed, a lower temperature will favor the process proceeding with the lower activation energy (high polymer). If polymodality has its origin in diffusion phenomena, cavitation can eliminate the formation of low oligomers. Eventually, a large excess of sodium may also influence the kinetics of coupling and favor the formation of high polymer.

The degradation of polysilanes in the presence of sodium indicates that the polymer is a kinetic product and is easily degraded to

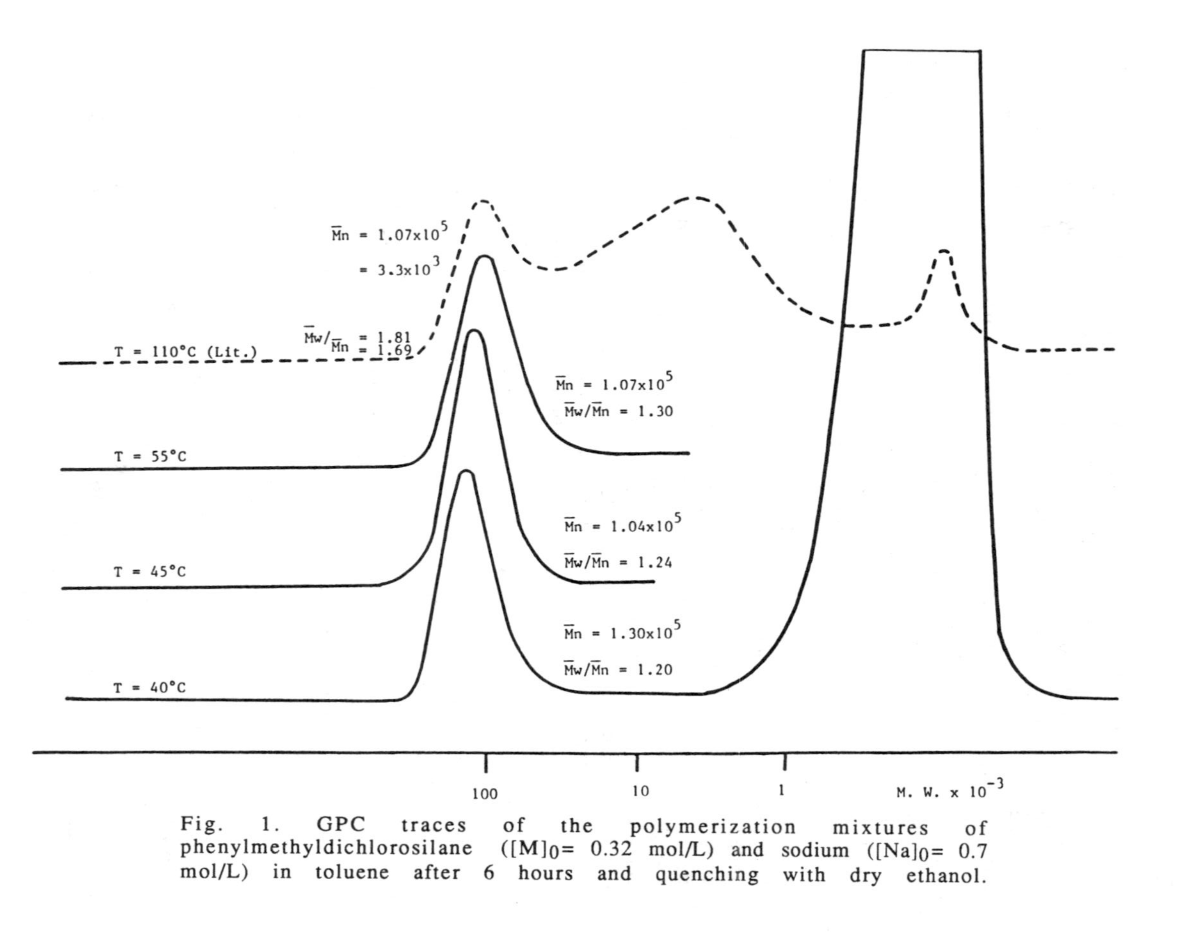

Fig. 1. GPC traces of the polymerization mixtures of phenylmethyldichlorosilane ($[M]_0$= 0.32 mol/L) and sodium ($[Na]_0$= 0.7 mol/L) in toluene after 6 hours and quenching with dry ethanol.

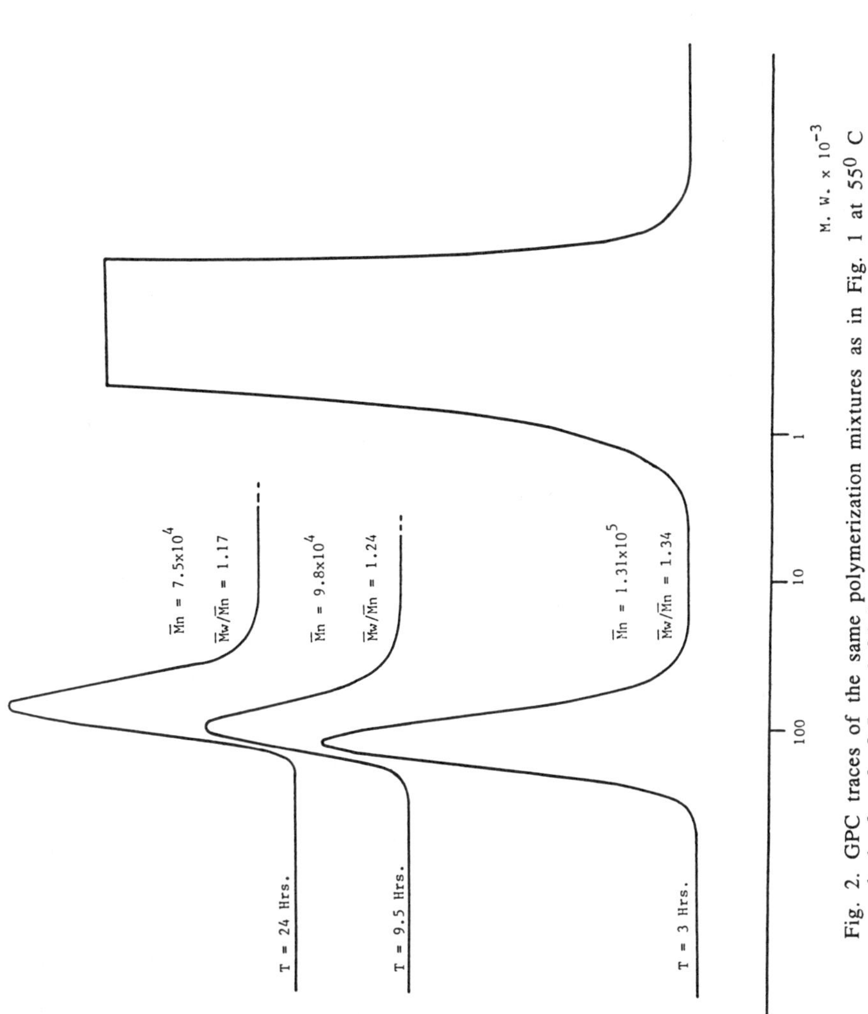

Fig. 2. GPC traces of the same polymerization mixtures as in Fig. 1 at 55^0 C quenched after 3, 9.5, and 24 hours.

thermodynamically more stable cyclic oligomers. Thus, some earlier experiments performed with soluble initiators (biphenyl radical anion) (11) in which only oligomers were formed may indicate very rapid polymerization as well as very rapid degradation. This means that isolation of the high polymer could have been possible but at shorter reaction times. We believe that the reproducibility of our results will enable more systematic kinetic and mechanistic studies of the reductive coupling of dichlorosilanes leading to the formation of high polymers.

Modification of Polysilanes

The presence of alkali metal during reductive coupling practically prevents the synthesis of polysilanes with groups which can be easily cleaved under basic conditions, such as pendant alkoxy or amino substituents. Polysilanes with pendant functional groups are not known. We have recently discoverd a very efficient method of functionalization of polysilanes.

Model reactions show that methyl, chloro, and aryl groups can be rapidly and quantitatively displaced by triluoromethanesulfonic acid:

$$R\text{-}Si(CH_3)_2\text{-}Si(CH_3)_2\text{-}R + 2HOSO_2CF_3 \rightarrow CF_3SO_2O\text{-}Si(CH_3)_2\text{-}Si(CH_3)_2\text{-}OSO_2CF_3 + 2RH$$

where $R=C_6H_5$, CH_3, Cl (2)

Bis 1,2-(trifluoromethanesulfonlyloxy)tetramethyldisilane is a stable and reactive compound towards different nucleophiles (14). It reacts rapidly with pyridine forming mono and (at a 1:2 ratio) disalts with pyridine. With secondary amines this compound forms the corresponding disilyldiamine. Dialkoxydisilanes were prepared in good yields in the reaction with different alcohols:

$$CF_3SO_2O\text{-}Si(CH_3)_2\text{-}Si(CH_3)_2\text{-}OSO_2CF_3 + 2ROH \xrightarrow[-BH]{B:} RO\text{-}Si(CH_3)_2\text{-}Si(CH_3)_2\text{-}OR \quad (3)$$

where RO-: CH_3O-, CF_3CH_2O-, $CH_2=CHCH_2O$-, $(CH_3)_3CO$-, etc

We used a similar reaction for the modification of polysilanes. Poly(phenylmethylsilylene) has been chosen as the model polysilane because cleavage of the aryl groups is much faster than that of the alkyl groups. The displacement reaction proceeds very rapidly at room temperature and immediately after addition of the acid the sharp signal of the benzene is found. ^{1}H-NMR spectra of the initial polymer indicate the presence of different iso-, syndio-, and atactic structures. The chemical shifts of the methyl groups after displacement of the phenyl groups by triflate groups move downfield approximately 1.2 ppm (cf. Fig. 3). A similar chemical shift of methyl groups is observed for mono and disilanes with triflate groups.

Polysilanes bearing triflate groups are very reactive and form an insoluble gel with a trace of moisture. The triflate group is hydrolized to silanol which rapidly condenses with the remaining triflate groups to a siloxane unit, linking intra or intermolecularly silicon atoms:

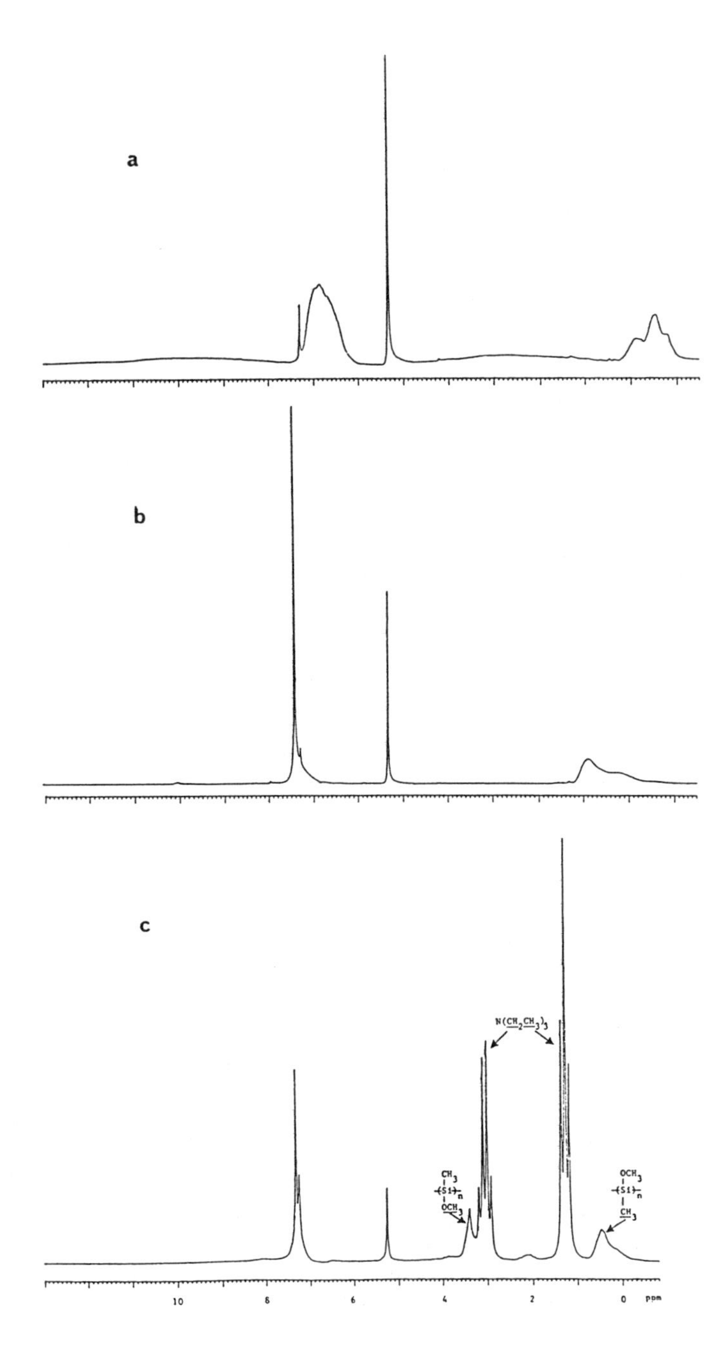

Fig. 3. ^{1}H-NMR spectra of poly(phenylmethylsilylene) ($[(SiPhMe)n]_0$=0.42 mol/L) (a), after reaction with triflic acid ($[CF_3SO_3H]_0$= 0.34 mol/L (b), and poly(methylmethoxysilylene) ($[(SiOMeMe)_n]_0$= 0.42 mol/L) (c) in $CDCl_3$ using CH_2Cl_2 as internal standard.

```
Ph                        CF3SO2O                    O-Si-...
|                           |                        | |...
...-(Si)n-...  + nHOSO2CF3 -->  ...-(Si)n-... + H2O -->  ...-(Si)n-...
|                           |                        |           (4)
CH3                        CH3                      CH3
```

Polymers with triflate groups react with alcohols to form alkoxysubstituted polysilanes. This reaction occurs readily in the presence of bases. The best results were obtained using triethylamine and hindered pyridine. In Fig. 3c the NMR spectrum of the reaction mixture containing the excess of triethylamine is shown, the methyl groups from the polymer chains absorb in the range typical for alkoxysilanes. Reaction in the presence of unsubstituted pyridine leads to the formation of insoluble polymer probably by attack at the p-C atom in the silylated pyridine.

Polysilanes with alkoxy groups are more light sensitive than conventional polysilanes. They degrade rapidly in the presence of light in agreement with the facile formation of silylene from dialkoxydisilanes. Properties of these polymers are currently being investigated.

Ring-Opening Polymerization

The formation of the linear polymer from the cyclic monomer requires a decrease of the free energy. Because usually entropy is lost during polymerization, the main driving force for the ring-opening process is the release of the angular strain upon conversion of the cycles to linear macromolecules. Thus, a majority of three- and four-membered rings can be readily and quantitatively converted into polymers.

Nonpolymerizability may have its origin either in thermodynamics or in kinetics. Some monomers, in spite of the strain, cannot be homopolymerized but readily copolymerize. 1-Benzyl-2,2-dimethylaziridine or tetramethyloxirane can serve as the examples (15). On the other hand unsuccessful polymerization can be ascribed to the lack of the initiator which could lead to the formation of active species capable of a significant decrease of the energy barrier for propagation. Therefore oxetanes, thietanes, azetidines, and aziridines cannot be polymerized by an anionic but only by a cationic mechanism(16).

Quite often in the ring-opening polymerization, the polymer is only the kinetic product and later is transformed to thermodynamically stable cycles. The cationic polymerization of ethylene oxide leads to a mixture of poly(ethylene oxide) and 1,4-dioxane. In the presence of a cationic initiator poly(ethylene oxide) can be almost quantitatively transformed to this cyclic dimer. On the other hand, anionic polymerization is not accompanied by cyclization due to the lower affinity of the alkoxide anion towards linear ethers; only strained (and more electrophilic) monomers can react with the anion.

We have attempted the polymerization of octaphenylcyclotetrasilane, which is the only commercially available strained cyclopolysilane.

We have used different anionic, cationic, and metathesis initiators but after the reaction of the initiator with one monomer molecule (initiation, k_i) no subsequent propagation was observed:

```
                         Ph Ph PhPh       (+Ph8Si4)         PhPhPhPhPhPhPhPh
                  ki     |  |  |  |          kp             |  |  |  |  |  |  |  |
BuLi + Ph8Si4 -> Bu-Si-Si-Si-Si-Li+          -->         Bu-Si-Si-Si-Si-Si-Si-Si-Si-Li+
                         |  |  |  |          <--            |  |  |  |  |  |  |  |
                         PhPh Ph Ph          kd             PhPhPhPhPhPhPhPh
```

(5)

This may suggest that either propagation is very slow due to the low reactivity of the derived species or that the equilibrium constant $K=k_p/k_d$ is very low which means that the cyclic tetramer is more stable than the polymer chain. This is apparently in agreement with the high yield (up to 40%) of cyclic tetramer; the yield of the polymer in the classical Kipping experiment of the coupling of diphenyldichlorosilane had to be very low because in addition to cyclic tetramer, large quantities of cyclic pentamer and hexamer were formed. Only very recently was existance of polysilanes with two aryl substituents proved by Miller who prepared soluble diaryl polysilanes with long p-alkyl substituents (17). Yields and stabilities of these polymers were not reported.

The interactions between bulky phenyl substituents in the polymer chain can give more steric hindrance than the deformation of the valency angles in the four membered ring. Similar interactions prevent the polymerization of 1,1-diphenylethylene and 2,2-diphenyloxirane (16). Thus, octaphenylcyclotetrasilane can be thermodynamically more stable than linear perphenylpolysilane and no initiator exists capable of converting this cycle to the linear polymer.

On the other hand, anions formed by the ring cleavage of octaphenylcyclotetrasilane may be very unreactive due to steric hindrances as well as to the formation of the tight silyl ion pair with alkali metals. We have attempted to decrease these interactions by using cryptands in order to better solvate the alkali metals and form the loose, separated ion pair. However, in addition to the rapid cleavage of one ring no propagation was observed. Using a 10 fold excess of monomer to butyl lithium in the presence of cryptand [2.1.1], only 10% of the monomer was converted to the linear chain; the rest did not react even after a longer time or at higher temperatures.

The polymerizabilty of cycles should increase for rings with smaller substituents. It has been reported that substitution of the chlorine atoms by alkyl groups in octachlorotetracyclosilane using methyl lithium led to the ring cleavage, indicating the possibility of polymerization. The sensitivity towards cationic or anionic systems should strongly depend on the structure of the substituents. Alkoxy groups may strongly stabilize positive charge and in a way similar to vinyl ethers can facilitate cationic polymerization. On the other hand, cyano groups should favor the anionic process. The facile displacement of phenyl groups by triflic acid was used as the first step in the synthesis of the four-membered cycles with alkoxy and cyano groups. During this substitution different isomers were formed and at the ratio 3:1 ($[HOSO_2CF_3]_0$:$[Ph_8Si_4]_0$) seventeen signals of methoxy groups were observed by ^{1}H NMR. The presence of different isomers in the puckered four-membered ring indicates small differences in reactivities of different aryl groups. The excess of the acid was observed by NMR at a ratio higher than 4:1 which showed difficulty in the formation of the geminal ditriflate. In the preliminary experiments, the ring-opening polymerization of octamethoxycyclotetrasilane led to a polymer with molecular weight $\overline{M}_n=10^4$ and high polydispersity. The conditions leading to better defined polymers are at present being investigated.

Acknowledgments. The financial support for this work by the Office of Naval Research is gratefully acknowledged.

Literature Cited

1. West, R. J. Organomet. Chem. 1986, 300, 327
2. Hiraoka, H., et al., U. S. Patent 4, 464,460 (1984)
3. Hofer, D. C., Miller, R. D., Willson, C. G. SPIE Adv. Resist. Tech. 1984, 469,16
4. Hofer, D. C. , Miller, R. D., Willson, C. G., Neureuther,A. ibid, 1984, 469, 108
5. Yajima, S., Hasegawa, Y., Hayashi, J., Ijmura, M. J. Mater. Sci. 1978, 13, 2569
6. Mazdyasni, K., West, R., David, L. D. J. Amer. Cer. Soc. 1978, 61, 504
7. West, R., Wolff, A. R., Petersen, D. J., J. Radiation Curing, 1986, 13, 35
8. West, R., David, L. D., Djurovich, P. I., Stearley, K. L., Srinivasan, K. S. V., Yu, H. J . Amer. Chem. Soc. 1981, 103, 7352
9. Kepler, R. G., Zeigler, J. M., Harrah, L. A., Kurtz, S. R. Bull. Am. Phys. Soc. 1983, 28, 362
10. Kajzar, F. , Messier, J., Rosilio,C. J. Appl. Phys., in press
11. Zeigler, J. M. Polymer Preprints 1986 ,27, 109
12. Zeigler, J. M. , Harrah, L. A. , Johnson, A. W. Polymer Preprints 1987, 28, 424
13. Jacobson, W. , Stockmayer, W. H. J. Chem. Phys. 1950, 18, 1600
14. Chen, Y. L., Matyjaszewski, K. J. Organomet. Chem., submitted
15. Penczek, S. , Kubisa, P., Matyjaszewski, K. Adv. Pol. Sci. 1980, 37,1
16. Penczek, S. , Kubisa, P., Matyjaszewski, K. Adv. Pol. Sci. 1986, 68/69, 1
17. Miller, R., Rabolt, J. R., Sooriyakumaran, R., Fickes, G.N., Farmer, B.L., Kuzmany, H. , Polymer Preprints 1987, 28, 422

RECEIVED September 1, 1987

Chapter 7

Polymerization of Group 14 Hydrides by Dehydrogenative Coupling

John F. Harrod

Department of Chemistry, McGill University, Montreal, Quebec H3A 2K6, Canada

The dehydrocoupling of organosilanes and germanes under the catalytic influence of titanocene and zirconocene derivatives is reviewed. Primary organosilanes generally give $RSiH_2$-terminated oligo-organosilylenes containing 10 to 20 silicon atoms, depending on the catalyst. Polymer structural assignments are based on a combination of ir, nmr and ms spectroscopies and on molecular weight studies using gpc and vpo. Primary germanes give three-dimensional gels with the titanium-based catalysts but secondary germanes give short chain linear oligomers. It is proposed that the polymerization proceeds by a repetitive insertion of silylene moieties into a metal-silicon bond.

The production of extremely long chains of carbon atoms is greatly facilitated by the common existence of compounds with metastable multiple carbon-carbon bonds. In addition to the highly favorable thermodynamics of polymerization of C=C and C≡C bonds, the initiation modes, which typically produce a propagating species with one active end, generally preclude the formation of rings by intramolecular coupling of the two chain ends. The thermodynamic stability of polymer relative to olefin is favorable for those cases where one of the carbons carries only hydrogen atoms. If both carbon atoms carry substituents, the ceiling temperature, above which the polymer is unstable relative to monomer, is generally too low for the polymer to be useful. The near absence of silicon analogs of the substituted olefins and acetylenes has precluded a parallel evolution of polysilylene chemistry. Indeed, the successful strategy for producing double bonds between elements of the third and lower periods has been to so encumber the atoms on either side of the double bond with bulky substituents that the molecule has no inclination to polymerize (1). The larger size of silicon compared to carbon makes a much higher steric encumbrance necessary to stabilize the monomer. The same considerations apply even more so to the heavier congeners of group 14.

0097-6156/88/0360-0089$06.00/0

Poly(dimethylsilylene) was first reported by Burkhard in 1949 (2), but the lack of solubility and general intractability of this material discouraged further studies at that time. The demonstration by Yajima and Hayashi that poly(dimethylsilylene) can be pyrolyzed to silicon carbide, and the subsequent development of silicon carbide fibres via this route, lead to a dramatic rise in interest in poly(organosilylenes) and their chemistry (3). More recently they have attracted increasing attention since potential uses in the areas of microlithography (4) and reprography (5) have been identified. Certain poly(organosilylenes) have also been shown to have unusual thermochromic behavior (6,7) and temperature-dependent transitions in chain conformation (8). These developments have been made possible by the application of modified Wurtz-Fittig-type coupling to the production of linear polysilylenes, as shown in Equation 1 .

$$n\ RR'SiCl_2 + 2n\ M \longrightarrow Cl{-}\!\!\left[\underset{R}{\overset{R'}{Si}}\right]_n\!\!{-}Cl + 2n\ MCl \qquad (1)$$

(M is a group 1 metal or alloy)

The contributions of West and co-workers in this area have been noteworthy (9). The Wisconsin group showed that the solubility of poly(dimethylsilylene) can be greatly enhanced by inclusion of a second substituent through copolymerization and that Equation 1 has considerable generality (10). It is now clear that poly(organosilylenes) with very high molecular weights (ca. 10^6) can be synthesised by this type of reaction. The polymers thus obtained have sufficed to allow the development of some new technologies and to point the way to others. A number of aspects of the Wurtz-Fittig-type coupling detract from its attractiveness as a commercially viable route to polysilylenes. Among the most serious difficulties are the poor control of molecular weight and polydispersity, production of low molecular weight cyclics, the hazards associated with handling hot, molten alkali metals, the limited tolerance of functional groups on the silicon to the reaction conditions and relatively high cost.

We have recently reported an alternative route to polysilylenes, involving the catalytic elimination of H_2 between two Si-H moieties (11,12,13). The reaction is homogeneously catalysed by titanocene and zirconocene derivatives and in principle should be easier to understand and control than the heterogeneous Wurtz-Fittig reaction. It is thus clear that these polymers provide at least an important complement to those made by the Wurtz-Fittig-type coupling and that improvements in the performance of catalysts could lead to a viable commercial route for the large scale production of polysilanes. In this paper a description of the progress we have made to date in understanding the reaction mechanism and characterizing the polymers will be described. The scope of the catalysis and some of its present limitations will also be discussed. Finally, some of the major questions that remain to be answered before this chemistry can be successfully applied to the general synthesis of polysilylenes will be addressed.

Dehydrogenative Coupling

The formation of an X-X bond by the extrusion of H_2 from two X-H containing molecules is a reaction of considerable generality. Until recently, such reactions have not been systematically exploited for the formation of polymers, although closely related reactions in which the thermodynamics of the process are enhanced by scavenging the hydrogen with oxygen (oxidative coupling) have been exploited to commercial advantage (14). Some of the reasons for the lack of progress in this area are the relatively low reactivity of many of the more interesting types of monomer, a lack of thermodynamic data which allow prediction of whether a reaction is feasible or not, and a lack of specificity in cases when there are several X-H bonds in the same molecule.

In the course of studying the reactions of Si-H compounds with dialkyltitanocenes, with a view to the synthesis of new hydridosilyltitanocene complexes, we adventitiously discovered that phenylsilane undergoes facile, quantitative dehydrogenative coupling to a linear poly(phenylsilylene) under the catalytic influence of dimethyltitanocene. The ease with which this reaction proceeds initially induced us to underestimate the significance of the observation.

Further studies quickly revealed that the rapid dehydrogenative coupling of primary organosilanes to oligomers and the slower coupling of secondary silanes to dimers can be effected under ambient conditions with compounds of the type Cp_2MR_2 (M = Ti, R = alkyl; M = Zr, R = alkyl or H)(11,12,13). None of the other metallocenes, metallocene alkyls, or metallocene hydrides of groups 4, 5 or 6 have shown any measurable activity for polymerization under ambient conditions, although vanadocene catalyzes the slow, stepwise oligomerization of phenylsilane in refluxing toluene (15). Excessive methylation of the Cp ligands deactivates the catalysts. For example, the mixed cyclopentadienyl-pentamethylcyclopentadienyl ($CpCp^*$) complexes of Ti and Zr are active, but the bis Cp^* complexes are not (16,17). However, the $Cp^*_2MR_2$ complexes, where M = Th or U are catalytically active; the former for the dimerization of primary silanes and the latter for their oligomerization (17). A problem with these compounds is that they are too reactive and have a tendency to react with the C-H bonds of substituents on the silicon (18). Bis(indenyl)dimethyltitanium does not catalyze the polymerization reaction, but does effect a slow stepwise oligomerization of phenylsilane to dimer and trimer at room temperature. The bis(indenyl)zirconium analog is active as a polymerization catalyst, giving essentially the same product as the bis(cyclo-pentadienyl) complex.

Replacement of one of the Cp groups in the titanocene or zirconocene-based catalysts by an alkyl group destroys the catalytic activity. It is thus clear that the constraints on catalytic activity are extremely severe.

A remarkable feature of the polymerization reactions is the absence of any evidence in nmr spectra of reaction mixtures for intermediate low molecular weight oligomers. This behavior is quite distinct from those systems mentioned above, which give dimers and trimers and it is believed that these represent two mechanistically distinct processes.

Normally, no small cyclopolysilanes are observed in these reactions. Two exceptions we have noted are the very slow reaction of phenylsilane under the influence of $Cp^*_2TiMe_2$ and the reaction of benzylsilane under the influence of dimethyltitanocene at very long reaction times. From both of these reactions we isolate a single isomer of the cyclohexasilane, in ca. 10 per cent yield in the case of the phenylsilane and ca. 60 per cent yield in the case of the benzylsilane. These isomers are believed to be the all-trans isomers. The phenyl derivative is identical to that previously reported by Hengge (19), the benzyl derivative is a new compound. In the case of the benzylsilane reaction, the cyclohexasilane is produced from polymer, following essentially complete conversion of the monomer to linear polysilane of about 10 silicon units average length. This observation is probably very important in understanding the factors that limit chain length.

The Polymers

The polymerization of primary organosilanes proceeds according to Equation 2. The rate of polymerization is strongly dependent on the

$$n\ RSiH_3 \longrightarrow H\!-\!\!\left[\!-\underset{H}{\overset{R}{Si}}-\!\right]_n\!\!-H \ + \ (n-1)\ H_2 \qquad (2)$$

steric demands of R. The relative rates for a number of different silane reactions, catalyzed by dimethyltitanocene and measured under co-hydrogenation conditions with cyclohexene as described below, is (20):

Phenyl (13) > p-Tolyl (10) > Phenylmethyl (4.6) > $PhSiD_3$ (3.6) >
Benzyl (1) > n-hexyl (1) > Cyclohexyl (0.5).

Cyclohexyl- and phenylmethylsilanes do not polymerize, but give dimer. With titanium-based catalysts the value of n is about 10 and does not vary very much with R or experimental conditions; with zirconium based catalysts, n can be as high as 20. Poly(phenyl- and p-tolysilylenes) produced with these catalysts are brittle glasses and poly(benzyl- and n-hexylsilylenes) are viscous oils. All of the polymers are atactic and highly soluble in most organic solvents. They have been shown by a variety of spectroscopic methods to be linear and SiH_2R terminated (21). The SiH_2R termini can be detected by the use of ^{29}Si-nmr using a DEPT pulse sequence. They are also evident from the presence of a strong SiH_2 infrared bending mode at about 910 cm^{-1}. The linear nature of the polymers and their degrees of polymerization have been determined using mass spectroscopy and particularly by comparing the mass spectrum of a linear poly(phenylsilylene), produced by dehydrogenative coupling with that reported by Hengge et al. (19) for hexaphenyl-cyclohexasilane. The data are shown in Table I and attention is drawn to the fact that the cyclic polymer, which does not separate into fragments with a single Si-Si bond cleavage, gives heavy ions in high abundance while the linear polymer, which is fragmented by a single Si-Si bond cleavage, gives heavy fragments in only very low abundance. The hexabenzylcyclohexasilane described above behaves in very similar fashion. The very sharp, conventional nmr spectra of these cyclic compounds also confirm the absence of small ring compounds in the normal polymer products, where they would be easily detectable.

The degrees of polymerization determined by mass spectroscopy are corroborated by vapor pressure osmometry and gel permeation chromatography studies (21).

The tertiary hydrogens of the polymer backbone show no detectable further reactivity under polymerization conditions but they can be hydrosilated using the classical Speier-type (22,23), or our new zirconium-based catalysts (see below) (17), to give polymers with fully, or partially substituted backbones. The presence of Si-H at every silicon atom and the attendant opportunity for further functionalization is one of the most interesting features of these new polymers. The potential for utilization of Si-H groups on poly(silylenes) for cross-linking and other modes of functionalization has already been recognized by West (8). Synthesis of Si-H functionalized polymers by Wurtz-Fittig-type coupling has been achieved by the copolymerization of dialkyldichlorosilanes with methyldichlorosilane. Great care must be exercised to maintain neutral conditions during the work-up in order to avoid reaction of the Si-H functions (8).

Cohydrogenation of Olefins

Thermodynamics only slightly favor Equation 1 and the copious evolution of H_2 can have undesirable effects on the course of the reaction. The inclusion of an internal olefin in the reaction mixture completely suppresses the evolution of hydrogen and increases the rate of polymerization. With titanocene-based catalysts the olefin undergoes hydrogenation but the polymerization, except for a rate increase, proceeds in the same way as in the absence of olefin (20). Cohydrogenation is useful for the synthesis of polymers from gaseous silanes, in particular SiH_4 and $MeSiH_3$, since it allows the progress of the reaction to be monitored by gas uptake. Another important advantage of cohydrogenation is that it can make Equation 1 much more thermodynamically viable because of the high heat of hydrogenation of olefins. In this respect the reaction resembles olefin hydrogenation-driven C-H bond activations studied by Crabtree (24).

Titanocene catalysts do not catalyze the hydrosilation of most internal olefins, although they can attach active olefins such as styrene, or norbornene to the growing polymer chain ends. The zirconocene-based catalysts, on the other hand, can be powerful hydrosilation catalysts and the remarkable copolymer synthesis shown in Equation 3 can be easily achieved under mild conditions (17).

With cyclohexene, polymerization occurs more rapidly than hydrosilation. After polymerization has proceeded to completion, there is a slow hydrosilation to introduce cyclohexyl groups onto the polymer chain, to a maximum extent of about 50 per cent of the Si-H groups. With more reactive olefins, such as styrene, hydrosilation occurs more rapidly than polymerization and the polymerization reaction is suppressed. As in the polymerization reaction, the reactivity of primary silanes is much greater than

$$2n\ PhSiH_3 + \frac{n}{2}\ \text{(cyclohexene)} \xrightarrow{Cp'_2ZrMe_2} H{+}\underset{H}{\overset{Ph}{Si}}-\underset{(C_6H_{11})}{\overset{Ph}{Si}}{+}_nH + (2n-1)\ H_2 \qquad (3)$$

that of secondary silanes and the product of hydrosilation of styrene by phenylsilane is almost entirely phenyl(1-phenylethyl)silane. This suggests that the incorporation of cyclohexyl groups into the polymer does not occur by simple hydrosilation. We currently favor a mechanism which involves cleavage of the preformed polymer chain by the catalyst to give an intermediate which is capable of yielding alkylsilane, but does not require direct breaking of the tertiary Si-H bond of the polymer chain.

Coupling of Germanes

Dimethyltitanocene is extremely active for the coupling of germanes (15). Even secondary germanes can be coupled rapidly to oligomers, but there seems to be a severe constraint on the chain length, as in the case of primary silane polymerization. Polymerization of diphenylgermane can be carried out under two different regimes using dimethyltitanocene as catalyst. Addition of freshly recrystallized dimethyltitanocene to an excess of diphenylgermane results in steady evolution of hydrogen at ca. 60°C, but the colour of the solution remains yellow until all of the germane is consumed. During this period, the formation of tetraphenyldigermane and small amounts of higher oligomers can be observed by nmr spectroscopy. When almost all of the monomer is consumed there is a dramatic change in colour to dark purple accompanied by a surge of gas. The resulting purple solution is much more active for the further dehydrogenative coupling of diphenylgermane and produces primarily octaphenyltetragermane. The dark purple product can be produced directly by reaction of dimethyltitanocene and diphenylgermane in a molar ratio of about 1:2. The structure of this dark coloured compound will be discussed further below.

Primary germanes undergo coupling to insoluble gels under the influence of dimethyltitanocene, presumably because the backbone hydrogens show sufficient activity to lead to crosslinking. Perhaps the most interesting aspect of this reaction is that it points to the possibility that similar reactions of silanes may be achievable with more active catalysts. Gels produced from coupling of primary silanes could have interesting electronic properties since they are the homologs of silicon monohydride, a material of considerable current interest to the electronics industry (25). In the presence of vanadocene, phenylgermane undergoes facile stepwise conversion to oligomers at about 60°C.

Results obtained with germanes also provide models for the kinds of reactions that may be occurring in the silane polymerization reaction as well. For example, we have succeeded in carrying out the reaction shown in Equation 4 (26). The analogous reactions with triphenylsilane, or triphenylstannane, were not

$$Cp_2Ti(CH_3)_2 + Ph_3GeH \longrightarrow Cp_2Ti(CH_3)(GePh_3) + CH_4 \qquad (4)$$

successfully demonstrated, but the fact that it occurs with the germane adds credence to the hypothesis that an analogous reaction is the first step in the reaction of silanes with dimethyltitanocene.

The Mechanism of Polymerization

The compounds 1, 2 and 3 have all been isolated from reactions of phenylsilane with either dimethyltitanocene (12) or dimethylzirconocene (13). All of the evidence points to the fact that these compounds are probably resting species and are not involved in the catalytic cycle. They do nevertheless give some indication of the complex series of reactions that transform the dimethylmetallocene to active catalyst.

Using the titanocene-catalyzed co-hydrogenation of cyclohexene, we have studied the kinetics of the polymerization of a number of primary silanes (20). The rate law was found to be:

$$\text{Rate} = k[\text{catalyst}]^{1/2}[RSiH_3]^{1/2}[C_6H_{12}]^0 \qquad (5)$$

We attribute the form of this rate law to be due to the pseudo-equilibrium 6. We refer to 6 as a pseudo-equilibrium, because it is in fact a steady state rather than a true equilibrium. If

$$Cp_2Ti(\mu\text{-}H)(\mu\text{-}H\text{-}SiHR)TiCp_2 + RSiH_3 \rightleftharpoons 2\ Cp_2Ti(H)(SiH_2R) \qquad (6)$$

1 4

compound 4 is a participant in the catalytic cycle, anything which changes the throughput rate of the cycle will alter the concentration of 4. Compounds 1 can be observed in solution by nmr and the R = Ph compound has been structurally characterized by X-ray analysis (12). A pseudo-equilibrium such as Equation 6 leads naturally to the rate law of Equation 5 if it is assumed that the equilibrium lies to the left and the rate of reaction is controlled by the unimolecular transformation of the titanium(IV)silylhydride, 4. The most plausible first step for the decomposition of 4 is an α-hydride elimination from the SiH_2R group, followed, or accompanied, by loss of H_2 from the complex to give a $Cp_2Ti{=}SiHR$ complex. It is then assumed that propagation occurs by some kind of repetitive insertion of the silylene into a Ti-Si bond. A possible mechanism of this kind is shown in Scheme 1. Two equally plausible routes for the propagation step are shown on the lower center and lower right of Scheme 1. This type of mechanism is attractive since it explains why secondary silanes will only dimerize.

A number of features of the silane and germane polymerization reactions show unequivocally that there are at least two independant mechanisms operating. This dichotomy is most evident in situations where the products of reaction can be compared for the induction period that precedes the formation of compound 1 and the products that are produced after the formation of 1 (12,15,20).

Table I. Mass Spectra of a Linear Oligo(phenylsilylene) and of Hexaphenylcyclohexasilane

Oligophenylsilylene			Hexaphenylcylcohexasilane[1]		
Ion	Mass	(abundance)	Ion	Mass	(abundance)
$Ph_5Si_5H_7$	532	(0.45)	$Ph_6Si_6H_6$	636	(30)
$Ph_5Si_4H_3$	499	(0.35)	$Ph_5Si_5H_5$	530	(20)
$Ph_4Si_5H_4$	452	(1.13)	$Ph_4Si_5H_4$	452	(50)
$Ph_4Si_4H_2$	422	(2.95)	Ph_4Si_4H	421	(40)
$Ph_3Si_4H_3$	346	(7.18)	$Ph_3Si_5H_5$	376	(16)
Ph_3Si_3H	316	(12.4)	Ph_3Si_4H	344	(75)
Ph_3Si_2	287	(27.0)	Ph_3Si_3	315	(20)
Ph_3Si	259	(24.0)	Ph_3Si	259	(60)
$Ph_2Si_3H_2$	240	(25.0)	Ph_2Si_3	238	(12)
Ph_2Si_2H	211	(42)	Ph_2Si_2	210	(16)
Ph_2SiH	183	(47)	Ph_2SiH	183	(80)
PhSi	105	(100)	PhSi	105	(100)

1. From ref. 1 .

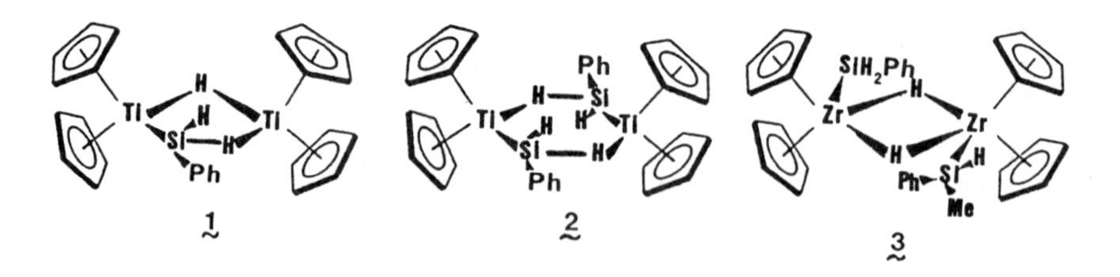

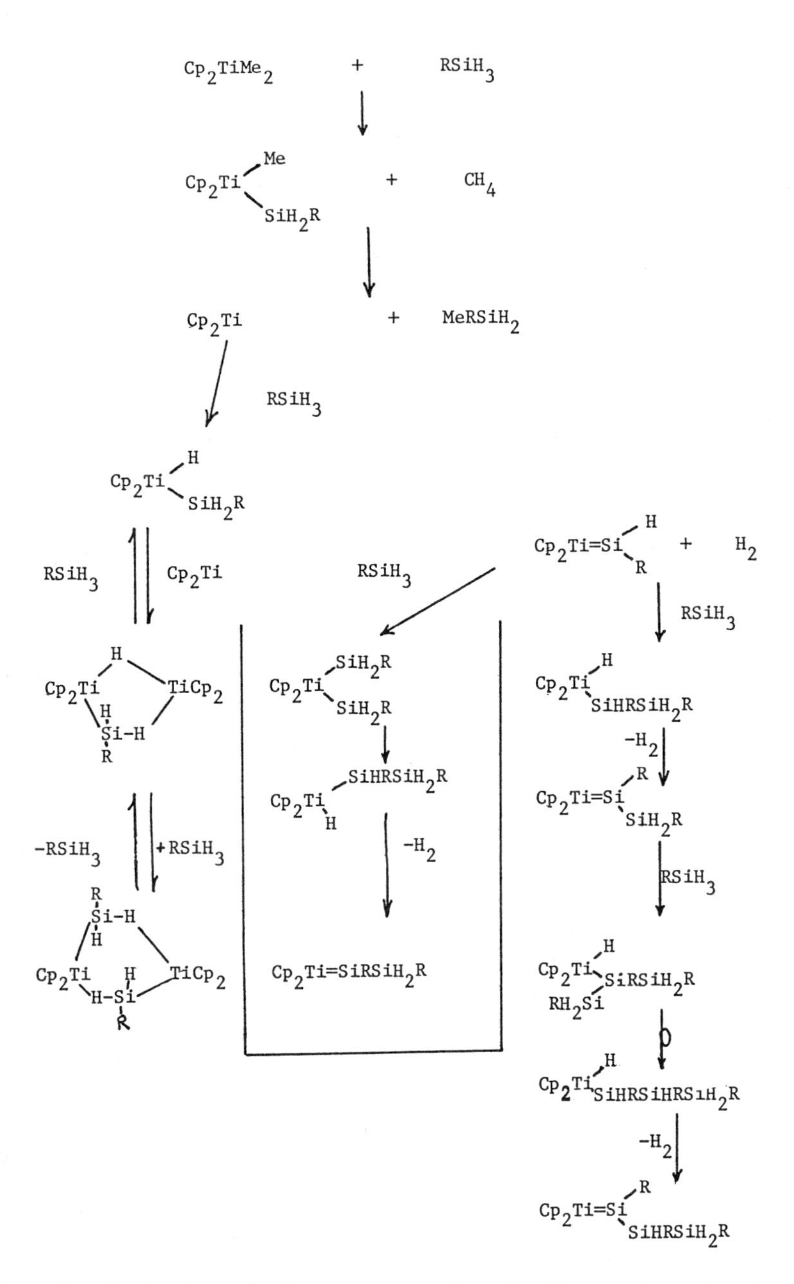

Scheme 1. Mechanism of titanocene catalyzed silane polymerization.

In the two regimes the products are quite different. For example, in the titanium catalyzed oligomerization of diphenylgermane described above, one catalytic regime seems to involve no gross reduction of the titanium(IV) and gives rise mainly to tetraphenyldigermane, the other involves the reduction of the titanium to give the diphenylgermyl analogue of 1, which is a powerful catalyst for the tetramerization of diphenylgermane. The characterization of the purple titanium intermediate as an analogue of 1 is based on the observation of an esr spectrum of the triplet state and on the similarity of its H-nmr spectrum to that of 1 (12).

A similar dichotomy was observed in the titanium catalyzed polymerization of primary silanes coupled to the hydrogenation of norbornene (20). At low catalyst concentration (ca. 0.004M), essentially complete conversion of norbornene to an equimolar mixture of norbornane and bis-phenylsilyl- (and/or 1,2-diphenyldisilyl)norbornane was observed. Under these conditions no evidence for reduction of titanium was obtained. At higher catalyst concentrations ($>$ 0.02M) rapid reduction of the dimethyltitanocene to 1 and 2 occurs and the catalytic reaction produces mainly polysilane ($\overline{DP}_n$ ca. 10) and norbornane in ca. 80 per cent yields, and silylated norbornanes in about 20 per cent yield.

Our present preferred hypothesis is that the reactions occurring in the regime where gross reduction of the titanium does not occur are metal catalyzed radical chain reactions. Oligomerization under these conditions proceeds by the stepwise coupling of silyl or germyl radicals. Following reduction of the titanium, we believe that polymerization occurs by some sort of rapid addition mechanism in which the intermediates are not observable because they are short lived, or because they are spectroscopically silent (e.g., cannot be seen in the nmr because they are paramagnetic). A likely candidate for the mechanism in this case is the rapid repetitive insertion of silylene moieties, produced by α-hydride elimination from 4, into a metal-silyl bond, as shown in Scheme 1.

At the present time we do not know what the termination reaction is. Given the fact that the polymers isolated from co-hydrogenation reactions do not have different molecular weight properties from those produced by simple polymerization, hydrogenolysis can be excluded as a likely termination reaction. This leaves spontaneous homolysis of a metal-Si bond, followed by hydrogen abstraction to neutralize the resulting silyl radical, or reductive elimination of a silyl with a hydride, or of two silyl ligands. There is nothing in the presently available information that allows us to discriminate between these alternatives and they must be considered as equally plausible. It is clear that any reaction that leads to polymer chain termination with $-SiH_2R$ groups will stop further chain growth since neither titanium, nor zirconium based catalysts are active for secondary silanes.

The formation of cyclohexasilanes and the zirconium catalyzed hydrosilation of poly(phenylsilylene), referred to above, both suggest that slow cleavage of polymer chains may occur in the presence of the catalysts. Such cleavage may also play an important role in limiting chain lengths and it will be the focus of further study.

Like many homogeneously catalyzed reactions, the overall cycle (or cycles) in these polymerization reactions probably contains too many steps to be easily analyzed by any single approach. Both kinetics and model compound studies have thrown light on some of the steps. However, as indicated above, many of the model compounds isolated from the reactions of primary silanes with metallocene alkyls and hydrides are too unreactive to explain the polymerization results.

Conclusions

The catalyzed dehydrogenative coupling of silicon and germanium hydrides has been achieved with high reaction rates and high conversions. This represents a major new route to the synthesis of Si-Si and of Ge-Ge bonds. The low degrees of polymerization of the products of primary organosilane coupling represent a serious limitation on their use in applications where mechanical strength is a pre-requisite. Further studies on the mechanism(s) of the reaction are being pursued, particularly with a view to understanding the nature of the chain termination process(es). It is possible that such knowledge will allow some control over the factors limiting chain length.

The selective reactions of terminal Si-H groups with appropriate couplers is a promising method for the conversion of our oligomers to high molecular weight materials and we are presently studying such reactions. It is likely that runs of twenty silicon atoms already exhibit many of the desirable features manifest by higher molecular weight materials and the coupling together of these chains can give them the dimensional stability and mechanical properties necessary for certain applications.

The principle of forming novel polymeric materials by dehydrogenative coupling is of considerable generality. The catalytic reactions of Si-H and OH are well know (27). Similar reactions with the heavier chalcogens might lead to some interesting new materials. Of even greater interest is the catalytic formation of high molecular weight poly(silazanes) by the elimination of H_2 between silanes and primary amines. This reaction has already been successfully carried out with the aid of transition metal complex catalysts (28).

Acknowledgments

The Natural Sciences and Engineering Research Council of Canada, the Fonds FCAR du Québec, the Dow Corning Corporation, and Esso Canada are all thanked for their financial support of this work. My collaborators, without whose experimental skills none of the work described above would have been done, are thanked and their contributions are recognized by citation in the references.

Literature Cited

1. Fink, J. M.; Michalczyk, M. J.; Haller, K.; West, R. J. Chem. Soc., Chem. Communs. 1983, 1010.

2. Burkhard, C. A. J. Am. Chem. Soc. 1949, 71, 963.
3. Yajima, S; Shishodo, T.; Kayano, H. Nature 1976, 264, 237.
4. Miller, R. D.; Hofer, D.; McKean, D. R.; Wilson, C. G.; West, R.; Trefonas, P. T. III ACS Symposium Series No. 266; American Chemical Society, Washington, DC, 1984, pp. 293-310.
5. Stolka, M.; Yuh, H.-J.; McGrane, K.; Pai, D. M. J. Polymer Sci., Part A, Polymer Chemistry 1986, 25, 823.
6. Trefonas, P.; Damewood, J. R.; West, R. Organometallics 1985, 4, 1318.
7. Harrah, L. A.; Ziegler, J. M. J. Polym. Sci., Polym. Lett. Ed. 1985, 23, 209.
8. Gobbi, G. C.; Fleming, W. W.; Sooriyakamuran, R.; Miller, R. D. J. Am. Chem. Soc. 1986, 108, 5624.
9. For an excellent review of the current status of organopolysilane chemistry, see West, R., J. Organometal. Chem. 1986, 300, 327.
10. Helmer, B; West, R. J. Organometal. Chem. 1982, 236, 21. Trefonas, P.; Djurovich, P. I.; Zhang, X.-H.; West, R.; Miller, R. D.; Hofer, D. J. Polym. Sci., Polym. Lett. Ed. 1983, 21, 819.
11. Aitken, C. T.; Harrod, J. F.; Samuel, E. J. Organomet. Chem. 1985, 279, C11.
12. Aitken, C. T.; Harrod, J. F.; Samuel, E. J. Am. Chem. Soc. 1986, 108, 4059.
13. Aitken, C. T.; Harrod, J. F.; Samuel, E, Can. J. Chem. 1986, 64, 1677.
14. Finkbeiner, H. L.; Hay, A. S.; White, D. M. Polymerization by oxidative coupling, in Polymerization Processes, Wiley-Interscience, New York, 1977, p. 537.
15. Malek, A. ; Harrod, J. F., unpublished results.
16. Aitken, C.; Harrod, J. F., unpublished results.
17. Barry, J.-P. ; Harrod, J. F., unpublished results.
18. The reactivity of organoactinides with C-H bonds has been extensively studied by Marks et al.; see e.g., Bruno, J. W.; Smith, G. M.; Marks, T. J.; Fair, C. K.; Schultz, A. J.; Williams, J. M. J. Am. Chem. Soc. 1986, 108, 40.
19. Hennge, E.; Lunzer, F. Monatheft. Chem. 1976, 107, 371.
20. Harrod, J. F. ; Yun, S. S. Organometallics, in press.
21. Aitken, C. T.; Harrod, J. F.; Gill, U. Can. J. Chem., in press.
22. Speier, J. L. Adv. Organomet. Chem. 1979, 17, 407.
23. Harrod, J. F.; Chalk, A. J. In Organic Synthesis via Metal Carbonyls, Wender, I., Pino, P. Eds.; Wiley: New York, 1977; Vol. 2, p. 673.
24. Crabtree, R. H.; Mihelcic, J. M.; Quirk, J. M. J. Am. Chem. Soc. 1982, 104, 107.
25. Stein, H. J.; Peercy, P. S. Appl. Phys. Lett. 1979, 34(9), 604. DeMeo, E. A.; Taylor, R. W. Science 1984, 224, 245.
26. Harrod, J. F.; Malek, A.; Rochon, F.; Melanson, R. Organometallics, in press.
27. See e.g., Sommer, L. H.; Lyons, J. E. J. Am. Chem. Soc. 1967, 89, 1521.
28. Blum, Y; Laine, R. M. Organometallics 1986, 5, 2081.

RECEIVED September 1, 1987

Chapter 8

Polysilylene Preparations

D. J. Worsfold

Division of Chemistry, National Research Council of Canada, Ottawa, Ontario K1A 0R9, Canada

The reductive coupling of dichlorosilanes with sodium in refluxing toluene has been studied with a view to determining the mechanism of the reaction. The reaction behaves as a chain reaction, rather than a simple polycondensation, in that high molecular weight material is formed near the start of the reaction. Also the formation of high molecular weight material is not dependant on equivalent proportions of reactants. The complex mixture of products of widely separated molecular weight can be interpreted by the simultaneous operation of two routes to polymer. One of these seems to be promoted by the presence of organometallic species, and could be via an anionic step chain growth. The reaction kinetics suggest that the growing chains have a comparatively long lifetime. A slow build up of rate suggest an initial formation of the growing chain ends at a rate dependant on the sodium surface area, but the subsequent chain propagation reaction is independant of surface area. Copolymerization experiments show a large difference in reactivity of aryl and alkyl substituted dichloromethylsilanes.

The discovery that soluble high molecular weight polysilanes may be prepared by the reductive coupling of dichlorodialkylsilanes by alkali metals (1,2) has led to considerable work on the properties of this interesting class of polymers (3,4,5). The preparation of the polymers leaves much to be desired as frequently the high polymer is only a minor product. Mechanistic studies of the reaction with a view to improving the relevant yields have been few (6). The major ones by Zeigler (7,8,9) showed that a silylene diradical was not involved in the reaction, and stressed the importance of polymer solvent interactions.

There appear to be three types of product, a low molecular weight product appearing to be small ring compounds (PI), an intermediate product (PII), and the usually desired high molecular weight polymer (PIII). Previous studies have concentrated only on PIII,

0097-6156/88/0360-0101$06.00/0

although PII could be useful for some purposes. However, to understand the reaction fully it is necessary to consider all three products, and also the possibility that more than one mechanism is operating. Three such products are formed in the anionic polymerization of cyclic siloxanes to polysiloxanes by a rearrangement mechanism, but interchanges between the three products did not appear as possible with the polysilanes.

Experimental

The preparations were carried out as usual in a three necked flask fitted with a stainless steel stirrer, condenser, and nitrogen inlet (10,11). The solvents and dichlorides were added by syringe. The sodium was added as a freshly cut block and, prior to the reaction, melted in the refluxing solvent and stirred to form sand. The overall surface area of the sodium could be controlled by the stirrer speed. When necessary the stirrer speed was measured by a tachometer.

Toluene, the major solvent, was stirred for three days with several portions of sulphuric acid, washed, dried, and stored over calcium hydride on a vacuum rack. The toluene was distilled out immediately prior to the reaction. The dichlorides were vacuum distilled at the time of the reaction into three fractions, and the middle fraction, about 60% of the total, used. The dichlorides were obtained from Petrach, stored in a nitrogen glove bag and handled by syringe.

The principal dichlorides used in this study were hexylmethyldichlorosilane (HMDS) and phenylmethyldichlorosilane (PMDS), some copolymerizations of the latter and dimethyldichlorosilane (DMDS) were also made. The usual practice was to add the dichloride (20 vol.%) in one portion to the refluxing reaction mixture at the start of the reaction. This usually entailed cooling the reaction mixture when PMDS was used to prevent a too vigorous reflux at the beginning of the reaction. In all cases the reaction turned dark purple and ultimately viscous.

When samples were removed during the course of the reaction, the reaction was terminated by adding to water, or to an alkyllithium, or in some cases to trimethylchlorosilane. The unreacted dichlorides could be determined by direct injection into the gas chromatograph (GC), or by the reaction product with an alkyllithium.

The reaction mixture at the end was treated carefully with water, and washed. The reaction mixture was analysed at this point by gel permeation chromatography (GPC). The high polymer could be isolated by precipitation in isopropylalcohol. The intermediate fraction PII was partly precipitated with the PIII or remained in solution according to the conditions of the reaction. Further fractions of PII could be isolated by precipitation in ethanol or methanol. PI was soluble in ethanol.

The molecular weight distributions were determined only in terms of the polystyrene equivalent from the polystyrene calibration of the GPC columns. This can only be regarded as semiquantitative, as the method determines only the relative hydrodynamic volumes. If the polysilane chain was significantly stiffer than polystyrene, the molecular weights would be estimated too high. If the chains are branched the molecular weights found would be too low. The errors can be upto a factor of 2, but this is small compared to the range covered. The complete reaction mixture was analysed to give the proportions of the three different molecular weight products (7).

RESULTS

Products

If samples were taken for the GPC and GC it was found that after all the dichloride was consumed that the product was distributed into the three peaks on the GPC PI, PII, PIII (Figure 1). Although high molecular weight material was present as soon as the dichloride was consumed, continued reflux with sodium caused changes in the molecular weight, particularly of PII (Table I). This was very

Table I. Change in MW PII of PMDS on Continued Contact with Sodium

Time (mins)	60	100	200	300	400
Mw x 10^{-3}	1.3	2.5	3.9	4.8	5.0

marked in the case of PMDS. This meant that when the reaction mixture was added to isopropyl alcohol to recover the polymer, the proportion of PII included was dependent on the time the reaction was left to reflux. Hence at short reflux times the polymer could appear monomodal as PIII, while at longer reflux times the precipitated polymer appeared mostly to be PII whereas the actual yield of PIII was the same. Hence the marked variations that appear in the literature of the proportions of the two peaks in the GPC of the precipitated product (7,8), could, in part, be an artifact of the time of reflux, and the unfortunate choice of isopropylalcohol as the precipitant. Isopropyl alcohol has a limiting solubility for polysilanes just in the range of molecular weights covered by PII.

Low Molecular Weight Product

The lowest molecular weight product could be isolated by its solubility in ethanol or methanol. It was found to have as its major product a cyclic pentamer. This was confirmed in the HMDS case from its ^{1}H and ^{13}C spectra which were consistant with its molecular formula, but containing many isomers. The mass spectrum gave a parent peak at 640.8 with an appropriate mix of isotopamers. The UV spectrum had a maximum at 261mu in agreement with the assignment of Watanabe et al. (12) for cyclic pentamers. The PMDS low MW product also gave ^{1}H and ^{13}C spectra in accord with a cyclic product, and the molecular weight determined by freezing point depression was 560 (600 for a pentamer).

In many polymerizations cyclic material is produced by a concurrent backbiting reaction as linear polymer is formed. For example dioxan is formed in the cationic polymerization of ethylene oxide to polyoxyethylene, and polyoxyethylene can be degraded to dioxan with acid catalysts. Thus it is possible that the formation of PI the low MW product in these polysilanes is an integral part of the polymerization process, and not a separate sidereaction. For example, Na/K alloy was reacted with HMDS in tetrahydrofuran (THF). The first product at short reaction time (1-2 min) was a polymer with a molecular weight up to 10^5, which rapidly degraded at longer times to the pentamer. Also high MW polymer from the reaction of HMDS in toluene with Na, on refluxing with K in THF degraded to the low M.W. cyclic material.

Chain Nature of Reaction

Although the coupling reaction of sodium with a dichloride to form a polymer would appear to be a form of a condensation reaction, the reaction appears more like a chain reaction. Normally a condensation polymerization will produce first dimers and trimers which then couple in turn to produce a rapid increase in molecular weight when all the monomer has reacted. Also high molecular weight material should form only when equimolar amounts of the reactants are present. In this reaction it is possible to isolate high molecular weight material from near the start of the reaction with molecular weight above 10^5, Figure 2. Moreover considerable high MW polymer is formed when either sodium or the dichloride is in excess, Figure 2.

This suggests that if the initiation reaction is a reaction between sodium and the dichloride, the subsequent chain growth reaction must be very much faster in order that long chains should form. There is some evidence that molecules containing sequences of silicon atoms react faster than those with single atoms (13). Nevertheless, the chain growth in general cannot be too rapid because the molecular weight of the high MW material increases during the course of the reaction, Table II, as dichloride is consumed indicating the chain carriers have a lifetime comparable to the reaction time.

TABLE II. PIII Formation with Time for HMDS

Time (mins)	10	20	33	180
HMDS used %	33	82	97	100
Yield PIII %	14	26	37	45
$M_w \times 10^{-5}$	2.0	3.0	4.5	4.5

Addition of Organometallics

Zeigler has shown by trapping experiments that the reaction intermediate was not a silylene diradical formed by the dehalogenation of the dichloride. This was confirmed here by using dimethoxydimethylsilane as a trapping reagent. Other possibilities for the reaction intermediate are chain end radicals or alkalimetal silane chainends. Also possible are radical anions as in the single electron transfer (SET) polymerization scheme described by Heitz (14). The latter scheme demonstrates several similarities to the present reaction particularly in the formation of higher molecular weight material near the start of polymerizations which would appear to be simple polycondensations.

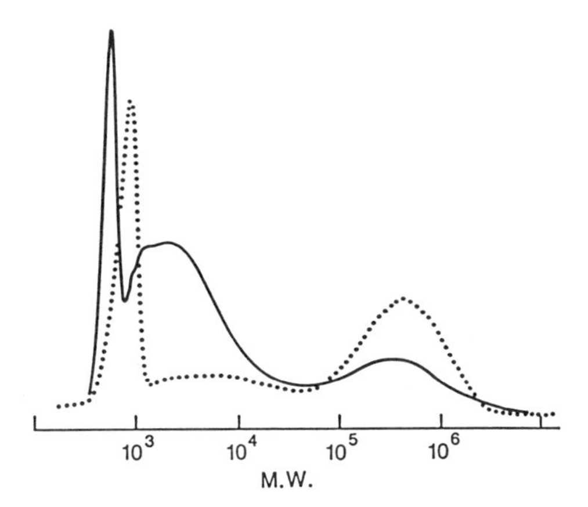

Figure 1.
GPC traces of complete products of the reductive coupling of dichlorosilanes with Na in refluxing toluene. •••• Hexylmethyldichlorosilane, —— phenylmethyldichlorosilane.

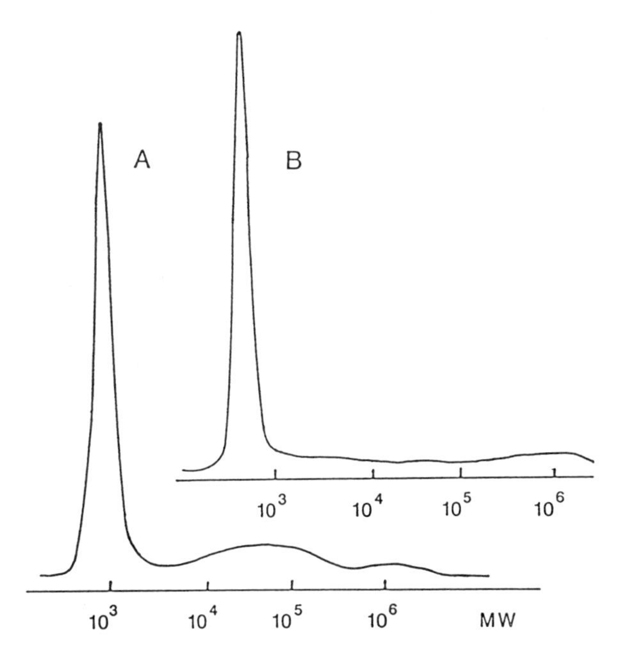

Figure 2.
GPC traces of products of the reductive coupling of hexylmethyldichlorosilane with Na in refluxing toluene. A, on a deficit of Na; B, with normal Na after 20% consumption of dichloride.

The role of ionic species in the reaction was studied by adding organometallic compounds at the start of the reaction. Either dibutylmercury or diphenylmercury were added to the sodium sand in refluxing toluene before the dichloride was added. A yellow color was produced indicating the reaction of the mercury compound with the sodium to give an organosodium compound. This color instantly disappeared on addition of the dichloride. With HMDS the final product had a lower MW for PIII, and usually a more complex GPC. With PMDS a very great increase in rate occurred, and most of the product was now PII, Figure 3. This product behaved in an identical fashion to the PII in the absence of added organometallic in that on continued reflux with sodium it increased in MW whilst remaining a constant fraction of the reaction product, Figure 3. After short reaction times the reaction solution could be separated from the sodium, washed with water, dried, and again refluxed with sodium. The product PII again would show an increase in molecular weight, although any of the original active chain ends could be expected to be destroyed by the washing.

The inference may be made that PII with PMDS and perhaps PIII with HMDS, are affected by the presence of organometallics, presumably in the role of producing more active chains. In the case of PMDS this then becomes the major reaction as seen by the increase in rate and the absence of PIII. The latter having perhaps a separate mechanism of formation. For HMDS the effect is smaller but perhaps the increase in chains is indicated by the decrease in MW of PIII. Although far from conclusive this is evidence for some intermediacy of alkalimetalsilane species in the production of these products, which in our hands were often the major polymeric species.

Attempts to promote radical reactions by adding azobisisobutyronitrile to HMDS polymerizations gave no marked effect.

Effect of Sodium Surface Area

Because of the heterogeneous nature of the reaction, the particle size, and hence the surface area of the sodium, might have some effect on the reaction. In these reactions the sodium was added as a block, melted, and stirred to give the dispersion. According to the speed of the stirrer the particle size varied from about 1mm to .1mm in diameter. This change caused a change in the ratio of the products of the reaction, particularly in the case of HMDS., Figure 4. In this case the large yield of the PIII compared to the low MW cyclics PI at low Na surface area, suggests the formation of the latter is promoted by the large surface area of the sodium. In the case of PMDS a relatively small effect was found.

Reaction Kinetics

The rate of consumption of PMDS was too fast to follow, however, this was not the case for HMDS, and the disappearance of dichloride was followed with time. The dependence of the reaction on the surface area of sodium is shown in Figure 5. There is an initial period of increasing rate, and this is very dependent on the sodium surface area. This induction period can be associated with an increase in the number of actively growing chains with comparatively long life-

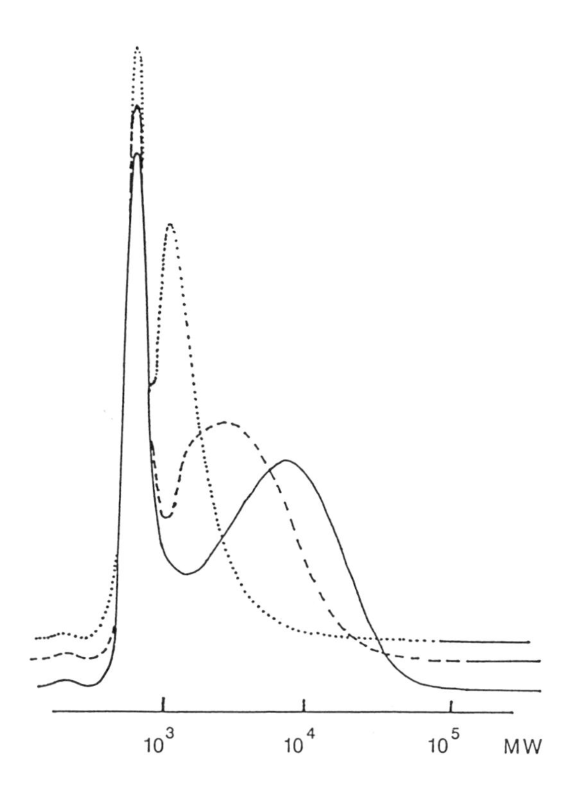

Figure 3.
GPC traces of products of the reductive coupling of phenylmethyldichlorosilane with Na in refluxing toluene in the presence of 3.4×10^{-2} M $HgBu_2$. After ••••• 5 mins, --- 120 min, —— 240 mins.

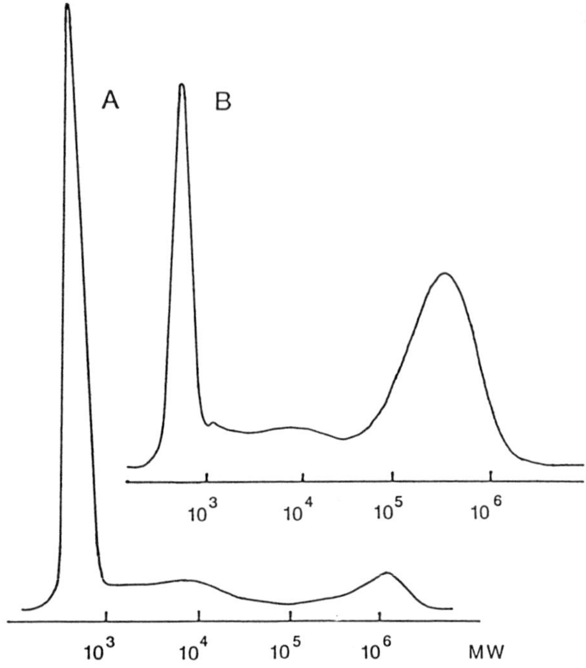

Figure 4.
GPC traces of products of the reductive coupling of hexylmethyldichlorosilane with Na in refluxing toluene. A, with rapid stirring; B, with slow stirring.

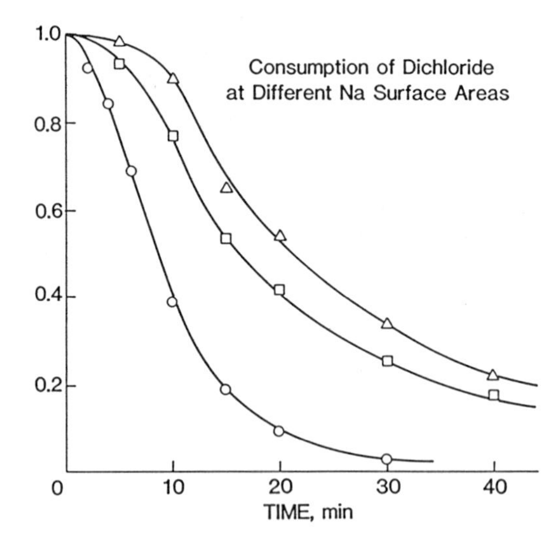

Figure 5.
Fraction of hexylmethyldichlorosilane remaining, plotted versus time, at various surface areas of sodium per mole dichloride Δ, 0.20; □ , 0.67; ○ , 4.64 meter^2.

times during this period. The later stages of the reaction are, however, relatively independent of the surface area of sodium (Table III).

Table III. Dependence of Maximum Rate on Sodium Surface Area

Surface Area / Mole Dichloride Meter^2	Max. Rate M/sec x 10^3
0.20	0.72
0.67	0.85
4.64	1.16

The rate decreases towards the end of the reaction which suggests the rate is dependent on the concentration of dichloride. This was confirmed by decreasing the concentration of HMDS by a factor of three, whilst keeping the sodium constant, when it was found that under comparable stirring conditions the fraction reaction curves nearly coincided. A slightly longer induction period with the lower monomer being consistant with a slower build up of the reaction centers at this concentration.

The apparent lack of dependence of the propagation reaction on the surface area of the sodium suggests that the reaction of a chlorine ended chain with sodium is probably fast and not the rate determining step. The rate determining process is probably the reaction of the sodium ended chain with the dichloride. This latter reaction is presumably not on the sodium surface because of the lack of dependence on the surface area. This is supported by the observation that if the sodium is allowed to settle part way through the reaction most of the polymer appears to be in the solution and not absorbed on the sodium surface via the longlived active chain ends.

Visual observation, however, suggests that some polymer is attached to the solid phase. This is particularly noticeable if water is added to the reaction mixture when the purple color is only difficultly discharged. The source of the color is not clear. It is possible that it is Na entrapped in the precipitating NaCl to give color centers as in the preparation of sodium alkyls. Another possibility is that it is due to a charge transfer reaction between the sodium and polymer or cyclic material. However, after isolation of the polymer it is not possible to regenerate the color in toluene solution with Na. The reactive chain end also might be colored as are the centers in living anionic polymerization. But if so this would indicate the attachment of the majority of the active centers to the solid surface because the supernatant liquid is only very slightly colored.

Effect of Reaction Termination Reagent

If the reaction has long lived intermediates, it is possible that the chain end could alternate between a chlorine ended and a sodium ended chain as a dichloride was added to give the chlorine end and then reacted further with sodium to regenerate the sodium end. On removing an aliquot the sodium is no longer in contact with the reaction mixture and the chain could be expected to attain the chlorine ended form. Then if the reaction was terminated with methyllithium in diethylether this would rapidly cap the end. If, however, water was added coupling of the chains could be expected leading to a higher molecular weight. This was indeed found. With HMDS, after 57% reaction on termination with CH_3Li or H_2O, PIII had MWs of 1.3 and $2.5x10^5$, after 84% reaction 2.4 and $3.1x10^5$, respectively. This confirms the long life of the active chains and the probability that it is the chlorine end at this time. It was found also that the polymer resided mostly in the solution, and not largely absorbed via its active end to the sodium surface.

Copolymerization

A few copolymerizations of PMDS and DMDS were carried out. By following the concentration of the dichloride it was possible to establish that PMDS reacts about four times faster than DMDS, Figure 6. This is reflected in the polymeric products of the reaction where at early stages of the reaction the polymer composition is far higher in PMDS than the feed. A feed ratio of $PhMe/Me_2$ of 0.75 gave a product ratio $PhMe/Me_2$ of 1.59. It is also apparent that in the later stages of the reaction only DMDS is present. This should lead to insoluble polymer if initiation and propagation both occur at this stage. In fact comparatively little insoluble polymer forms, and this is further evidence for the long life of the active polymer chains. In the later stages of the reaction much DMDS adds as a dimethylsilyl rich block onto the existing phenylmethylsilyl blocks and remains in solution. The resulting polymers would have a graded composition along the chain as the composition of the feed changes.

Conclusions

Much is left not undestood in this reaction. It would appear that more than one route leads to the higher MW material, most likely two to account for the appearance of the two higher MW peaks in the GPC. The low molecular weight cyclics are perhaps the products of a simultaneous growth and backbiting mechanism in one of these routes. One of the mechanisms shows some evidence of ionic or alkalimetalsilyl intermediates, and this maybe also the route to the cyclics because this is promoted in THF solution. This reaction would then form most of the products and be the major factor deciding the kinetics of the reaction.

The reaction kinetics in the case of HMDS suggests a two stage mechanism for the major dichloride consuming reaction. First a slow Na surface dependent build up of long lived active centers then a propagation step in which the Na surface is not rate determining. The reaction which occurs on the surface must be fast.

The other mechanism, if in fact there is one, is perhaps via a radical anion as per the SET mechanism suggested by Heitz (14,15) for other polymerizations. The ability of the long silicon chain to delocalize electrons would assist in stabilizing either radical anions or straight anionic species. A simple radical is preferred by Zeigler (9) for the formation of the highest MW fraction in the polymerization of PMDS, a product for whose route no evidence has accrued in this work.

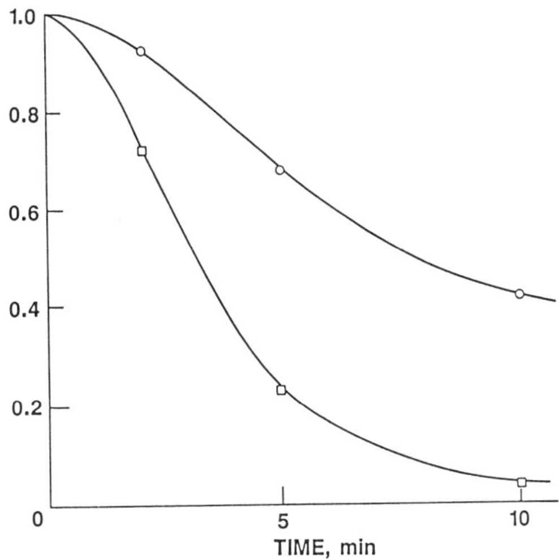

Figure 6.
Fraction of phenylmethyldichlorosilane (□) and dimethyldichlorosilane (O) remaining as a function of time in a copolymerization.

Literature Cited

1. Wesson, J.P.; Williams, T.C. J. Polym. Sci., Polym. Chem. Ed., 18, 959 (1980).
2. Trujillo, R.E. J. Organometal. Chem., 198, C27 (1980).
3. West, R.; David, L.D.; Djurovich, P.I.; Stearley, K.L.; Srinivasan, K.S.V.; Yu, H. J. Amer. Chem. Soc., 103, 7352 (1981).
4. Harrah, L.A.; Zeigler, J.M. J. Polym. Sci., Polym. Let. Ed., 23, 209 (1985).
5. Hofer, D.C.; Miller, R.D.; Miller, C.G. SPIE Vol. 469, 16, Advances in Resist Technology (1984).
6. West, R. J. Organometal. Chem., 300, 327 (1986).
7. Miller, R.D.; Hofer, D.; McKean, D.R.; Willson, C.G.; West, R.; Trefonas, P.T. Materials for Microlithography, ACS Symp. Series 226, 293 (1984).
8. Zeigler, J.M. Polym Preprints, 27(1), 109 (1986).
9. Zeigler, J.M. Polym Preprints, 28(1), 424 (1987).
10. Wesson, J.P.; Williams, T.C. J. Polym. Sci., Polym. Chem. Ed., 17, 2833 (1979).
11. Xing-Hua Zhang, R.; West, J. Polym. Sci., Polym. Chem. Ed., 22, 159 (1984).
12. Watanabe, H.; Muraoka, T.; Kageyama, M.; Yoskizumi, K.; Nagai, Y. Organometallics, 141, 3 (1984).
13. Weyenberg, D.R.; Atwell, W.H. Pure Appl. Chem., 19, 343 (1969).
14. Heitz, W. Makromol. Chem., Macromol. Symp., 4, 35 (1986).
15. Koch, W.; Risse, W.; Heitz, W. Macromol. Chem., Suppl., 12, 105 (1985).

RECEIVED September 23, 1987

Chapter 9

Characterization of Copolydiorganosilanes with Varying Compositions

Samuel P. Sawan, Yi-Guan Tsai, and Horng-Yih Huang

Polymer Science Program, Department of Chemistry, University of Lowell, Lowell, MA 01854

Soluble polydiorganosilane homo and copolymers have recently shown great potential in such areas as precursors for the preparation of silicon carbide fibers (1), as photoinitiators in alkene polymerization (2), as photoconductors (3), and as positive or negative self-developing photoresists for photolithographic applications (4). A number of copolydiorganosilane copolymers have been reported recently (5) in which the copolymer contained equal amounts of both monomers in the feed. In this work seven copolymers, including; dimethyl-co-cyclohexylmethyl, cyclohexylmethyl-co-n-propylmethyl, cyclohexylmethyl-co-n-propylmethyl, phenylmethyl-co-isopropylmethyl, phenylmethyl-co-n-propylmethyl, phenylmethyl-co-cyclohexylmethyl and n-propylmethyl-co-isopropylmethyl silane with varying percentages of the respective comonomers in the feed have been prepared and characterized.

The previous discovery of polydimethylsilane by Burkhard in 1949 (6) indicated that this material was insoluble in common organic solvents, highly crystalline and readily decomposed without melting when heated above 250°C. By introducing larger organo-substituents onto the silicon backbone, polysilane polymers were shown to readily dissolve in organic solvents and show melting behavior common to many amorphous carbon backboned polymers. With these properties, polysilanes can be molded, casted and potted into various shapes or even drawn into fibers.

Polysilane polymers possess several rather remarkable properties. Some of the unique properties of these inorganic polymers result from the delocalization of sigma electrons in their uninterrupted silicon backbone with a concurrent photoablative sensitivity to UV radiation. Several polysilanes have been characterized and all show a strong UV absorption. The maximum absorption wavelength in their UV spectra is strongly dependent on the organo-substituents (5), molecular weight and the composition of

0097-6156/88/0360-0112$06.00/0

copolymers. Also, the absorption maxima may be temperature dependent for some polysilanes such as poly(di-n-hexylsilane) (7).

The photochemical degradation of high molecular weight polysilanes are quite unusual and is the key to potential uses of these polymer. A photolytic cascade mechanism has been proposed in which both silylene and silyl radical generation occur in the photodegradation (8). Photoscission and photocrosslinking both may occur when the polymer is exposed to UV light and is dictated by the composition of the polysilane. The rate of photodegradation is also strongly dependent on the substituents and copolymer composition.

Although considerable interest exists in the polysilanes the number of systems that have been investigated is limited. Thus, a need to prepare and characterize additional polysilane copolymers is readily apparent. It is expected that systematic investigations, as presented in this report, will yield additional and valuable insight into the unique behavior of this class of polymers.

Experimental

Polydiorganosilane copolymers were synthesized by adding 2.2 moles of a sodium dispersion in a light oil (Aldrich) at a constant rate (320 meq/min) into a toluene solution which contained a total of 1 mole of diorganodichlorsilane monomers.

$$n\ \mathrm{Cl{-}Si(R_1)(R_2){-}Cl} + m\ \mathrm{Cl{-}Si(R_3)(R_4){-}Cl} \xrightarrow[\text{Toluene},110^{o}\mathrm{C}]{2.2(n+m)\ \mathrm{Na}} \mathrm{-(Si(R_1)(R_2))_n-(Si(R_3)(R_4))_m-}$$

All monomers (Petrarch) used in this study were carefully fractionally distilled prior to use. The condensation polymerization was carried out at approximately 110°C and allowed to reflux for 4 hours after the addition of the sodium metal was completed. The high molecular weight copolymers were obtained after two precipitations from toluene into ethyl acetate.

Molecular weights of the copolymers were determined by gel permeation chromatography (GPC) with four µ-styragel (Waters) columns calibrated using polystyrene standards. Chloroform was used as the eluate at a flow rate 1.5 ml/min. An LKB-2140 Ultraviolet Photodiode Array detector was used to detect the polymer with a scan range from 190 to 370 nm.

Proton and carbon-13 nuclear magnetic resonance (NMR) spectra were recorded on a IBM Instruments 270 MHz NMR Spectrometer on 6-8 weight percent solutions in deuterated chloroform. Ultraviolet spectra were recorded on an IBM Ultraviolet Spectrophotometer Model 9420 using chloroform solutions containing 2×10^{-5} g/ml of the copolymers.

Results and Discussion

The preparation of high molecular weight polymers requires careful attention to the purity of the reagents employed and the reaction conditions used. A typical yield for high molecular weight polydiorganosilane using the Wurtz reaction ranged from 10 to 35%. All polymers prepared in this study were found to be soluble in chloroform, toluene and tetrahydrofuran and could be drawn into fibers or cast into films.

Gel permeation chromatographic (GPC) analyses showed all samples to be bimodal in their molecular weight distribution except for c-hex/n-pro (50/50) and c-hex/i-pro (25/75), copolymers 5 and 9, respectively. Molecular weights were calculated from a calibration curve prepared using polystyrene standards without hydrodynamic correction for the silane polymers. The calculated molecular weights ranged from several tens of thousands up to several millions. Table I shows the molecular weights observed for each mode in the bimodal distributions. Although no attempt has been made to determine the hydrodynamic correction factor for these polymers, a previous report (9) indicates that the actual weight average molecular weights for polysilanes are higher than calculated by a factor of 2 to 3.

The lowest molecular weight observed for all polymers prepared in this study was for copolymer 16, a copolymer prepared using 75 mole % cyclohexyl methyl and 25 mole % phenyl methyl silane. The low molecular weight observed for this copolymer may be due to the two bulky side groups of the monomers (phenyl and cyclohexyl) sterically interfering with propagation during the polymerization.

All copolymers show strong UV absorption maxima ranging from 304 to 340 nm as shown in Table I and Figure 1. The absorbance maxima were found to be dependent on the comonomers employed and the composition of the copolymers. Copolymers which contain the phenyl group on the polymer chain show a greater red shift in their UV spectrum such that the λ_{max} appears near 340 nm. This red shifting due to aryl substitution has been reported by others (10) and is due to σ-π electron resonance of the phenyl group and silicon backbone. It is interesting to note that as the amount of phenyl monomer content is reduced in the copolymer the λ_{max} is observed to decrease. This is probably due to less conjugative interaction of the phenyl substituent with the silicon backbone.

As shown in Table I, cyclohexyl and isopropyl substituents also raised the absorption maxima to longer wavelengths as compared with linear alkyl substituted polysilanes. The red shifting behavior of cyclohexyl substituted polysilanes has been attributed to an extended trans conformation of the polymer chain due to the bulky cyclohexyl side groups (11). The isopropyl group also appears to sterically affect chain conformation such that polymers containing this residue show red shifting in their UV spectra. This effect is evidenced in an examination of the UV spectra of copolymers prepared from n-propyl methyl and i-propyl methyl silane monomers. As the content of the isopropyl residue is increased from 25 to 75 mole %, the λ_{max} is observed to increase from 307.4 to 312.4 nm, respectively (Figure 2).

The change in the maximum absorption wavelength of the

Table I. Properties of Copolydiorganosilanes

Sample Number	Monomer Feed Composition ---A---		---B---		Mole % A in A + B	$\overline{Mw} \times 10^{-3}$	λ_{max} (nm)
1	c-Hexyl	Me	Me	Me	75	1100 8	307.6
2	c-Hexyl	Me	Me	Me	50	1300 14	307.8
3	c-Hexyl	Me	Me	Me	25	1100 7	304.4
4	c-Hexyl	Me	n-Propyl	Me	75	840 7	317.2
5	c-Hexyl	Me	n-Propyl	Me	50	600	313.4
6	c-Hexyl	Me	n-Propyl	Me	25	1100 5	308.4
7	c-Hexyl	Me	iso-Propyl	Me	75	960 5	323.4
8	c-Hexyl	Me	iso-Propyl	Me	50	1000 5	323
9	c-Hexyl	Me	iso-Propyl	Me	25	1200	320.8
10	Phenyl	Me	n-Propyl	Me	75	1300 1	340
11	Phenyl	Me	n-Propyl	Me	50	910 3	336
12	Phenyl	Me	n-Propyl	Me	25	850 7	309.6
13	Phenyl	Me	iso-Propyl	Me	75	1300 8	340
14	Phenyl	Me	iso-Propyl	Me	50	980 4	339.8
15	Phenyl	Me	iso-Propyl	Me	25	740 4	333.6
16	Phenyl	Me	c-Hexyl	Me	75	54 8	340
17	Phenyl	Me	c-Hexyl	Me	50	150 8	340.6
18	Phenyl	Me	c-Hexyl	Me	25	180 4	326.2
19	n-propyl	Me	iso-Propyl	Me	75	1400 34	307.4
20	n-Propyl	Me	iso-Propyl	Me	50	1700 14	308.2
21	n-Propyl	Me	iso-Propyl	Me	25	1200 10	312.4
22	n-Propyl	Me			100	1300 23	307

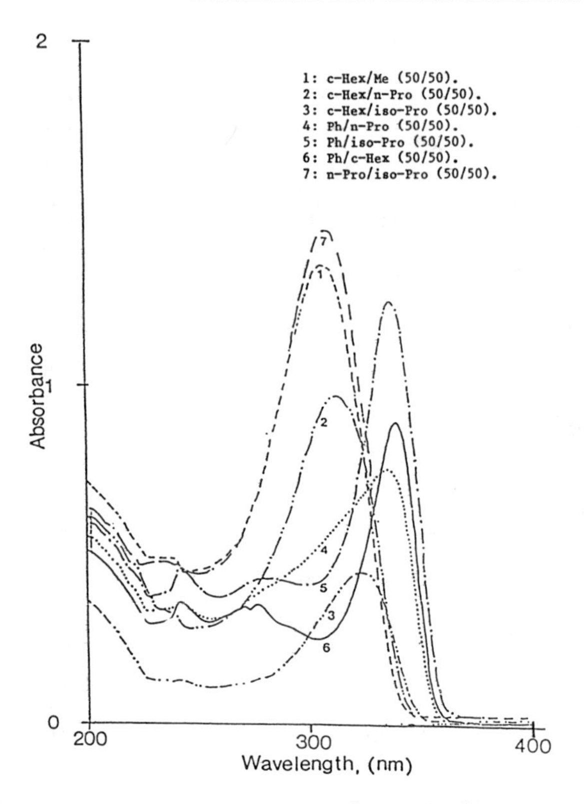

Figure 1. UV Spectra for Representative Polysilane Copolymers.

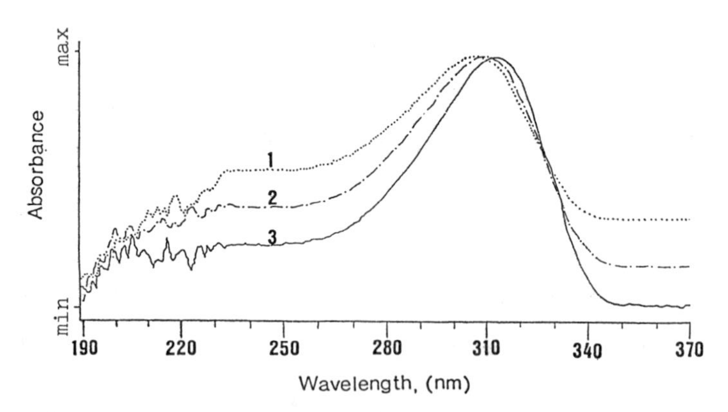

Figure 2. The UV Spectra of Poly(n-propyl-methyl-co-iso-propyl-methylsilane) Copolymers with Similar Molecular Weights Having the Following Compositions: (1) 75/25, (2) 50/50 and (3) 25/75.

copolymers also depends upon the molecular weight of the copolymer. The UV absorption spectra for copolymer 5, c-hex/n-pro (50/50), with different molecular weights are shown in Figure 3. These spectra result from the UV analysis of the various fractions that pass through the photodiode array detector during the chromatographic analysis. The λ_{max} for polymer fractions with molecular weights of 1040, 52, 11 and 5 x 10^3 were found at 314, 313.5, 308.2 and 304 nm, respectively. A similar observation for the wavelength dependent shift in λ_{max} as a function of molecular weight has been reported for linear silane oligomers (12,13) in which the silicon chain acts as a $\sigma-\sigma^*$ or $\sigma-\pi$ chromophore.

The proton decoupled carbon 13 NMR spectra for three poly(cyclohexylmethyl-co-isopropylmethyl) copolymers are shown in Figure 4. The backbone methyl group is observed as occurring between -4 and -1 ppm and consists of multiple resonances which are due to polymer microstructure. Multiple resonances are also observed for the methyl and tertiary carbon of the isopropyl group and for the methine carbon of the cyclohexyl group. Microstructural assignments for these resonances remain to be made. It has also been found that increasing the bulky character of the substituent yielded broader resonance peaks in the carbon-13 NMR spectra.

One additional peak occurring at 29.5 ppm was always apparent in the ^{13}C NMR spectra of all copolymers prepared in this study. Proton coupled ^{13}C NMR spectra reveal a triplet for this resonance with a coupling constant of 127 Hz (Figure 5). Such NMR spectral results are consistent with an additional methylene resonance in the copolymers. Evidence in the literature suggests that the condensation reaction may proceed via a radial pathway (14). Thus, a methylene group may be readily generated during the condensation polymerization due to silyl radical abstraction of a methyl group hydrogen atom.

Fourier transform infrared spectroscopic examination of these polymers also supports this conclusion. As can be observed in Figure 6, a peak occurring at 2086 cm^{-1} is seen in all the copolysilanes. This absorbance has been assigned to a Si-H stretching absorption (15). Such a moiety is expected if silyl radical abstraction of a hydrogen occurs. Examination of the infrared spectra for a $Si-CH_2-Si$ vibrational peak which should be located between 1000 and 1100 cm^{-1} is inconclusive due to the presence of a multitude of absorbances in this region.

Polydiorganosilanes undergo photodegradation readily when exposed to mid or deep UV radiation. In this study, the photodegradation was carried out by exposing copolysilane solutions (5 x 10^{-5} g/ml in chloroform) to a low pressure mercury UV light source and recording the UV spectra of such solutions as a function of the exposure time. As shown in Figure 7 for copolymer 20, i.e. n-pro/i-pro (50/50), the λ_{max} is observed to shift to shorter wavelengths with increasing exposure time. The shift in the maximum absorbance wavelength is accompanied by a change in the extinction coefficient for the photoproduct formed during irradiation. Such bleaching spectra are very similar to spectra reported for cyclosilanes with different ring sizes (16). The cyclosilanes show an increase in their extinction coefficients and a shift to longer

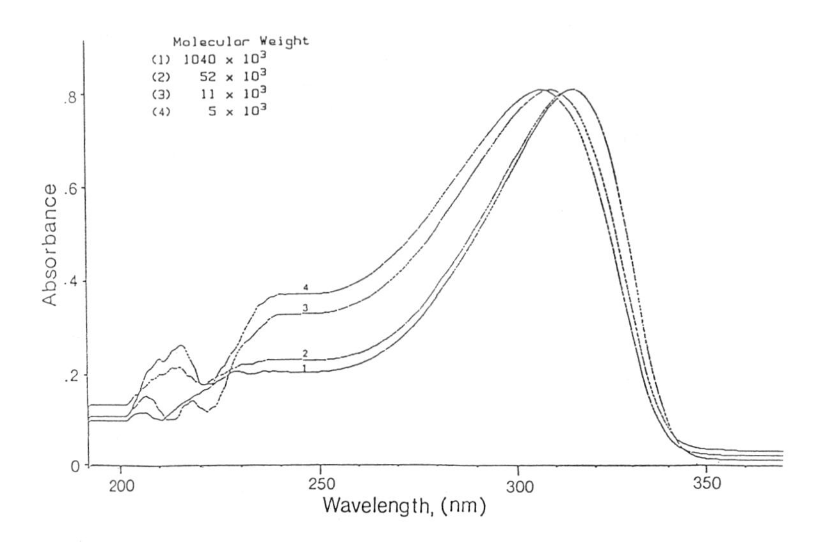

Figure 3. UV Spectra for Poly(cyclohexyl-methyl-co-iso-propyl-methyl silane) (50/50) with Different Molecular Weights.

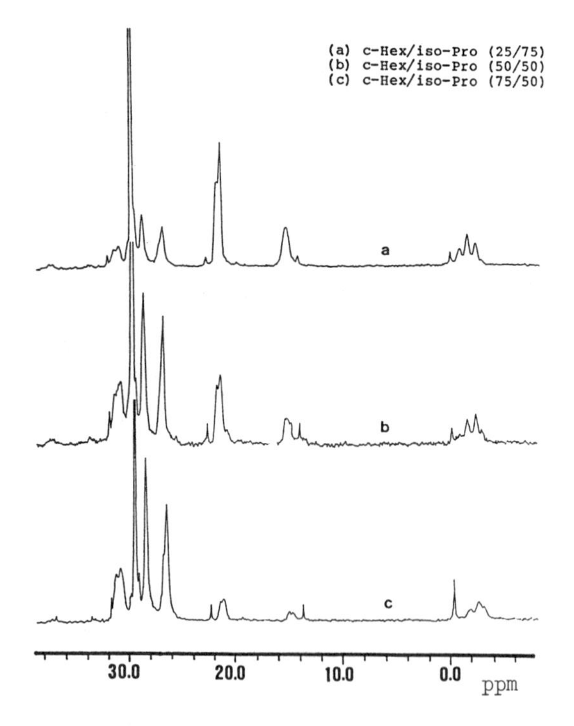

Figure 4. 67 MHz Carbon-13 NMR Spectra for Poly(cyclohexyl-methyl-co-n-propyl-methylsilane) Copolymers.

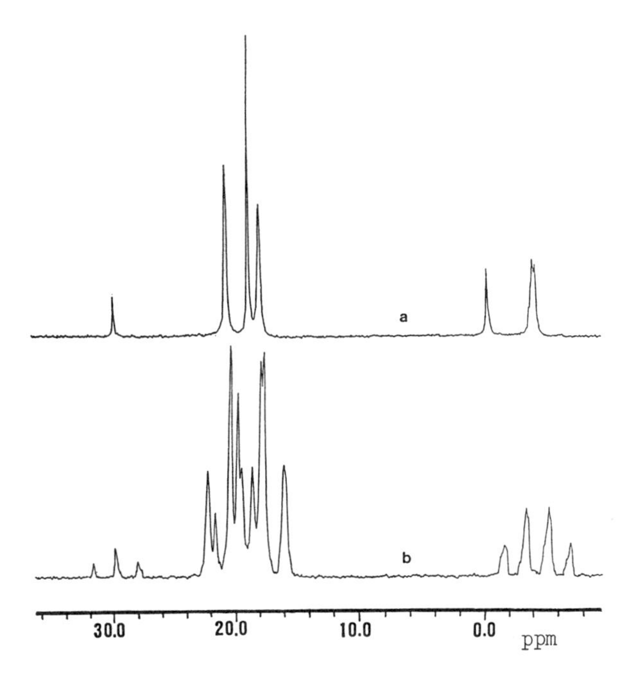

Figure 5. 67 MHz Carbon-13 NMR Spectra for Poly(n-propyl-methylsilane). (a) Proton Decoupled (b) Proton Coupled.

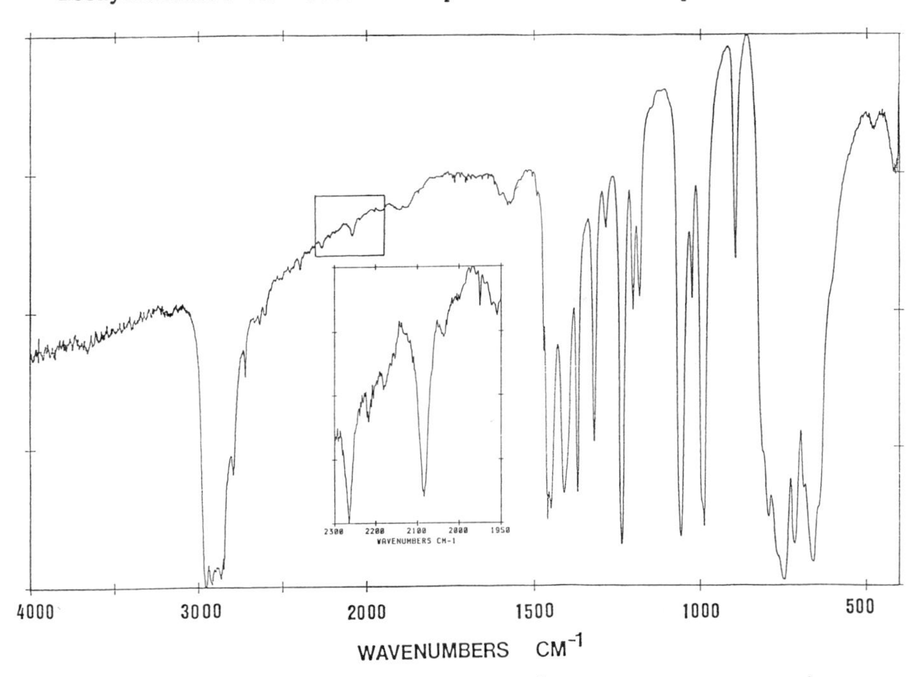

Figure 6. FT-IR Spectrum for Poly(n-propyl-methylsilane).

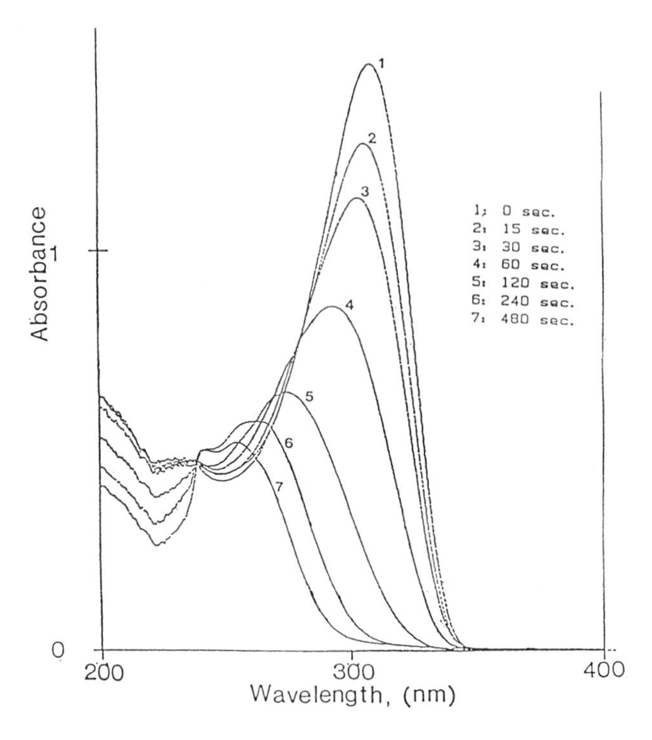

Figure 7. UV Spectra for Poly(n-propyl-methyl-co-iso-propyl-methylsilane) (50/50) as a Function of UV Exposure Time.

wavelengths in their λ_{max} as the ring size is increased. These UV spectral results indicate that the polydiorganosilanes are cleaved into small fragments, perhaps cyclic, during irradiation.

The rate of photodegradation of the high molecular weight species was examined by plotting the intensity at the original maximum absorption wavelength as a function of the exposure time. As shown in Figures 8 and 9, such plots yield a straight line with varying slopes depending upon the copolymer composition. Copolymers containing the phenyl group showed the highest solution degradation rates of all copolymers studied. The photodegradation rate was found to increase directly with the content of phenyl residues in the copolymers. Copolymers which contain a high ratio of bulky groups, such as cyclohexyl and isopropyl, show a higher degradation rate as compared to linear alkyl substituted polysilanes.

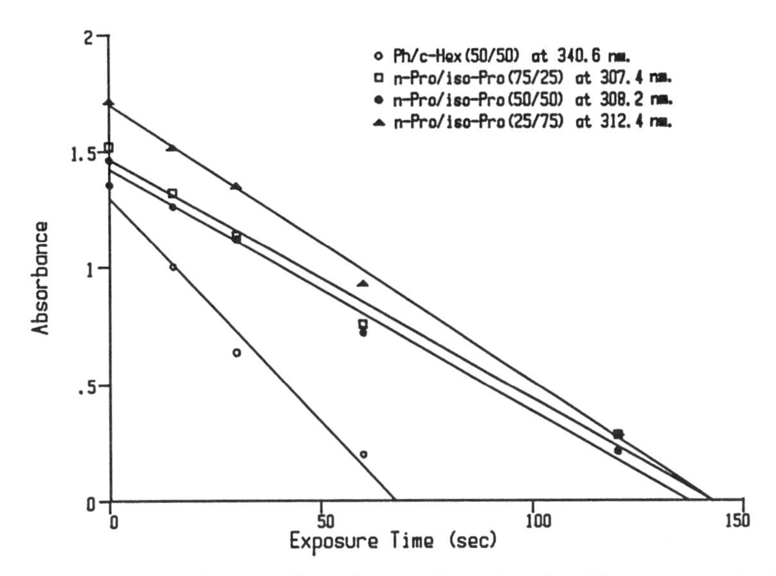

Figure 8. Change in UV Absorbance for Copolysilanes at Their λ_{max} as a Function of UV Exposure Time.

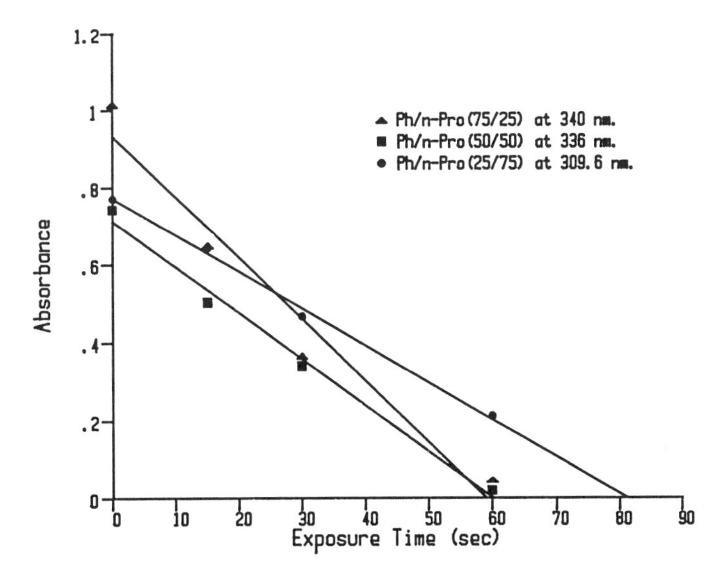

Figure 9. Change in UV Absorbance for Poly(phenyl-methyl-co-n-propyl-methylsilane) Copolymers at Their Respective λ_{max} as a Function of UV Exposure Time.

Acknowledgments

The authors wish to thank Digital Equipment Corporation for financial support for this work and Dr. Collins Conley for his technical assistance.

Literature Cited

1. Yajima, S. Am. Ceram. Soc. Bull. 1983, 62, 893.
2. West, R.; Maxka, J.; Sinclair, R.; Cotts, P. Presented at the 193rd ACS National Metg., Sym. on Organomet. Polym., Denver, 1987.
3. Kepler, R.G.; Zeigler, J.M.; Harrah, L.A.; Kurtz, S.R. Bull. Am. Phys. Soc. 1993, 28, 362.
4. Zeigler, J.M.; Harrah L.A.; Johnson, A.W. SPIE Adv. Resist. Tech. 1985, 539, 166.
5. Zhang, X.H.; West, R. J. Polym. Sci., Polym. Lett. Ed. 1985, 23, 479; J. Polym. Sci., Polym. Chem. Ed. 1984, 22, 159.
6. Burkhard, C.A. J. Am. Chem. Soc. 1949, 71, 963.
7. Miller, R.D.; Hofer, D.; Rabolt, J.; Fickes, G.N. J. Am. Chem. Soc. 1985, 107, 2172.
8. Trefonas III, P.; West, R.; Miller, R.D. J. Am. Chem. Soc. 1985, 107, 2737.
9. West, R. J. Organomet. Chem. 1986, 300, 327.
10. Pitt, C.G.; Carey, R.N.; Toren, E.C. Jr. J. Am. Chem. Soc. 1972, 94, 3806.
11. Trefonas III, P.; Djurovich, P.I.; Zhang, X.H.; West, R. J. Polym. Sci., Polym. Lett. Ed. 1983, 21, 819; J. Polym. Sci., Polym. Lett. Ed. 1983, 21, 823.
12. MacKay, K.M.; Watt, R. Organometal. Chem. Rev. 1969, 4, 137.
13. Gilman, H.; Atwell, W.H.; Schwebke, G.L. J. Organometal. Chem. 1964, 2, 369.
14. Zeigler, J.M. A.C.S. Polym. Prep. 1986, 27(1), 109.
15. Yajima, S.; Hasegawa, Y.; Hayashi, J.; Iimura, M. J. Mate. Sci. 1978, 13, 2569.
16. West, R. Pure Applied Chem. 1982, 54, 5, 1041.

RECEIVED September 1, 1987

POLYSILAZANES AND POLYSILAZOXANES

Chapter 10

Synthetic Routes to Oligosilazanes and Polysilazanes

Polysilazane Precursors to Silicon Nitride

Richard M. Laine [1], Yigal D. Blum, Doris Tse, and Robert Glaser [2]

Inorganic and Organometallic Chemistry Programs, SRI International, Menlo Park, CA 94025

Organometallic polymer precursors offer the potential to manufacture shaped forms of advanced ceramic materials using low temperature processing. Polysilazanes, compounds containing Si-N bonds in the polymer backbone, can be used as precursors to silicon nitride containing ceramic materials. This chapter provides an overview of the general synthetic approaches to polysilazanes with particular emphasis on the synthesis of preceramic polysilazanes. The latter part of the chapter focusses on catalyzed dehydrocoupling reactions as a route to polysilazanes.

Non-oxide ceramics such as silicon carbide (SiC), silicon nitride (Si_3N_4), and boron nitride (BN) offer a wide variety of unique physical properties such as high hardness and high structural stability under environmental extremes, as well as varied electronic and optical properties. These advantageous properties provide the driving force for intense research efforts directed toward developing new practical applications for these materials. These efforts occur despite the considerable expense often associated with their initial preparation and subsequent transformation into finished products.

The expense in preparation derives from the need for high purity products. The cost in fabricating three-dimensional shapes of SiC, Si_3N_4 or BN arises from the need to sinter powders of these materials at high temperatures (>1500°C) and often under high pressure. Coatings of these materials are at present produced by chemical or

[1]Current address: Department of Materials Science and Engineering, University of Washington, Seattle, WA 98195
[2]On leave from the Department of Chemistry, Ben Gurion University, Beer Sheva, Israel

0097-6156/88/0360-0124$06.00/0

physical vapor deposition (CVD or PVD). In each instance, the process is equipment and energy intensive and, has numerous limitations. If inexpensive methods for the synthesis and fabrication of non-oxide ceramics can be developed, then the number of useful applications could be enormous.

One potential solution to these problems, suggested some 20 years ago by Chantrell and Popper (1), involves the use of inorganic or organometallic polymers as precursors to the desired ceramic material. The concept (2) centers on the use of a tractable (soluble, meltable or malleable) inorganic precursor polymer that can be shaped at low temperature (as one shapes organic polymers) into a coating, a fiber or as a matrix (binder) for a ceramic powder. Once the final shape is obtained, the precursor polymer can be pyrolytically transformed into the desired ceramic material. With careful control of the pyrolysis conditions, the final piece will have the appropriate physical and/or electronic properties.

The commercialization of Nicalon (Nippon Carbon Co.), a SiC based ceramic fiber fabricated by pyrolysis of fibers of polycarbosilane (3), has sparked intensive efforts to synthesize polymer precursors of use in the fabrication of Si_3N_4 and BN fibers, coatings and binders. In this regard, oligo- and polysilazanes have recently been shown to offer considerable potential as precursors to Si_3N_4. The objective of this chapter is to provide an overview of the general methods for synthesizing oligo- and polysilazanes and to show how some of these routes have been used to synthesize oligo- and polysilazane preceramic polymers. In the latter part of the chapter, we will emphasize our own work on transition metal catalyzed dehydrocoupling reactions.

General Synthetic Approaches

Historically, oligo- and polysilazane synthetic chemistry has passed through three stages of development with quite differing goals. Initial efforts, beginning with the work of Stock and Somieski in 1921, were directed simply towards the preparation and classification of the general properties of polysilazanes (4,5). The commercial success of polysiloxanes or silicones in the fifties and sixties prompted studies on the synthesis of polysilazanes as potential analogs (without much success). As mentioned above, current interest derives from their use as silicon nitride preceramic polymers.

Reactions (1)-(5) illustrate known methods for forming silicon-nitrogen bonds of potential use in the formation of oligo- and polysilazanes. The most common method of forming silazanes is via ammonolysis or aminolysis as shown in reaction (1) (4,5):

$$R_3SiCl + 2R'_2NH \longrightarrow R_3SiNR'_2 + R'_2NH_2^+Cl^- \qquad (1)$$

Si-N bonds can also be formed by a dehydrocoupling reaction catalyzed by alkali (6-11) or transition metals (12):

$$R_3SiH + R'_2NH \xrightarrow{catalyst} R_3SiNR'_2 + H_2 \quad (2)$$

In rare instances, one can also obtain displacement of oxygen by nitrogen as illustrated by reaction (3) (13). Reaction (3) proceeds despite the fact that the Si-O bond is 25-30 kcal/mole stronger than the Si-N bond (14).

$$R_3Si\text{-}OH + R'_2NH \longrightarrow R_3SiNR'_2 + H_2O \quad (3)$$

Verbeek and Winter have used a deamination/condensation reaction to form new Si-N bonds as exemplified by reaction (4) (15,16).

$$2R_2Si(NHR')_2 \xrightarrow{\Delta} (R'NH)R_2SiN(R')SiR_2(NHR') + R'NH_2 \quad (4)$$

It has been suggested that reaction (4) passes through an $R_2Si{=}NR'$ intermediate, although no substantiation exists (17,18).

Van Wazer (19) has described the use of a redistribution reaction, illustrated in (5), to form linear oligomers from cyclic species:

$$2R_2SiCl_2 + \text{cyclo-}[R_2SiNH]_3 \longrightarrow ClR_2SiNHSiR_2Cl + Cl\text{-}[R_2SiNH]_2\text{-}SiR_2Cl \quad (5)$$

With the exception of reaction (3), the utility of all of the above reactions for the synthesis of oligo- and polysilazanes has been examined in varying detail. In the following paragraphs, we attempt to examine the pertinent studies in each area especially as it relates to the synthesis of preceramic polymers.

Polymerization by Ammonolysis and Aminolysis

The simplicity of the ammonolysis/aminolysis of dihalohalosilanes, reactions (6) and (7), made them the original method of choice for the synthesis of oligo- and polysilazanes, especially because of the analogy to the hydrolytic synthesis of polysiloxanes. The ammonolysis of dihalosilanes, reaction (6), has been found to be extremely sensitive to steric factors (6).

$$R^1RSiCl_2 + 3xNH_3 \longrightarrow -[R^1RSiNH]_x- + 2xNH_4{}^+Cl^- \quad (6)$$

For example, with $R^1 = R = H$, reaction (6) gives only oligosilazanes; with $R^1 = CH_3$ and $R = H$, both cyclic and oligosilazanes are formed; and with $R^1 > CH_3$ and $R = H$ then one obtains mostly cyclotri- and cyclotetrasilazanes as the products.

Aminolysis of dihalosilanes, reaction (7), is also extremely sensitive to the steric bulk of the amine (20). For $R^1 = CH_3$,

$$xR^1RSiCl_2 + 3xR'NH_2 \longrightarrow -[R^1RSiNR']_x- + 2xR'NH_3^+Cl^- \quad (7)$$

$R = CH_3$ and $R' \geqslant Et$, the products are predominantly the cyclotri- and cyclotetrasilazanes and/or simple disilazanes, e.g., $R^1RSi(NHR')_2$. For reactions where $R^1 = H$ or CH_3, $R = H$ and $R' = CH_3$, the products can be mostly linear oligosilazanes depending on conditions (see below) (21-23). The preference for cyclotri- and cyclotetrasilazanes, in these reactions, is not surprising given that the hydrolysis of dihalosilanes also leads to cyclotri- and cyclotetrasiloxanes.

The search for silicon nitride precursors has renewed interest in ammonolysis as a route to polysilazanes. In our own work, we have followed up on early studies by Seyferth and Wiseman (20,21) on the preparation of $-[H_2SiNMe]_x-$ from $MeNH_2$ and H_2SiCl_2. We find that the molecular weight of this oligomer is greatly affected by the reaction conditions. Originally, Seyferth and Wiseman were able to produce mixtures of oligomers of $-[H_2SiNMe]_x-$ and the cyclotetramer (22). Removal of the cyclotetramer by distillation gave a polymer with Mn ≃ 600 Daltons (x = 10). We have now succeeded in preparing quantities of $-[H_2SiNMe]_x-$ with Mn ≈ 1000-1200 (x = 18-20) with less than 10% cyclomers (23). The resulting oligosilazane can be used directly as a precursor for transition metal catalyzed dehydrocoupling polymerization (see below).

Arai et al. (24) have devised a novel approach to ammonolysis wherein the silyl halide is modified by complexation to pyridine (py) prior to ammonolysis:

$$RHSiCl_2 + 2py \longrightarrow RHSiCl_2 \cdot 2py \quad (8)$$

$$RHSiCl_2 \cdot 2py + NH_3 \longrightarrow NH_4Cl \text{ (or py.HCl)} + -[RHSiNH]_x- \quad (9)$$

$$R = Me \text{ or } H$$

The oligosilazanes formed via this approach give higher molecular weights than obtainable by direct ammonolysis. When R = H, the product is a polydispersed polymer with a weight average molecular weight (M_w) of ≈ 1300 D. The polymer is not stable and crosslinks to form an intractable product in a short time in the absence of solvent. Pyrolysis of this material gives an 80% ceramic yield of silicon nitride mixed with silicon. Silicon is found even when pyrolysis is conducted under an ammonia atmosphere. The tentatively proposed oligomer structure is (25):

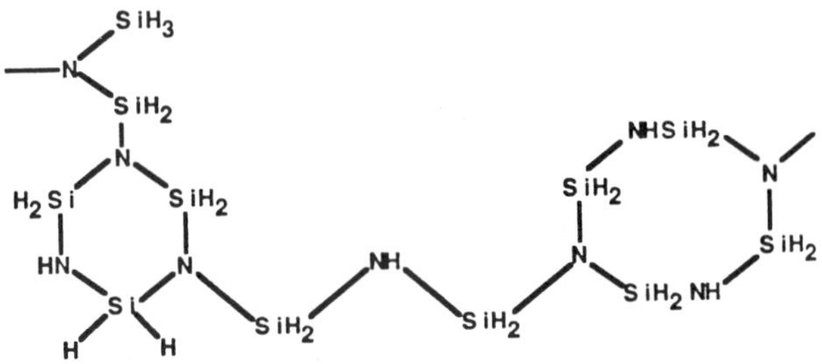

Of interest is the fact that IR and NMR evidence point to the presence of SiH_3 groups in the oligosilazane. The SiH_3 groups must arise as a consequence of a redistribution process, although further study remains to verify these results. When R = Me, the resulting polymer, $-[MeHSiNH]_x-$, can have molecular weights of 1700 D. Pyrolysis of this material gives high ceramic yields. The composition of the ceramic product was not reported.

Ring Opening Polymerization

Base and acid catalyzed ring opening polymerization of cyclotrisiloxane, $-[Me_2SiO]_3-$ [e.g. reaction (10)], is a well-known method of generating high molecular weight polysiloxanes.

$$RO- + x \ \ulcorner[Me_2SiO]_3\urcorner \longrightarrow RO-[Me_2SiO]_{3x}- \qquad (10)$$

RO- = alkoxide or hydroxide

A number of groups have attempted to develop the analogous reaction for cyclotri- and cyclotetrasilazanes. Ring opening polymerization represents an alternative to direct ammonolysis even though the cyclomer precursors are normally made by ammonolysis or aminolysis.

Thus, Andrianov et al. (26) attempted to catalyze polymerization of a number of alkyl and alkyl/aryl cyclosilazanes using catalytic amounts of KOH or other strong bases at temperatures of up to 300°C. In general, the reactions proceed with evolution of NH_3, hydrocarbons and the formation of intractable, crosslinked, brittle products even at low temperatures. Contrary to what is observed with cyclotrisiloxanes, no evidence was found for the formation of linear polysilazanes. Copolymerization of mixtures of cyclosilazanes and cyclosiloxanes gave somewhat more tractable polymers with less evolution of hydrocarbons or ammonia, however very little was done to characterize the resulting materials.

An explanation for the lack of formation of linear polysilazanes in base catalyzed ring opening is suggested by one of Fink's publications (7). Fink observes that oligosilazane amides will undergo fragmentation to simple amides under the conditions employed by Andrianov:

$$R_3SiN(Li)SiR_2N(SiR_3)_2 \longrightarrow LiN(SiR_3)_2 + 0.5(R_3SiNSiR_3)_2 \qquad (11)$$

Presumably, any type of strong base will react with cyclosilazanes to form amido species. If amido species are extremely susceptible to fragmentation then reactions such as reaction (11) would promote depolymerization rather than polymerization. Consequently, the only way polymeric species can form would be through a condensation process that leads to trisubstituted nitrogens and fused rings.

Rochow et al. (27-28) have examined acid catalyzed ring opening polymerization as an approach to the formation of polysilazanes. They find that heating cyclotri- and cyclotetrasilazanes with ammonium halide catalysts at temperatures of 160°C for 6-8 h results in a condensation polymerization process that evolves ammonia and leads to the formation of waxy polysilazanes. Analytical results suggest that these polysilazanes consist of rings linked by silyl bridges as illustrated by the following structure:

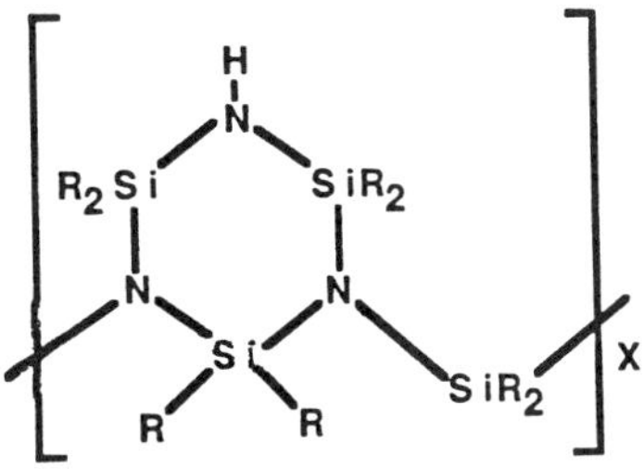

Polymers of this type were found to have the highest molecular weight, $M_n \approx$ 10K D, of any polysilazanes produced to date. Unfortunately, no one has examined the utility of these compounds as preceramic polymers.

Rochow et al. (29) also report that heating tri- and cyclotetrasilazanes under ammonia leads to the formation of linear oligodimethylsilazanes, reaction (12); however, they were never able

$$NH_3 + \ulcorner[Me_2SiNH]_3\urcorner \longrightarrow NH_2\text{-}[Me_2SiNH]_x\text{-}H \qquad (12)$$

to obtain conversions much greater than 10% and molecular weights were always on the order of $M_n \approx$ 1200 D.

Ring opening polymerization can also be catalyzed by transition metals as shown in reaction (13) (12,30). The process appears to involve hydrogenolysis of the Si-N bond, given that reaction (13)

$$(Me_3Si)_2NH + \ulcorner[Me_2SiNH]_4\urcorner \xrightarrow{Ru_3(CO)_{12}/135^{\circ}C/1h/H_2} Me_3Si\text{-}[Me_2SiNH]_x\text{-}SiMe_3 \qquad (13)$$

requires only 1h to equilibrate in the presence of as little as 1 atm of H_2, and more than 30h without a source of hydrogen. The value of x in $Me_3Si-[Me_2SiNH]_x-SiMe_3$ varies depending on the relative proportions of the capping agent, $(Me_3Si)_2NH$, to the tetracyclomer. Because the catalyst indiscriminantly cleaves Si-N bonds in both the tetracyclomer and the product oligomers in reaction (13), equilibration prevents the growth of high molecular weight species.

If reaction (13) is carried out under H_2 in the absence of capping agent then oligomers with $M_n \approx 2000$ D can be isolated following distillation of the volatile cyclomers. We believe that in this case the product polymers are both condensed ring systems and hydrogen capped linear oligomers.

None of the above described ring opening polymerization methods has, as yet, proved useful for the formation of polysilazane preceramic polymers. However, Si-N bond cleavage and reformation, as it occurs in reaction (13), is probably responsible in part for the curing or thermoset step in transition metal catalyzed dehydrocoupling polymerization of hydridosilazanes (31), as described below.

Deamination/Condensation Polymerization Reactions

Reaction (4) was the first reaction successfully used to synthesize preceramic polysilazanes. Verbeek et al. found that fusible polysilazane resins could be produced by pyrolysis of bis- or trisalkylaminosilazanes (or mixtures) (15,16):

$$R_2Si(NHMe)_2 \xrightarrow{\Delta\ 200\text{-}800^{\circ}C} MeNH_2 + \overline{(R_2SiNMe)_3} + \text{polymer} \qquad (14)$$

$$RSi(NHMe)_3 \xrightarrow{\Delta\ 520^{\circ}C/3h} MeNH_2 + \text{polymer} \qquad (15)$$

$$R = Me \text{ or Phenyl}$$

The cyclotrisilazane (R = Me) produced in reaction (14) is recycled at 650°C [by reaction with $MeNH_2$, the reverse of reaction (14)] to increase the yield of processible polymer. Physicochemical characterization of this material shows it to have a softening point at 190°C and a C:Si ratio of 1:1.18. Filaments 5-18 μm in diameter can be spun at 315°C. The precursor fiber is then rendered infusible by exposure to air and transformed into a ceramic fiber by heating to 1200°C under N_2. The ceramic yield is on the order of 54%; although, the composition of the resulting amorphous product is not reported. The approach used by Verbeek is quite similar to that employed by Yajima et al. (13) in the pyrolytic preparation of polycarbosilane and its transformation into SiC fibers.

Reaction (15) has recently been studied in greater detail by a group at Marshal Space Flight Center (31-33). Wynne and Rice propose (2) the following structure for the polysilazane produced when R = Me:

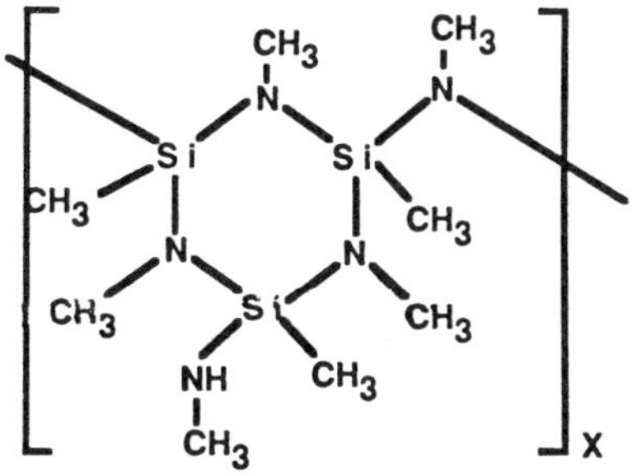

The Marshall group has optimized reaction (14) to obtain a polysilazane with $M_w \approx 4000$ Daltons which can be hand drawn to give 10-20 μm preceramic fibers. These fibers are then rendered infusible by exposure to humid air and pyrolyzed to give fibers with the same ceramic yields, 55^+%, as found by Verbeek et al. The ceramic products are mainly amorphous SiC and Si_3N_4 with some SiO_2 (a consequence of the humidity treatments).

Unfortunately, the chemistry of the deamination/condensation process is poorly understood; thus, little can be said about the mechanisms and kinetics of the amine elimination step, the nature of the "silaimine" (7,8) intermediate or the condensation step. It seems reasonable to predict that if one could learn to control the relative rates for these two steps, more control could be exerted over the condensation process and the properties of the precursor polymer.

Si-Cl/Si-N Redistribution Polymerization Reactions

In the past five years, researchers at Dow Corning have made a concentrated effort to explore and develop the use of Si-Cl/Si-N redistribution reactions as a means of preparing tractable, polysilazane precursors to Si_3N_4 (34-42). Initial work focussed on the reaction of chlorosilanes and chlorodisilanes with hexamethyldisilazane:

$$MeSiCl_3 + (Me_3Si)_2NH \longrightarrow Me_3SiCl + Polymer \quad (16)$$

$$MeCl_2SiSiMeCl_2 + (Me_3Si)_2NH \longrightarrow Me_3SiCl + Polymer \quad (17)$$

Surprisingly, pyrolysis of these polysilazane polymers gave mostly SiC rather than the expected Si_3N_4. It was later found that the use of $HSiCl_3$ in place of $MeSiCl_3$ provides useful precursors to Si_3N_4 (42):

$$HN(SiMe_3)_2 + HSiCl_3 \longrightarrow Me_3SiCl + -[HSi(NH)_{1.5}]_x[HSiNH(NHSiMe_3)]_y- \quad (18)$$

$M_n \approx$ 3,500 Daltons

$M_w \approx$ 15,000 Daltons

The sequence of redistribution reactions leading to the polymer shown in reaction (18) is illustrated by the following:

$$HSiCl_3 + (Me_3Si)_2NH \longrightarrow Me_3SiCl + Me_3SiNHSiHCl_2 \quad (19)$$

$$Me_3SiNHSiHCl_2 + (Me_3Si)_2NH \longrightarrow Me_3SiCl + (Me_3SiNH)_2SiHCl \quad (20)$$

$$(Me_3SiNH)_2SiHCl + (Me_3Si)_2NH \rightleftharpoons Me_3SiCl + (Me_3SiNH)_3SiH \quad (21)$$

$$2(Me_3SiNH)SiHCl \longrightarrow (Me_3Si)_2NH + Me_3SiNHSiHClNHClHSiNHSiMe \quad (22)$$

The polymer produced in reaction (18) can be spun to give 15-20 μm fibers. These fibers must then be made infusible prior to being pyrolyzed in order to obtain high quality ceramic fibers. The curing process involves vapor phase reaction of the spun precursor with $HSiCl_3$. This final redistribution reaction removes residual Me_3Si-groups, reduces carbon content and crosslinks the precursor polysilazane. After appropriate processing the resulting amorphous ceramic fiber consists of 96% Si_3N_4, 2% C and 2% O and the overall ceramic yield is approximately 45-55%.

Catalytic Dehydrocoupling--Dehydrocyclization Reactions

Catalytic dehydrocoupling, as shown in reaction (3), was pioneered by Fink (6-10) and later studied by Andrianov et al (43). Although Fink described the synthesis of the first polysilazanes, (9) reaction (23), this route to oligo- and polysilazanes remained unexplored

$$H_2NRNH_2 + 2R_2SiH_2 \xrightarrow{\text{strong base}} H_2 + -[RN(\mu\text{-}R_2Si)_2N]_x- \quad (23)$$

until the work of Seyferth and Wiseman, (11,20-22,44,45) twenty years later.

Seyferth and Wiseman find that cyclomers and oligomers of the type $-[MeSiHNH]_x-$, produced by ammonolysis of $MeSiHCl_2$, will react with a strong base, eg KH, to undergo "dehydrocyclization", reaction (24), akin to reaction (23). The resulting products are soluble, tractable, sheetlike polymers that can be spun into fibers and give extremely high ceramic yields upon pyrolysis.

$$-[MeSiHNH]_x- \xrightarrow{1\% KH}$$
$x \approx 4.9$

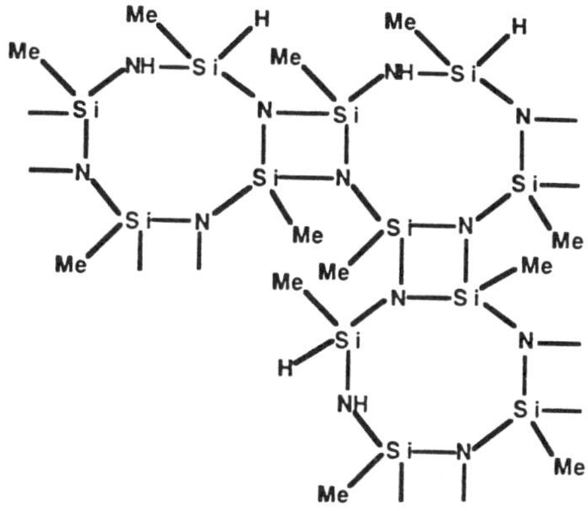

(24)

This structure derives from the formula, $(MeSiHNH)_{0.39}(MeSiHNMe)_{0.04}(MeSiN)_{0.57}$, established by both NMR and combustion analysis.

Pyrolysis of a typical polysilazane produced according to reaction (24), under N_2, gives ceramic yields of 85%. Based on chemical analysis the ceramic composition appears to be a mixture of Si_3N_4, SiC, carbon and possibly some SiO_2 in a 0.88:1.27:0.75:0.09 ratio. If the pyrolysis is carried out under NH_3 then carbon can be eliminated from the ceramic product giving essentially pure Si_3N_4. Seyferth suggests that NH_3 displaces the Me-Si groups via nucleophilic attack.

The mechanism of the dehydrocyclization reaction is not completely understood. Two alternatives were proposed (11), one which proceeds via a silaimine mechanism, exemplified by reaction (25), and one wherein ring closure occurs by nucleophilic displacement of hydrogen, reaction (26).

$$2R_2Si(H)\text{-}NR'(Li) \xrightarrow{-LiH} 2[R_2Si{=}NR'] \rightarrow R_2Si(\mu\text{-}NR')_2SiR_2 \quad (25)$$

$$2R_2Si(H)\text{-}NR'(Li) \xrightarrow{-LiH} [R_2Si\text{-}NR'\text{-}SiR_2\text{-}NR'(Li)]$$

$$\xrightarrow{-LiH} R_2Si(\mu\text{-}NR')_2SiR_2 \quad (26)$$

Modeling studies where R = R' = Me provide evidence in favor of a silaimine intermediate; however, efforts to trap the silaimine species were unsuccessful. Thus, the support for a specific ring closure mechanism remains inconclusive.

Transition Metal Catalyzed Dehydrocoupling Polymerization Reactions

Transition metal catalyzed dehydrocoupling, reaction (27), offers an alternative to the alkali metal dehydrocyclization process

$$R_2SiH_2 + R'NH_2 \xrightarrow{\text{catalyst}} H_2 + -[R_2SiNR']_x- \quad (27)$$

and provides routes to different types of oligo- and polysilazanes (12,46-50). We have examined, in a preliminary fashion, the kinetics and mechanisms involved in reaction (27) using reaction (28) (50):

$$Et_3SiH + R'NH_2 \xrightarrow{Ru_3(CO)_{12}/70^oC/THF} H_2 + Et_3SiHNR' \quad (28)$$

We find that the rates of reaction for the various amines examined in (28) are governed by an extremely complex set of equilibria. For example, when R' = n-Pr, n-Bu or s-Bu, the rate of reaction exhibit first order dependence on $[Et_3SiH]$ at constant amine concentration. However, the rate of reaction exhibits inverse non-linear dependence on $[n\text{-}PrNH_2]$ and $[n\text{-}BuNH_2]$, but positive non-linear dependence on $[s\text{-}BuNH_2]$ at constant $[Et_3SiH]$. Furthermore, if R' = t-Bu, then the rate of reaction is almost independent of both $[t\text{-}BuNH_2]$ and $[Et_3SiH]$. Studies of the rate dependence on catalyst concentration for reaction (28) where $R'NH_2$ is $n\text{-}BuNH_2$ reveal relative catalyst activities that are inversely dependent on $[Ru_3(CO)_{12}]$. Similar studies with $R'NH_2$ = $t\text{-}BuNH_2$ reveal that the rate of reaction is linearly dependent on $[Ru_3(CO)_{12}]$. Piperidine is unreactive under the reaction conditions studied.

Based on our previous work with $Ru_3(CO)_{12}$/amine catalytic systems, we can propose mechanisms that account for these observations; although, additional work will have to performed to validate some of our assumptions. The rate vs reactant concentration studies suggest that there are three different rate determining steps in the catalytic cycle for reaction (28), depending on the steric demands of the amine. In the cases where the rate of reaction is inversely dependent on $[RNH_2]$, one can assume that the amine is not participating in the rate determining step or there are competing reactions where the dominant reaction is inhibition. The conclusion then is that the rate determining step is probably oxidative addition of Et_3SiH to the active catalyst. The inverse dependence can be interpreted in light of Kasez et al's work (51) on the low temperature reactions of primary amines with $Ru_3(CO)_{12}$, and our work on amine/$Ru_3(CO)_{12}$ interactions (52,53).

Kaesz et al. have shown that simple, primary amines (e.g. $MeNH_2$) will react with $Ru_3(CO)_{12}$ to form μ-acetamido ligands at temperatures as low as -15°C. We find that simple primary, secondary and tertiary alkyl amines will react with $Ru_3(CO)_{12}$ at temperatures of 70-150°C to undergo catalytic deuterium for hydrogen exchange reactions on the hydrocarbon groups and transalkylation (52). We have found that a variety of aliphatic and aromatic amines will react with $M_3(CO)_{12}$ (M = Ru or Os) by binding through the nitrogen coincident with oxidative addition of an alpha C-H bond to form (μ^2-iminium)$HM_3(CO)_{10}$ and (μ^2-iminium)$_2H_2M_3(CO)_9$ complexes (53). We suggest that with n-$PrNH_2$ and n-$BuNH_2$ there is successful competition between amine and silane for sites of coordinative unsaturation on the active catalyst species.

With regard to the s-$BuNH_2$ results, it is likely that steric bulk, especially at the tertiary C-H alpha to the NH_2- group, limits its ability to bind to the active catalyst site and therefore it cannot compete with Et_3SiH, although it can still function as a reactant. If this logic is correct, then we can also suggest that Si-N bond formation probably occurs by nucleophilic attack directly on the silicon moiety bound to metal rather than through initial ligation at the metal followed by reaction. Because the rate of silazane formation is dependent on both [Et_3SiH] and [s-$BuNH_2$] we suggest that the rate limiting step in this instance is the formation of the Si-N bond. On changing amine from s-$BuNH_2$ to t-$BuNH_2$, there is a significant reduction in the reaction rate. As discussed above, the fact that this rate is almost completely independent of changes in both [Et_3SiH] and [t-$BuNH_2$] suggests that catalyst activation must become the rate determining step.

An alternate rationale to the above cycle can be proposed if the amine serves as both a spectator ligand and reactant in the catalytic cycle. In the case of n-$PrNH_2$ and n-$BuNH_2$, the complex containing the spectator ligand can add a second amine (causing inhibition) or it can react with silane. Inhibition may include stabilization of the cluster towards fragmentation given the stability of the (μ^2-iminium)$HM_3(CO)_{10}$ and (μ^2-iminium)$_2H_2M_3(CO)_9$ complexes (53). In this regard, we find that piperidine will react with $Ru_3(CO)_{12}$ to form very stable bis(piperidino)cluster complexes (53). If we attempt to carry out reaction (28) using piperidine, a sterically undemanding secondary amine, we observe no reaction. If piperidine is acting as a ligand to totally deactivate by inhibition, then addition of piperidine to reaction (28) run with n-$BuNH_2$ should inhibit or totally poison the reaction. In fact, the addition of more than one equivalent of piperidine per equivalent of catalyst served only to slightly accelerate the reaction rather than inhibit it. Thus, we must conclude that the spectator ligand concept does not appear to be valid. These results support the following very general mechanism for dehyrocoupling as catalyzed by M = $Ru_3(CO)_{12}$:

$$M + RNH_2 \rightleftharpoons M(RNH_2)$$

$$M + Et_3SiH \longrightarrow Et_3SiMH$$

$$Et_3SiMH + RNH_2 \longrightarrow Et_3SiNHR + MH_2$$

$$MH_2 \rightleftharpoons M + H_2$$

There appear to be considerable steric demands involved in formation of the Si-N bond in reaction (28). Support for the importance of steric constraints in dehydrocoupling comes from studies on the synthesis of oligosilazanes from $PhSiH_3$ and NH_3. When reaction (29) is run at 60°C, NMR and elemental analysis

$$PhSiH_3 + NH_3 \xrightarrow{Ru_3(CO)_{12}/60^{o}C/THF} H_2 + H\text{-}[PhSiHNH]_x\text{-}H \quad \text{viscous oil, } M_n \approx 1000 \tag{29}$$

confirm that the resulting oligomer is essentially linear (49). In order for this to occur, chain growth must occur by stepwise addition at the $PhSiH_2$ end caps without competition from reaction at the interior Si-H bonds. This implies that steric selectivity affects the facility of the oxidative addition step. If the oligomer produced in reaction (29) is treated with additional NH_3, but at 90°C, higher molecular weight oligomers can be formed but, NMR and elemental analysis now indicate the formation of pendant NH_2 groups and imino crosslinks as shown in reaction (30).

$$H\text{-}[PhSiHNH]_x\text{-}H + NH_3 \xrightarrow{Ru_3(CO)_{12}/90^{o}C/16h} H_2 + \text{-}[PhSiHNH]_{x-y}[Ph\overset{\overset{NH_{0.5}}{|}}{S}iNH]_y\text{-} \quad \text{solid, } M_n \approx 1400 \tag{30}$$

The structural changes that occur at 90°C are again indicative of steric constraints. We find no evidence that the N-H bonds in the $-[PhSiHNH]_x-$ backbone participate in the crosslinking process observed in (30) which is in keeping with our observation concerning piperidine's lack of reactivity. We also observe similar behavior for reactions of n-hexylsilane with NH_3 (23).

The following reactions were performed to demonstrate the utility of transition metal dehydrocoupling for the synthesis of different types of polysilazanes and a novel polysiloxazane:

$$HMe_2SiNHSiMe_2H + NH_3 \xrightarrow{Ru_3(C)_{12}/60^{o}C/THF} H_2 + \text{-}[Me_2SiHN]_x\text{-} \quad Mn \approx 2000 \text{ (volatiles distilled off)} \tag{31}$$

$$xHMe_2SiNHSiMe_2H + NH_3 + yPhSiH_3 \xrightarrow{Ru_3(CO)_{12}/60^{o}/THF} H_2 + [HMe_2SiNHSiMe_2H]_x[PhSiHNH]_y \tag{32}$$

$$HMe_2SiOSiMe_2H + NH_3 \xrightarrow{Ru_3(CO)_{12}/60^o/THF} H_2 + -[Me_2SiHNMe_2SiO]_x- \quad (33)$$
$$Mn \approx 5000\text{-}7000$$

Preceramic Polymers by Dehydrocoupling

We have also explored the use of the dehydrocoupling reaction for the synthesis of preceramic polysilazanes starting from the precursor $MeNH-[H_2SiNMe]_x-H$. As described above, $MeNH-[H_2SiNMe]_x-H$ is produced by aminolysis of H_2SiCl_2 with $MeNH_2$ under conditions where $x \approx 18\text{-}20$ ($Mn \approx 1150$).

Seyferth and Wiseman reported that oligomers of $MeNH-[H_2SiNMe]_x-H$ where $x \approx 10$ gave a 38% ceramic yield when pyrolyzed to 900°C (21,22). The ceramic product appeared to be mostly silicon nitride. Because $MeNH-[H_2SiNMe]_x-H$ has N-H bonds as end caps and internal Si-H bonds, the possibility of forming chain extended and/or branched polymers using the dehydrocoupling reaction exists. We find that treatment of this precursor, as in reaction (34), does lead to species with higher

$$MeNH-[H_2SiNMe]_x-H \xrightarrow{Ru_3(CO)_{12}/90^oC/THF} H_2 + \text{polymer} \dashrightarrow \text{gel} \dashrightarrow \text{resin} \quad (34)$$

molecular weights. Depending on the reaction conditions and time, it is possible to produce tractable, processable polymers with visco-elastic properties useful for making ceramic coatings, fibers and three dimensional objects (when used as a binder for silicon nitride powder). With increased reaction times or higher temperatures more crosslinking occurs in reaction (34), and the resulting product first turns into a gel and then into an intractable resin.

Figure 1 shown below illustrates the changes in molecular weight and dispersion as reaction (34) proceeds (49). The bimodal distribution suggests that there is more than one mechanism for polymer growth. At least one mechanism, if not more, leads to gelation as evidenced by the disparity between $M_n \approx 2300$ and $M_w \approx 25K$ at 65 h. This disparity is typical of a gelation via branching process (54).

Table I lists the molecular weights and viscoelastic properties for the precursors and selected polymers produced in reaction (34). It also contains the ceramic yields obtained on pyrolysis to 900°C and the composition of the ceramic product. The salient features of the results summarized in Table I are: (1) The molecular weight and the structure (extent of branching and/or

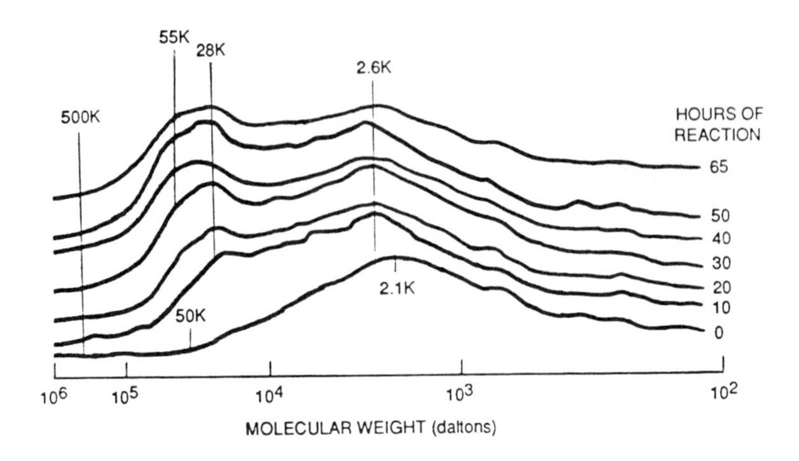

Figure 1. Size exclusion chromatography of $Ru_3(CO)_{12}$ catalyzed polymerization of MeNH-$[H_2SiNMe]_x$-H as a function of time. Polystyrene standards used for calibration.

Table I. Pyrolysis Studies on MeNH-$[H_2SiNMe]_x$-H Oligomers and Polymers

Oligomer	Mn (GPC)	Viscosity (poise)	Ceramic Yield (% at 900°C)	Si_3N_4 (Percent)
-$[H_2SiNMe]_x$- x = 10	600-700	1	40	80-85
-$[H_2SiNMe]_x$- x = 19	1150	5	45-50	80-85
-$[H_2SiNMe]_x$- $Ru_3(CO)_{12}$/90 C/THF or 30h	1560	18	60-65	80-85
-$[H_2SiNMe]_x$- $Ru_3(CO)_{12}$/90°C/THF for 65h	1620	100	65-70	80-85

crosslinking) of the oligo- or polysilazane play a role in the total ceramic yield; (2) Catalytic dehydrocoupling can be used to increase the overall ceramic yield and modify the viscoelastic properties of the precursor polymer; and, (3) The composition of the ceramic product (83% Si_3N_4 and 17%C for all precursors) appears to be defined by the monomer unit and appears to be independent of molecular weight or viscoelastic properties. This latter point is extremely important to synthetic chemists in that it strongly suggests that one can use custom-designed "molecular building blocks" to synthesize precursors to a wide range of known materials and possibly to some that are unknown.

Concluding Remarks

It is intriguing to note that good synthetic routes to tractable, high molecular weight (>20K) polysilazanes are still not available despite the extensive efforts and progress made in the last five years and the facility with which the analogous polysiloxanes are prepared. Part of the problem may be that the reaction mechanisms for the above discussed approaches are still poorly understood. We would like to suggest here some additional reasons for this lack of success.

Consider the effects of the N-R group on the structure and reactivity of polysilazanes as susceptible to hydrogen bonding effects. These effects alone should favor ring closure over the respective siloxane analogs. Moreover, for R = H, a new type of depolymerization reaction, analogous to reaction (4), is available as illustrated in reaction (35). Thus, reaction (35) could also contribute to the

$$H\text{-}[R_2SiNH]_{x-y}[R_2SiNH]_y\text{-} \overset{\Delta}{\longleftrightarrow} \text{-}[R_2SiNH]_{x-y}\text{-}H + [R_2Si{=}N][R_2SiNH]_{y-1}\text{-}H \qquad (35)$$

instability of oligosilazanes. In fact, the polysiloxazane produced in reaction (33) cannot rearrange via reaction (35) which may explain why we are able to produce this polymer with such high M_n (5000-7000 D). For the case where R > Me, it has already been shown that steric effects work in favor of ring closure (21-23).

Another likely but heretofore unrecognized problem with the common ammonlysis/aminolysis method of polysilazane synthesis is that the byproduct will always contain Cl^- salts. Walsh finds that the Si-Cl and Si-N bond strengths are equivalent (about 100 kcal/mole) (14). Given the well known ability of F^- to promote cleavage and rearrangement of bonds at silicon, it may be that Cl^- exhibits similar properties with the relatively weak Si-N bonds (compared to Si-O bonds); albeit, to a lesser degree than F^-. Support for this idea comes from the identification of SiH_3 oligomer caps formed during the ammonolysis of $H_2SiCl_2{\cdot}2py$, see reaction (9) above (24). The redistribution reaction necessary to obtain the SiH_3 groups could readily be promoted by Cl^- and represents another mechanism whereby growth of high molecular weight polysilazanes is prevented. If redistribution reactions to form chain capping species are promoted by Cl^-, then processes which avoid ammonolysis may provide the key to making high molecular weight polysilazanes. The results shown in the Figure suggest that catalytic dehydrocoupling may offer the opportunity to surmount this problem if appropriately selective catalysts can be developed.

In the area of preceramic polysilazanes, sufficient progress has been made to produce precursors for silicon nitride fibers, coatings and as binders for silicon nitride powder. However, particular problems still remain to be solved particularly with regard to reducing impurity levels and improving densification during pyrolysis.

Acknowledgments

We would like to thank Dr. Andrea Chow and Mr. Richard Hamlin for their contributions in the characterization of the polymer species and for discussions concerning the implications of these results. We gratefully acknowledge support for this research from the Strategic Defense Sciences Office through Office of Naval Research Contracts N00014-84-C-0392 and N00014-85-C-0668.

Literature Cited

1. Chantrell, P. G. and Popper, E. P.; "Special Ceramics", E. P. Popper Ed.; New York; Academic (1964) pp. 87-102.
2. For recent reviews see a. Wynne, K. J. and Rice, R. W.; Ann. Rev. Mater. Sci. (1984) 14, 297. b. Wills, R. R.; Mark, R. A.; Mukherjee, S. A.; Am. Ceram. Soc. Bull. (1983) 62, 904-915. c. Rice, R. R.; Cer. Bull. (1983) 62, 889-892.
3. Yajima, S.; Shishido, T.; Kayano, H.; Nature (London) (1976), 264, 237.
4. Stock, A. and Somieski, K.; Ber. Dtsch. Chem. Ges.(1921) 54, 740 see also ref 5.
5. a. Brewer, S. D. and Haber, C. P.; J. Am. Chem. Soc. (1948) 70, 361. b. Osthoff, R. C. and Kantor, S. W.; Inorg. Synth. (1957) 5, 61.
6. Fink, W.; Angew. Chem. Inter. Nat. Ed. (1966) 5, 760-776.
7. Fink, W.; Chem. Ber. (1963) 96, 1071-1079.
8. Fink, W.; Helv.Chim. Acta. (1964) 47, 498-508.
9. Fink, W.; Helv. Chim. Acta. (1964) 49, 1408-1415.
10. Fink, W.; Helv. Chim. Acta. (1964) 51, 954.
11. Wiseman, G. H.; Wheeler, D. R.; and Seyferth, D.; Organomet. (1986) 5, 146-152.
12. Blum, Y. D. and Laine, R. M.; Organomet. (1986) 5, 2081.
13. Gilman, H.; Hofferth, B.; Melvin, H. W. and Dunn, G. E.; J. Am. Chem. Soc.(1950) 72, 5767-5768.
14. Walsh, R.; Acc. Chem. Res. (1981) 14, 246-252.
15. Verbeek, W.; U. S. Patent No. 3,853,567 Dec 1974.
16. Winter, G. and Verbeek, W.; and Mansmann, M.; U. S. Patent No. 3,892,583 Jul 1975.
17. a. Parker, D. R. and Sommer, L. H.; J. Am. Chem. Soc. (1976) 98, 618-620. b. Parker, D. R. and Sommer, L. H.; J. Organomet. Chem. (1976) 110, C1-C4. For other studies on the Si=N bond see the following ref. and ref 7.

18. Elseikh; M. and Sommer, L. H.; J. Organomet. Chem. (1980) 186, 301-308.
19. a. Van Wazer, J. R.; Moedritzer, L.; Moedritzer, K. and Groves, W.; U. S. Patent No. 3,393,218 July, 1968. b. see also Wannagat, U.; Pure and Appl. Chem. (1966) 13, 263-279.
20. Seyferth, D. and Wiseman, G. H.; "Ultrastructure Processing of Ceramics, Glasses and Composites", L. L. Hench and D. R. Ulrich, Eds. (1984) pp. 265-275.
21. Wiseman, G. H.; Ph.D. Dissertation, M. I. T., Aug. 1984.
22. a. Seyferth, D.; Wiseman, G. H. and C. Prud'homme, J. Am. Ceram. Soc.(1983) 66, C13-C14. b. Seyferth, D. and Wiseman, G. H.; Am. Chem. Soc. Poly. Div. Prprt. (1984) 25, 10-12.
23. Laine, R. M.; Blum, Y. D.; Dodge, A. unpublished results.
24. Arai, M. and Isoda, T.; Japan Kokai Tokkyo Koho JP. 61 89,230.
25. Arai, M.; Sakurada, S.; Isoda, T. and Tomizawa, T.; Am. Chem. Soc. Polymer Div. Polym. Prpts. (1987) 27, 407.
26. a. Zhdanov, A. A.; Kotrelev, G. V.; Kazakova, V. V. and Redkozubova, Ye. P.; Polym. Sci., U.S.S.R. (1985) 27, 1593-1600. b. Andrianov, K. A.; Ismailov, B. A.; Konov, A. M. and Kotrelev, G. V. ; J. Organomet. Chem. (1965) 3, 129-137. c. Andrianov, K. A.; Kotrelev, G. V.; Kamaritski, B. A.; Unitski, I. H. and Sidorova, N. I.; J. Organomet. Chem. (1969) 16, 51-62.
27. Rochow, E. G.; Mon. Chem. (1964) 95, 750-765.
28. Kruger, C. R. and Rochow, E.; J. Poly. Sci. A, (1964) 2, 3179-3189.
29. Redl, G. and Rochow, E. G; Angew. Chem. (1964) 76, 650.
30. Zoeckler, M. T. and Laine, R. M.; J. Org. Chem. (1983) 48, 2539.
31. Chow, A. W.; Hamlin, R. D.; Blum, Y. and Laine, R. M.; submitted to J. Pol. Sci.
32. a. Penn, B. G.; Daniels, J. G.; Ledbetter, III, F. E.; and Clemons, J. M.; Poly. Eng. and Sci. (1986) 26, 1191-1194. b. Penn, B. G.; Ledbetter III, F. E.; Clemons, J. M. and Daniels, J. G.; J. Appl. Polym. Sci.(1982) 27, 3751-3761.
33. Penn, B. G.; Ledbetter III, F. E. and Clemons, J. M.; Ind. Eng. Chem. Process. Des. Dev. (1984) 23, 217-220.
34. Gaul Jr, J. H.; U. S. Patent No. 4,340,619 Jul 1982.
35. Baney, R. H.; Gaul, J. H.; U. S. Patent 4,310,651 1982.
36. Gaul Jr, J. H.; U. S. Patent No. 4,395,460 Jul 1983.
37. Gaul Jr, J. H.; U. S. Patent No. 4,312,970 Jan 1985.
38. Cannady, J. P.; U. S. Patent No. 4,535,007 Aug 1985.
39. Cannady, J. P.; U. S. Patent No. 4,543,344 Sep 1985.
40. Bujalski, D. R.; Eur. Patent Appl. EP 175,382 Mar 1986.
41. Baney, R. H.; Bujalski, D. R.; Eur. Patent Appl. EP 175,383 Mar 1986.
42. Legrow, G. E.; Lim, T. F.; Lipowitz, J. and Reaoch, R. S.; "Better Ceramics Through Chemistry II", Mat. Res. Symp. Proc. Vol. 73, Brinker, C. J.; Clark, D. E.; and Ulrich, D. R.; Eds. (1986) pp 553-558.
43. Andrianov, K. A.; Shkol'nik, M. I.; Syrtsova, Zh. S.; Petrov, K. I.; Kopylov, V. M.; Zaitseva, M. G. and Koroleva, E. Z.; Dokl. Akad. Nauk. SSSR (1975) 223, 347-350.
44. Seyferth, D.; Prud'Homme, C. C.; Wiseman, G. H.; U. S. Patent 4,397,828 Aug 1984.
45. Seyferth, D. and Wiseman, G. H.; U. S. Patent 4,482,669 Nov. 1984.

46. Blum, Y. D.; Laine, R. M.; Schwartz, K. B.; Rowcliffe, D. J.; Bening, R. C.; and Cotts, D. B. "Better Ceramics Through Chemistry II" Mat. Res. Soc. Symp. Proc., Brinker, C. J.; Clark, D. E. and Ulrich, D. R. Eds. (1986) 73, pp 389-393.
47. Schwartz, K. B.; Rowecliffe, D. J.; Blum, Y. D. and Laine, R. M.; "Better Ceramics Through Chemistry II" Mat. Res. Soc. Symp. Proc., Brinker, C. J.; Clark, D. E. and Ulrich, D. R. Eds. (1986) 73 pp 407-412.
48. Laine, R. M. and Blum, Y. D.;U. S. Patent 4,612,383 Sept 1986.
49. Laine, R. M.; Blum, Y. D.; Hamlin, R. D.; Chow, A.; in "Ultrastructure Processing of Glass, Ceramics and Composites II" D. D. Mackenzie and D. R. Ulrich, Eds., J. Wiley and Sons, (1987) in press.
50. Biran, C.; Blum, Y. D., Laine, R. M.; Glaser, R. and Tse, D. S.; manuscript submitted for publication.
51. Lin, Y. C.; Knobler, C. B. and Kaesz, H. D.; J. Am. Chem. Soc. (1981) 103, 1216-1218 and references therein.
52. Wilson, Jr., W. B. and Laine, R. M.; J. Am. Chem. Soc. (1985) 107, 361-368.
53. Eisenstadt, A.; Giandomenico, C.; Fredericks, M. F.; and Laine, R. M.; Organomet. (1985) 4, 2033.
54. Billmeyer, Jr., F. W.; "Textbook of Polymer Science" 2nd Ed. John Wiley and Sons, N. Y., N. Y., 1971 pp. 272-278.

RECEIVED September 1, 1987

Chapter 11

Organosilicon Polymers as Precursors for Silicon-Containing Ceramics

Recent Developments

Dietmar Seyferth, Gary H. Wiseman, Joanne M. Schwark, Yuan-Fu Yu, and Charles A. Poutasse

Department of Chemistry, Massachusetts Institute of Technology, Cambridge, MA 02139

Following general comments about the preceramic polymer approach to the preparation of ceramic materials, we describe the preparation of a novel polysilazane by dehydrocyclodimerization of the ammonolysis product of CH_3SiHCl_2, or coammonolysis products of CH_3SiHCl_2 with other chlorosilanes, the conversion of such polymers to ceramic products and their use in upgrading Si-H containing organosilicon polymers, e.g., methylhydrogenpolysiloxane, to more useful ceramic precursors.

Silicon-containing ceramics include the oxide materials, silica and the silicates; the binary compounds of silicon with non-metals, principally silicon carbide and silicon nitride; silicon oxynitride and the sialons; main group and transition metal silicides, and, finally, elemental silicon itself. There is a vigorous research activity throughout the world on the preparation of all of these classes of solid silicon compounds by the newer preparative techniques. In this report, we will focus on silicon carbide and silicon nitride.

Silicon carbide, SiC [1] and silicon nitride, Si_3N_4 [2], have been known for some time. Their properties, especially high thermal and chemical stability, hardness, high strength, and a variety of other properties have led to useful applications for both of these materials.

The "conventional" methods for the preparation of SiC and Si_3N_4, the high temperature reaction of fine grade sand and coke (with additions of sawdust and NaCl) in an electric furnace (the Acheson process) for the former and usually the direct nitridation of elemental silicon or the reaction of silicon tetrachloride with ammonia (in the gas phase or in solution) for the latter, do not involve soluble or fusible intermediates. For many applications of these materials this is not necessarily a disadvantage (e.g., for the application of SiC as an abrasive), but for some of the more recent desired applications soluble or fusible (i.e., processable) intermediates are required.

The need for soluble or fusible precursors whose pyrolysis will give the desired ceramic material has led to a new area of macromolecular science, that of preceramic polymers [3]. Such polymers are needed for a number of different applications. Ceramic powders by themselves are

0097–6156/88/0360–0143$06.00/0

difficult to form into bulk bodies of complex shape. Although ceramists have addressed this problem using the more conventional ceramics techniques with some success, preceramic polymers could, in principle, serve in such applications, either as the sole material from which the shaped body is fabricated or as a binder for the ceramic powder from which the shaped body is to be made. In either case, pyrolysis of the green body would then convert the polymer to a ceramic material, hopefully of the desired composition. In the latter alternative, shrinkage during pyrolysis should not be great.

Ceramic fibers of diverse chemical compositions are sought for application in the production of metal-, ceramic-, glass- and polymer-matrix composites [3c]. The presence of such ceramic fibers in a matrix, provided they have the right length-to-diameter ratio and are distributed uniformly throughout the matrix, can result in very considerable increases in the strength (i.e., fracture toughness) of the resulting material. To prepare such ceramic fibers, a suitable polymeric precursor is needed, one which can be spun by melt-spinning, dry-spinning, or wet-spinning techniques [4] into fibers which then can be pyrolyzed (with or without a prior cure step).

Some materials with otherwise very useful properties such as high thermal stability and great strength and toughness are unstable with respect to oxidation at high temperatures. An example of such a class of materials is that of the carbon-carbon composites. If these materials could be protected against oxidation by infiltration of their pores and the effective coating of their surface by a polymer whose pyrolysis gives an oxidation-resistant ceramic material, then one would have available new dimensions of applicability of such carbon-carbon composite materials.

In order to have a useful preceramic polymer, considerations of structure and reactivity are of paramount importance. Not every inorganic or organometallic polymer will be a useful preceramic polymer. Some more general considerations merit discussion. Although preceramic polymers are potentially "high value" products if the desired properties result from their use, the more generally useful and practical systems will be those based on commercially available, relatively cheap starting monomers. Preferably, the polymer synthesis should involve simple, easily effected chemistry which proceeds in high yield. The preceramic polymer itself should be liquid or, if a solid, it should be fusible and/or soluble in at least some organic solvents, i.e., it should be processable. Its pyrolysis should provide a high yield of ceramic residue and the pyrolysis volatiles preferably should be non-hazardous and non-toxic. In the requirement of high ceramic yield, economic considerations are only secondary. If the weight loss on pyrolysis is low, shrinkage will be minimized as will be the destructive effects of the gases evolved during the pyrolysis.

There are important considerations as far as the chemistry is concerned. First, the design of the preceramic polymer is of crucial importance. Many linear organometallic and inorganic polymers, even if they are of high molecular weight, decompose thermally by formation and evolution of small cyclic molecules, and thus the ceramic yield is low. In such thermolyses, chain scission is followed by "back-biting" of the reactive terminus thus generated at a bond further along the chain. Thus high molecular weight, linear poly(dimethylsiloxanes) decompose thermally principally by extruding small cyclic oligomers, $(Me_2SiO)_n$, n = 3,4,5 . . . When a polymer is characterized by this type of thermal decomposition, the ceramic yield will be low and it will be necessary to convert the linear polymer structure to a highly cross-linked one by suitable chemical reactions prior to its pyrolysis. In terms of the high ceramic yield requirement, the ideal preceramic polymer is one which has functional substituent groups which will give an efficient thermal cross-linking process so that on pyrolysis non-volatile, three-dimensional networks (which lead to maximum weight retention) are formed. Thus, preceramic polymer design requires the introduction of reactive or potentially reactive functionality.

In the design of preceramic polymers, achievement of the desired elemental composition in the ceramic obtained from them (SiC and Si_3N_4 in the present cases) is a major problem. For instance, in the case of polymers aimed at the production of SiC on pyrolysis, it is more usual than not to obtain solid residues after pyrolysis which, in addition to SiC, contain an excess either of free carbon or free silicon. In order to get close to the desired elemental composition, two approaches have been found useful in our research: (1) The use of two comonomers in the appropriate ratio in preparation of the polymer, and (2) the use of chemical or physical combinations of two different polymers in the appropriate ratio.

Preceramic polymers intended for melt-spinning require a compromise. If the thermal cross-linking process is too effective at relatively low temperatures (100-200°C), then melt-spinning will not be possible since heating will induce cross-linking and will produce an infusible material prior to the spinning. A less effective cross-linking process is required so that the polymer forms a stable melt which can be extruded through the holes of the spinneret. The resulting polymer fiber, however, must then be "cured", i.e., cross-linked, chemically or by irradiation, to render it infusible so that the fiber form is retained on pyrolysis. Finally, there still are chemical options in the pyrolysis step. Certainly, the rate of pyrolysis, i.e., the time/temperature profile of the pyrolysis, is extremely important. However, the gas stream used in the pyrolysis also is of great importance. One may carry out "inert" or "reactive" gas pyrolyses. An example of how one may in this way change the nature of the ceramic product is provided by one of our preceramic polymers which will be discussed in more detail later in this paper. This polymer, of composition $[(CH_3SiHNH)_a(CH_3SiN)_b]_m$, gives a black solid, a mixture of SiC, Si_3N_4, and some free carbon, on pyrolysis to 1000°C in an inert gas stream (nitrogen or argon). However, when the pyrolysis is carried out in a stream of ammonia, a white solid remains which usually contains less than 0.5% total carbon and is essentially pure silicon nitride. At higher temperatures (>400°C), the NH_3 molecules effect nucleophilic cleavage of the Si-C bonds present in the polymer and the methyl groups are lost as CH_4. Such chemistry at higher temperatures can be an important and sometimes useful part of the pyrolysis process.

The first useful organosilicon preceramic polymer, a silicon carbide fiber precursor, was developed by S. Yajima and his coworkers at Tohoku University in Japan [5]. As might be expected on the basis of the 2 C/1 Si ratio of the $(CH_3)_2SiCl_2$ starting material used in this process, the ceramic fibers contain free carbon as well as silicon carbide. A typical analysis [5] showed a composition 1 SiC/0.78 C/0.22 SiO_2. (The latter is introduced in the oxidative cure step of the polycarbosilane fiber).

The Yajima polycarbosilane, while it was one of the first, is not the only polymeric precursor to silicon carbide which has been developed.

Another useful system which merits mention is the polycarbosilane which resulted from research carried out by C.L. Schilling and his coworkers in the Union Carbide Laboratories in Tarrytown, New York [6]. More recently, a useful polymeric precursor for silicon nitride has been developed by workers at Dow Corning Corporation [7].

New Silicon-Based Preceramic Polymer Systems: Recent Research at M.I.T.

In earlier work [8], we have developed a process for the preparation of useful preceramic polymers using commercially available CH_3SiHCl_2 as the starting material. In the initial step, this chlorosilane was treated with ammonia to give oligomeric, mostly cyclic $[CH_3SiHNH]_n$ ($n \sim 5$). This product, itself not a useful preceramic material, was subjected to the base-catalyzed dehydrocyclodimerization (DHCD) reaction (eq. 1) in which the species present in the $[CH_3SiHNH]_n$ oligomer mixture are linked together via cyclodisilazane units.

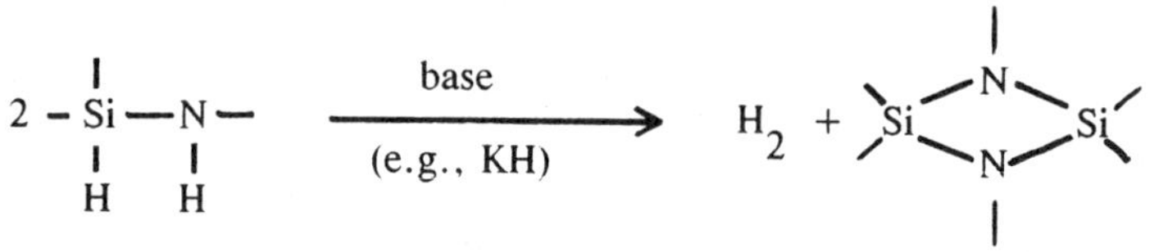

$$2\ {-}\underset{H}{\overset{|}{Si}}{-}\underset{H}{\overset{}{N}}{-} \xrightarrow[\text{(e.g., KH)}]{\text{base}} H_2 + \text{cyclo-}(Si_2N_2) \tag{1}$$

The repeating unit in the $[CH_3SiHNH]_n$ cyclics is 1.

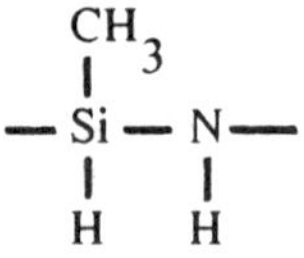

$$-\underset{H}{\overset{CH_3}{Si}}-\underset{H}{N}-$$

1

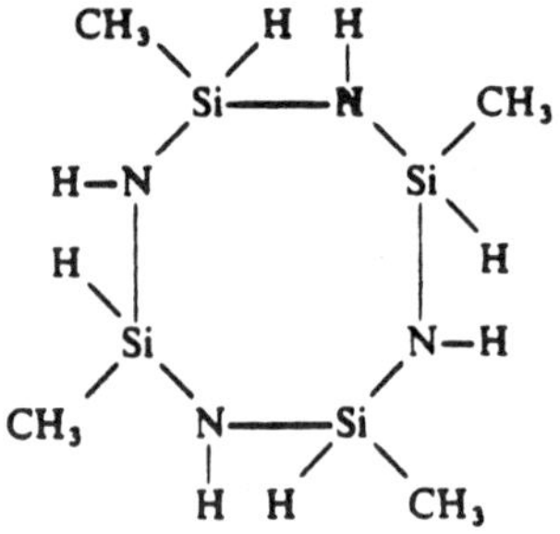

2

For example, the cyclic tetramer is the 8-membered ring compound 2. On the basis of equation 1, the adjacent NH and SiH groups provide the functionality which permits the molecular weight of the $[CH_3SiHNH]_n$ cyclics to be increased. Thus, $[CH_3SiHNH]_n$ cyclics will be linked together via four-membered rings as shown in 3.

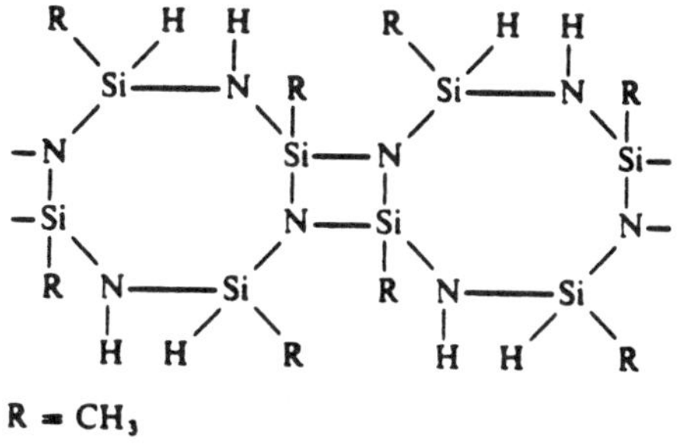

R = CH_3

3

We believe that a sheet-like network polymer (which, however, is not flat) having the general functional unit composition $[(CH_3SiHNH)_a(CH_3SiN)_b(CH_3SiHNK)_c]_n$ results. When hydrogen evolution ceases, a "living" polymer with reactive silylamide functions is present. On reaction of the latter with an electrophile (CH_3I or a chlorosilane), the neutral polysilazane is formed and can be isolated in the form of an organic-soluble white solid when the reaction is carried out in THF solution. Depending on solvent and reaction conditions, the molecular weight of this product is between 800 and 2500 g/mol. The molecular weight can be approximately doubled by using a difunctional chlorosilane (such as $ClMe_2SiCH_2CH_2SiMe_2Cl$) to quench the "living" polymer. Such polysilazanes, when pyrolyzed in a stream of argon, leave behind a black ceramic residue equivalent to 80-85% of the weight of the originally charged polymer. Analysis gave % Si, C and N values which could be rationalized in terms of a composition of (by weight) 67% Si_3N_4, 28% SiC and 5% C. The volatiles evolved during the pyrolysis consisted of H_2 and CH_4 and a trace of NH_3. Initial evaluation of the polysilazane shows it to have promise in three of the main potential applications of preceramic polymers: in the preparation of ceramic fibers and of ceramic coatings and as a binder for ceramic powders.

This polysilazane undergoes thermal crosslinking too readily to permit melt-spinning, but it can be dry-spun. It was clear that if melt-spinning was to be successful we would have to prepare a less reactive polysilazane with fewer reactive Si(H)-N(H) groups. To achieve this, we have studied polysilazanes derived by the dehydrocyclodimerization of the products of coammonolysis of CH_3SiHCl_2 and $CH_3(Un)SiCl_2$, where Un is an unsaturated substituent such as vinyl, CH_2=CH, or allyl, $CH_2CH=CH_2$. These groups were used because a melt-spun fiber requires a subsequent cure step to render it infusible. If this is not done, the fiber would simply melt on pyrolysis and no ceramic fiber would be obtained. The vinyl and allyl groups provide C=C functionality which can undergo UV-catalyzed Si-H additions, reactions which after melt-spinning in principle should lead to extensive further crosslinking without heating.

To provide the required monomers, we ammonolyzed mixtures of CH_3SiHCl_2 and $CH_3(CH_2=CH)SiCl_2$ and CH_3SiHCl_2 and $CH_3(CH_2=CHCH_2)SiCl_2$, respectively. The $[(CH_3SiHNH)_x(CH_3(Un)SiNH)_y]_n$ oligomers were prepared using $CH_3SiHCl_2/CH_3(Un)SiCl_2$ ratios of 3, 4 and 6, although other ratios can be used. This gives materials which still contain many $(CH_3)Si(H)$-$N(H)$- units, so polymerization by means of the process of eq. 1 still will be possible. However, dilution of the ring systems of the cyclic oligomers with $CH_3(Un)Si$ units will decrease the cross-linking which will occur on treatment with the basic catalyst.

The coammonolysis of CH_3SiHCl_2 and $CH_3(Un)SiCl_2$ in the indicated ratios was carried out by the procedure as described for the ammonolysis of CH_3SiHCl_2 [8]. A mixture of cyclic oligomeric silazanes is to be expected in these reactions, $[(CH_3(H)SiNH)_x(CH_3(Un)SiNH)_y]_n$, with more than one ring size present. In each preparation, the soluble, liquid products were isolated and used in the base-catalyzed polymerizations. When a ratio $CH_3SiHCl_2/CH_3(Un)SiCl_2$ of x:1 (x>1) was used in the ammonolysis reaction, the $CH_3(H)Si/CH_3(Un)Si$ ratio in the soluble ammonolysis product usually was somewhat less than x:1.

The addition of the various ammonolysis products to suspensions of catalytic amounts of KH in dry tetrahydrofuran resulted in hydrogen gas evolution with formation of a clear solution. Thermal treatment was followed by treatment with methyl iodide to "kill" the "living" polymeric potassium silylamide present in solution. The resulting silazane generally was obtained in high (>90%) yield and these products, always white powders, were soluble in organic solvents such as hexane, benzene, toluene and THF. Their average molecular weights were in the 800-1200 range. The presence of unchanged vinyl and allyl groups was proven by their IR and ^1H NMR spectra. A typical elemental analysis led to the empirical formula $SiC_{1.5}NH_{4.4}$ which could be translated into a composition $[(CH_3SiHNH)(CH_3[CH_2{=}CH]SiNH)_{2.3}]_x$.

These polysilazanes, upon pyrolysis to 1000°C under argon or nitrogen, gave black ceramic materials in good yield (73-86%, by weight). Analysis of the ceramic produced by such pyrolysis of one of these polysilazanes gave a composition 67.5 wt.% Si_3N_4, 22.5 wt.% SiC, and 10.0 wt.% unbound carbon. Another such ceramic sample, obtained by pyrolysis of another polysilazane, had the composition 68.8 wt.% Si_3N_4, 21.7 wt.% SiC, and 9.9 wt.% C.

The possibility of "curing" fibers pulled from polysilazanes of the type prepared here was demonstrated in the following experiment: Fibers were pulled from a concentrated syrup of a polysilazane derived from a 4:1 molar ratio $CH_3SiHCl_2/CH_3(CH_2{=}CH)SiCl_2$ coammonolysis product by DHCD in THF. Some fibers were pyrolyzed without any further treatment. These melted in large part, leaving very little in the way of ceramic fibers. Other fibers were subjected to UV irradiation for 2 hours. These, on pyrolysis under argon, did not melt and ceramic fibers were obtained. This polysilazane then is a good candidate for melt-spinning.

In order to obtain a SiC/Si_3N_4 mixture rich in Si_3N_4 by the preceramic polymer route, one requires a polymer which is richer in nitrogen than the CH_3SiHCl_2 ammonolysis product, $(CH_3SiHNH)_n$. In designing such a preceramic polymer, one would like to retain the facile chemical and thermal cross-linking system which the Si(H)-N(H)-unit provides. We have found that the coammonolysis of CH_3SiHCl_2 and $HSiCl_3$ serves our purposes well. For $HSiCl_3$, ammonolysis introduces three Si-N bonds per silicon atom, so the ammonolysis product of $CH_3SiHCl_2/HSiCl_3$ mixtures will contain more nitrogen than the ammonolysis product of CH_3SiHCl_2 alone.

In order to define the optimum system, we have investigated the ammonolysis of $CH_3SiHCl_2/HSiCl_3$ mixtures in various ratios in two solvents, diethyl ether, Et_2O, and tetrahydrofuran, THF. CH_3SiHCl_2/-$HSiCl_3$ mol ratios of 6, 3 and 1 were examined.

In both solvents, the 6:1 and 3:1 ratios produced polysilazane oils with molecular weights in the range 390-401 g/mol and 480 g/mol, respectively. When a 1:1 reactant ratio was used, waxes of somewhat higher (764-778 g/mol) molecular weights were obtained in both solvents. In the 1:1 reaction carried out in Et_2O the yield of soluble product was only 40%, but in THF it was nearly quantitative.

The oils produced in the 6:1 and 3:1 reactions in Et_2O appeared to be stable on long-term storage at room temperature in the absence of moisture (e.g., in the inert atmosphere box). However, the waxy product of 1:1 (Et_2O) reactions and all the coammonolysis products prepared in THF formed gels (i.e., became insoluble) after 3-4 weeks at room temperature, even when stored in a nitrogen-filled dry box.

The pyrolysis of the coammonolysis products was studied. The 6 CH_3SiHCl_2/1 $HSiCl_3$ ammonolysis product would be the least cross-linked since it contains the least amount of trifunctional component and, as expected, low ceramic yields were obtained on pyrolysis of these products. Pyrolysis of the 3:1 products gives increased ceramic yields, while pyrolysis of the most highly cross-linked 1:1 ammonolysis products gives quite good ceramic yields, 72% for the product prepared in Et_2O; 78% for that prepared in THF.

All of the ammonolysis products were submitted to the KH-catalyzed dehydrocyclodimerization reaction in order to obtain more highly cross-linked products that would give higher ceramic yields on pyrolysis. In all cases, the standard procedure [8] was used (1% KH in THF, followed by quenching with CH_3I (or a chlorosilane)). In every reaction, the product was a white solid which was produced in virtually quantitative yield. The proton NMR spectra of these products, as expected, showed an increase in the $SiCH_3/SiH$ + NH proton ratio, while the relative SiH/NH ratio was unchanged. In all cases, the molecular weights of the solid products were at least double that of the starting ammonolysis product, so the desired polymerization had occurred. Although the increase in molecular weight in these DHCD reactions is not great, any increase is useful for further processing. The reactions bring the advantage that the oils are converted to more easily handleable solids.

Pyrolysis of the white solids obtained in these KH-catalyzed dehydrocyclodimerization reactions (under argon from 50-950°C) produced black ceramic residues, with the exception of the 1:1 THF ammonolysis-derived solid which left a brown residue. The ceramic yields were excellent (all greater than or equal to 82%, with the highest being 88%).

Analysis of bulk samples of the ceramic materials produced in the pyrolysis of the various KH-catalyzed dehydrocyclodimerization products showed that our goal of a higher Si_3N_4/SiC ratio has been achieved: for the 1:1 ammonolysis product-derived polymers, 86% Si_3N_4, 8% SiC and 5% C (THF ammonolysis) and 83% Si_3N_4, 11% SiC and 6% C (Et_2O ammonolysis); for the 3:1 and 6:1 ammonolysis product-derived polymers: 77% Si_3N_4, 18-19% SiC and 4-5% C (Et_2O ammonolysis) and 74% Si_3N_4, 20% SiC and 5-6% C (THF ammonolysis).

These polymers may be used in the preparation of quite pure silicon nitride if the pyrolysis is carried out in a stream of ammonia (a reactive gas) rather than under nitrogen or argon. The ammonia reacts with the polymer at higher temperatures to cleave methyl groups from silicon (eq. 2), so that essentially all carbon is lost. Thus pyrolysis of the dehydrocyclodimerization product of the 1:1 (THF) ammonolysis

$$-\overset{|}{\underset{|}{Si}}-CH_3 + NH_3(g) \longrightarrow -\overset{|}{\underset{|}{Si}}-NH_2 + CH_4 \qquad (2)$$

product to 1000°C in a stream of ammonia gave a white ceramic residue in high yield which contained only 0.29% C, the remainder being silicon nitride.

Other commercially available $RSiCl_3$ compounds are CH_3SiCl_3 (a cheap by-product of the Direct Process) and $CH_2{=}CHSiCl_3$ (also an inexpensive starting material) and both were included in this study. In both cases, 6:1, 3:1 and 1:1 $CH_3SiHCl_2/RSiCl_3$ ammonolysis products were prepared and submitted to the dehydrocyclodimerization procedure. The ceramic yields obtained on pyrolysis of the resulting polymers were high. In the CH_3SiHCl_2/CH_3SiCl_3 experiments: 78-86%. In all cases, a black ceramic residue resulted when the pyrolysis to 1000°C was carried out in a stream of argon. As expected, the carbon content (in the form of SiC and free C) was higher than that of the $CH_3SiHCl_2/HSiCl_3$-derived ceramics: 12-18% SiC, up to 9.5% C. Nonetheless, higher Si_3N_4 contents than those obtained when CH_3SiHCl_2 is used alone (~ 67%)) were obtained: ~ 76-80% Si_3N_4.

To produce a ceramic material containing only Si_3N_4, the white solid polysilazane derived from DHCD of the oil obtained by ammonolysis of 6:1 CH_3SiHCl_2/CH_3SiCl_3 was pyrolyzed in a stream of ammonia (to 1000°C). A <u>white</u> ceramic residue containing only 0.36% by weight C resulted.

With 6:1, 3:1 and 1:1 $CH_3SiHCl_2/CH_2{=}CHSiCl_3$ ammonolysis products DHCD gave white solids whose pyrolysis resulted in increased carbon content and decreased Si_3N_4 (vs. the CH_3SiHCl_2/CH_3SiCl_3 examples): ~ 69-73% Si_3N_4, 9-13% SiC, 12-18% C (by weight).

Our research also has been directed at SiC precursors. It began with an examination of a potential starting material in which the C:Si ratio was 1, the ratio desired in the derived ceramic product. Available methylsilicon compounds with a 1 C/1 Si stoichiometry are CH_3SiCl_3 and CH_3SiHCl_2. The latter, in principle, could give $[CH_3SiH]_n$ cyclic oligomers and linear polymers on reaction with an alkali metal. In practice, the Si-H linkages also are reactive toward alkali metals. Thus, mixed organochlorosilane systems containing some CH_3SiHCl_2 have been treated with metallic potassium by Schilling and Williams [9]. It was reported that the CH_3SiHCl_2-based contribution to the final product was $(CH_3SiH)_{0.2}(CH_3Si)_{0.8}$, i.e., about 80% of the available Si-H bonds had reacted. Such Si-H reactions lead to cross-linking in the product, or to formation of polycyclic species if cyclic products are preferred. Nevertheless, we have used this known reaction for CH_3SiHCl_2 with an alkali metal as an entry to new preceramic polymers.

When the reaction of CH_3SiHCl_2 with sodium pieces was carried out in tetrahydrofuran medium, a white solid was isolated in 48% yield. This solid was poorly soluble in hexane, somewhat soluble in benzene, and quite soluble in THF. Its ^1H NMR spectrum (in $CDCl_3$) indicated that extensive reaction of Si-H bonds had occurred. The $\delta(SiH)/\delta(SiCH_3)$

integration led to a constitution $[(CH_3SiH)_{0.4}(CH_3Si)_{0.6}]_n$. Here the CH_3SiH units are ring and chain members which are not branching sites; the CH_3Si units are ring and chain members which are branching sites. In our reactions it is expected that mixtures of polycyclic and linear (possibly cross-linked) polysilanes will be formed. (Attempts to distill out pure compounds from our preparations were not successful. Less than 10% of the product was volatile at higher temperatures at 10^{-4} torr.). The ceramic yield obtained when the $[(CH_3SiH)_{0.4}(CH_3Si)_{0.6}]_n$ polymer was pyrolyzed (TGA to 1000°C) was 60%; a gray-black solid was obtained whose analysis indicated a composition 1.0 SiC + 0.49 Si.

The reaction of methyldichlorosilane with sodium in a solvent system composed of six parts of hexane and one of THF gave a higher yield of product which was soluble in organic solvents. Such reactions give a colorless oil in 75 to over 80% yield which is soluble in many organic solvents. In various experiments the molecular weight (cryoscopic in benzene) averaged 520-740 and the constitution (by ^1H NMR) $[(CH_3SiH)_{0.76}(CH_3Si)_{0.24}]_n$ to $[(CH_3SiH)_{0.9}(CH_3Si)_{0.1}]_n$. This less cross-linked material (compared to the product obtained in THF alone) gave much lower yields of ceramic product on pyrolysis to 1000°C (TGA yields ranging from 12-27% in various runs). Again, the product was (by analysis) a mixture of SiC and elemental silicon, 1.0 SiC + 0.42 Si being a typical composition. These results are not especially promising, and it was obvious that further chemical modification of the $[(CH_3SiH)_x(CH_3Si)_y]_n$ products obtained in the CH_3SiHCl_2/Na reactions was required [10].

A number of approaches which we tried did not lead to success, but during the course of our studies we found that treatment of the $[(CH_3SiH)_x(CH_3Si)_y]_n$ products with alkali metal amides (catalytic quantities) serves to convert them to materials of higher molecular weight whose pyrolysis gives significantly higher ceramic yields. Thus, in one example, to 0.05 mol of liquid $[(CH_3SiH)_{0.85}(CH_3Si)_{0.15}]$ in THF was added, under nitrogen, a solution of about 1.25 mmol (2.5 mol%) of $[(CH_3)_3Si]_2NK$ in THF. The resulting red solution was treated with methyl iodide. Subsequent nonhydrolytic workup gave a soluble white powder in 68% yield, molecular weight 1000, whose pyrolysis to 1000°C gave a ceramic yield of 63%.

The proton NMR spectra of this product showed only broad resonances in the Si-H and Si-CH_3 regions. In the starting $[(CH_3SiH)_x(CH_3Si)_y]_n$ material, the observed proton NMR integration ratios, $SiCH_3$/SiH, ranged from 3.27-3.74. This ratio was quite different in the case of the product of the silylamide-catalyzed process, ranging from 8.8 to 14. Both Si-H and Si-Si bonds are reactive toward nucleophilic reagents. In the case of the alkali metal silylamides, extensive structural reorganization, involving both Si-H and Si-Si bonds of the $[(CH_3SiH)_x(CH_3Si)_y]_n$ polysilane, that results in further cross-linking, must have taken place.

While these silylamide-catalyzed reactions provided a good way to solve the problem of the low ceramic yield in the pyrolysis of $[(CH_3SiH)_x$-$(CH_3Si)_y]_n$, the problem of the elemental composition of the ceramic product remained (i.e., the problem of Si/C ratios greater than one) since only catalytic quantities of the silylamide were used.

As noted above, KH-catalyzed polymerization of the CH_3SiHCl_2 ammonolysis product gives a polymeric silylamide of type $[(CH_3SiHNH)_a$-$(CH_3SiN)_b(CH_3SiHNK)_c]$. In a typical example, a = 0.39, b = 0.57, c = 0.04, so there is only a low concentration of silylamide functions in the polymer. This polymeric silylamide reacts with electrophiles other than methyl iodide, e.g., with diverse chlorosilanes, and it has been isolated and analyzed. Since it is a silylamide, we expected that it also would react with $[(CH_3SiH)_x(CH_3Si)_y]_n$ polysilane-type materials. Not only would it be expected to convert the latter into material of higher molecular weight, but it also would be expected to improve the Si/C ratio (i.e., bring it closer to 1). As noted above, pyrolysis of $[(CH_3SiHNH)_a(CH_3SiN)_b(CH_3SiHNK)_c]_n$ gives a ceramic product in 80-85% yield containing Si_3N_4, SiC and excess carbon. Thus, combination of the two species in the appropriate stoichiometry, i.e., of $[(CH_3SiH)_x(CH_3Si)_y]_n$ and $[(CH_3SiHNH)_a(CH_3SiN)_b(CH_3SiHNK)_c]_n$, and pyrolysis of the product (which we will call a "graft" polymer) after CH_3I quench could, in principle, lead to a ceramic product in which the excess Si obtained in pyrolysis of the former and the excess C obtained in the pyrolysis of the latter combine to give SiC. Accordingly, experiments were carried out in which the two polymer systems, $[(CH_3SiH)_x(CH_3Si)_y]_n$ and the "living" polymer–silylamide, $[(CH_3SiHNH)_a(CH_3SiN)_b(CH_3SiHNK)_c]_n$, were mixed in THF solution in varying proportions (2.4:1 to 1:2 mole ratio) and allowed to react at room temperature for 1 h and at reflux for 1 h. (Such experiments were carried out with the $[(CH_3SiH)_x(CH_3Si)_y]_n$ materials prepared in hexane/THF as well as with those prepared in THF alone.) After quenching with methyl iodide, nonhydrolytic workup gave a new polymer in nearly quantitative yield (based on weight of material charged). The molecular weight of these products was in the 1800–2500 range. Their pyrolysis under nitrogen gave ceramic products in 74–83% yield. Thus, the reaction of the two polymer systems gives a new polymer in close to quantitative yield, which seems to be an excellent new preceramic polymer in terms of ceramic yield.

In an alternative method of synthesis of $[(CH_3SiH)_x(CH_3Si)_y]$-$[(CH_3SiHNH)_a(CH_3SiN)_b]$ "combined" polymers, the polysilyl amide was generated in situ in the presence of $[(CH_3SiH)_x(CH_3Si)_y]_n$. This, however, gave materials that were somewhat different. In one such experiment, a mixture of $(CH_3SiHNH)_n$ cyclics (as obtained in the ammonolysis of CH_3SiHCl_2 in THF) and the $[(CH_3SiH)_x(CH_3Si)_y]_n$ material (x = 0.76; y = 0.26) in THF was treated with a catalytic amount of KH. After the reaction mixture had been treated with methyl iodide, the usual workup gave an 89% yield of hexane-soluble white powder, molecular weight ~2750. On pyrolysis, this material gave a 73% yield of a black ceramic.

The "combined" polymer prepared in this way ("in situ polymer") was in some ways different from the "combined" polymer prepared by the first method ("graft" polymer). Principal differences were observed in their proton NMR spectra and in the form of their TGA curves. This suggests that the two differently prepared polymers have different structures. It is likely that in the "in situ" preparation intermediates formed by the action of KH on the $(CH_3SiHNH)_n$ cyclics are intercepted by the reaction with the $[(CH_3SiH)_x(CH_3Si)_y]_n$ also present before the $[(CH_3SiHNH)_a(CH_3SiH)_b(CH_3SiHNK)_c]_n$ polymer (which is the starting reactant used in the "graft" procedure) has a chance to be formed to the extent of its usual molecular weight. Thus, less of the original CH_3SiHNH protons are lost and/or more of those of the $[(CH_3SiH)_x$-$(CH_3Si)_y]_n$ system are reacted.

The TGA curves of the "graft" polymer and the "in situ" polymer are different as well. Noteworthy in the former is a small weight loss between 100°C and 200°C, which begins at around 100°C. This initial small weight loss occurs only at higher temperature (beginning at ~175°C) in the case of the "in situ" polymer. This difference in initial thermal stability could well have chemical consequences of importance with respect to ceramics and both kinds of polymers may be useful as preceramic materials.

Further experiments showed that the "combined" polymers may be converted to black ceramic fibers. Pyrolysis of pressed bars of the "combined" polymer to 1000°C gave a black product of irregular shape (74-76% ceramic yield). In other experiments, SiC powder was dispersed in toluene containing 20% by weight of the "combined" polymer. The solution was evaporated and the residue, a fine powder of SiC with the "combined" polymer binder, was pressed into bars and pyrolyzed at 1000°C. A ceramic bar (6% weight loss, slightly shrunk in size) was obtained.

The ceramic products obtained in the pyrolysis of the "combined" polymers have not been studied in detail, but some of them have been analyzed for C, N, and Si. The compositions of the ceramic materials obtained cover the range 1 Si_3N_4 + 3.3 to 6.6 SiC + 0.74 to 0.85 C. Thus, as expected, they are rich in silicon carbide and the excess Si which is obtained in the pyrolysis of the $[(CH_3SiH)_x(CH_3Si)_y]_n$ materials alone is not present, so that objective has been achieved. By proper adjustment of starting material ratios, we find that the excess carbon content can be minimized [11].

The "living polymer" intermediate in our polysilazane synthesis is useful in "upgrading" other Si-H containing polymers for application as precursors for ceramic materials. Another example of this application is provided by such "upgrading" of methylhydrogenpolysiloxane, $[CH_3Si(H)O]_n$. Such a linear polymer, average molecular weight 2000-5000 (vendor data), on pyrolysis under argon to 1000°C, left a black ceramic residue of only 13%.

In this study, experiments were carried out with a $[CH_3Si(H)O]_m$ prepared using conditions under which the yield of the cyclic oligomers (m = 4, 5, 6, ...) is maximized [12], as well as with the commercial $[CH_3Si(H)O]_m$ polymer of higher molecular weight, with presumably high linear content.

In one approach, the polymeric silylamide was prepared as described above (using the product of CH_3SiHCl_2 ammonolysis in THF) and to this "living" polymer solution were added slowly the $[CH_3Si(H)O]_m$ oligomers (high cyclic content). An immediate reaction with some gas evolution occurred. The resulting clear solution was treated with CH_3I to react with any remaining silylamide or silanolate units. Silylamide/siloxane weight ratios of 1:1 and 1:5 were used. In both cases the polymeric product was an organic-soluble white solid of moderate (1700 and 2400, respectively) average molecular weight. In both cases pyrolysis under nitrogen or argon gave high char yields (78% and 76%, respectively, by TGA). Pyrolysis to 1000°C of a bulk sample of the 1:1 by weight polymer gave a black solid in 80% yield. Elemental analysis indicated a "composition": 1 SiC + 0.84 Si_3N_4 + 2.17 SiO_2 + 2.0 C. (This is not meant to reflect the composition in terms of chemical species present. No doubt, rather than separate Si_3N_4 and SiO_2, silicon oxynitrides are present). On the other hand, pyrolysis under gaseous ammonia gave a <u>white</u> ceramic solid in 78.5% yield. It is suggested that here also high temperature nucleophilic cleavage of $Si-CH_3$ bonds by NH_3 occurred and that the ceramic product is a silicon oxynitride. This was confirmed by analysis. The white solid contained less than 0.5% carbon.

In an alternate approach to this "graft" procedure, the "in situ" procedure was used. A mixture (~1:1 by weight) of the CH_3SiHCl_2 ammonolysis (in THF) product, $[CH_3SiHNH]_m$, and the CH_3SiHCl_2 hydrolysis product, $[CH_3Si(H)O]_n$, in THF, was added to a suspension of a catalytic amount of KH in THF. Hydrogen evolution was observed and a clear solution resulted. After quenching with CH_3I, further work-up gave the new polymer, a soluble white powder, average molecular weight 1670. Pyrolysis to 1000°C gave a black ceramic solid in 84% yield (by TGA). Pyrolysis of bulk sample under argon yielded a black ceramic (73%). Analysis indicated the "composition": 1 SiC + 1.03 Si_3N_4 + 1.8 SiO_2 + 2.63 C.

In place of the (mostly) cyclic $[CH_3SiHO]_n$ oligomers, a commercial methylhydrogenpolysiloxane (Petrarch PS-122) of higher molecular weight, presumably mostly linear species, may serve as the siloxane component. When a 1:1 by weight ratio of the preformed polymeric silylamide and the $[CH_3Si(H)O]_n$ polymer was used and the reaction mixture was quenched with CH_3I, the usual work-up produced a soluble white solid, molecular weight 1540. Pyrolysis under argon to 1000°C gave a black ceramic material (77% yield by TGA). Pyrolysis of a bulk sample yielded a black solid (73% yield) whose analysis indicated a "composition": 1 SiC + 1.5 Si_3N_4 + 3.15 SiO_2 + 3.6 C. Application of the "in situ" procedure [1:1 by weight ratio of $(CH_3SiHNH)_m$ and $[CH_3Si(H)O]_n$ gave a soluble white powder, average molecular weight 1740, pyrolysis yield (TGA) 88% (black solid). The fact that the pyrolysis of these polymers under a stream of ammonia gives white solids, silicon oxynitrides which contain little, if any, carbon, in high yield is of interest. The ceramics applications of these silicon oxynitride precursor systems are receiving further study by Yu and Ma of Universal Energy Systems [13].

The chemistry which is involved in the "graft" and "in situ" procedures and the structures of the hybrid polymers which are formed remain to be elucidated. However, there is no doubt that these procedures are useful ones. We have used them also to form new and useful hybrid preceramic polymers from the Yajima polycarbosilane which contains a plurality of $[CH_3Si(H)CH_2]$ units [14].

Conclusions

We have described new routes to useful preceramic organosilicon polymers and have demonstrated that their design is an exercise in functional group chemistry. Furthermore, we have shown that an organosilicon polymer which seemed quite unpromising as far as application is concerned could, through further chemistry, be incorporated into new polymers whose properties in terms of ceramic yield and elemental composition were quite acceptable for use as precursors for ceramic materials. It is obvious that the chemist can make a significant impact on this area of ceramics. However, it should be stressed that the useful applications of this chemistry can only be developed by close collaboration between the chemist and the ceramist.

Acknowledgments

The work reported in this paper was carried out with generous support of the Office of Naval research and the Air Force Office of Scientific Research. The results presented here derive from the Ph.D. dissertations of Gary H. Wiseman and Joanne M. Schwark and the postdoctoral research of Dr. Yuan-Fu Yu and Dr. Charles A. Poutasse. I thank these coworkers for their skillful and dedicated efforts.

Literature Cited

1. Gmelin Handbook of Inorganic Chemistry, 8th Edition, Springer-Verlag: Berlin, Silicon, Supplement Volumes B2, 1984, and B3, 1986.
2. Messier, D. R.; Croft, W. J. in "Preparation and Properties of Solid-State Materials," Vol. 7, Wilcox, W. R., ed.; Dekker: New York, 1982, Chapter 2.
3. a. Wynne, K. J.; Rice, R. W. Ann. Rev. Mater. Sci. (1984) 14, 297. b. Rice, R. W. Am. Ceram. Soc. Bull. (1983) 62, 889. c. Rice, R. W. Chem. Tech. (1983) 230.
4. Billmeyer, F. W. Jr., Textbook of Polymer Chemistry, 2nd edition, Wiley: New York, 1984, Chapter 18.
5. Yajima, S. Am. Ceram. Soc. Bull. (1983) 62, 893.
6. Schilling, C. L., Jr.; Wesson, J. P.; Williams, T. C. Am Ceram. Soc. Bull. (1983) 62, 912.
7. LeGrow, G. E.; Lim, T. F.; Lipowitz, J.; Reaoch, R. S. in "Better Ceramics Through Chemistry II," edited by C. J. Brinker, D. E. Clark and D. R. Ulrich, Materials Research Society, Pittsburgh, 1986, pp. 553-558.
8. a. Seyferth, D.; Wiseman, G. H. J. Am. Ceram. Soc. (1984) 67, C-132. b. U. S. Patent 4,482,669 (Nov. 13, 1984). c. Seyferth, D.; Wiseman, G. H. in "Ultrastructure Processing of Ceramics, Glasses and Composites," 2, edited by L. L. Hench and D. R. Ulrich, Wiley: New York, 1986, Chapter 38.
9. a. Schilling, C. L., Jr.; Williams, T. C., Report 1983, TR-83-1, Order No. AD-A141546; Chem. Abstr. 101 196820p. b. U. S. Patent 4,472,591 (Sept. 18, 1984).
10. Very much the same study, with the same results, was carried out by Sinclair and Brown-Wensley at 3M prior to our investigation. This work came to our attention when the U.S. patent issued: Brown-Wensley, K. A.; Sinclair, R. A. U. S. patent 4,537,942 (Aug. 27, 1985).
11. This "combined polymer approach" has been patented: Seyferth, D.; Wood, T. G.; Yu, Y.-F; U. S. patent 4,645,807 (Feb. 24, 1987).
12. Seyferth, D.; Prud'homme, C.; Wiseman, G. H. Inorg. Chem. (1983) 22, 2163.
13. Yu, Y.-F.; Ma, T.-I in "Better Ceramics Through Chemistry II," edited by C. J. Brinker, D. E. Clark and D. R. Ulrich, Materials Research Society, Pittsburgh, 1986, pp. 559-564.
14. Seyferth, D. and Yu, Y.-F., U. S. patent 4,650,837 (Mar. 17, 1987).

RECEIVED October 23, 1987

Chapter 12

NMR Characterization of a Polymethyldisilylazane

A Precursor to Si–C–N–O Ceramics

Jonathan Lipowitz, James A. Rabe, and Thomas M. Carr[1]

Dow Corning Corporation, Midland, MI 48640

^{29}Si and ^{13}C NMR spectroscopy was used to characterize a polymethyldisilylazane polymer, a precursor to Si-C-N-O ceramics. Polymer is prepared by reaction of mixed chloromethyldisilanes with hexamethyldisilazane. The polymer is a low MW ($\bar{M}_n$ ~2000), glassy oligomer with a polycyclic, cage-like structure and a broad MW distribution. ^{29}Si and ^{13}C NMR spectra give only broad signals. A simplified polymerization reaction using *sym*-tetrachlorodimethyldisilane was followed by ^{29}Si NMR to give insight into development of polymer structure. The broad NMR signals were shown to result from a multitude of chemical environments about the Si atoms which develop early in the reaction, not from restricted motion of Si atoms in the polymer molecules or from ^{14}N quadrupole broadening.

Ceramics can be prepared by pyrolysis of various organosilicon polymers. The advantages of ceramic formation from polymers include the ability to prepare shapes difficult to achieve by conventional powder processing methods, such as films and fibers; the use of lower processing temperatures than in conventional methods; the ability to achieve very high purity because reagents can be purified by well-established methods such as distillation and recrystallization; the ability to vary ceramic composition by variation of polymer composition; and the ability to create unique metastable ceramic compositions which cannot be achieved by conventional processing (3). This technology has been reviewed recently by several authors (4-11).

[1]Current address: Technical Center, Owens Corning Fiberglas, Granville, OH 43023

0097-6156/88/0360-0156$06.00/0

One of the main advantages of the polymer route to ceramics is the preparation of ceramic fibers, a shape difficult to achieve by other methods. Ceramic fiber-based composites are becoming an increasingly important group of structural materials (12, 13).

A ceramic fiber with Si-C-N-O composition can be prepared by melt-spinning, cure and pyrolysis of a polymethyldisilylazane polymer precursor (14, 15), which is the reaction product of a mixture of 50 mol % 1,1,2,2- tetrachloro-1,2-dimethyldisilane (Ia), 40 mol % 1,1,2-trichloro- 1,2,2-trimethyldisilane (Ib) and 10 mol % 1,2-dichloro-1,1,2,2- tetramethyldisilane +1,1-dichloro-1,2,2,2-tetramethyldisilane (Ic).

This mixture is reacted with excess hexamethyldisilazane (HMDZ) to generate oligomeric species (polymethyldisilylazane "polymer") with the approximate composition II, as determined by elemental analysis and M_n from GPC analysis,

$$[Me_{2.6}Si_2(NH)_{1.5}(NHSiMe_3)_{0.4}Cl_{0.15}]_{\sim 13} \qquad (II)$$

along with Me_3SiCl and NH_4Cl as by-products. The volatile species Me_3SiCl and HMDZ are removed by gradually increasing temperature to about 250° C during reaction.

The polymer is a glassy material with a highly crosslinked presumably cage-like polycyclic structure and is readily soluble in aromatic and aliphatic hydrocarbons. Number average molecular weights ($\overline{M}_n$) of ca. 2000 are found by GPC. Polymers are reactive with both oxygen and moisture and must be handled in a dry and inert atmosphere. A broad MW distribution is found by GPC, suggesting that a variety of structures are present. The mixture of chlorodisilane monomers and their high average functionality (f=3.4) suggest a complex highly crosslinked structure.

The intent of the effort described in this paper was to obtain insight into the structure of this complex polymer and to characterize the structures present using NMR.

Experimental

Approximately 1 g polymer and 0.06 M $Cr(acac)_3$ were dissolved in $CDCl_3$ to prepare solutions for ^{29}Si and ^{13}C NMR spectroscopy. NMR spectra were run on a Varian XL-200 FT-NMR instrument. To aid in obtaining quantitative data, the solution was doped with 0.06 M chromium acetylacetonate [$Cr(acac)_3$)] to remove possible signal artifacts resulting from long spin-lattice relaxation times (T_1's) and the nuclear Overhauser effect, well-known features associated with ^{29}Si and ^{13}C NMR spectroscopy. This permits quantitative signal acquisition. From the literature (16) and additional work done in this laboratory, it was expected that $Cr(acac)_3$ would be an inert species. A solution of HMDZ (2.04 g, 12.67 mmole), 1,1,2,2-tetrachloro-1,2-dimethyldisilane, Ia (0.99 g, 4.34 mmole), Cr $(acac)_3$ (0.09 g, 0.3 mmole), in $CDCl_3$ to 5.0 ml volume was prepared and ^{29}Si NMR data were sampled at 10 minute intervals.

1,1,2,2-tetrachloro-1,2-dimethyldisilane was prepared by the method of Sakurai et al. (17). Hexamethyldisilazane enriched to 99% ^{15}N was prepared by a Ni-catalyzed reaction of trimethylsilane with 99% enriched $^{15}NH_3$. Reaction was carried out in a 5 L flask on a vacuum line using 0.67 atm (0.14 mole) Me_3SiH (PCR Research Chemicals, Gainesville, FL) and 0.33 atm (0.07 mole) $^{15}NH_3$ (Stohler Isotope Chemicals, Waltham, MA) and 25 g Ni powder, 99.9%, 500-600 μm, spherical (Johnson Matthey, Inc., Seabrook, NH). Ni powder was washed with reagent grade THF and vacuum degassed before use. Reaction conditions are similar to that described in (18). The bottom section of the flask containing Ni powder was heated 96 hrs at 170° C. Distillation gave 7.1 g (0.044 mole) $(Me_3Si)_2{}^{15}NH$ of 98% purity by gc analysis. The only impurity found was hexamethyldisiloxane. A nmr experiment to follow formation of products and loss of reactants was performed just as described for unlabeled HMDZ. Formation of products and loss of reactants was virtually identical to that shown in Figure 5.

Results

To gain insight into the polymer structure, ^{13}C NMR spectroscopy was performed. From examination of the broad signals in Figure 1 it can be seen that this experimental approach yields little information. There is little additional information obtained from examination of the broad signals in a ^{29}Si NMR spectrum (Figure 2) of this polymer, except for the observation of residual Me_3SiCl by-product at ca. + 32 ppm and Me_3SiNH groups at ca. +2 ppm. ^1H NMR shows only $Si-CH_3$ groups. It was hoped that the broad, featureless ^{29}Si NMR spectrum could be simplified by the use of a pure disilane starting material. Since the polymer itself is highly cross-linked and the pure, symmetrical 1,1,2,2-tetrachloro-1,2-dimethyldisilane is likely to produce a simpler polymer of similar structure, it was chosen for detailed study. Polymer prepared from the sym-tetrachlorodisilane exhibits similar properties to that from the mixture (I) of chlorodisilanes. Indeed, the sym-tetrachlorodisilane is the predominant molecular species in the mixture of disilanes (I) ised. The ^{29}Si NMR spectrum of the polymer that results from the reaction of HMDZ with sym-tetrachlorodimethyldisilane ($Cl_2MeSi-SiMeCl_2$), looks very much like that of polymer prepared from the mixture of dichlorosilanes (Figure 2). There is no observable fine structure that would lead to an increased understanding of the chemical structure of this simpler oligomer.

The broad, featureless nature of the ^{29}Si NMR signal is likely due to one of three phenomena that are well-known to cause such effects in NMR spectroscopy: a multiplicity of chemical and physical environments about the silicon atoms; restricted segmental motion in polycyclic or cage-like molecules; or line broadening due to ^{29}Si magnetic interactions with the numerous ^{14}N quadrupoles. To obtain information on the reason for NMR signal broadening and insight into the chemical nature of these pre-ceramic polymers, attempts were made to follow the reaction of the sym-tetrachlorodisilane and HMDZ by ^{29}Si NMR spectroscopy.

From inspection of Figure 3, which is the ^{29}Si NMR spectrum of the reaction mixture after 10 minutes, it is seen that reaction is quite rapid and that a significant amount of the trimethylsilazane monoadduct of sym-tetrachlorodisilane has formed, along with Me_3SiCl by-product and a small amount of the sym-1,2- diadduct. As reaction proceeds, rapid consumption of starting materials with resultant

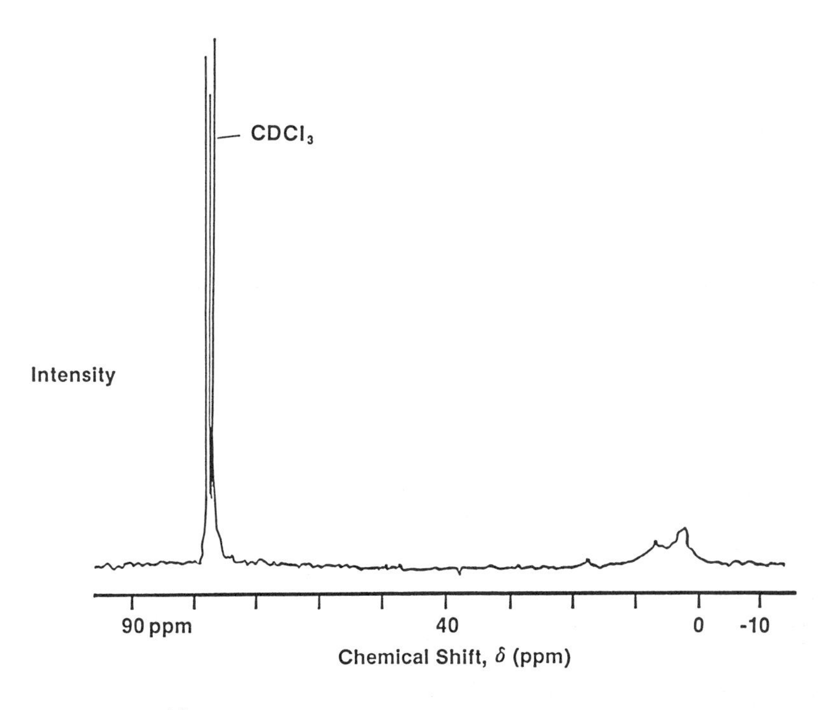

Figure 1. ^{13}C NMR Spectrum of Polymethyldisilylazane Polymer in $CDCl_3$.

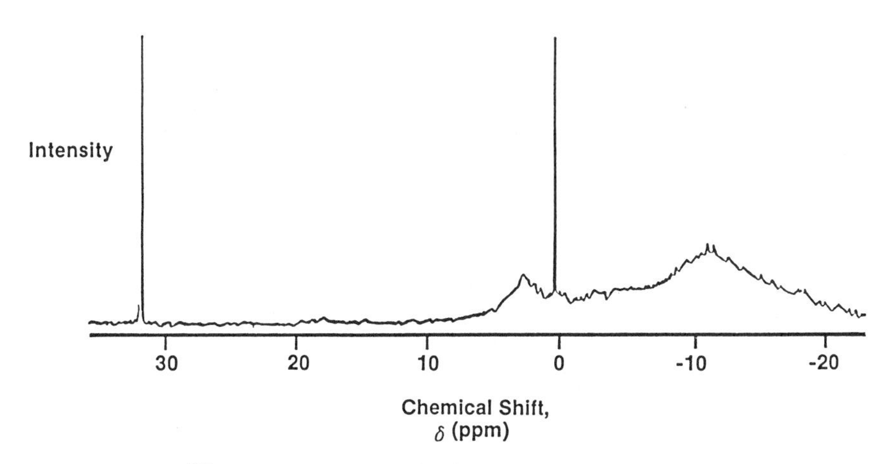

Figure 2. ^{29}Si NMR Spectrum of Polymethyldisilylazane Polymer in $CDCl_3$; 0.06 M in $Cr(acac)_3$.

formation of monoadduct and diadduct species is seen. About 40 minutes after initial mixing, formation of the triadduct species is observed, which remains as a minor component throughout the course of the reaction. A compilation of the ^{29}Si NMR chemical shifts of these low molecular weight species is shown in Table I.

After 80 minutes it was observed that the only significant changes were the near-total consumption of the $[MeCl_2Si]_2$ starting material and the monoadduct species, along with increased diadduct formation. In addition, triadduct content is at its maximum concentration, with only small amounts of this strained species with $[Me_3Si]_2$-SiMe- linkages formed. From inspection of Figure 4, obtained after 10 hours, it can be seen that, along with Me_3SiCl and HMDZ species, appreciable amounts of higher order oligomers with $[\equiv SiNH]_2SiMe$-type linkages are formed.

These observations lead to the following conclusions regarding the reaction scheme of $[MeCl_2Si]_2$ and HMDZ.

The initial step involves the fast ligand exchange reaction of HMDZ with $[MeCl_2Si]_2$ to create the monoadduct and Me_3SiCl by-product. A similar fast reaction occurs at the other Si atom to create the symmetrical 1,2-diadduct and additional Me_3SiCl. Steric hindrance causes formation of two trimethylsilylazane groups on a silicon atom to yield $[Me_3SiNH]_2$-SiMe-type species to be slow. This would account for the presence of only minor amounts of triadduct and the fact that the 1,1-diadduct is not observed. The formation of higher order species than triadduct involves self-reaction of monoadduct and diadduct species as well as reaction between these species to create a multiplicity of linear and cyclic species with Si-Si-NH-Si-Si-type linkages. No single higher order species than triadduct is formed in sufficient concentration to stand out from the multiplicity of species produced.

The validity of the conclusion regarding the reaction scheme can be verified from Figure 5, which plots relative moles of each species *vs* time. There is a rapid rise in monoadduct and diadduct content initially. Shortly after initial mixing, a rapid decrease in monoadduct with a significantly slower decline in diadduct content is seen. Triadduct is slow to form and slow to react. A rapid decrease in HMDZ content followed by a gradual increase in HMDZ is found due to the following condensation reactions:

$$\equiv Si\text{-}NH\text{-}SiMe_3 + Me_3SiNH\text{-}Si\equiv \xrightarrow{slow} \equiv Si\text{-}NH\text{-}Si\equiv + [Me_3Si]_2NH \quad (III)$$

It is observed that the long-term increase in HMDZ is significantly less than the short-term decrease in the initial reaction period. This is significant because, in terms of the proposed reaction scheme (Figure 6), it would be difficult to envision formation of polycyclic or cage-like structures without regeneration of HMDZ that is nearly equal in magnitude to that of the initial consumption of HMDZ, long before larger molecules of restricted segmental motion are produced. However, broad signals are forming as shown by the reduction in total integrated intensity of sharp signals.

No silicon resonance from a molecule with more than 5 Si atoms (triadduct) can be distinguished from the multitude of Si environments produced. With this evidence, it would seem that broad

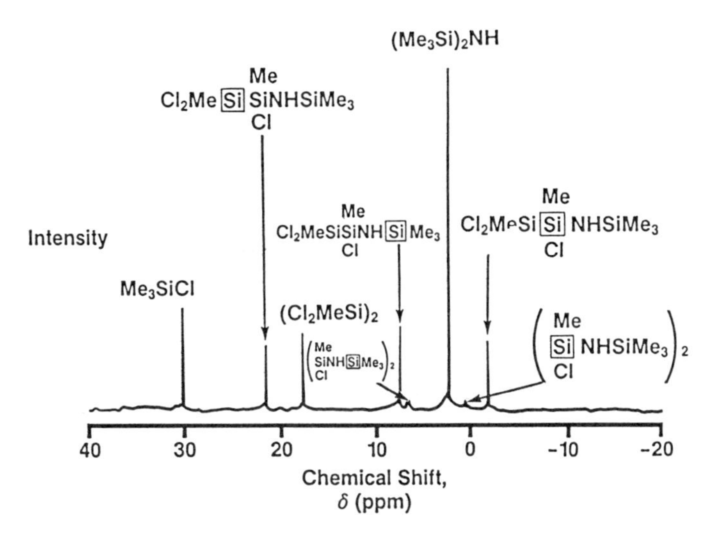

Figure 3. ^{29}Si NMR Spectrum of 10-Minute Reaction Product of $(Me_3Si)_2NH$, 2.5 M, and $(MeCl_2Si)_2$, 0.87 M, in $CDCl_3$; 0.06 M in $Cr(acac)_3$.

TABLE I. ^{29}Si NMR Chemical Shifts in the Reaction of $(Me_3Si)_2NH$ and $(MeCl_2Si)_2$

Starting Materials	δ (^{29}Si) in ppm Relative to Me_4Si	
$(Me_3Si)_2NH$	+ 2.24	
$(MeCl_2Si)_2$	+ 17.45	
By-Product		
Me_3SiCl	+ 30.08	
Monoadduct		
	+ 7.48	(a)
$Me_3Si_aNH\text{-}ClMeSi_b\text{-}Si_cCl_2Me$	− 1.91	(b)
	+ 21.61	(c)
1,2-Diadduct		
	+ 6.60	(a)
$[Me_3Si_a\text{-}NHClMeSi_b]_2$	− 2.20	(b)*
Triadduct		
	+ 3.42	(a)
$[Me_3Si_aNH]_2\text{-}MeSi_b\text{-}Si_cMeCl\text{-}NHSi_dMe_3$	− 17.65	(b)
	− 3.97	(c)
	+ 5.65	(d)

* Doublet observed (ca. 0.1 ppm separation) due to the presence of erythro and threo diastereomers.

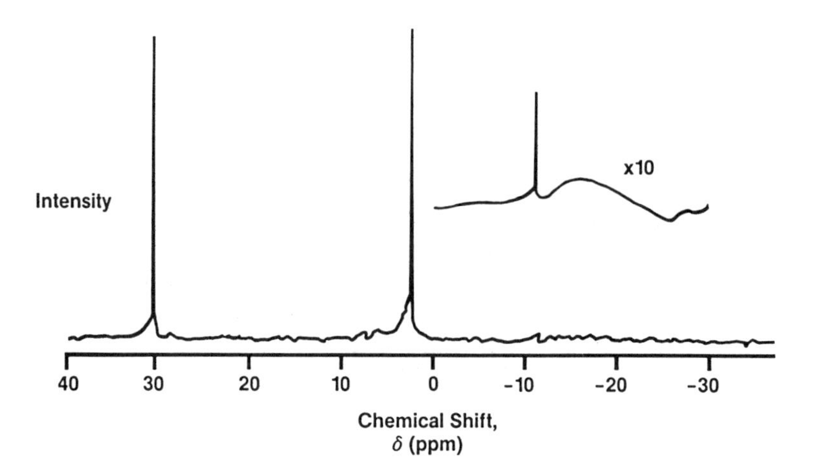

Figure 4. ^{29}Si NMR Spectrum of 10-Hour Reaction Product of $(Me_3Si)_2NH$ and $(MeCl_2Si)_2$ in $CDCl_3$; 0.06 M in $Cr(acac)_3$.

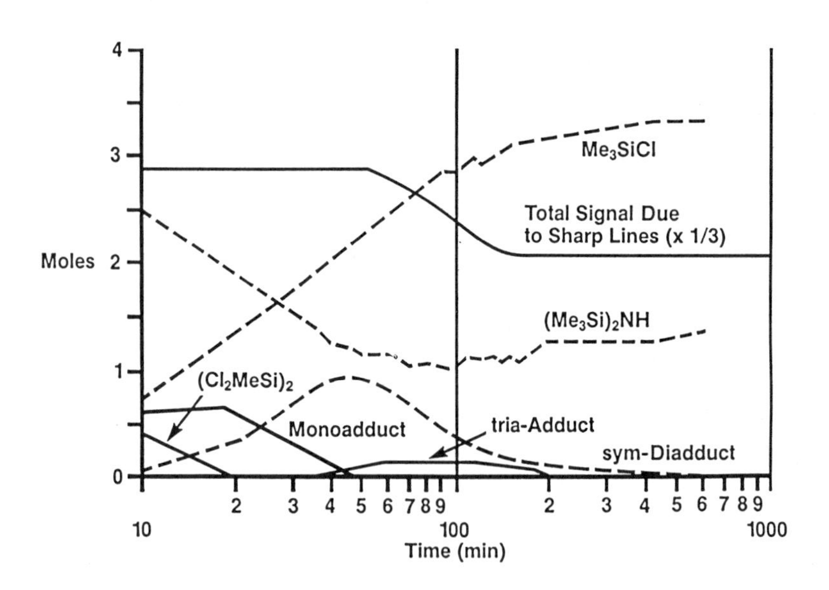

Figure 5. Reaction of $(Me_3Si)_2NH$ and $(MeCl_2Si)_2$ in $CDCl_3$.

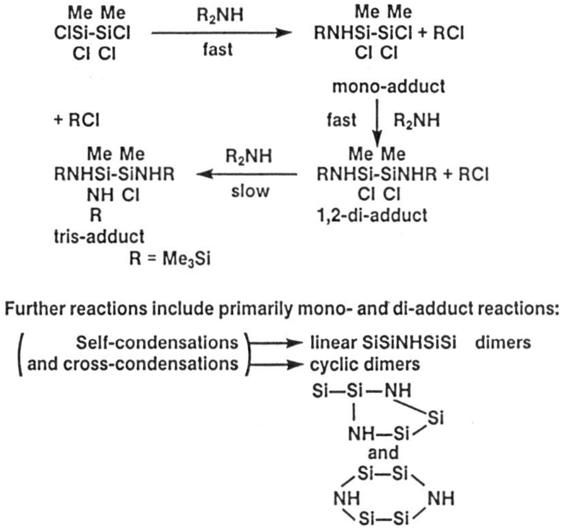

Me Me
ClSi-SiCl —R_2NH, fast→ RNHSi-SiCl + RCl
Cl Cl

Me Me
RNHSi-SiCl + RCl
Cl Cl
mono-adduct

↓ fast, R_2NH

Me Me
RNHSi-SiNHR + RCl
Cl Cl
1,2-di-adduct

←R_2NH, slow—

+ RCl
Me Me
RNHSi-SiNHR
NH Cl
R
tris-adduct

R = Me_3Si

Further reactions include primarily mono- and di-adduct reactions:

(Self-condensations and cross-condensations) → linear SiSiNHSiSi dimers
→ cyclic dimers

Si—Si—NH—Si—Si—NH (five-membered ring)

and

Si—Si—NH—Si—Si—NH (six-membered ring)

Figure 6. Reaction Scheme.

signals are not due to restricted segmental motion in cage-like moieties. Instead, it would appear that either ^{14}N quadrupolar interactions or production of numerous Si-N species would cause this lack of spectral detail. Substitution of 99% ^{15}N-enriched HMDZ, which has no quadrupolar nuclei, in the reaction mixture results in the same signal breadth. This indicates that the reaction of HMDZ and $[MeCl_2Si]_2$ results in a continuum of species that cannot be resolved by NMR techniques at the present time. The ^{15}N chemical shifts are shown in Table II.

TABLE II. ^{15}N-NMR Chemical Shifts

Compound	^{15}N Chemical Shift*
$(Me_3Si)_2{}^{15}NH$	-358.4
$Me_3Si{}^{15}NHSiClMeSiCl_2Me$ (monoadduct)	-351.4
$(Me_3Si{}^{15}NHSiClMe)_2$ (1,2-diadduct)	-350.3
$Me_3Si{}^{15}N_aHSiClMeSiMe({}^{15}N_bHSiMe_3)_2$ (triadduct)	-349.2 (a) -352.2 (b) doublet

* ppm relative to CH_3NO_2

Conclusions

As a result of the above work, the following has been achieved. A proposed reaction scheme for formation of polymethyldisylazane oligomer, a ceramic precursor, has been developed, giving some insight into structural features of this type of resin-like molecule. In addition, the cause for broad featureless signals in the NMR spectroscopy of these polymers has been determined to be the multiplicity of environments about the Si atoms which develop early in the reaction before appreciable polymerization occurs.

Acknowledgments

This work was made possible by support from the Defense Advanced Research Projects Agency (DARPA) and the Air Force Wright Aeronautical Laboratories (AFWAL) under Contract F33615-83-C-5006 to Dow Corning Corporation. The authors gratefully acknowledge the encouragement of the Contract managers S. Wax (DARPA) and A. Katz (AFWAL).

Literature Cited

3. Lipowitz, J., Freeman, H. A., Chen, R. T., and Prack, E. R. Adv. Ceram. Mater. 1987, 2, 121.
4. Rice, R. W. Am. Ceram. Soc. Bull. 1983, 62, 889.
5. Yajima, S. ibid., 893.
6. West, R., David, L. D., Durovich, P. I., Yu, H., and Sinclair, R. ibid., 899.
7. Wills, R., Markle, R. A., and Mukherjee, S. P. ibid., 904.
8. Schilling, Jr., C. L., Wesson, J. P., and Williams, T. C. ibid., 912.
9. Walker, Jr., B. E., Rice, R. W., Becher, P. F., Bender, B. A., and Coblenz, W. S. ibid., 916.
10. Baney, R. H. Division of Industrial and Engineering Chemistry, I. Structural, Electronic and Refractory Ceramics, ACS Symposium on Materials in Emerging Technologies, 1985.
11. K. J. Wynne and R. W. Rice. An. Rev. Mater. Sci. 1984, 14, 297.
12. Mah, T., Mendiratta, M. G., Katz, A. P. and Mazdiyasni, K. S. Am. Ceram. Soc. Bull. 1987, 66, 304.
13. Marshall, D. B. and Ritter, J. E. ibid. 1987, 66, 309.
14. Gaul, Jr., J. H. U.S. Patent 4 340 619, 1982.
15. Baney, R. H. In Ultrastructure Processing of Ceramics, Glasses, and Composites. Ed. Hench, L. L. and Ulrich, D. R. Wiley, J. New York, 1984, Chapter 20.

16. Harris, R. K. Kennedy, J. D., McFarlane, W. In NMR and the Periodic Table. Ed. Harris, R. K. and Mann, B. E., Academic Press, London, 1978, 309-340.
17. Sakurai, H., Watanabe, T. and Kumada, M. J. Organometal. Chem. 1967, 7,15.
18. Kotzsch, H. J., Draese, R., and Vahlensieck, H. J. U.S. Patent 4 115 427, 1978.

RECEIVED September 1, 1987

Chapter 13

Silicon–Nitrogen-Containing Rings and Polymers

Z. Lasocki, B. Dejak, J. Kulpinski, E. Lesniak, S. Piechucki, and M. Witekowa

Institute of Polymers, Technical University, Lodz, Poland

It has been of interest to explore the possibilities of synthesizing useful polymers from the following types of silazane (Si-N) containing ring compounds: cyclic bis (silyl) imidates, N-acylcyclosilazoxanes, N-arylcyclosilazoxanes and N,N′-diarylcyclodisilazanes. The imidates and acylcyclosilazoxanes are readily hydrolyzed and decomposed on heating. The interesting internal mobility of these molecules will be discussed. Arylcyclosilazoxanes undergo anionic, ring-opening polymerization and copolymerization with cyclosiloxanes to yield linear polysilazoxanes of remarkable thermal stability. Another method of improving the thermal stability of polysiloxanes is based on the polycondensation of functional cyclodisilazane oligomers with α, ω-dihydroxypolysiloxanes. High temperature reactions of diarylcyclodisilazanes with phenylenediamines yield a variety of cyclolinear polysilazane oligomers and polymers with aromatic spacing groups. Their structure and properties will be discussed.

In a research program to explore the possibilities of synthesizing useful polymeric materials from silazane (Si-N) containing ring compounds, reactions of the following ring systems were studied:

(I) cyclic silylamides of various ring sizes,
(II) cyclosilazoxanes; six- and eight-membered,
(III) cyclodisilazanes

Formation of polymer chains or networks from these compounds can be conceived either as a ring-opening polymerization process or a polyreaction involving exo-cyclic functional groups.

Cyclic Bis(silylamides): Structure and Rearrangements

In an attempt to prepare polymers of type **Ic** or **Id** from acid amides and 1,3-dichlorodisiloxanes or 1,5-dichlorotrisiloxanes, it was found that bimolecular cyclization prevailed over polycondensation and cyclic bis(silylamides) in the amido (**Ia**) or imidato (**Ib**) isomeric forms were exclusively formed in high yields

0097–6156/88/0360–0166$06.00/0

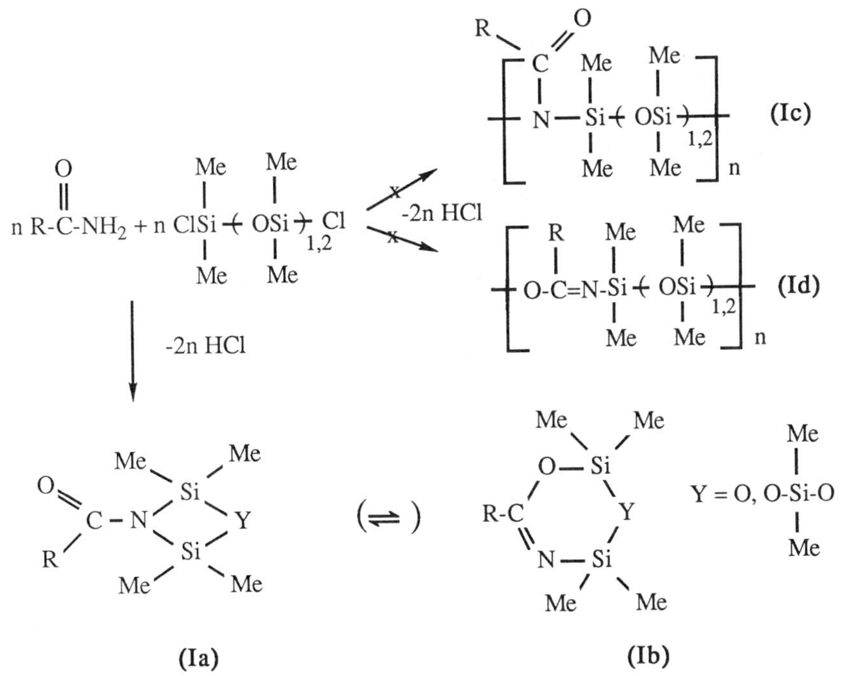

(Ic)

(Id)

(Ia)

(Ib)

Scheme 1

(1). This reaction bears a close similarity to the one discovered by J.F. Klebe who found that acid amides with dialkyldichlorosilanes yield disilaoxadiazines (2). The interesting intramolecular mobility of these compounds was studied by ^1H NMR spectroscopy and the mechanism of internal silyl migrations within their molecules arose some discussion in the literature (3,4).

The present cyclic bis(silylamides) are a family of very reactive compounds, sensitive to moisture and heat, readily hydrolyzed to form amides and cyclosiloxanes and readily decomposed at moderate temperatures with elimination of nitrile. However, many attempts to produce from them linear polymers by ring-opening polymerization failed. In a few cases compounds **Ia** and **Ib** were found to be interconvertible, e.g. for R = Me and Y = O the two isomers in benzonitrile, pyridine or carbon tetrachloride are found to be in a tautomeric equilibrium, which was quantitatively studied by ^1H NMR (5). The tautomer in the amido form , N-acetyltetramethylcyclodisilazoxane, is the main component in the tautomeric mixture at -30°C, but its signals have almost completely disappeared from the spectrum at 90°C, where the imidato isomeric form, 2,4,6-pentamethyl-2,4-disila-1,3,5-dioxazine, predominates. The amido tautomer is the first example of a stable four-membered silazoxane ring structure; it represents an intermediate case between the long known disilazane ring, reviewed in 1968 by W. Fink (6) and cyclodisiloxane, recently synthesized by R. West et al. (7). An isomeric ring structure to the cyclodisilazoxane, the disiloxazetidine, has been quite recently reported (8).

Alkyl substituents R in **I** and particularly R = H tend to stabilize the amido form of the molecule, whereas electronegative groups (R = CCl_3, p-XC_6H_4; X =

H, NO_2, Cl, OCH_3) favor the imidato isomer, e.g. the reaction product of trichloroacetamide with 1,5-dichlorohexamethyltrisiloxane exists in the eight-membered ring imidato form (**Ib**, R = CCl_3, Y = $OSiMe_2O$). The ring in that form, however, does not lose its internal mobility and two out of the three methylsilyl singlets in the ^{1}H NMR and ^{13}C NMR spectra undergo coalescence at 46°C ($\Delta G^{\ddagger}$ = 17.1 kcal/mole) and 90°C ($\Delta G^{\ddagger}$ = 18.5 kcal/mol), respectively. This behavior is best explained as the result of an intramolecular silyl migration resulting in a degenerate rearrangement and involving a transannular attack of nitrogen on silicon, formation of an unstable intermediate in the amido six-membered ring form to afford finally an identical structure with the initial molecule except for the respective positions of silicons (**a**) and (**b**), which have been interchanged (9).

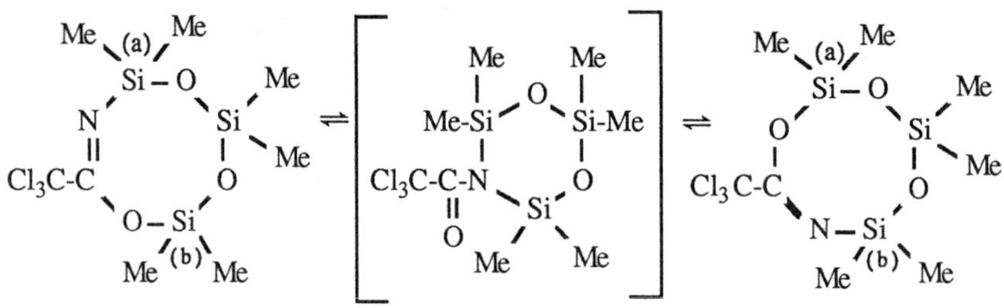

The dynamic NMR behavior of the chloroacetamide derivative of hexamethyltrisiloxane (**I**, R = $ClCH_2$, Y = $OSiMe_2O$) offers a good example of the structural versatility of cyclic bis(silylamides) (10). The ^{13}C NMR spectrum of this compound at -50°C is consistent with the amido form of the molecule and contains singlets at (δ ppm) 174.21 (C = O), 45.11 (CH_2Cl), 2.47 and 0.79 ($SiCH_3$). At -10°C new signals, which can be assigned to the imidato isomer come into sight at (δ ppm) 157.24 (C = N) and 46.41 (CH_2Cl). Three additional weak singlets in the methylsilyl region of the spectrum are also observed. At 10°C the carbonyl signal has disappeared and coalescence phenomena are apparent in both chloromethyl and methylsilyl regions. Finally, at 68°C the spectrum is consistent with the sole existence of the imidato form of the molecule, which undergoes a degenerate rearrangement (i.e. sharp singlets at 45.63, 1.23 and 0.97 ppm, δ) (10).

^{29}Si NMR spectroscopy is a reliable tool for the structure determination of siloxane derivatives, since the chemical shifts can be assigned to definite ring sizes and hence to the isomeric forms of the compounds being examined. For instance, in the spectrum of the hexamethyltrisiloxane derivative of p-chlorobenzamide (**I**, R = p-ClC_6H_4, Y = $OSiMe_2O$) at -32°C there are three singlets at (δ ppm) -12.82 (**a**), -15.01 (**b**) and -21.03 (**c**); signals (**a**) and (**b**) undergo coalescence at 23°C yielding one peak at -14.67 ppm δ, while the chemical shift of silicon (**c**) is almost unaffected. In the spectra of model six-membered rings, the chemical shifts are at (δ ppm) -4.45 (2Si, N-Si-O) and -8.66 (1Si, O-Si-O) for N-phenylhexamethylcyclotrisilazadioxane and -5.78 (2Si, N-Si-O) and -9.81 (1Si, O-Si-O) for N-acetylhexamethylcyclotrisilazadioxane. On the other hand, it is known that on passing from hexamethylcyclotrisiloxane to octamethylcyclotetrasiloxane there is a diamagnetic shift from (δ ppm) -9.2 to -20.0 (11). Therefore, the examined compound can be safely assigned the eight-membered imidato ring structure **Ib** (9).

To summarize the effects of the nature of groups R and Y on the structure of bis(silylamides) (**I**) it can be concluded that relative increase of the π character in the carbon-nitrogen bond is probably responsible for the relative stabilization of the amido form of the molecules by hydrogen and methyl substituents R. This effect is pronounced in the trisiloxane derivatives of the amides due to the absence of considerable strain in the six-membered rings and the conformation lability of the eight-membered structures.

Silazoxane and Silazane Linear and Cyclo-linear Polymers

Introductory Remarks. In contrast with the popularity and usefulness of the polysiloxane chains, which constitute the structural backbone of silicones, the knowledge of polymers based on the silazane unit is still limited. In the sixties there was still some hope of the possibility of producing long chain polysilazane molecules and a number of laboratories were active in seeking convenient methods for their synthesis (e.g. see review by Aylett (12)).

Although the energy of the silicon-nitrogen bond is ca. 30 kcal/mole lower than that of the siloxane linkage, the extensive (p-d) π bonding tends to increase the SiN bond strength. Indeed, some low molecular weight silazanes were found to be stable to high temperatures; for example, the four-membered cyclodisilazanes, which are among the most thermostable organosilicon-nitrogen compounds; they decompose over 500°C in spite of the obviously high ring strain in their molecules (13). This raised hope that polysilazanes would surpass silicones in thermo-stability. However, the polarity of the silicon-nitrogen bond and, particularly the basicity of nitrogen, make these compounds sensitive to hydrolysis in acidic media, thus limiting their general utility.

It has become fairly evident that for most silazane systems the equilibrium between rings and chains lies well to the side of the ring compounds and the concentration of linear oligomers is very low. VanWazer and Moedritzer have shown that equilibration of N-trimethylhexamethylcyclotrisilazane with dichlorodimethylsilane yields mainly the cyclic trimer. The concentration of linear oligomers decreases rapidly with increasing chain lengths and the dimer, 1,3-dichlorotetramethyldisilazane, is the main linear component at equilibrium (14). In a similar equilibration study of octamethylcyclotetrasiloxane with dichloro-dimethylsilane it was found that the amount of rings at equilibrium is negligibly low (15). The reluctance of silazanes to build long chain molecules arises, therefore, from thermodynamic grounds.

Varying the substituents on silicon and nitrogen does little to change the situation. Hydrogen on silicon, however, does appear to promote silazane chains formation (12,16). It has been recently claimed that unsubstituted cyclosilazanes can be polymerized by heat to yield soluble, high molecular weight polysilazanes, which were further pyrolyzed to β-Si_3N_4 and β-SiC ceramics (17). However, the linear chain structure of the polymers was not established and some decomposition of the material during polymerization was observed.

The reasons for the high thermodynamic stability of the silazane rings with respect to chains are not clear. Quantum mechanical calculations have shown that $p\pi$ - $p\pi$ interactions between non-bonded nitrogen atoms are stronger in silazane rings relative to similar interactions between oxygens in cyclosiloxanes, which results in relatively high N-N bond orders (18). It has been suggested that there are larger angular strains, in cyclosiloxanes. Moreover, the Si-O bond energy is higher in linear structures and increases with chain length, the reverse being true for the Si-N energy in cyclic and linear polysilazanes (19). Whatever the reason, it is an experimental fact that cyclosilazanes of various ring sizes (four-, six- and eight) are reluctant to polymerize by a ring-opening process although they readily undergo rearrangements with ring expansions and contractions.

In the present research program, which is directed to synthesizing useful polymers from silazane containing ring compounds, three general pathways were followed:

- the first consists of using preformed silazoxane rings to produce silazoxane chains by ring-opening polymerization,
- the second is based on a combination of disilazane rings with polysiloxane chains to obtain silazane-siloxane block copolymers,
- the third takes advantage of the susceptibility of silazanes to nucleophilic attack by amines and to transamination, which with bifunctional reagents, may result in polyreaction and formation of chains containing alternating silazane and organic units.

Silazoxane Linear Polymers and Copolymers

Several sets of six- and eight-membered N-arylcyclosilazoxanes (**IIa-e**) have been prepared from aniline and dichlorosiloxanes:

(IIa) 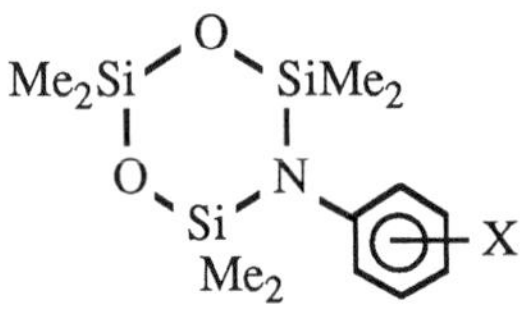 **(IIb)**

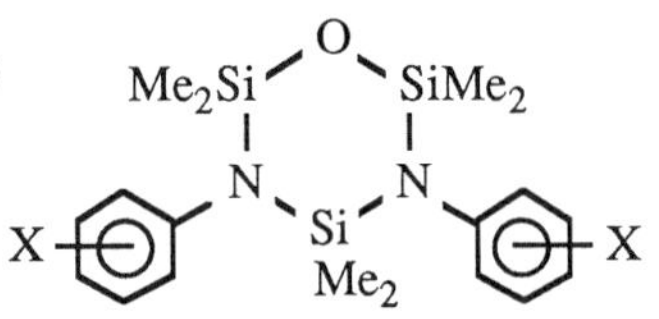

X = H; m,p-Me; m,p-OMe; m,p-Cl; m,p-Br X = p-Me; p-OMe; m-Cl; m,p-Br

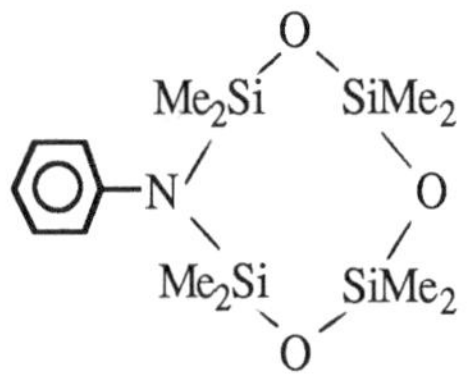

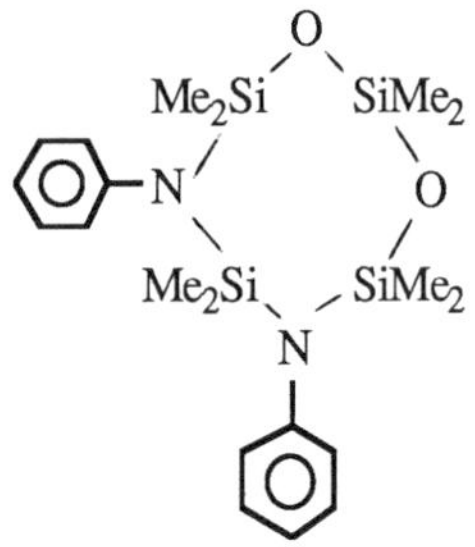

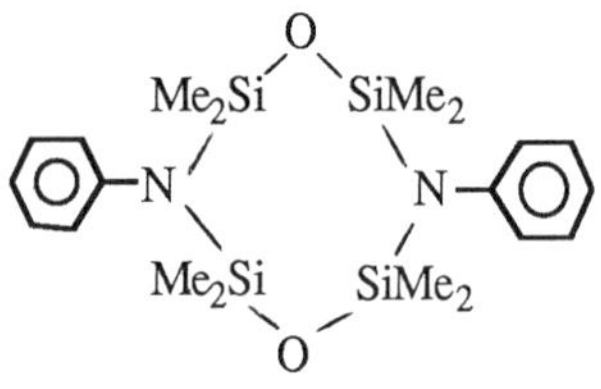

(IIc) (IId) (IIe)

For the preparation of **IIa** and **IIb** see (20). Compounds **IIc-e** were prepared in reasonably good yields from aniline and 1,7-dichlorooctamethyltetrasiloxane (**IIc**), bis-(phenylamino)-dimethylsilane and 1,5-dichlorohexamethyltrisiloxane (**IId**), and aniline and 1,3-dichlorotetramethyldisiloxane (**IIe**). All syntheses were carried out in DMF and in the presence of stoichiometric amounts of triethylamine as hydrogen chloride acceptor (21).

Rings **IIa, b** and **c** were found to polymerize in bulk with DMSO as an activator and various catalytic amounts of the anionic initiator, potassium poly(dimethylsiloxane)diolate (for polymerization of **IIa** and **b**, see (22)). The latter is well known as an efficient initiator of cyclosiloxane polymerization. Thus initiation is believed to be an elementary reaction of ring cleavage at the siloxane bond. The polymerization is a slow process. After 20 to 60 hours at 120 to 160°C in vacuo or under nitrogen and after chain stopping with an excess of chlorotrimethylsilane, liquid polymers of molecular weights ($\bar{M}_n$, osmometric) ranging from 1,000 to 6,500 were formed in 50 to 80% yield. The linear silazoxane structure of the polymers was established by spectroscopic (IR, NMR) methods (22).

The polymerization reached an equilibrium of chains with rings and, surprisingly, a crystalline fraction (up to 40% (w/w)) was always present in the mixture. The solid was identified as the six-membered cyclodiazasiloxane **IIb** in all experiments, including those of polymerization of rings containing a single silazane unit (**IIa** and **c**). Polymerization of eight-membered **IIc** yielded additionally some **IId**. These observations are summarized in the scheme below:

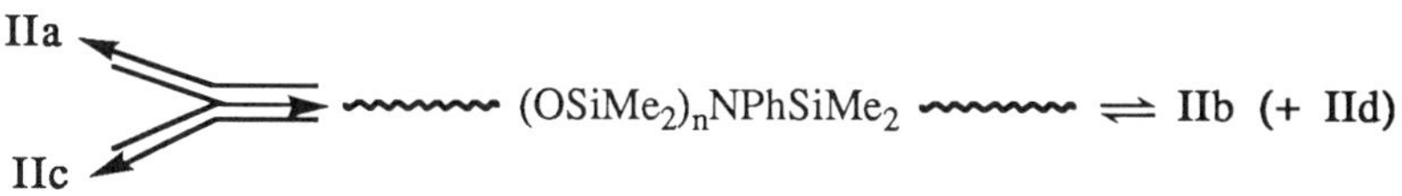

$$\text{IIa},\ \text{IIc} \rightleftharpoons \sim\!\!\sim (OSiMe_2)_nNPhSiMe_2 \sim\!\!\sim \rightleftharpoons \text{IIb (+ IId)}$$

The mechanism of polymerization of **IIa**, taking account of the formation of **IIb**, can be rationalized in the following way:

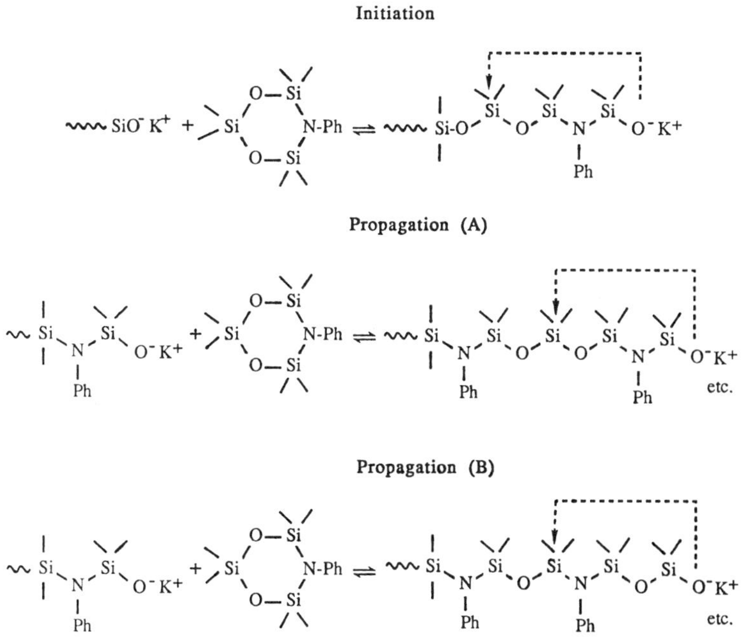

Initiation

$$\sim\!\!\sim SiO^- K^+ + \text{(IIa)} \rightleftharpoons \sim\!\!\sim Si\text{-}O\text{-}Si\text{-}O\text{-}Si\text{-}N(Ph)\text{-}Si\text{-}O^- K^+$$

Propagation (A)

$$\sim Si\text{-}N(Ph)\text{-}Si\text{-}O^- K^+ + \text{(IIa)} \rightleftharpoons \sim\!\!\sim Si\text{-}N(Ph)\text{-}Si\text{-}O\text{-}Si\text{-}O\text{-}Si\text{-}N(Ph)\text{-}Si\text{-}O^- K^+ \quad \text{etc.}$$

Propagation (B)

$$\sim Si\text{-}N(Ph)\text{-}Si\text{-}O^- K^+ + \text{(IIa)} \rightleftharpoons \sim\!\!\sim Si\text{-}N(Ph)\text{-}Si\text{-}O\text{-}Si\text{-}N(Ph)\text{-}Si\text{-}O\text{-}Si\text{-}O^- K^+ \quad \text{etc.}$$

Chain Transfer

(IIb)

Propagation (A) will result in the expected ordered sequence of siloxane and silazane units in the polymer chain. However, nucleophilic attack by the silanolate anion (probably ion pair) on the ring silicon adjacent to nitrogen (propagation (B)) will yield a chain containing some silazane units separated by a single siloxane unit. Chain transfer at the site shown in the scheme will produce an unstable amide anion, which, by a fast back-biting reaction, will yield the cyclic diazasiloxane, **IIb**.

The anionic mechanism of polymerization of cyclosilazoxanes is also supported by the fact that compounds **IIa** and **IIc** are able to copolymerize with cyclosiloxanes in similar conditions to their polymerization. Copolymerization of **IIa** with hexamethylcyclotrisiloxane (D_3) at 130°C *in vacuo* yielded polymers of reasonably high molecular weight. However, on increasing the comonomer ratio **IIa**:D_3 from 0.8:1 to 2:1, the molecular weight was found to decrease from 80,000 to 16,000, the highest ratio of comonomer units found in the chain being 1.25:1 (22). This effect is also reflected by the fall of the intrinsic viscosity of polymer solutions in toluene upon increasing the comonomer ratio. The linear structure of the polymers containing both siloxane and silazane units in their chains was well established by spectroscopic methods, e.g. the ^{1}H NMR spectrum contains only the peaks attributable to the phenyl groups and the two different positions of the methyl-silyl protons, the latter giving approximately the expected integration (22). Copolymerizations of eight-membered **IIc** with octamethylcyclotetrasiloxane yielded polymers whose $\bar{M}_n$ never exceeded 7,200 (21).

From these results it is evident that there is a definite limiting number of silazane units that can be introduced into a siloxane chain by copolymerization of silazoxane and siloxane rings. The six-membered and eight-membered cyclosilazoxanes containing two silazane bonds (**IIb, d** and **e**) do not copolymerize with cyclosiloxanes by the anionic process, although **IIb** is able to polymerize alone. The ultimate member of the six-membered ring series, N-triphenylhexamethylcyclotrisilazane (23) neither homopolymerizes nor copolymerizes with cyclosiloxanes.

The silazoxane copolymers are viscous liquids similar in all respects to silicone oils. Surprisingly, their thermal stability was found to be much higher. Polydimethylsiloxane chains depolymerize completely at ca. 350°C leaving practically no residue. In the case of N-phenylpolysilazoxanes the highest rate of weight loss is at 550°C, as revealed by a sharp peak on the DTG line for a copolymer of **IIa** with D_3 (1:1 molar ratio, $\bar{M}_n = 70{,}000$). At 800°C there remains up to 40% (w/w) of a solid residue which does not change on further heating and is probably composed of silica, silicon carbide and silicon nitride (22). Therefore, the mechanism of chain splitting must be different from that of polysiloxane.

However, it follows from isothermal kinetic studies of degradation of these type of polymers, that random scission of the polymer chain followed by complete depolymerization of the kinetically active fragments, a mechanism established by M. Zeldin et al. for trimethylsilyl end-blocked polydimethylsiloxanes (24), is also valid here to quite a large extent of conversion (25). Volatile compounds are produced with 5-10% of the initial copolymer weight remaining as a solid glassy residue (25). Thus there is some controversy between the conclusions concerning the close similarity of thermal degradation mechanisms on the one hand and the experimental facts which suggest fundamental differences in the thermal behavior of polysiloxane and polysilazoxane chains, on the other. The reasons for the obviously higher thermostability of N-phenylsilazoxane linear polymers are not clear at the moment and the problem requires further studies.

Substituting phenyl groups on nitrogen in all cyclosilazoxanes (**IIa-e**) was intended to increase the resistance to hydrolysis of the monomers and polymers by decreasing the basicity of their molecules. A kinetic study of methanolysis of cyclosilazoxanes **IIa** and **b** has shown that the reaction is inhibited by bases. Methanolysis in slightly acidic acetate buffer solution proceeds fairly fast (26,27). Pseudo-first-order rate constants were measured using the UV technique and partial catalytic rate constants for all the acidic species present in the solution (methoxonium ion, undissociated acetic acid and methanol molecules) were determined. It was found that general acid catalysis operates in this reaction. This fact, together with other kinetic tests of mechanism such as substituent effects on the benzene ring, activation parameters and solvent kinetic isotope effects suggest a mechanism involving concerted proton transfer to the substrate and nucleophilic attack on silicon by a solvent molecule (26,27). Since proton transfer to electronegative elements such as oxygen or nitrogen is usually fast and diffusion controlled, it follows that in the present case the nitrogen basicity must be very low as a result of extensive (p-d) π conjugation with both adjoining silicon atoms coupled with the electron withdrawing effect of the phenyl group.

Siloxane-cyclodisilazane Block Copolymers (29)

Polycondensation of difunctional oligomeric cyclodisilazanes with α,ω-dihydroxypolysiloxanes proved to be another successful attempt to synthesize silazane modified polysiloxanes of improved thermostability. The cyclodisilazane oligomers **IIIa** (see scheme below) are readily available from common and cheap materials, dichlorodimethylsilane and ammonia (6). The chlorine atoms in these compounds can be easily exchanged to other functional groups, such as amino, hydroxyl, etc.

W. Fink (13) and K.A. Andrianov (28) and others reported a series of attempts to combine the cyclodisilazane monomers and oligomers into a polymeric system with siloxanes. In most of these syntheses some ring cleavage and crosslinking of the linear block copolymer occurred due to the susceptibility of the disilazane ring to nucleophilic attack by the functional groups of the second component of the reaction mixture. The only fully successful attempt seems to have been the polycondensation of the lithium salts of hydroxycyclosilazanes or hydroxysiloxanes with chlorofunctional comonomers (13).

In the present work it has been found that the side reactions resulting in cleavage of the disilazane ring can be largely avoided if comparatively long polysiloxanediol chain molecules are made to react with the oligomeric dichlorocyclodisilazanes where the hydrogen chloride formed is carefully and continuously neutralized by an amine. The idealized polycondensation of cyclodisilazane oligomer **IIIa**, composed of an average number of 8.36 rings, with commercial α,ω-dihydroxypolydimethylsiloxanes of 100, 150 and 200 siloxane units in the average chain is represented below, as well as the structure of the cyclodisilazane-siloxane linear block copolymer which is formed. The reaction was carried out in benzene at 80°C for 20 hr and followed by evaporating the solvent and heating the residue at 200°C/0.2 torr for 2 hr. The product is a wax-like material completely soluble in benzene, toluene and some other solvents. The intrinsic viscosity of toluene solutions is increased on passing from the initial polysiloxanes to the copolymers; i.e., from $[\eta] = 0.113 - 0.159$ to $[\eta] = 0.225 - 0.325$. The IR spectrum of the copolymer contains all the bands attributable to the silazane ring. It can be concluded that the polycyclodisilazane blocks have been preserved and the copolymer has a linear structure. On the other hand, crosslinking and gel formation occurred when equivalent amounts of oligomeric dimethylsiloxane-α,ω-diols were taken as components; e.g. on polycondensation of **IIIa** with 1,7-dihydroxyoctamethyltetrasiloxane rubber-like, insoluble material was formed.

$Cl\text{-}SiMe_2\text{-}[N(SiMe_2)_2N\text{-}SiMe_2]_x\text{-}Cl + HO\text{-}SiMe_2\text{-}O \sim [SiMe_2\text{-}O]_n \sim SiMe_2\text{-}OH$

(IIIa)

$\xrightarrow[C_6H_6]{Et_3N}$ (-Et_3NHCl)

$\bar{x} = 8.36$

$\bar{n} = 100,150,200$

$HO\text{-}SiMe_2\text{-}O \sim SiMe_2\text{-}O\text{-}SiMe_2\text{-}[N(SiMe_2)_2N\text{-}SiMe_2]_x\text{-}O \sim Si\text{-}O\text{-}Si\text{-}[N(SiMe_2)_2N\text{-}SiMe_2]_x\text{-}O \sim SiMe_2 \sim$

The copolymers have a remarkably improved thermostability with respect to the polysiloxanediols. While the polysiloxane homopolymers undergo 15% weight loss at 350-360°C, as determined by the thermogravimetric experiments, the same percent weight loss is observed at 430-510°C for various samples of cyclosilazoxane-siloxane block copolymers obtained from equivalent proportions of the reagents. However, increased stability by ca. 50°C (using the same test) was also evident for a sample prepared with 3% (w/w) of cyclosilazane oligomer, which corresponds to 25% of the equivalent amount. Thus, copolymerization of polysiloxane diols with dichlorocyclodisilazanes is an efficient method of modifying polysiloxanes to achieve considerably higher thermostability of the material.

Polysilazanes with Aromatic Spacing Groups (30)

The susceptibility of cyclodisilazanes to nucleophilic attack by aromatic amines has also been used to prepare silazane containing polymers. Polysilazane cyclo-linear chains with aromatic spacing groups, synthesized by polycondensations of difunctional cyclodisilazanes with bis-phenols and N,N′-diorganosilane diamines, have been reported (13).

In the present work a series of N,N′-diphenyltetramethylcyclodisilazanes with substituents on the benzene ring were prepared from bis(phenylamino)silanes (23) (for spectroscopic data, see (31)); and, the parent compound **IIIb** was treated with aniline and phenylene-diamines in the melt at 200-300°C.

IIIb with aniline yields bis(phenylamino)tetramethyldisilazane **IVb** (see scheme below) in a reversible reaction whose equilibrium is shifted to the right with decreasing temperature.

With m-phenylenediamine, the unsymmetrical linear disilazane **IVb′** is formed which undergoes recyclization to the unsymmetrical cyclodisilazane **IIIb′** The following set of equilibria is thus established.

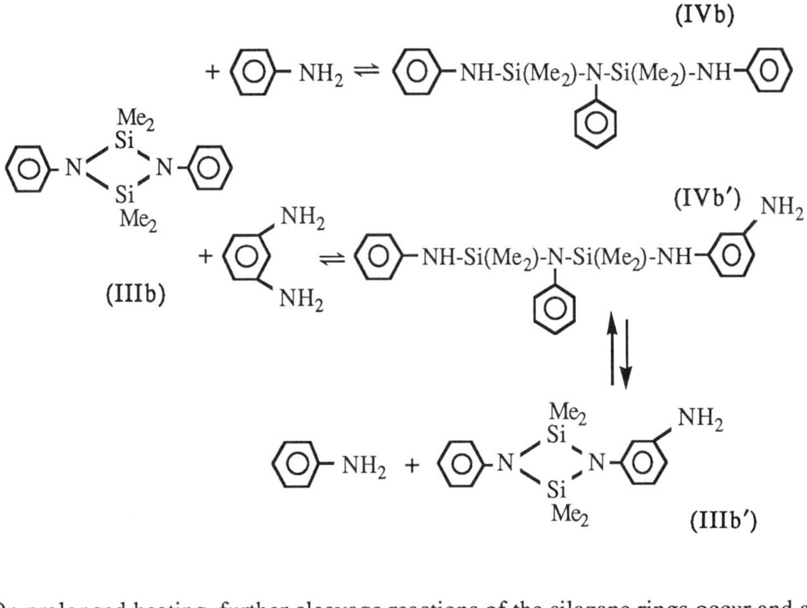

(IVb)

(IIIb) + $C_6H_5NH_2$ ⇌ C_6H_5-NH-Si(Me_2)-N(C_6H_5)-Si(Me_2)-NH-C_6H_5

(IVb′)

(IIIb) + m-$C_6H_4(NH_2)_2$ ⇌ C_6H_5-NH-Si(Me_2)-N(C_6H_5)-Si(Me_2)-NH-C_6H_4-NH_2

⇅

$C_6H_5NH_2$ + C_6H_5-N<($SiMe_2$)$_2$>N-C_6H_4-NH_2 (IIIb′)

On prolonged heating, further cleavage reactions of the silazane rings occur and a fraction, consisting of linear oligomers of the idealized structure shown below, was collected as a resinous residue after evaporating the lower boiling material. The molecular weight of the oligomeric fraction is ca. 1000 and the structure is established by IR and NMR spectroscopy and elemental analysis.

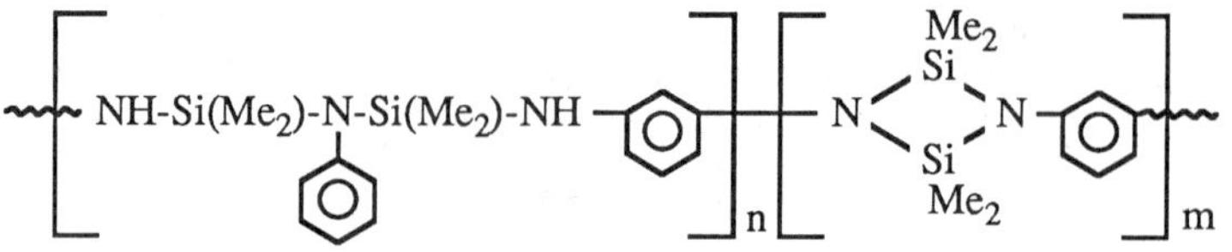

~ NH-Si(Me2)-N-Si(Me2)-NH]n [N (Si Me2)2 N]m ~

Similar results were obtained when o-phenylenediamine was reacted with **IIIb**. A range of low boiling compounds were separated from the reaction mixture after 10-12 hr heating at 220°C. These have been identified as diazadimethylsilaindane **Vb** (m.p. 76-78°C; lit. (32) m.p. 75-83°C) and bis(phenylamino)dimethylsilane **VIb** (m.p. 46-48°C; lit. (33) m.p, 45 ± 2°C). They are probably formed by thermal decomposition of the intermediate unsymmetrical disilazane **IVb"**. At 190-225°C/0.3 torr, a crystalline fraction with

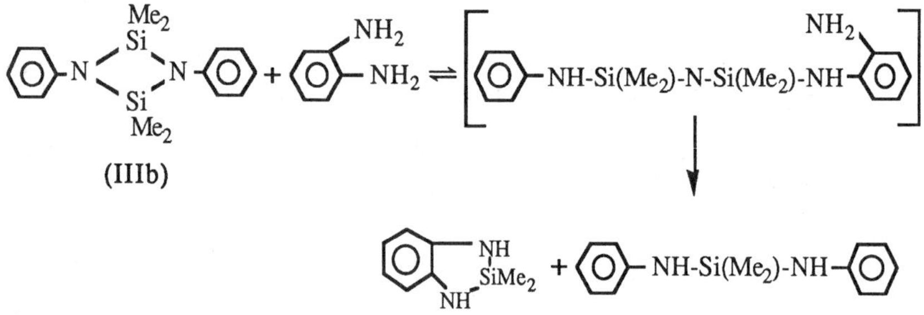

Me2 Si N N Si Me2 + NH2 NH2 ⇌ [NH-Si(Me2)-N-Si(Me2)-NH NH2]

(IIIb)

NH SiMe2 NH + NH-Si(Me2)-NH

(Vb) (VIb)

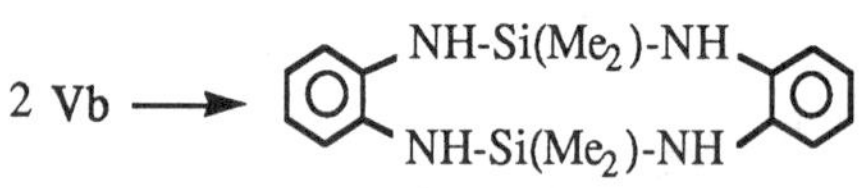

2 Vb ⟶ NH-Si(Me2)-NH NH-Si(Me2)-NH (VIIb)

VIIb + VIb ⟶ Me2 Si N N Si Me2 NH NH Si Me2 (VIIIb)

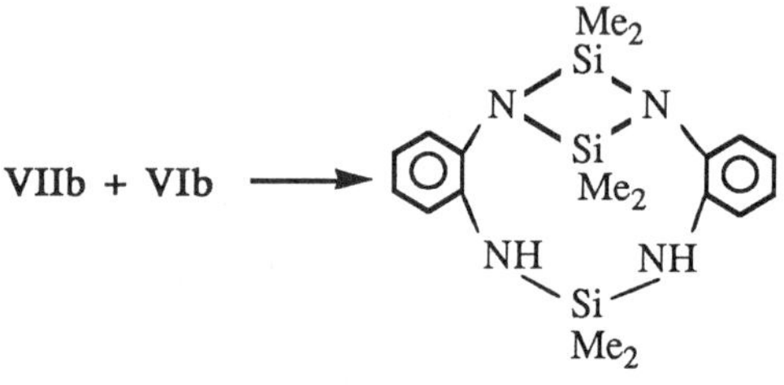

m.p. 209-211°C was collected. The analytical data for this compound suggest structure **VIIIb**, (for $C_{18}H_{28}N_4Si_3$: $M_{calc.}$ 384.7, m/e 384.7; calc %: C, 56.19; H, 7.34; N, 14.56; Si, 21.90; found %: C, 56.22; H, 7.56; N, 15.19; Si, 21.79; IR (υ, cm^{-1}): 3381 s (NH), 970, 890 s (Si-N-Si); ^1H NMR (δ, ppm, $SiMe_4$ int. standard) 0.51 d (18 H) $SiCH_3$, 3.82 s (2 H) NH, 6.38-6.75 m (8 H) H_{Ar}.

VIIIb is formed by dimerization of the diazasilaindane, **Vb**, and subsequent reaction of the cyclic dimer, **VIIb**, with bis(phenylamino)-dimethylsilane, **VIb**, as it was shown in separate experiments starting from **Vb** and **VIb**. It was subsequently found that **VIIIb** is formed in fairly good yield whenever o-phenylenediamine is brought into reaction with difunctional silanes, such as dichlorodimethylsilane, bis(diethylamino)dimethylsilane and hexamethyl-cyclotrisilazane.

Some negligible amounts of linear oligomers, obtained as a non-volatile, solid brown residue, were also formed on reaction of **IIIb** with o-phenylene-diamine; these were almost completely soluble in toluene. No high molecular fraction was detected among the reaction products.

When **IIIb** is heated with p-phenylenediamine at 200-300°C under conditions similar to the other phenylenediamines, mainly insoluble, probably crosslinked, oligomers (rubber-like solid) are formed. On the other hand, p-phenylenediamine with bis(diethylamino)dimethylsilane at 180°C yielded a toluene-soluble solid whose molecular weight and elemental analysis agree well with the probable cyclophane structure **IX**; for $C_{32}H_{48}N_8Si_4$: $M_{calc.}$ 657.16, found $M_{cryoscopic}$ 648, m/e 657.16; calc. %: C, 58.49; H, 7.36; N, 17.05; Si, 17.10; found %: C, 58.40; H, 7.42; N, 17.10; Si, 17.16. The structure of this compound is being further investigated.

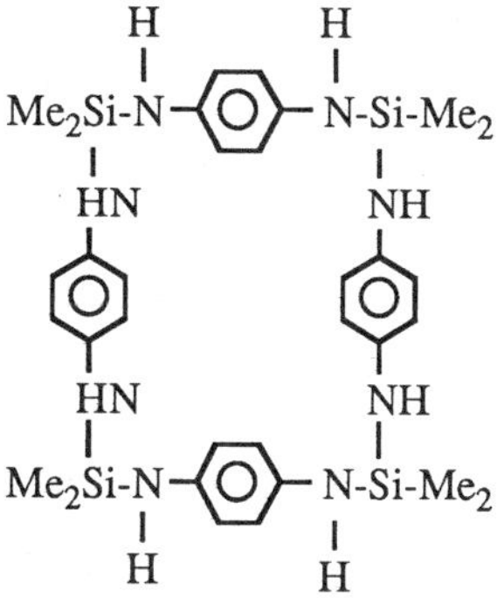

H H
Me_2Si-N–C₆H₄–N-Si-Me_2
HN NH
HN NH
Me_2Si-N–C₆H₄–N-Si-Me_2
H H

(IX)

Thus, the tendency of silazanes to form ring structures, including macro-rings, is extraordinary.

Conclusion

Cyclosilazanes are found to be reluctant to polymerize by the ring-opening process, probably for thermodynamic reasons. On the other hand, six- and eight-membered silazoxane rings are able to undergo anionic polymerization under similar conditions to those which have been widely used for cyclosiloxane polymerization provided there is no more than two silazane units in the cyclic monomer. They can also copolymerize with cyclosiloxanes; however, the chain length of the linear polymer formed is substantially decreased with increasing proportion of silazane units.

The thermostability of siloxane-silazane copolymers of both random and block structure is found to be much higher (i.e. 100-200°C) with respect to polysiloxanes. This effect is brought about by introducing only a few silazane entities into the polymer chain. The reasons for the effect are not clear and the mechanism of thermal degradation of polysilazoxanes will require further experimental studies.

Acknowledgments

This work was supported in part by the Polish Academy of Sciences, Programs CPBP 03-13 and 03-14. I am sincerely grateful to Ms. Sharon Fricke (IUPUI) for typing and producing the manuscript.

Literature Cited

1. Dejak, B.; Lasocki, Z. J. Organometal. Chem. 1972, 44, C39.
2. Klebe, J. F. J. Amer. Chem. Soc. 1968, 90, 5246.
3. Boer, F. P.; vanRemoertere, F. P. J. Amer. Chem. Soc. 1970, 92, 801.
4. Klebe, J. F. Acc. Chem. Res. 1970, 3, 299.
5. Dejak, B.; Lasocki, Z. J. Organometal. Chem. 1983, 246, 151.
6. Fink, W. Helv. Chim. Acta 1968, 51, 1011.
7. West, R.; Fink, M. J.; Michl, J. Science 1981, 214, 1343.
8. Gilette, G. and West, R. presented at the 19th Organosilicon Symposium, Lousiana State University, Baton Rouge, LA, April 26-27, 1985.
9. Dejak, B.; Lasocki, Z.; Jancke, H. Bull. Pol. Acad. Sci. Chem. 1985, 33, 275.
10. Dejak, B.; Lasocki, Z. unpublished results.
11. Engelhardt, G.; Magi, G.; Lipmaa, E. J. Organometal. Chem. 1973, 54, 115.
12. Aylett, B. J. Organomet. Chem. Rev. 1968, 3, 151.
13. Fink, W. J. Paint. Technol. 1970, 42, 220.
14. VanWazer, J. R.; Moedritzer, K. J. Chem. Phys. 1964, 41, 3122.
15. Moedritzer, K. Organomet. Chem. Rev. 1966, 1, 179.
16. Redl, G.; Rochow. E.G. Angew Chem. 1964, 76, 650.
17. Matsumoto, M.; Niwata, K.; Tanaka, S. presented at the 7th International Symposium on Organosilicon Chemistry, Gunma Symposium, September, 1984.
18. Kirichenko, E. A.; Ermakov, A. J.; Samsonova, I. N. Zh. Fiz. Khim. U.S.S.R. 1977, 51, 2506.
19. May, L. A.; Rumba, G. Ya. Latv. P.S.R. Zinat. Akad. Vest., Khim. Ser., 1964, 1, 23.
20. Lasocki, Z.; Witehowa, M., Syn. React. Inorg. Metal-Org. Chem. 1974, 4, 231.
21. Lasocki, Z.; Witekowa, M. unpublished results.
22. Lasocki, Z.; Witekowa, M. J. Macromol. Sci. Chem. 1977, All, 457.
23. Kulpinski, J.; Lasocki, Z.; Piechucki, S. Bull. Pol. Acad. Sci. Chem. (in press).
24. Zeldin, M.; Quian, B.; Choi, S. J. J. Polym. Sci., Polym. Chem. Ed. 1983, 21, 1361.
25. Kang, D. W.; Rajendran, D. P.; Zeldin, M. J. Polym. Sci., Polym. Chem. Ed. 1986, 24, 1085.
26. Lasocki, Z.; Witekowa, M. J. Organometal. Chem. 1984, 264, 49.
27. Lasocki, Z.; Witekowa, M. J. Organometal. Chem. 1986, 311, 17.
28. Andrianov, K. A.; Emelianov, U. N. Usp. Khim. 1977, 66, 2066.
29. Kulpinski, J.; Lasocki, Z.; Lesniak, E. unpublished results.
30. Kulpinski, J.; Lasocki, Z. unpublished results.
31. Albanov, A. I.; Voronkov, M. G.; Dorokhova, V. V.; Kulpinski, J.; Larin, M. F.; Lasocki, Z.; Piechucki, S.; Brodskaia, E. I.; Pestunovich, V. A. Izv. Akad. Nauk U.S.S.R. 1982, 1781.
32. Wieber, M.; Schmidt M. Z. f. Naturforschung 1963, 18b, 849.
33. Anderson, H. H. J. Amer. Chem. Soc. 1951, 73, 5802.

RECEIVED October 13, 1987

POLYSILOXANES

Chapter 14

Recent Advances in Organosiloxane Copolymers

J. D. Summers, C. S. Elsbernd, P. M. Sormani[1], P. J. A. Brandt[2], C. A. Arnold, I. Yilgor[3], J. S. Riffle, S. Kilic, and J. E. McGrath[4]

Department of Chemistry and Polymer Materials and Interfaces Laboratory, Virginia Polytechnic Institute and State University, Blacksburg, VA 24061

Organosiloxane copolymers have been of great interest over the past several years due to the somewhat unique characteristics imparted by the siloxane blocks. The excellent UV and oxidative resistance of these structures together with their outstanding thermal stability and wide temperature use range make them candidates as modifiers for a multitude of applications. Organosiloxane materials have now been prepared which demonstrate a variety of interesting properties such as atomic oxygen resistance, biocompatibility, high gas permeabilities, and hydrophobic surfaces. In this paper, the synthesis and properties of functional polysiloxane oligomers which are the precursors for the copolymers are discussed. Additionally, a summary and review of the preparation and characteristics of the organosiloxane copolymers is given.

The general theme of our research focuses on multiphase copolymer systems obtained via living polymers, engineering polymers (both thermoplastics and thermosets) and polysiloxane systems. Organosiloxane copolymers have been of interest for a number of years due to the unusual characteristics of polysiloxanes, such as their thermal and UV stability, low glass transition temperature, very high gas permeability and, especially, their low surface energy characteristics (1-2). In the polydimethylsiloxane chain, 1, many workers have demonstrated that the methyl groups are oriented toward the air or vacuum surface. This provides a hydrophobic, low surface energy characteristic, which has a number of secondary but important applications including enhanced biocompatibility and more recently, resistance to atomic oxygen and oxygen plasmas. Our current work on

[1]Current address: Experimental Station, E. I. du Pont de Nemours and Company, Wilmington, DE 19898
[2]Current address: Specialty Chemical Division, 3M Center, St. Paul, MN 55144
[3]Current address: Mercor, Inc., Berkeley, CA 94710
[4]Correspondence should be addressed to this author.

0097-6156/88/0360-0180$06.00/0

siloxane chemistry includes the preparation of functional oligomers both by equilibration processes employing the commonly available cyclic tetramer, D_4, and to some extent, living polymerizations which utilize the lithium siloxanolate initiated polymerization of the cyclic trimer, D_3 (1-4). The novel oligomers are then incorporated into high performance block, segmented and graft copolymers.

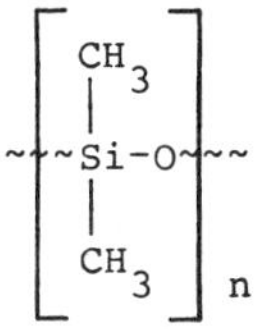

$$\sim\!\!\left[\sim \mathrm{Si(CH_3)_2{-}O} \sim\right]_n\!\!\sim \qquad \mathbf{1}$$

Functional Polysiloxane Oligomers

For equilibration processes, one must synthesize both oligomers and what are termed dimers, or disiloxanes. Our primary interest is in the utilization of these functional oligomers for the synthesis of both linear block or segmented copolymers, and also surface modified, oughened networks such as the epoxy and imide systems (3-27). The generalized structure of the oligomers of interest is shown in Scheme 1.

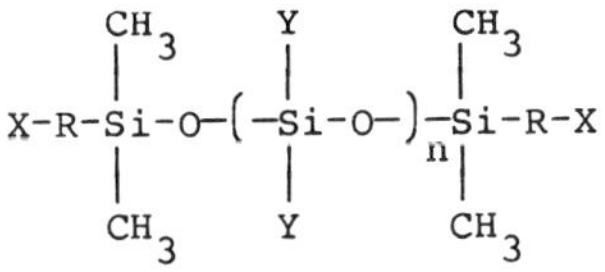

$$\mathrm{X{-}R{-}Si(CH_3)_2{-}O{-}(-Si(Y)_2{-}O-)_n{-}Si(CH_3)_2{-}R{-}X} \qquad \text{Variables: X, R, Y, n}$$

Scheme 1. General Structure of an α,ω-Difunctional Polysiloxane.

The overall synthesis of functionally terminated oligomers involves equilibration of the cyclic tetramer in the presence of a functional disiloxane as illustrated in Scheme 2. In this case, one utilizes a catalyst which can be either an acidic or basic moiety. A basic catalyst (for example) attacks the silicon of the tetramer and initiates the ring-opening polymerization. However, in the presence of the disiloxane, the polymerization undergoes what is effectively a very useful chain transfer reaction. This occurs because, like the monomer and growing chain, the silicon-oxygen-silicon bond of the disiloxane is sufficiently polar that it is attacked by the active ionic species and, thus, takes part in the equilibration process. On the other hand, one takes advantage of the fact that the silicon-alkyl (or aryl) bond is more covalent in nature and, thus, stable during the polymerization. The equilibration process is generalized in Scheme 3 wherein "D_4" refers to the cyclic, tetrameric starting material and "MM" denotes a disiloxane with a terminal group associated with each "M". This shorthand nomenclature system is one traditionally utilized by siloxane chemists. As one can note from Scheme 3, the dimer will be incorporated into the chain and the chain ends will be identical with the terminal units of "MM" or, as we have called it, the disiloxane. This entire concept is well-known and, in

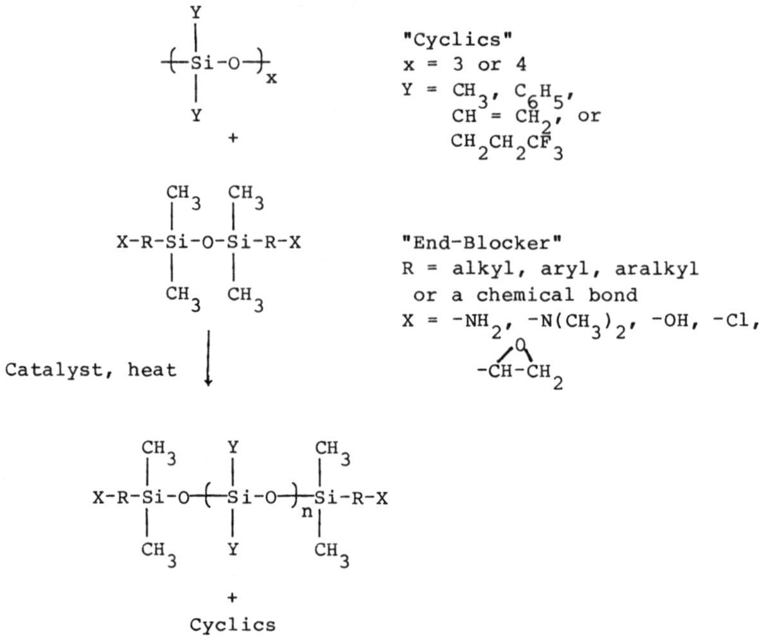

$$\left(\begin{array}{c} Y \\ | \\ Si\text{-}O \\ | \\ Y \end{array}\right)_x$$

"Cyclics"
x = 3 or 4
Y = CH_3, C_6H_5, $CH = CH_2$, or $CH_2CH_2CF_3$

+

$$X\text{-}R\text{-}Si(CH_3)_2\text{-}O\text{-}Si(CH_3)_2\text{-}R\text{-}X$$

"End-Blocker"
R = alkyl, aryl, aralkyl or a chemical bond
X = $-NH_2$, $-N(CH_3)_2$, -OH, -Cl, $-CH\text{-}CH_2$ (epoxide, with O bridging)

Catalyst, heat ↓

$$X\text{-}R\text{-}Si(CH_3)_2\text{-}O\left(Si(Y)_2\text{-}O\right)_n Si(CH_3)_2\text{-}R\text{-}X$$

+
Cyclics

Scheme 2. General Synthesis of "End-Blocked" Polysiloxanes via Equilibration Processes.

$$\sim\sim\sim D_x^{(-)} + D_4 \longrightarrow \sim\sim\sim D_{(x+4)}^{(-)}$$

$$\sim\sim\sim D_x^{(-)} + MM \longrightarrow MD_xM$$

$$\sim\sim\sim MD_xM + MD_yM \xrightarrow{\text{cat.}} MD_{(x+w)}M + MD_{(y-w)}M$$

Scheme 3. Equilibration Polymerization Processes (Shown utilizing base catalysis).

fact, methyl terminated polydimethylsiloxane fluids and oils have been prepared in this way for some time. However, the additional feature which we have focused on is to design the endgroups to be organofunctional.

Many oligomers with various functionalities have been prepared in our laboratory (4). However, the amino terminated species has been studied most extensively due to its wide utility as a component of a large number of segmented copolymers (e.g. imides, amides, ureas, etc.). In order to prepare functional oligomers of this type, one must first prepare the disiloxane. One route to this was pointed out some years ago by Saam and Speier (28). They showed that it was possible to react allylamine with a protecting reagent such as hexamethyldisilazane to produce the protected amine material, plus ammonia which could be removed by distillation. The protected allyamine group undergoes hydrosilylation with dimethylchlorosilane in the presence of various platinum catalysts (such as chloroplatinic acid) to produce mostly the chlorosilane intermediate. Purification of this material by vacuum distillation is desirable. In order to prepare the disiloxane, one may hydrolyze the chlorosilane to produce first a silanol, which readily undergoes dehydration to produce a stable siloxane bond. At the same time, water is able to easily hydrolyze the labile silicon-nitrogen bond to regenerate the desired primary amine group. Many other methods have been reported for the preparation of analogous disiloxanes. For example, unsaturated nitriles can be hydrosilylated and subsequently reduced to produce an amine group. Alternatively, allylamine can also be protected via imine formation. The general area is of great interest now since it has been demonstrated that the available linear siloxane segmented copolymers have a variety of interesting properties. A number of other reactive endgroups have been prepared such as secondary amines, aromatic amines, silylamines, phenolic hydroxyls, epoxides, alkenes, and silanes (4). In addition to the dimethylsiloxane repeat unit in the backbone, it is possible to produce a number of other important structures, such as methyl-vinyl, diphenyl, trifluoropropyl-methyl, and methyl-silane. All of these have been demonstrated to be useful for various reasons in our laboratory and elsewhere.

The synthetic process for oligomeric species is to react the aminopropyl-functional disiloxane with the cyclic tetramer as illustrated in Scheme 4.

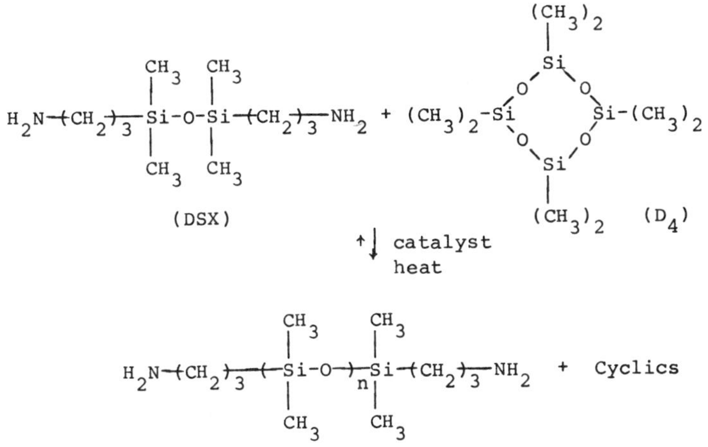

$$H_2N{-}(CH_2)_3{-}Si(CH_3)_2{-}O{-}Si(CH_3)_2{-}(CH_2)_3{-}NH_2 \quad (DSX) \quad + \quad [(CH_3)_2SiO]_4 \quad (D_4)$$

$$\xrightarrow[\text{heat}]{\text{catalyst}}$$

$$H_2N{-}(CH_2)_3{-}(Si(CH_3)_2{-}O)_n{-}Si(CH_3)_2{-}(CH_2)_3{-}NH_2 \; + \; \text{Cyclics}$$

- ⊙ Cyclics are removed via vacuum stripping
- ⊙ Catalyst must be neutralized for maximum stability
- ⊙ $\bar{M}n$ is controlled at equilibrium via initial $[D_4]/[DSX]$ ratio
- ⊙ Other endgroups possible with appropriate DSX and catalyst

Scheme 4. Synthesis of Difunctional Bis-(Aminopropyl) Polydimethylsiloxane Via An Equilibration Polymerization Process.

Here one may choose a suitable catalyst, typically a potassium or tetramethylammonium siloxanolate. The process then goes through a series of equilibration steps (as discussed earlier) to produce the ring-chain equilibrium between the linear oligomer and the cyclic species. Cyclic species are predominantly the tetrameric cyclic, which can be removed by vacuum distillation. In the case of the preferred quaternary ammonium catalysts, one may decompose the catalyst by simply heating the reaction mixture briefly to about 150°C. This will deactivate the catalyst, as indeed was pointed out a number of years ago by Gilbert and Kantor (29). Effectively, it is possible to obtain a stable, linear, functionalized oligomer free of cyclics. This can be verified via chromatography (e.g. GPC and/or HPLC). In the case of a potassium siloxanolate catalyst, one must neutralize the equilibrated product (for example, with ion exchange systems) to achieve maximum stability. We have already demonstrated that it is possible to synthesize a variety of molecular weights. The number average molecular weight at equilibrium is governed by the initial molar ratio of the cyclic tetramer to the disiloxane. In general, the DSX incorporates more readily using the quaternary ammonium catalyst.

Some characteristics of aminopropyl terminated polydimethylsiloxane oligomers are illustrated in Scheme 5.

$\bar{M}n$ (g/mole)	n	Tg°C
600	6	-115
1000	11	-118
1800	22	-121
2400	30	-123
3800	50	-123

$$H_2N\text{-}(CH_2)_3\text{-}(\underset{CH_3}{\overset{CH_3}{Si}}\text{-}O)_n\text{-}\underset{CH_3}{\overset{CH_3}{Si}}\text{-}(CH_2)_3\text{-}NH_2$$

Scheme 5. Characteristics of Aminopropyl Terminated Polydimethylsiloxane Oligomers Synthesized by a Base Catalyzed Equilibration Process.

The number average molecular weights of the oligomers can be obtained via titration of the amine endgroups. They can also be estimated by ^{1}H-NMR up to reasonably high molecular weights, by comparing the integrated ratio of the methylene protons at the end of the chain to the silicon methyls in the backbone. The titration values for the endgroups and the number average molecular weight determined by proton NMR agree quite nicely in the case of the polydimethylsiloxane system described bearing α,ω-aminopropyl groups. Hydroxyalkyl terminated oligomers are not so simple. Problems can occur if the hydroxyl group interacts with the growing species to produce a silicon-oxygen-carbon bond. In this case, one may find effectively lower -OH concentrations than anticipated. In fact, it may be necessary to also block the hydroxyl group during the equilibration.

Organosiloxane Copolymers

During the past few years (4, 15-27, 30-36), we have prepared block or segmented copolymers from a variety of hard segments. In particular, we have focused on polyureas, polyurethanes, polyamides and polyimides. In addition, we have utilized these oligomers to modify epoxy networks. Thus, the equilibration procedure works extremely well from a synthetic point of view. Although relatively little (1,7,37) has been reported with respect to the detailed kinetics and mechanisms involved in these overall processes, we have begun to explore this important aspect and some of our results have been recently reported (38,39). In this paper, a review of the utilization of polysiloxane oligomers for the preparation of soluble, high performance segmented polyimide-siloxane copolymers and various elastomeric copolymers will be included.

Much of the early effort on siloxane systems has been summarized in reference 9. In the synthesis of organosiloxane copolymers it should be noted that one has the choice of introducing either a silicon-oxygen-carbon link or a silicon-carbon link between the two dissimilar segments. In small molecules, quite distinctly different hydrolytic stabilities are observed. For example, one often uses silicon-oxygen-carbon bonds as protecting groups which are stable under anhydrous neutral or basic conditions, but which quickly revert when treated with aqueous acidic environments. The benefit of the silicon-carbon bond is that it is significantly more hydrolytically stable than the silicon-oxygen-carbon bond. However, Noshay, Matzner and coworkers (9) showed that the silicon-oxygen-carbon bond in hydrophobic high molecular weight copolymers was much more stable than would have been predicted from small molecule considerations. Apparently the hydrophobic environment protects the potentially hydrolyzable group.

The chemistry for synthesizing functional polysiloxane oligomers appropriate for producing block copolymers with silicon-carbon links between the blocks has been reviewed earlier in this monograph. Representative structures have been depicted in Scheme 1. One generally prepares the Si-C linked functionalities by conducting hydrosilylation reactions between Si-H groups and the corresponding unsaturated, functional endgroup. In some cases, as previously described, the endgroup must either first be protected or subsequently further reacted after hydrosilylation, in order to finally produce the desired functionality. Polysiloxanes, **2**, appropriate for producing Si-O-C units between the blocks in copolymers are shown below. The copolymerization reaction involves

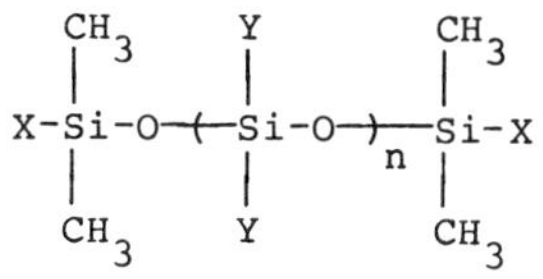

$$X\text{-}Si(CH_3)_2\text{-}O\text{-}\left(Si(Y)_2\text{-}O\right)_n\text{-}Si(CH_3)_2\text{-}X \qquad \mathbf{2}$$

$X = N(CH_3)_2$, OCH_2CH_3, Cl, $O\text{-}C(=O)\text{-}CH_3$

$Y = CH_3$, C_6H_5, $CH{=}CH_2$, $CH_2CH_2CF_3$

attack of a nucleophile on the terminal silicon atom of the siloxane with either concerted or subsequent elimination of the functional group "X".

Noshay and coworkers developed an interesting route for the preparation of perfectly alternating siloxane poly(arylene ether sulfone) copolymers via reaction of hydroxyl groups with silylamines ($X = N(CH_3)_2$ in 2) which is reviewed in reference 9. A variety of different hard segments were incorporated into the resulting copolymers. These workers observed that, in fact, the processibility of these copolymers was very dependent upon the differential solubility parameter between the hard segment and the non-polar polydimethylsiloxane block (14). In some cases, the microphase separation was apparently so well developed that the materials developed an extremely high viscosity in the melt, rendering them essentially non-processible. This phenomenon occurred even though the materials were linear, and indeed, solvent-castable from a variety of appropriate organic liquids. Although this viscosity effect is general with microphase separated systems, it is particularly pronounced for the organosiloxane materials (14). Despite some of the melt fabrication problems, the organosiloxane systems produced by the silylamine-hydroxyl reaction (9) produced interesting, perfectly alternating copolymers.

Polycarbonate (15-18) and aromatic polyester (19,20) copolymers produced using this same general route have been investigated in our laboratories. The chemistry is illustrated in Scheme 6. The perfectly alternating copolymers developed very uniform morphology (17) reminiscent of the triblock styrene-diene materials. Typical stress-strain and dynamic mechanical behavior is shown in Figures 1 and 2. By contrast, if the copolymers were prepared by either a randomly coupled route or via an <u>in-situ</u> generation of the hard block (16,17), the morphology was not regular and, indeed, the physical properties of both elastomeric and rigid compositions were quite different from those of the perfectly alternating systems. The aromatic polyester/siloxane copolymers synthesized in our laboratories included a series of siloxanes which were themselves comprised of "blocky" structures. Both diphenylsiloxane sequences as well as trifluoropropyl-methylsiloxane units were combined with dimethylsiloxane units. These copolymers are of considerable interest as multiphase damping materials (19) due to the fact that two, three or even more relaxations can be designed into these copolymers as a function of block length and composition. Indeed, significant loss peaks ranging from -120 to well over 200°C have been achieved (19,20).

Recent work has focused on a variety of thermoplastic elastomers and modified thermoplastic polyimides based on the aminopropyl end functionality present in suitably equilibrated polydimethylsiloxanes. Characteristic of these are the urea linked materials described in references 22-25. The chemistry is summarized in Scheme 7. A characteristic stress-strain curve and dynamic mechanical behavior for the urea linked systems in provided in Figures 3 and 4. It was of interest to note that the ultimate properties of the soluble, processible, urea linked copolymers were equivalent to some of the best silica reinforced, chemically crosslinked, silicone rubber materials described in the literature. Various other fundamentally oriented studies which describe the kinetics and mechanisms of the synthesis of these oligomers are provided in the references herein. A number of other hard segments such as the imides and amides have also been investigated. One was able to show a good correlation

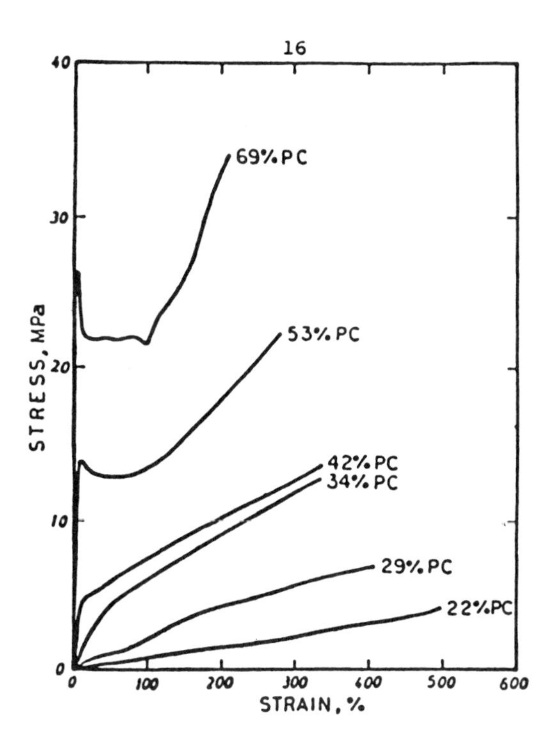

Figure 1. Stress-strain curves of polycarbonate-polydimethylsiloxane block copolymers (Crosshead Speed: 5 cm/min). (Reproduced from Refs. 15, 18. Copyright 1980, 1984 American Chemical Society.)

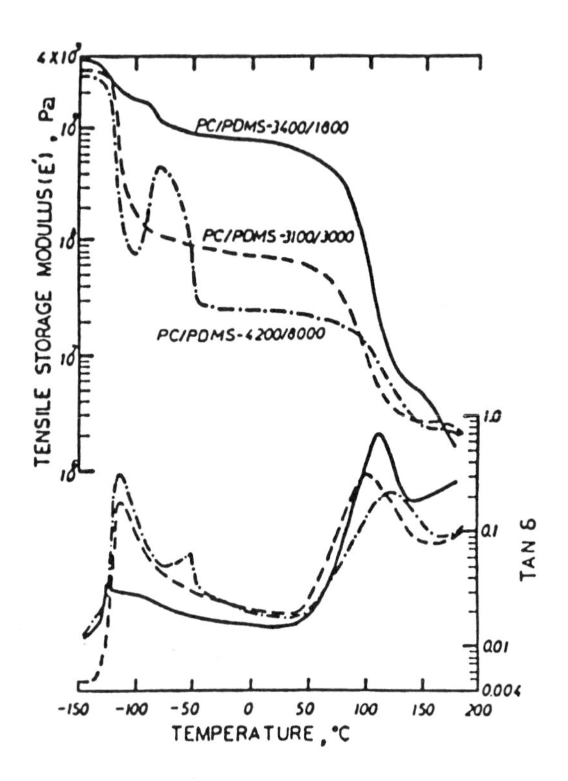

Figure 2. Storage moduli and loss tangents versus temperature for three typical block copolymers: Frequency: 11 Hz; $\bar{M}_n$ of blocks expressed in g/mole. (Reproduced from Refs. 15, 18. Copyright 1980, 1984 American Chemical Society.)

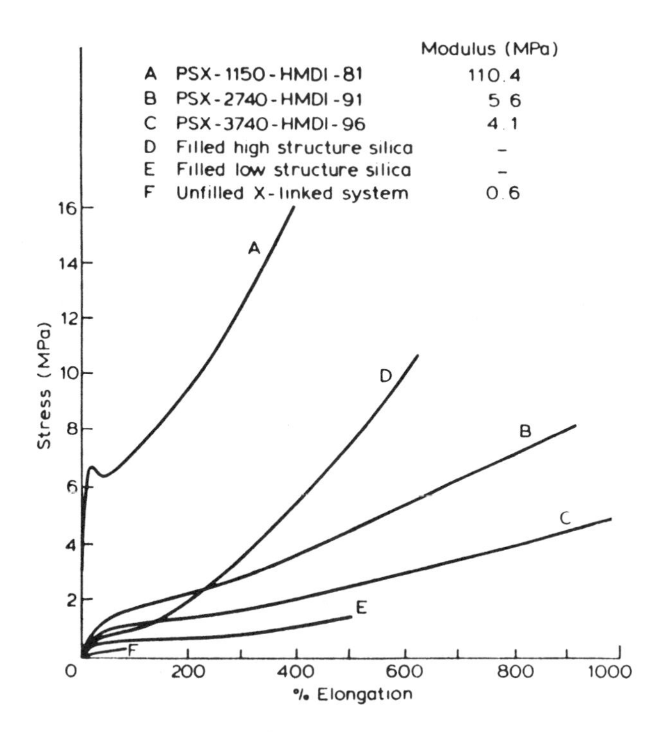

Figure 3. Stress versus % elongation behavior for siloxane-urea segmented copolymers from H-MDI as a function of molecular weight of oligomer used and the hard segment content at 25 °C. (Reproduced with permission from Ref. 25. Copyright 1984 IPC Business Press, Ltd.)

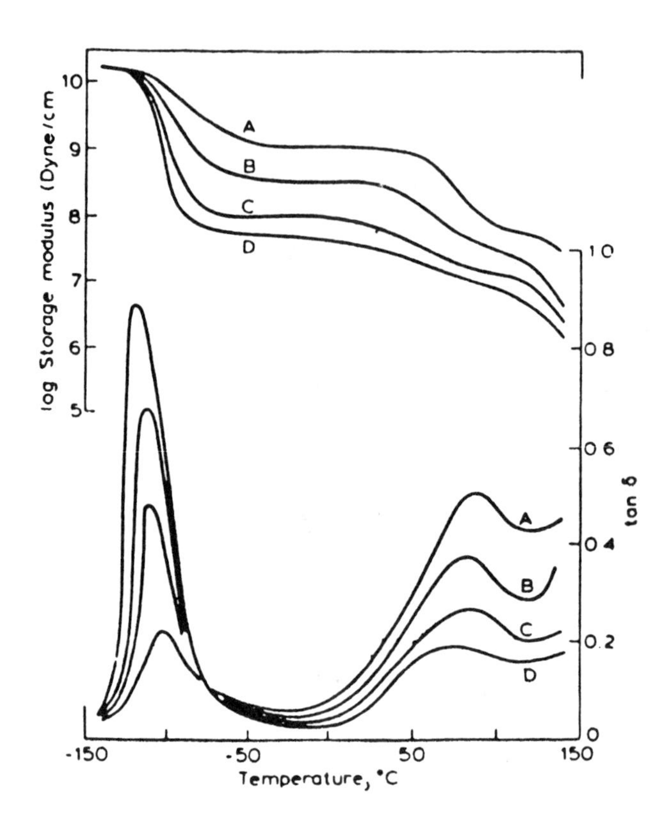

Figure 4. Dynamic mechanical behavior of siloxane-urea copolymers prepared from H-MDI. Curve A: PSX-1 50-HMDI-81; curve B: PSX-1770-HMDI-87; curve C: PSX-770-HMDI-91; curve D: PSX-3680-HMDI-94. (Reproduced with permission from Ref. 25. Copyright 1984 IPC Business Press, Ltd.)

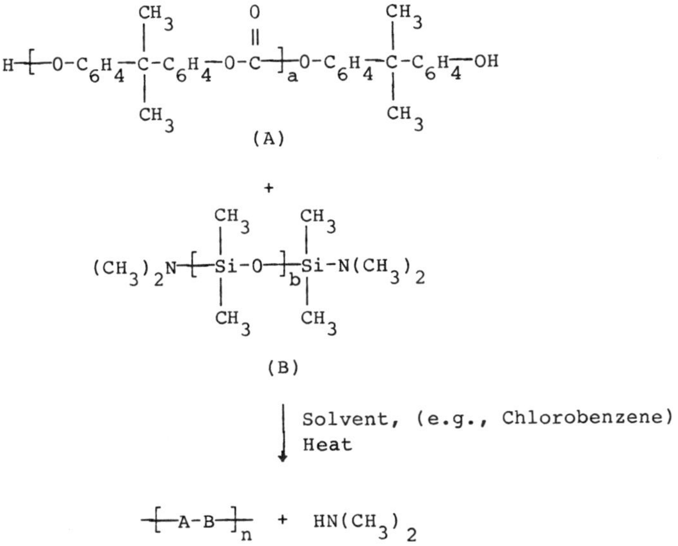

$$H{+}O{-}C_6H_4{-}\underset{CH_3}{\overset{CH_3}{C}}{-}C_6H_4{-}O{-}\overset{O}{\overset{\|}{C}}{+}_a O{-}C_6H_4{-}\underset{CH_3}{\overset{CH_3}{C}}{-}C_6H_4{-}OH$$

(A)

+

$$(CH_3)_2N{+}\underset{CH_3}{\overset{CH_3}{Si}}{-}O{+}_b\underset{CH_3}{\overset{CH_3}{Si}}{-}N(CH_3)_2$$

(B)

↓ Solvent, (e.g., Chlorobenzene)
Heat

$${+}A{-}B{+}_n \; + \; HN(CH_3)_2$$

Scheme 6. Perfectly Alternating Polycarbonate-Polydimethylsiloxane Block Copolymers.

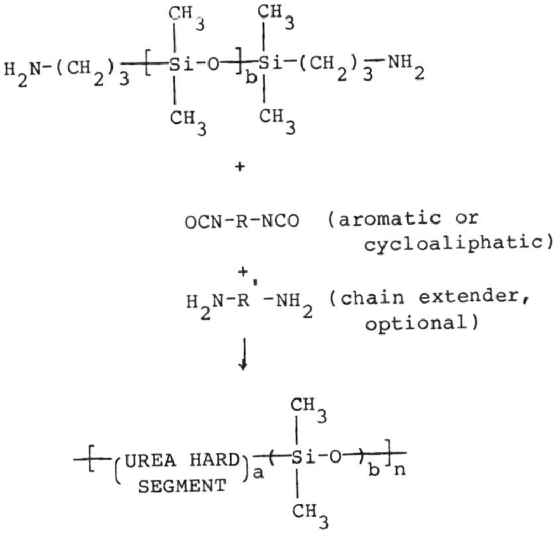

$$H_2N{-}(CH_2)_3{+}\underset{CH_3}{\overset{CH_3}{Si}}{-}O{+}_b\underset{CH_3}{\overset{CH_3}{Si}}{-}(CH_2)_3{-}NH_2$$

+

OCN-R-NCO (aromatic or cycloaliphatic)

+

$H_2N{-}R'{-}NH_2$ (chain extender, optional)

↓

$${+}(\text{UREA HARD SEGMENT})_a{(}\underset{CH_3}{\overset{CH_3}{Si}}{-}O{)}_b{+}_n$$

Scheme 7.

between the intensity of the hydrogen bond formation and the ultimate mechanical properties at room temperature. In this regard, the urea linked systems were preferred. On the other hand, thermal stability was best for the polyimide systems.

Brief discussion of the novel surface properties in these elastomers is appropriate. The bulk microphase separation characteristics of thermoplastic elastomers has been studied in great detail and is very important. However, less information is known about the surface structure in these two-phase materials. A number of investigators have been interested in these areas. Some of the first work, described in reference 26, focused on the (at that time) somewhat unique siloxane enhancement at the air/vacuum surface in pure copolymers and even in homopolymer-block copolymer blends. Essentially, the polydimethylsiloxane displays very low surface energy which provides a thermodynamic driving force for migration to the air or vacuum interface. In contrast to the homopolymer, the siloxane segment is chemically linked and cannot macroscopically phase separate. X-ray photoelectron spectroscopy studies (XPS or ESCA) conducted demonstrated that the surface structure was predominantly siloxane, even when the bulk siloxane compositions were relatively low. An important criterion, though, was the development of microphase structure in the bulk, which apparently frees the siloxane microphase to more easily migrate to the air or vacuum interface. Many other studies of this phenomenon have appeared in the literature since that time. In the cases of the siloxane modified polyimides, discussed briefly below, it is of great interest in both the area of oxygen plasma processing and atomic oxygen resistance in outer space applications. In the presence of atomic oxygen, apparently the siloxane structure on the surface is converted to an organosilicate-type, ceramic-like material which provides protection during the etching process. Further discussion of this phenomenon is provided in references 32 and 40-43.

Although most of the multi-block or segmented thermoplastic elastomer studies have utilized the readily available cyclic tetramer (D_4) structure as a starting monomer (4), there has also been interest in materials derived from the cyclic trimer (D_3). The cyclic trimer has the advantage that one can produce predictable molecular weights with narrow distributions, due to the ability for the cyclic trimer to produce living lithium siloxanolate chain ends. This has been described by a number of authors (9). Recently, polymers based upon t-butylstyrene and D_3 (44) have been studied. These materials, again, rely somewhat upon the lower solubility parameter associated with the t-butylstyrene group relative to styrene. As a result, the subsequent block polymers that were prepared process easily and may lead to attractive specialty elastomeric and rigid copolymers.

As mentioned earlier, siloxanes impart a number of beneficial properties to polymeric systems into which they are incorporated, including enhanced solubility, resistance to degradation in aggressive oxygen environments, impact resistance and modified surface properties. These particular advantages render polysiloxane-modified polyimides attractive for aerospace, microelectronic and other high performance applications (40-43).

The thermal-mechanical properties of siloxane-modified polyimides are a function of the weight fraction of incorporated siloxane, molecular weight of the siloxane blocks and chemical architecture of both the siloxane and imide segments. Thus, copolymers with high concentrations of incorporated siloxane (>50%), where the siloxane is the continuous phase, behave as thermoplastic elastomers, whereas lower siloxane concentrations result in more rigid materials which behave essentially as modified polyimides. At high polyimide compositions, the upper glass transition value and most mechanical properties approach those of the unmodified controls. However, the surface structure can still be strongly dominated by the low surface energy polysiloxane microphase.

The great difference in solubility parameters of the siloxane and imide segments is a driving force for microphase separation, particularly when higher molecular weight siloxane oligomers are incorporated. Additionally, because the siloxane component possesses a relatively low surface energy, it will migrate to the air or vacuum interface, yielding a surface dominated by the siloxane component. This effect is observed even for low (~5 weight %) levels of siloxane incorporation. In aggressive oxygen environments, the surface siloxane segments convert to a ceramic-like silicate (SiO_2) which provides a protective outer coat to the bulk material. The conversion of polysiloxane to silicon dioxide in oxygen plasma has been documented within the literature of the electronics industry as well as in our laboratory.

The copolymers investigated were largely based upon 3,3',4,4'-benzophenone tetracarboxylic dianhydride (BTDA), the meta-substituted diamine 3,3'-diaminodiphenyl sulfone (DDS) and aminopropyl-terminated polydimethylsiloxane oligomers of various molecular weights in the range of 950 to 10,000 grams per mole. The incorporated siloxane oligomer has been varied from 5 to 70 weight percent. Conversion of the poly(amic acid) to the fully imidized polyimide was accomplished by two different techniques. The first, and most common, route was via bulk thermal imidization with the loss of water. In this case, temperatures near the glass transition temperature of the fully imidized product must be employed in order to achieve complete imidization. In the second method, imidization was accomplished at lower temperatures in the range of 150 to 170°C by means of a solution imidization procedure (42,43). Although in both cases, tough, transparent, flexible and soluble films were obtained. The choice of imidization method affects material properties. The solution imidized materials are markedly more soluble than the bulk imidized systems, for example, but their thermal and mechanical properties do not vary significantly. The polysiloxane-amic acid intermediates were prepared as previously described (42,43) then converted to the fully imidized segmented copolymers by either the bulk or solution imidization route. The bulk-thermal imidization technique involved casting the amic acid solutions onto a substrate, then removing solvent in a vacuum oven at 100°C for several hours. The films were then thermally cycled at 100, 200 and 300°C for an hour at each temperature in a forced air convection oven to complete the cyclization. The solution imidization procedure has been further described by Summers, et. al. (42,43). A representative structure for the poly(imide siloxane) copolymers is depicted in Figure 5.

Figure 5. Structural Representation of Poly(Imide-Siloxane) Segmented Copolymers (PSX = Polydimethylsiloxane).

One of the major goals of this endeavor was to solubilize the normally intractable polyimides by the incorporation of siloxane segments, and, optionally, by solution imidization. Solubilities of a series of siloxane-modified polyimide copolymers were evaluated in a variety of solvents as indicated in Table I. Copolymer solubility was found to be a function of the siloxane oligomer concentration

Table I. Solubilities of Bulk (B) and Solution (S) Imidized Poly(Imide Siloxane) Segmented Copolymers

	Control		20% PSX		40% PSX		60% PSX	
Solvent	B	S	B	S	B	S	B	S
N-methylprrolidinone (NMP)	I	S	M	S	M	S	S	S
Dimethylacetamide (DMAc)	I	S	M	S	M	S	S	S
Tetrahydrofuran (THF)	I	I	I	I	I	S	S	S
Methylene Chloride (CH_2Cl_2)	I	I	I	I	I	S	S	S

Key: S = Soluble; M = Marginally soluble; I = Insoluble

as well as the method of imidization. The solution imidized copolymers and also the solution imidized control (no siloxane) were all soluble in dipolar, aprotic solvents such as N-methylpyrollidinone (NMP) and dimethylacetamide (DMAc). The solubility of the solution imidized copolymers and control in NMP or DMAc ranged from ≈10 weight percent for the control to ≈20 weight percent for the copolymers. At 40 weight percent siloxane, the solution imidized copolymers were soluble in a variety of alternative solvents, including tetrahydrofuran, methylene chloride and diglyme. The bulk imidized materials were less soluble than those which were solution imidized, requiring at least 10 weight percent siloxane at solids concentrations of less than 5 percent to achieve solubility in NMP and DMAc. The bulk imidized control was not soluble in dipolar, aprotic solvents. Intrinsic viscosities of the copolymers ranged from 0.50 to 0.85 dl/g, indicating that high molecular weight copolymers were obtained by both imidization techniques over the entire composition and segment molecular weight range.

Values of the upper glass transition temperatures of the siloxane modified polyimides were found to be a function of both the level of incorporated siloxane as well as the siloxane molecular weight (Table II). The upper transition temperature of the solution

Table II. Upper Glass Transitions of Poly(Imide Siloxane) Segmented Copolymers

WT. % PSX	PSX $\bar{M}n$	IMIDIZATION METHOD	$[\eta]$ (NMP, 25°C)	T_g (°C)
CONTROL	-----	BULK	----	272
CONTROL	-----	SOLN	1.36	265
10	900	BULK	0.62	256
10	900	SOLN	0.63	251
10	2100	BULK	0.78	261
10	2100	SOLN	0.73	260
10	5000	BULK	0.71	264
10	10000	BULK	0.73	266
20	900	BULK	0.78	246
20	900	SOLN	0.67	240
20	2100	BULK	0.60	258
20	2100	SOLN	0.57	252
20	5000	BULK	0.51	262
40	900	BULK	0.55	225
40	900	SOLN	0.58	218

imidized materials was found to be only slightly depressed from its bulk imidized analogue. Generally, the upper transition temperature increased with greater siloxane oligomer molecular weight and with decreasing siloxane incorporation. In many cases, the copolymers' upper transition temperature was depressed only slightly relative to that of the control, indirectly indicating that good microphase separation was achieved. The lower temperature siloxane transition is difficult to detect by DSC for low levels of siloxane incorporation, i.e., <20 weight percent. At greater levels of incorporation, however, the transition is detected by both DSC and DMTA within the range -117 to -123°C, representing 20 to 50 weight percent siloxane, respectively. Initial transmission electron micrographs of the solution imidized copolymers indicate siloxane microphase formation for a 20 weight percent siloxane copolymer with a 2000 molecular weight siloxane segment. The domain size is relatively small, approximately 30 Å, and phase contrast is somewhat poor.

In summary, it has been possible to prepare a variety of organosiloxane copolymers which demonstrate useful mechanical properties but also provide a variety of other interesting properties such as hydrophobic character, high gas permeability, thermal and UV stability, atomic and oxygen plasma resistance, and biocompatibility. The synthesis and characterization of these materials is continuing.

Acknowledgments

Portions of this research were supported by the ARO, AFOSR, DARPA and NASA and this is gratefully acknowledged.

Literature Cited

1. Wright, P. V. In Ring Opening Polymerization Vol. 2; Ivin K.; Saegusa T., Eds.; Elsevier, New York, 1984; pp 1055-1123.
2. Noll, W. The Chemistry of Silicones; Academic Press: New York, 1968.
3. McGrath, J. E. In Ring Opening Polymerization: Kinetics, Mechanisms and Synthesis; McGrath, J. E., Ed.; ACS Symposium Series, No. 286; American Chemical Society: Washington, DC, 1985; pp 1-22.
4. Yilgor, I.; McGrath, J. E. Advances in Organosiloxane Copolymers; Springer-Verlag: Heidelberg, in press.
5. McGrath, J. E. Ring Opening Polymerization: Kinetics, Mechanisms and Synthesis; ACS Symposium Series No. 286; American Chemical Society: Washington, DC, 1985.
6. Yilgor, I.; Riffle, J. S.; McGrath, J. E. In Reactive Oligomers; Harris, P., Ed.; ACS Symposium Series, No. 282; American Chemical Society: Washington, DC, 1985; pp 161-174.
7. Sormani, P. M.; Minton R. J.; McGrath, J. E. In Ring Opening Polymerization: Kinetics, Mechanisms and Synthesis; McGrath, J. E., Ed., ACS Symposium Series, No. 286; American Chemical Society: Washington, DC, 1985; pp 147-160.
8. McGrath, J. E.; Riffle, J. S.; Yilgor, I.; Banthia, A. K.; Wilkes, G. L. In Initiation of Polymerization, ACS Symposium Series No. 212; American Chemical Society: Washington, DC, 1983; pp 145-173.
9. Noshay, A.; McGrath, J. E. Block Copolymers: Overview and Critical Survey; Academic Press: New York, 1977.
10. McGrath, J. E.; Matzner, M.; Noshay, A.; Robeson, L. M. In Encyclopedia of Polymer Science and Technology; Supplement 2; Wiley: New York, 1977; pp 129-158.
11. McGrath, J. E. In Block Copolymers; Meier, D. J., Ed.; MMI Press: Midland, MI, 1983; pp 1-12.
12. Matzner, M.; Noshay, A.; Robeson, L. M.; Merriam, C. N.; Barclay, R., Jr.; McGrath, J. E. Appl. Polymer Symposia 973, 22, 143-156.
13. McGrath, J. E. J. Chem. Ed., 1981, 58, 914-921.
14. Matzner, M.; Noshay, A.; McGrath, J. E. Trans. Soc. Rheology 1977, 21/22, 273-290.
15. Tang, S.; Meinecke, E.; Riffle, J. S.; McGrath, J. E. Rubber Chem. and Tech. 1980, 54(5), 1160.
16. Riffle, J. S.; Freelin, R. G.; Banthia, A. K.; McGrath, J. E. J. Macromol. Sci.-Chem. 1981 A15(5), 967-998.

17. Ward, T. C.; Sheehy, D. P.; Riffle, J. S.; McGrath, J. E. Macromolecules 1981, 14(6), 1791-1797.
18. Tang, S.; Meinecke, E.; Riffle, J. S.; McGrath, J. E. Rub. Chem. and Tech. 1984, 57(1), 184.
19. Andolino-Brandt, P. J.; McGrath, J. E. Proc. 30th National SAMPE Mtg , 1985, pp 959-971.
20. Andolino-Brandt, P. J.; Elsbernd, C. S.; McGrath, J. E. J. Polym. Sci. 1987, in press.
21. McGrath, J. E. Pure and Applied Chemistry 1983, 55(10), 1573-85.
22. Yilgor, I.; Riffle, J. S.; Wilkes, G. L.; McGrath, J. E. Polymer Bulletin 1982, 8, 535-542.
23. Tyagi, D.; Yilgor, I.; Wilkes, G. L.; McGrath, J. E. Polymer Bulletin 1982, 8, 543-550.
24. Yilgor, I.; Steckle, W. P., Jr.; Tyagi, D.; Wilkes, G. L.; and McGrath, J. E. Polymer, London 1984, 25(12), 1800-1806.
25. Tyagi, D.; Yilgor, I.; McGrath, J. E.; Wilkes, G. L. Polymer, London 1984, 25(12), 1806-1814.
26. Dwight, D. W.; Beck, A.; Riffle, J. S.; McGrath, J. E. Polym. Prepr. 1979, 20(1), 702-706; Macromolecules 1987, in press.
27. Johnson, B. C.; Summers, J. D.; McGrath, J. E. J. Polym. Sci. 1987, in press.
28. Saam, J. C.; Speier, J. L. J. Org. Chem. 1959, 24, 119.
29. Gilbert, A. R.; Kantor, S. W. J. Polym. Sci. 1959, 40, 35.
30. Yilgor, I.; Yilgor, E.; Eberle, J.; Steckle, W. P., Jr.; Johnson, B. C.; Tyagi, D.; Wilkes, G. L.; McGrath, J. E. Polym. Prepr. 1983, 24(1), 167-170.
31. Yilgor, I.; Yilgor, E.; Johnson, B. C.; Eberle, J.; Wilkes, G. L.; McGrath, J. E. Polym. Prepr. 1983, 24(2), 78-80.
32. Johnson, B. C. Ph.D. Thesis, VPI&SU, Blacksburg, VA, 1984.
33. Yilgor, I.; Lee, B.; Steckle, W. P., Jr.; Riffle, J. S.; Tyagi, D.; Wilkes, G. L.; McGrath, J. E. Polym. Prepr. 1983, 24(2), 35-37.
34. Johnson, B. C.; Yilgor, I.; McGrath, J. E. Polym. Prepr. 1984, 25(2), 54-56.
35. Tran, C. Ph.D. Thesis, VPI&SU, Blacksburg, VA, 1985.
36. Yorkgitis, E. M.; Eiss, N. S., Jr.; Tran, C.; Wilkes, G. L.; McGrath, J. E. Adv. Polym. Sci. 1985, 72, 79.
37. Grubb, W. T.; Osthoff, R. C. JACS 1955, 77, 1405-1411.
38. Sormani, P. M. Ph.D. Thesis, VPI&SU, Blacksburg, VA, 1986.
39. McGrath, J. E.; Sormani, P. M.; Elsbernd, C. S.; Kilic, S. Makromol. Chem. 1987, in press.
40. Johnson, B. C.; Yilgor, I.; McGrath, J. E. Polym. Prepr. 1984, 25(2), 54-56;
41. Johnson, B. C.; Yilgor, I.; McGrath, J. E. J. Polymer Sci. 1987, in press.
42. Summers, J. D.; Arnold, C. A.; Bott, R. H.; Taylor, L. T.; Ward, T. C.; McGrath, J. E. Polym. Prepr. 1986, 27(2), 403.
43. Summers, J. D. Ph.D. Thesis, VPI&SU, Blacksburg, VA, 1987.
44. Smith, S. D. Ph.D. Thesis, VPI&SU, Blacksburg, VA, 1987.

RECEIVED October 23, 1987

Chapter 15

Polysiloxanes Functionalized with 3-(1-Oxypyridinyl) Groups

Catalysts for Transacylation Reactions of Carboxylic and Phosphoric Acid Derivatives

Martel Zeldin, Wilmer K. Fife, Cheng-xiang Tian, and Jian-min Xu

Department of Chemistry, Indiana University–Purdue University at Indianapolis, Indianapolis, IN 46223

Poly(methyl 3-(1-oxypyridinyl)siloxane) was synthesized and shown to have catalytic activity in transacylation reactions of carboxylic and phosphoric acid derivatives. 3-(Methyldichlorosilyl)pyridine (**1**) was made by metallation of 3-bromopyridine with n-BuLi followed by reaction with excess $MeSiCl_3$. **1** was hydrolyzed in aqueous ammonia to give hydroxyl terminated poly(methyl 3-pyridinylsiloxane) (**2**) which was end-blocked to polymer **3** with $(Me_3Si)_2NH$ and Me_3SiCl. Polymer **3** was N-oxidized with m-$ClC_6H_4CO_3H$ to give **4**. Species **1-4** were characterized by IR and ^1H NMR spectra. MS of **1** and thermal analysis (DSC and TGA) of **2-4** are discussed. 3-(Trimethylsilyl)-pyridine 1-oxide (**6**), 1,3-dimethyl-1,3-bis-3-(1-oxypyridinyl) disiloxane (**7**) and **4** were effective catalysts for conversion of benzoyl chloride to benzoic anhydride in CH_2Cl_2/aqueous $NaHCO_3$ suspensions and for hydrolysis of diphenyl phosphorochloridate in aqueous $NaHCO_3$. The latter had a $t_{1/2}$ of less than 10 min at 23°C. These results were compared to those obtained with pyridine 1-oxide (**9**) and poly(4-vinylpyridine 1-oxide) (**8**) as catalysts.

The catalytic efficiency of enzymes has not, as yet, been matched by synthetic materials. Nature has endowed living organisms with unique macromolecules which, through precise composition, structure and solvent interactions, provide reactive centers that have the ability to control the rate, specificity and stereochemistry of an enormous variety of chemical transformations. Although scientists have been aware of this fact for some time, it is only recently with the rapid development of polymer chemistry that chemists have acquired the methodologies necessary to synthesize multifunctional long-chain molecules with catalytically active substituents. Moreover, there have been very few reports of efforts to complement catalytically active groups with binding domains and segments which can effect the structural and/or substrate affinity-solubility characteristics of the polymer. As with enzymes, it is the precise number, arrangement and physico-chemical behavior of the relevant active centers that need to be controlled and systematically modified to optimize reaction specificity, stereochemistry and rate. At the present time there are no functionalized synthetic polymers which combine all the above attributes or even approach enzyme performance characteristics, although polyethylenimines (1), cryptates (2) and

0097–6156/88/0360–0199$06.00/0

cyclodextrins (3), to some extent, offer potential. To our knowledge none of these has found utility in industrial organic chemistry.

Recent work by Fife et.al.(4) has shown that pyridine 1-oxide and poly[4-vinylpyridine 1-oxide] are effective nucleophilic catalysts for transacylation reactions of carboxylic and phosphoric acid derivatives. This class of reaction is among the most important and widely used in industrial and biochemistry. Fife has demonstrated that, in a phase-managed process, the catalysts cause the high yield ($\geq$ 80%) synthesis of symmetric and unsymmetric organic anhydrides in well-stirred organic solvent/water suspensions from an acid chloride in the organic phase and a carboxylate ion in the aqueous phase in as little as 10 min at ambient temperature. The phase-transfer process is depicted in Scheme 1 in which the catalyst is acylated in CH_2Cl_2 and the intermediate is carried into the water where reaction with a carboxylate ion generates the acid anhydride. Since the rate of anhydride diffusion into the organic phase is considerably faster than hydrolysis, cessation of agitation isolates phases and desired product. It has also been determined that there is selective transport of more hydrophobic carboxylate ions (e.g., p-toluate vs. isobutyrate) at the mixed solvent interface which suggests an interfacial component in the process. In order to examine more closely the question of hydrophobic binding, catalytic behavior, interfacial activity and transport facility, we have designed an organic soluble-water insoluble polymeric catalyst (viz. polysiloxanes) which can effect transacylation at the mixed solvent interface without the catalyst partitioning into the aqueous medium.

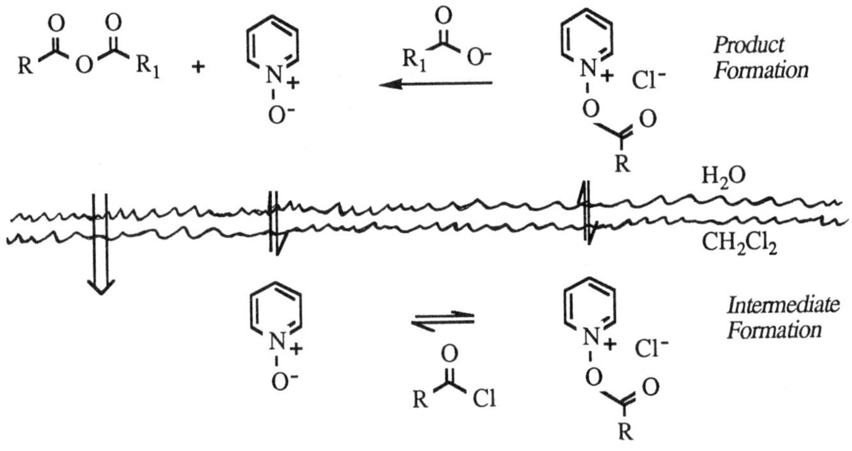

Scheme 1.

The feasibility of bonding pyridinyl groups to silicon which contains a hydrolytically sensitive functional group has recently been demonstrated (5-7). 2-Fluoro-3-(dimethylchlorosilyl)pyridine and 3-fluoro-4-(dimethylchlorosilyl)pyridine as well as 2-, 3-, and 4-(dimethylchlorosilyl)pyridine were prepared by the reaction of the corresponding lithiopyridines with excess Me_2SiCl_2. Hydrolysis of the pyridinyl substituted chlorosilanes gave disiloxanes which were insoluble in water. In the present report we will describe extension of this work to include pyridinyl dichlorosilanes which can be hydrolyzed to polysiloxanes. These polymers can be N-oxidized and the resultant derivatives have been shown to be effective hydrophobic transacylation catalysts.

Results and Discussion

Synthesis and Spectroscopic Characterization. 3-(Methyldichlorosilyl)-pyridine (1) was synthesized by the reaction of 3-lithiopyridine, prepared in situ from 3-bromopyridine and n-butyllithium, with a large excess of methyltrichlorosilane at -76° (Scheme 2). The product is a colorless distillable liquid which is soluble in aromatic, aliphatic and chlorinated hydrocarbons and extremely sensitive to moisture and protonic solvents.

The infrared spectrum of **1** exhibits a pair of intense absorptions at 505 cm^{-1} and 550 cm^{-1} characteristic of symmetric and asymmetric stretching vibrations, respectively, of the $SiCl_2$ group (8). Bands at 710(s)/760(vs) cm^{-1} and 455(m) cm^{-1} are consistent with r- and t-ring vibrations, respectively, associated with Si-C ring modes observed in phenylsilanes (9). The ^{1}H-NMR spectrum of **1** contains a singlet at 0.60 ppm assigned to Me_3Si protons and a complex pattern of peaks in the aromatic region corresponding to the pyridinyl ring protons (viz. δ 8.57, d/d, H^2; 8.33, d/d, H^6; 7.60 d/t, H^4 and 6.88, d/d/d, H^5). The integrated areas of the multiplets and comparison of the pattern with other 3-substituted pyridine derivatives are consistent with the assignments (10).

The mass spectrum of **1** exhibits polyisotopic ion clusters with relative abundances diagnostic of a dichloro derivative for the molecular ion $M^{+\bullet}$ (m/e: 195, 193, 191) and fragment ions $(M\text{-}CH_3)^+$ (m/e: 180, 178, 176) and $(M\text{-}Py)^+$ (m/e: 117, 115, 113). The appearance of Py^+ (m/e: 78) indicates a competition in fragmentation

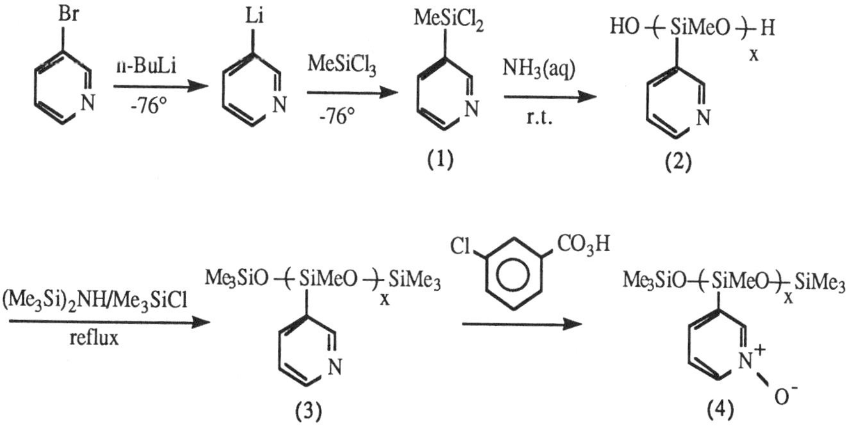

Br
n-BuLi
-76°
N
Li
MeSiCl3
-76°
N
MeSiCl2
NH3(aq)
r.t.
N
(1)
HO (SiMeO) H
x
N
(2)
(Me3Si)2NH/Me3SiCl
reflux
Me3SiO (SiMeO) SiMe3
x
N
(3)
Cl
CO3H
Me3SiO(SiMeO) SiMe3
x
N+
O-
(4)

Scheme 2.

in which ion current can be carried either by the resonance stabilized aromatic ring or the electropositive metalloid. Conspicuous by its absence, but common to all silyl-substituted pyridines and their oxide derivatives, is the $(M\text{-}HCN)^{+\bullet}$ which is characteristic of ring fragmentation elimination evident in MS of 3-substituted pyridines (11).

Hydrolysis of **1** is carried out at room temperature by the slow addition of aqueous NH_3 at 0°C (Scheme 2). The resulting hydroxyl end-functional polysiloxane (2) is dispersed as fine droplets in the aqueous medium and can be separated by centrifugation. When neat, **2** is a colorless fluid with molecular weight ranging from

1×10^3 to 1×10^4 depending primarily on the length of time the polymer is heat-dried (e.g. < 100°C) under vacuum. Upon standing at room temperature for several days or heating for shorter periods of time above 100°C, the fluid becomes a rubbery semi-solid characteristic of a high molecular weight polymer resulting from chain-end condensation.

The IR spectrum of **2** displays a strong, broad band in the 1050-1100 cm^{-1} region characteristic of the Si-O-Si stretching vibration in siloxanes (12). Additionally a strong broad band centered at 3,200 cm^{-1} due to retained water is superimposed on terminal OH groups. Attempts to reduce or completely remove the water with conventional drying agents or by azeotropic distillation with organic solvents were unsuccessful.

The ^1H-NMR spectrum of **2** in $CDCl_3$ (Figure 1) exhibits broad unresolved resonances in the aromatic region similar to those found in the monomer. Broad signals with lack of resolution are consistent with magnetic non-equivalence of the methyl group protons resulting from a mixture of triad tacticities.

Species **2** was end-blocked to give **3** and simultaneously dehydrated by refluxing with a mixture of $(Me_3Si)_2NH$ and Me_3SiCl (Scheme 2). Product **3** is a clear colorless viscous fluid which remains unchanged on standing or upon heating under vacuum or in air and is soluble in organic solvents but immiscible with water.

The IR spectrum of **3** differs from **2** by the appearance of new bands at 845 cm^{-1}/870 cm^{-1} which are characteristic of the end-block $Si(CH_3)_3$ group (13). The intensity of the bands decrease with increase in polymer molecular weight thus supporting the assignment. There is also a significant decrease in the OH absorption confirming an anhydrous polymer.

The ^1H-NMR spectrum of **3** is given in Figure 1. The principal difference between the spectrum of **2** and **3** is the appearance of a singlet at -0.07 ppm in **3** which has been assigned to the terminal trimethylsiloxyl protons. The relative area of the peak decreases with increase in molecular weight. Thus, if the molecular weight is not too large (i.e. < 10,000), the ^1H-NMR spectrum is a convenient method for determining the average degree of polymerization (viz. $A^{Me}/A^{Me_3} = 1.4$; $\bar{D}_p = 8.3$ $\bar{M}_n = 1{,}300$). The NMR data agree well with other methods of molecular weight determination.

When **3** is treated with m-chloroperoxybenzoic acid oxidation at the ring nitrogen occurs (Scheme 2). Product **4** is a pale yellow viscous fluid which is soluble in organic solvents but immiscible with water having a partition ratio of 24:1 in a CH_2Cl_2/H_2O medium. The IR spectrum of **4** gives direct evidence for the presence of the $\overset{+}{N}-\overset{-}{O}$ bond with a characteristic stretching vibration at 1250 cm^{-1} (14). The band is intense and broad thus masking the much sharper CH rocking modes in the same region. It is noteworthy that a new medium intensity absorption occurs in **4** at 920 cm^{-1}, which is a window in **3**. This band is analogous to the one assigned to the $\overset{+}{N}-\overset{-}{O}$ deformation mode at 800-900 cm^{-1} reported by Kireev et al. (14) and Shindo (15).

The ^1H-NMR spectrum of a low molecular weight oligomer of **4** is given in Figure 1. A salient feature of the spectrum is the appearance of two rather than three unresolved signals in the aromatic region. Similar spectral changes (i.e. the merging of $H^{3,4}$ protons and the upfield shift of $H^{2,6}$ protons by ~ 0.5 ppm) upon N-oxidation have been observed with other 3-substituted pyridines (10). The position of the high-field methyl proton signal in **4** is comparable to that found in **3**.

Thermal Analysis of 2, 3 and 4. The TGA of **2** and **3** (x = 10) under argon is given in Figure 2. Both polymers have approximately the same temperature at which

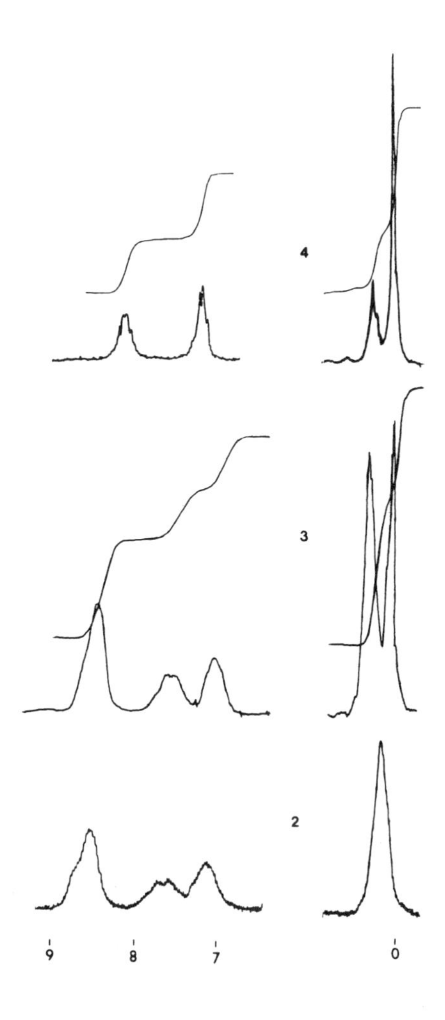

Figure 1. 60 MHz ^{1}H NMR Spectra of Hydroxy Terminated-(**2**), Trimethylsiloxyl End-Capped-(**3**) and Trimethylsiloxyl End-Capped/N-oxidized-(4) Poly(methyl 3-pyridinylsiloxane).

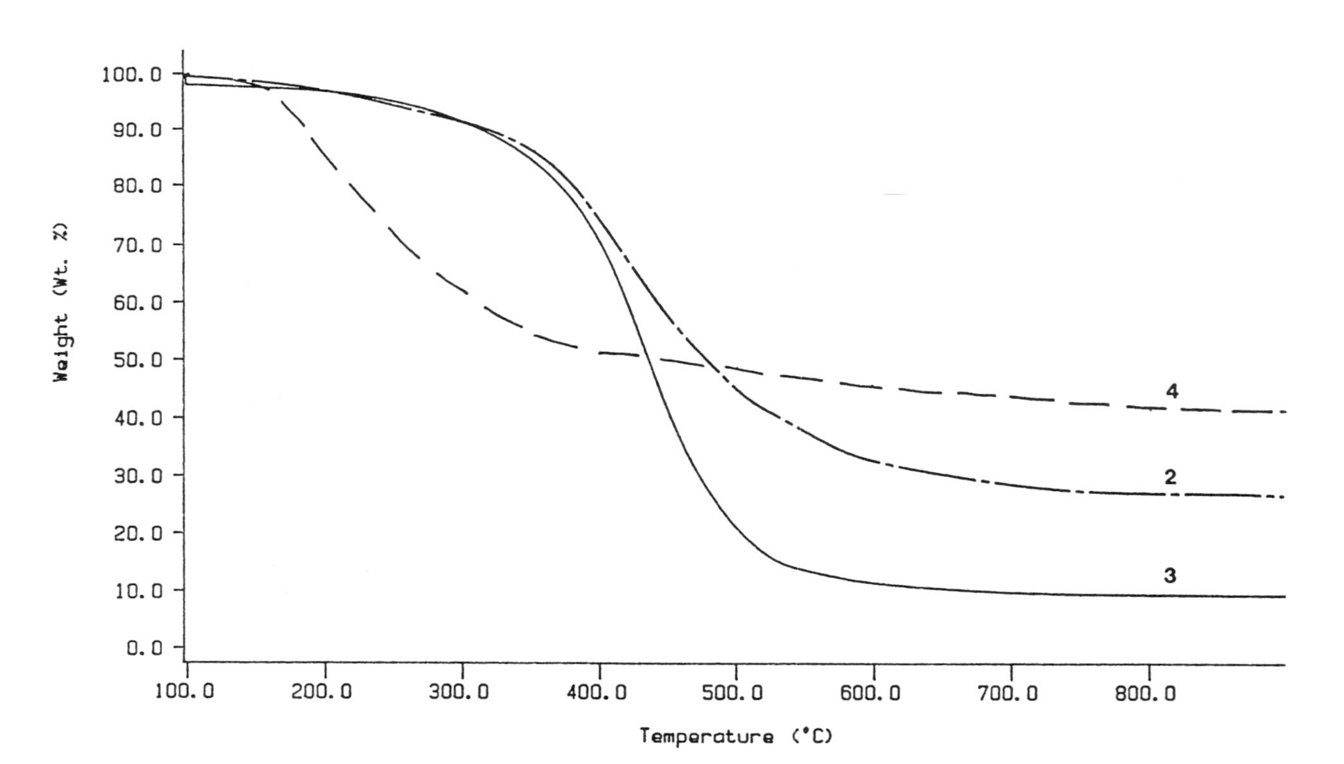

Figure 2. TGA of Hydroxy Terminated-(2), Trimethylsiloxyl End-Capped-(3) and Trimethylsiloxyl End-Capped/N-oxidized-(4) Poly(methyl 3-pyridinyl-siloxane); Heating Rate: 40°C/min, Under Ar.

the onset of weight loss occurs, presumably in part by depolymerization (16, 17). No attempt, thus far, has been made to identify the volatile products. If consistent with other polysiloxanes (e.g. Me_2, Me/Ph), cyclic oligomers are formed. The larger residue fraction for the hydroxy terminated species (28% vs. 9%) may be the result of crosslinking by pyridinyl decomposition perhaps involving end-group (nucleophilic) attack at the 2/6 positions of the ring. The TGA of **4**, however, indicates a lower onset of weight loss accompanied by a higher (40%) amount of residue. This suggests that the 1-oxide plays a significant role not only in chain scission but also in the crosslinking process.

The order of stability from DSC data is **3** > **2** ≈ **4** as judged from the exotherm maximum. In order to determine the nature of the degradation process, DSC experiments were carried out on **3** in which heating was terminated at different points in the heating program; i.e. prior to, at and post exothermic maximum. With the exception of the samples heated above 500°C, the residues from **3** were clear viscous yellow fluids soluble in CH_2Cl_2. The IR spectrum of the soluble residues were almost identical to the unheated material, the only change being the appearance of a new weak band at 940 cm^{-1} which suggests some N-oxidation by ambient air. Indeed, the band is not present when precautions are taken to exclude oxygen from the sample holder and the analyzer. The combination of SiC, SiOSi and SiC_3 vibrations, increased viscosity and solubility in organic solvents of the residues suggests that heating to 450°C promotes principally redistribution polymerization with little, if any, decomposition and crosslinking. In contrast, if **3** is heated above 500°C, the residue is a black insoluble solid with a relatively featureless IR spectrum containing broad bands in the SiOSi and SiC regions. The absence of ring modes and the appearance of a sharp band at 2350 cm^{-1}, which corresponds to the SiH vibration, implies that degradation involves crosslinking by pyridine ring decomposition and H migration. The latter is a well known pathway in mass spectrometry of organosilanes (18). Clearly, **3** displays remarkable thermal stability for its molecular weight. Further studies of the thermal properties and degradation process of **2-4** are underway and will be reported elsewhere.

Catalysis in Transacylation Reactions. The principal objective of the study was to evaluate **4** as an effective organic soluble lipophilic catalyst for transacylation reactions of carboxylic and phosphoric acid derivatives in aqueous and two-phase aqueous-organic solvent media. Indeed **4** catalyzes the conversion of benzoyl chloride to benzoic anhydride in well-stirred suspensions of CH_2Cl_2 and 1.0 M aqueous $NaHCO_3$ (Equations 1-3). The results are summarized in Table 1 where yields of isolated acid, anhydride and recovered acid chloride are reported. The reaction is believed to involve formation of the poly(benzoyloxypyridinium) ion intermediate (**5**) in the organic phase (Equation 1) and **5** then quickly reacts with bicarbonate ion and/or hydroxide ion at the interphase to form benzoate ion (Equation 2 and 3). Apparently most of the benzoate ion is trapped by additional **5** in the organic layer or at the interphase to produce benzoic anhydride (Equation 4), an example of normal phase-

$$PhC(=O)Cl + 4 \xrightarrow[CH_2Cl_2]{} -\!(Si(Me)(C_5H_4N^{+}\!-\!O\!-\!C(=O)Ph)\!-\!O)\!-\ Cl^{-} \quad (1)$$

(5)

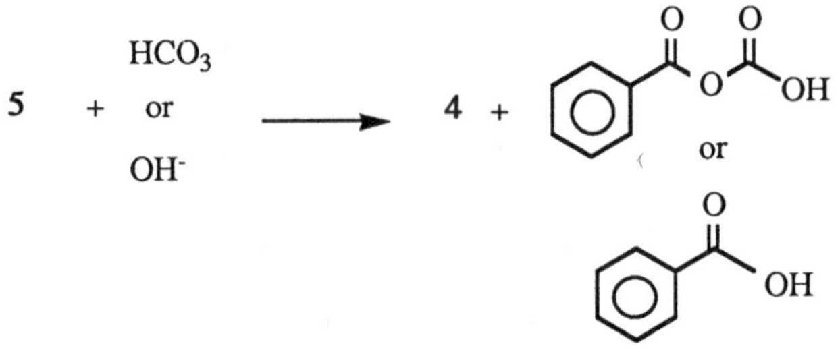

$$5 + HCO_3 \text{ or } OH^- \longrightarrow 4 + \text{(benzoyl–O–C(O)OH) or (benzoic acid, C(O)OH)} \quad (2)$$

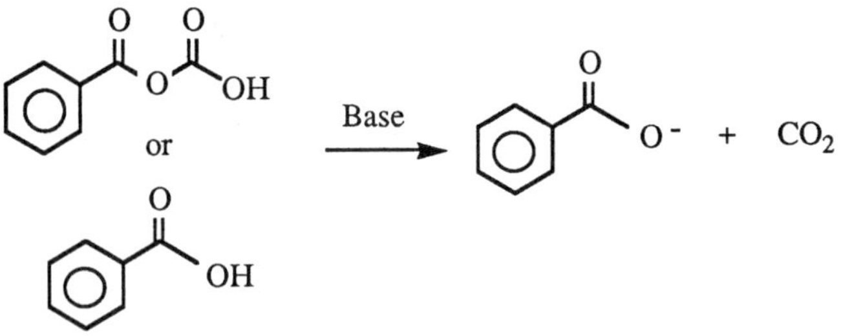

$$\text{(benzoyl–O–C(O)OH) or (benzoic acid)} \xrightarrow{\text{Base}} \text{(benzoate, } O^-) + CO_2 \quad (3)$$

$$5 + \text{(benzoate, } O^-) \longrightarrow 4 + \text{(benzoic anhydride)} \quad (4)$$

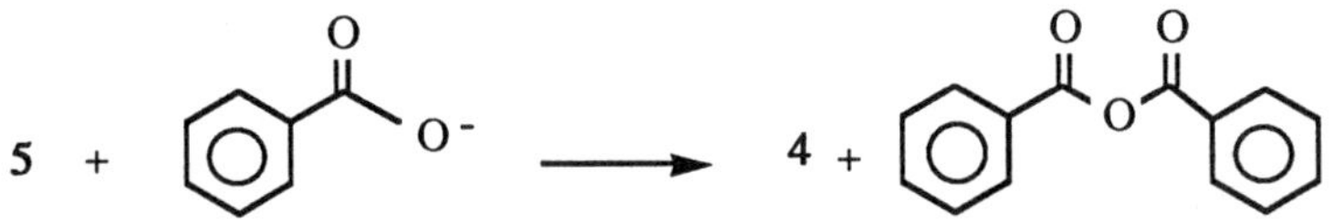

transfer catalysis (19). The same result is observed for catalysis by 3-(trimethylsilyl)-pyridine 1-oxide (**6**), 1,3-dimethyl-1,3-bis-3-(1-oxypyridinyl) disiloxane (**7**) and poly(4-vinyl pyridine 1-oxide) (**8**). Interestingly, catalysis by pyridine 1-oxide (**9**) furnishes nearly equal amounts of anhydride and hydrolysis product. This has been taken as evidence for a high proportion of inverse phase-transfer catalysis (1a), i.e., reaction between 1-benzoyloxypyridinium ion and bicarbonate ion in the aqueous phase. It is particularly significant that water-soluble (**8**) and organic-soluble (**4**, **6** and **7**) catalysts behave quite similarly in two-phase media. This argues for a symmetry-like behavior in catalysis at the interphase of the water-organic solvent suspension.

Hydrolysis of diphenyl phosphorochloridate (DPPC) in 2.0 M aqueous sodium carbonate is also believed to be a two-phase process. DPPC is quite insoluble in water and forms an insoluble second phase at the concentration employed (i.e. 0.10 M). It seems highly significant that the hydrophobic silicon-substituted pyridine 1-oxides (**4**, **6**, **7**) are much more effective catalysts than hydrophilic **8** and **9**. In fact, **4** is clearly the most effective catalyst we have examined for this reaction ($t_{1/2} < 10$ min). Since derivatives of phosphoric acids are known to undergo substitution reactions via nucleophilic addition-elimination sequences (20) (Equation 5), we believe that the initial step in hydrolysis of DPPC occurs in the organic phase. Moreover, the

$$(RO)_2P(=O)\text{-}X + :Nu \rightleftharpoons (RO)_2P(O^-)(X)(^+Nu) \rightleftharpoons (RO)_2P(=O)\text{-}Nu + :X \quad (5)$$

Table 1. Catalysis in Transacylation Reactions[a]

Benzoyl Chloride in CH_2Cl_2 — 1.0 M $NaHCO_3$

Catalysts (0.10 equiv) [b]	*Acid (%)*	*Anhydride (%)*	*Recovered Acid Chloride (%)*
-	0.4	0.0	> 90.0
(9)	42.0	49.4	1.6
(8)	8.3	81.0	0.7
(6)	4.9	91.6	1.5
(7)	8.1	79.1	2.0
(4)	5.0	89.3	3.0

Diphenyl Phosphorochloridate in 2.0 M $NaHCO_3$

Catalysts (0.10 equiv) [b]	*Acid (%)*	*Recovered Acid Chloride (%)*
-	14.2	72.6
(9)	22.4	71.9
(8)	39.2	47.3
(6)	72.3	22.9
(7)	66.0	23.5
(4)	85.9	10.7

[a] Reaction mixtures contained either 4 mMole of benzoyl chloride or 1 mMole of diphenyl phosphorochloridate. Reactions were run for 30 minutes at 22-23°C.

[b] **9**, Pyridine 1-oxide; **8**, Poly(4-vinylpyridine 1-oxide); **6**, 3-(Trimethylsilyl)pyridine 1-oxide; **7**, 1,3-Dimethyl-1,3-bis-3-(1-oxypyridinyl)disiloxane; **4**, Me_3SiO end-blocked poly(methyl 3-(1-oxypyridinyl)siloxane).

relatively high catalytic activity of **4** provides strong evidence for the importance of association or binding between the hydrophobic catalyst and DPPC prior to the first catalytic step.

It is noteworthy that polysiloxanes, long recognized and utilized because of their inertness and hydrophobicity, are now shown to be important platforms for chemical catalysis even in water containing media provided they are appropriately functionalized. Moreover, to our knowledge this is the first example in the published literature of a polysiloxane serving as a catalyst in organic synthesis (21).

The importance of hydrophobic binding interactions in facilitating catalysis in enzyme reactions is well known. The impact of this phenomenon in the action of synthetic polymer catalysts for reactions such as described above is significant. A full investigation of a variety of monomeric and polymeric catalysts with nucleophilic sites is currently underway. They are being used to study the effect of polymer structure and morphology on catalytic activity in transacylation and other reactions.

Acknowledgment

Acknowledgement is made to Reilly Tar and Chemical Corporation and the Dow Corning Corporation for partial support of this research and to the National Science Foundation for funds to purchase the thermal analyzer (grant CHE 84-10776). J.-m. X. and C.-x. T. express their sincerest gratitude to the Synthetic Research Institute of Tianjin (PRC) for leaves-of-absence to carry out the research.

Liturature Cited

1. (a) Hirl, M.A.; Gamson, E.P.; Klotz, I.M. J. Am. Chem. Soc. **1979**, 101, 6020. (b) Delaney, E.J.; Ward, L.E.; Klotz, I.M. J. Am. Chem. Soc. **1982**, 104, 799.
2. Lehn, J.-m. Acc. Chem. Res. **1978**, 11, 49.
3. Bender, M.; Kamiyama, M. Cyclodextrin Chemistry. Reactivity and Structure Concepts in Organic Chemistry; Springer-Verlag: Berlin, **1978**; Vol. 16.
4. (a) Fife, W.K.; Xin, Y. J. Am. Chem. Soc. **1987**, 109, 1278. (b) Fife, W.K.; Zhang, Z.-d. J. Org. Chem. **1986**, 51, 3744. (c) Fife, W.K.; Zhang, Z.-d. Tetrahedron Lett. **1986**, 27, 4933. (d) Fife, W.K.; Zhang, Z.-d. ibid. **1986**, 27, 4937.
5. Zeldin, M.; Xu, J.-m. J. Organomet. Chem. **1987**, 320, 267.
6. Zeldin, M.; Xu, J.-m.; Tian, C.-x. ibid. **1987**, 326, 341.
7. Zeldin, M.; Xu, J.-m.; Tian, C.-x.; Polymer Preprints **1987**, 28(1), 417.
8. Smith, A.L. Spectrochim. Acta, **1960**, 16, 87.
9. Maslowsky, E., Jr. Vibrational Spectra of Organometallic Compounds; Wiley-Interscience: New York, **1977**, p. 65-66, 410-411.
10. Pouchert, C.J.; Campbell, J.R. The Aldrich Library of NMR Spectra; Aldrich Chemical Co. Inc., Milwaukee, **1974**; Vol IX.
11. Budzikiewicz, H.; Djerassi, C.; Williams, D.H. Mass Spectrometry of Organic Compounds; Holden-Day Inc.: San Francisco, **1967**, Chapt. 20.
12. See ref. 9, p. 111-116.
13. Smith, A.L. Analysis of Silicones; J. Wiley and Sons: New York, **1974**.
14. Kireev, G.V.; Leont'ev, V.B.; Kurbatov, Y.V.; Otroshchenko, O.S.; Sadykov, A.S. Izv. Akad. Nauk SSSR, Ser. Khim. **1980**, 5, 1034.
15. Shindo, H. Chem. and Pharm. Bull. (Tokyo) **1958**, 6, 117.
16. Grassie, N.; Francey, K.F.; MacFarlane, I.G. Polym. Degrad. and Stability, **1980**, 2, 67; (b) Grassie, N.; Francey, K.F.; MacFarlane, I.G. ibid., **1980**, 2, 53.
17. Kang, D.W.; Rajendran, G.P.; Zeldin, M. J. Polym. Sci.: Part A: Polym. Chem. **1986**, 24, 1085.
18. Spalding, T.R. in "Mass Spectrometry of Inorganic and Organometallic Compounds," Litzow, M.R.; Spalding, T.R. Ed., Elsevier Scientific Publishing Co., **1973**; Chapter 7.
19. Dehmlow, E.V.; Dehmlow S.S. Phase Transfer Catalysis; 2nd ed.; Verlag Chemie; Weinheim, **1983**.
20. Jencks, W.P. Catalysis In Chemistry And Enzymology; McGraw-Hill, New York, **1969**.
21. A patent (FR 2474891; see CA 96:124298c) assigned to B.A. Ashby, GE Company, claims a platinum-siloxane complex which catalyzes hydrosilation of vinylsiloxanes or SiOH groups in the preparation of silicone resins and rubbers.

RECEIVED September 1, 1987

Chapter 16

Photochemical Behavior of Organosilicon Polymers Bearing Phenyldisilanyl Units

Mitsuo Ishikawa and Kazuo Nate

Department of Applied Chemistry, Faculty of Engineering, Hiroshima University, Saijo-cho, Higashi-Hiroshima 724, Japan; and Production Engineering Research Laboratory, Hitachi Ltd., Totsuka, Yokohama 244, Japan

The synthesis and photochemical behavior of polysiloxanes containing phenyl(trimethylsilyl)siloxy units and p-(disilanylene)phnylene polymers have been reported. Irradiation of thin liquid films of polysiloxanes containing phenyldisilanyl units, and both phenyldisilanyl units and alkenyl groups (vinyl, allyl, and butenyl) with a Xe-Hg lamp afforded transparent films which are insoluble in common organic solvents. Irradiation of poly[p-(1,2-diethyldimethyldisilanylene)phenylene] and poly[p-(1,2-dimethyldiphenyldisilanylene)phenylene] with a low-pressure mercury lamp in solutions resulted in homolytic scission of silicon-silicon bonds, followed by the formation of silicon-carbon unsaturated compounds and hydrosilanes. Irradiation of thin solid films of p-(disilanylene)phenylene polymers in air led to the formation of photodegradation products involving Si-OH and Si-O-Si groups. Lithographic applications of a double-layer system of the latter polymers are also described.

The π-electron system-substituted organodisilanes such as aryl-, alkenyl-, and alkynyldisilanes are photoactive under ultraviolet irradiation, and their photochemical behavior has been extensively studied (1). However, much less interest has been shown in the photochemistry of polymers bearing π-electron substituted disilanyl units (2-4). In this paper, we report the synthesis and photochemical behavior of polysiloxanes involving phenyl(trimethylsilyl)-siloxy units and silicon polymers in which the alternate arrangement of a disilanylene unit and a phenylene group is found regularly in the polymer backbone. We also describe lithographic applications of a double-layer system of the latter polymers.

0097-6156/88/0360-0209$06.00/0

Polysiloxanes Containing Phenyldisilanyl Units

In 1975, we found that irradiation of pentamethylphenyldisilane with a low-pressure mercury lamp leads to the transient formation of a silene. In the presence of a trapping agent such as alcohol, the silene thus formed reacts with alcohol to give addition products, while in the absence of the trapping agent, it undergoes polymerization to give nonvolatile substances (5).

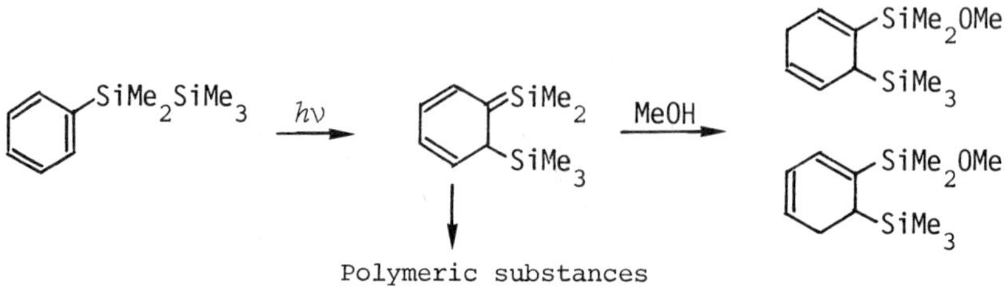

$PhSiMe_2SiMe_3 \xrightarrow{h\nu}$ (silene: $=SiMe_2$, $SiMe_3$) $\xrightarrow{MeOH}$ ($SiMe_2OMe$, $SiMe_3$ adducts)

↓

Polymeric substances

We initially thought that if polysiloxanes which are incorporated with phenyldisilanyl units and hydroxy groups in the polymer chain could be prepared, they would undergo photocrosslinking on exposure to ultraviolet radiation. In order to learn whether such polysiloxanes undergo photocrosslinking, we synthesized siloxane copolymers involving phneyl(trimethylsilyl)siloxy units in the polymer backbone, and studied their photochemical behavior (6).

Polysiloxanes (1)-(4) bearing phenyldisilanyl units could be readily prepared by copolymerization of 1,3,5-triphenyl[tris(trimethylsilyl)]cyclotrisiloxane and cyclopolysiloxanes in the presence of a catalytic amount of an intercalation compound prepared from graphite and potassium metal in a ratio of 8:1, in high yields. The polymers 1-4 thus obtained are colorless greaselike, and are soluble in common organic solvents.

$$[Me_3SiSi(Ph)O]_3 + (Me_2SiO)_4 + [R(Me)Si)]_{3\ or\ 4}$$

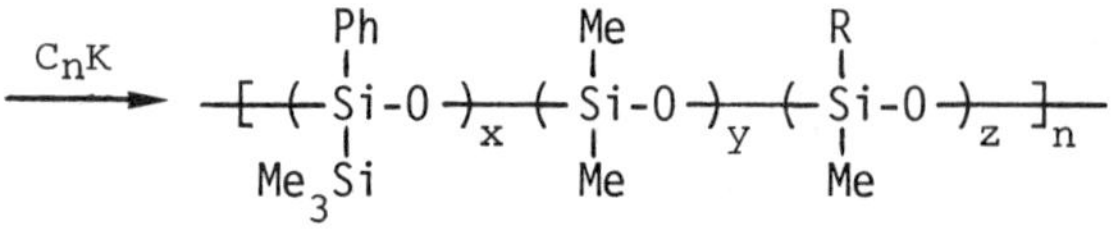

$$\xrightarrow{C_nK} -[(Si(Ph)(SiMe_3)-O)_x-(Si(Me)(Me)-O)_y-(Si(R)(Me)-O)_z]_n-$$

1, z = 0

2, R = CH_2=CH-

3, R = CH_2=$CHCH_2$-

4, R = CH_2=$CHCH_2CH_2$-

The UV spectrum of the polymer 1 ($\overline{Mn}$, 28,000) exhibits a characteristic absorption at 235 nm, indicating the presence of phenyldisilanyl units in the molecule. It is well known that pentamethylphenyldisilane shows a characteristic absorption at 231 nm (7-9), but trimethylphenylsilane exhibits an absorption at 211 nm. The polymers 2-4 also show strong absorption in the similar region, and photoactive under UV-irradiation.

Irradiation of a hexane solution of the polymer 1 with a 15-W low-pressure mercury lamp under a nitrogen atmosphere, produced the insoluble crosslinked polymer whose IR spectrum shows a weak absorption at 2150 cm^{-1} attributed to the stretching vibration of an Si-H bond.

In order to learn more about the photocrosslinking process, we synthesized 1,1-bis(trimethylsiloxy)-1-phenyl(trimethyl)disilane (5) as a model compound and examined its photochemical behavior in solutions. Compound 5 could readily be prepared by cohydrolysis of 1,1-dichloro-1-phenyl(trimethyl)disilane with a large excess of chlorotrimethylsilane in high yield.

$$Cl_2(Ph)SiSiMe_3 \quad + \quad 2Me_3SiCl \longrightarrow Me_3Si\text{-}O\text{-}Si(Ph)(SiMe_3)\text{-}O\text{-}SiMe_3$$

5

Unlike pentamethylphenyldisilane, the photolysis of compound 5 in the presence of methanol in hexane gave no products arising from the reaction of a silene with methanol, but polymeric substances were obtained as main products, in addition to small amounts of bis(trimethylsiloxy)phenylsilane (6) (4%) and bis(trimethylsiloxy)-phenylmethoxysilane (7) (5%).

$$5 \xrightarrow[MeOH]{h\nu} Me_3Si\text{-}O\text{-}Si(Ph)(H)\text{-}O\text{-}SiMe_3 \quad + \quad Me_3Si\text{-}O\text{-}Si(Ph)(OMe)\text{-}O\text{-}SiMe_3$$

6 7

\+ Polymeric substances

Irradiation of the compound 5 in isopropylbenzene under similar consitions afforded m- and p-[bis(trimethylsilyl)phenylsilyl]isopropylbenzene (8) (m:p=1.4:1) and m- and p-(trimethylsilyl)isopropylbenzene (9) (m:p=3:1) in 45 and 24% yields, respectively, together with 9% yield of hydrosilane 6. The formation of products 8 and 9 may be best understood in terms of homolytic aromatic substitution of silyl radicals produced by homolytic scission of an Si-Si bond in compound 5, while hydrosilane 6 can be explained by hydrogen abstraction of the bis(trimethylsiloxy)phenylsilyl radical (Scheme 1).

On the basis of the photochemical reaction of the model compound 5, it seems likely that the photochemically generated silyl radicals play an important role in the photocrosslinking of the polymer 1. First, photochemical bond scission of the Si-Si bonds in the polymer chains takes place to give silyl radicals. The resulting radicals undergo homolytic aromatic substitution to give phenylene linkages, or abstract hydrogen from methylsilyl groups to form silylmethylene radicals which would result in the formation of silylmethylene linkages (Scheme 2).

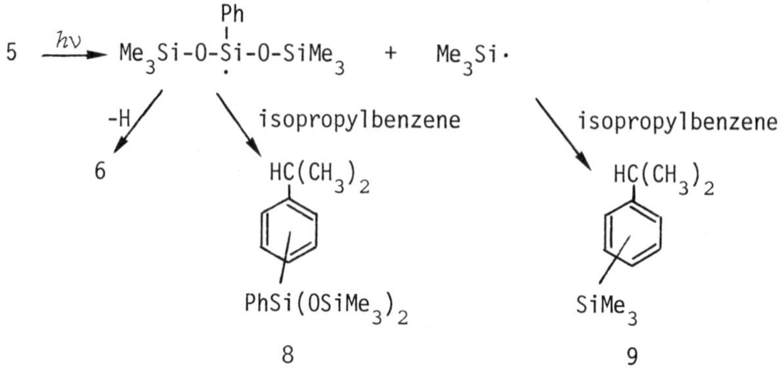

$$5 \xrightarrow{h\nu} Me_3Si\text{-}O\text{-}\dot{Si}(Ph)\text{-}O\text{-}SiMe_3 + Me_3Si\cdot$$

-H → 6; isopropylbenzene → 8 ($HC(CH_3)_2$-C₆H₄-$PhSi(OSiMe_3)_2$); $Me_3Si\cdot$ + isopropylbenzene → 9 ($HC(CH_3)_2$-C₆H₄-$SiMe_3$)

Scheme 1.

$$\sim Si(Ph)(SiMe_3)\text{-}O\text{-}SiMe_2\text{-}O\sim \xrightarrow{h\nu} \sim \dot{Si}(Ph)\text{-}O\text{-}SiMe_2\text{-}O\sim + \cdot SiMe_3$$

$\sim \dot{Si}(Ph)\text{-}O\text{-}SiMe_2\text{-}O\sim$ + $\sim Si(C_6H_5)(SiMe_3)\text{-}O\text{-}SiMe_2\text{-}O\sim$ → cyclohexadienyl radical adduct (H) → $\sim Si(Ph)(\text{-}O\text{-}SiMe_2\text{-}O\sim)\text{-}C_6H_4\text{-}Si(SiMe_3)\text{-}O\text{-}SiMe_2\text{-}O$ + $\cdot H$

$$Me_3Si\cdot + \sim O\text{-}Si(Me)(CH_3)\text{-}O\text{-}SiMe_2\text{-}O\sim \longrightarrow \sim O\text{-}Si(Me)(\dot{C}H_2)\text{-}O\text{-}SiMe_2\text{-}O\sim + Me_3SiH$$

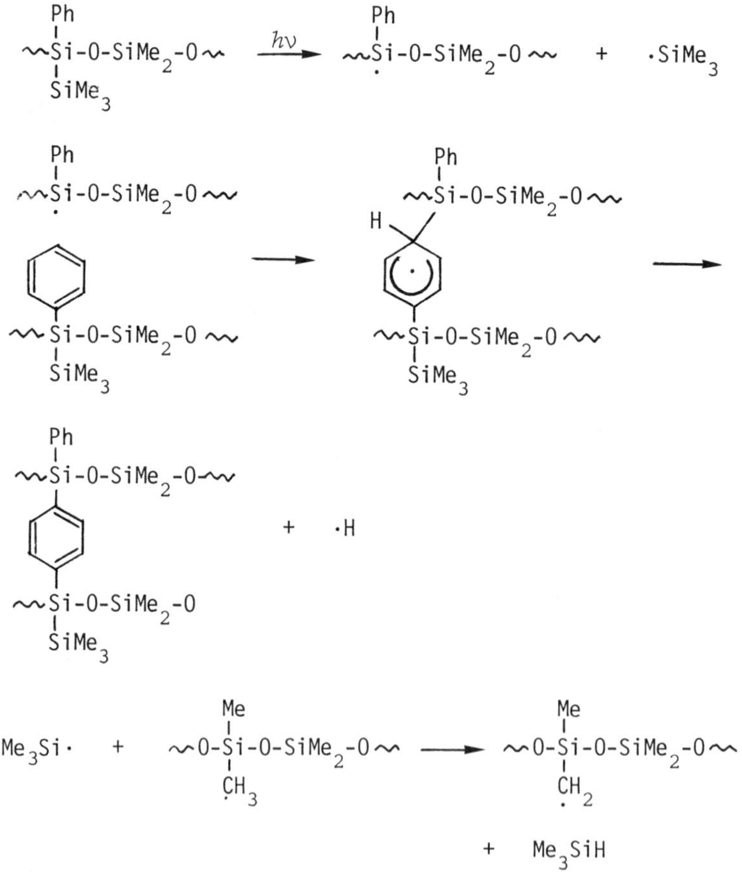

Scheme 2.

```
        Me                                 Me
        |                                  |
~~O-Si-O-SiMe2-O~~              ~~O-Si-O-SiMe2-O~~
                                           |
 ——→    CH2                ——→             CH2
        .                                  |
        .                                  |
~~O-Si-O-SiMe2-O~~              ~~O-Si-O-SiMe2-O~~
        |                                  |
        Ph                                 Ph
```

When a thin liquid film with a thickness of approximately 2 μm prepared by spin coating of a 15% benzene solution of polymer 1 was irradiated with a 500-W Xe-Hg lamp for 300 s in air, a transparent solid film was obtained. The UV spectrum of this solid film shows that an absorption at 235 nm due to phenyldisilanyl units vanishes after UV-irradiation (Figure 1). This clearly indicates that photolytic cleavage of silicon-silicon bonds leading to the crosslinking occurred. Similar photolysis of the thin liquid films under a nitrogen atmosphere again afforded transparent solid films whose UV spectra show no absorption at 235 nm due to phenyldisilanyl units.

We have investigated the photochemical behavior of polymer 1 under various conditions in air and found that UV irradiation of the thin liquid films with a thickness of less than 10 μm, indeed produced transparent solid films. However, when the films with a thickness of 100 μm were irradiated with a mercury lamp, crosslinking leading to the solid films occurred only on the surface of the films, but inside remained as liquid after prolonged irradiation. In thses cases, tha surface of the films was found to be slightly opaque. Therefore, most of the light would not be transmitted to the inside of the films.

Since crosslinking of the polymer occurs as the result of the initial formation of silyl radicals, the siloxane polymer containing both phenyldisilanyl units and functional groups which undergo radical polymerization should produce solid material whatever the thickness of the films. To ascertain this, we have examined the photochemical behavior of the polymers 2-4.

Irradiation of thin films prepared by spin coating of a 15% benzene solution of polymers 2 and 3 on a quartz plate with a mercury lamp in air gave solid films. The absorption at 235 nm observed in the UV spectra of the starting films again disappeared after UV irradiation, as observed in the similar photolysis of 1, indicating that the radical scission of silicon-silicon bonds occurred. IR spectra of the resulting films show that intensity of the bands at 1595 cm^{-1} for polymer 2 and at 1640 cm^{-1} for polymer 3, due to the C=C stretching frequencies decreased and a broad band at 1725 cm^{-1} attributed to the C=O stretching frequencies was observed. Similar irradiation of the thin films of 2 and 3 under a nitrogen atmosphere also afforded transparent solid films whose IR spectra, however, show no C=O stretching frequencies. Intensity of the bands due to the C=C stretching frequencies in the resulting films decresed to a certain extent.

All of the photocrosslinking described here begin with surface of the films. The photocrosslinking leading to the solid films under a nitrogen atmosphere would proceed by the mechanism analogous

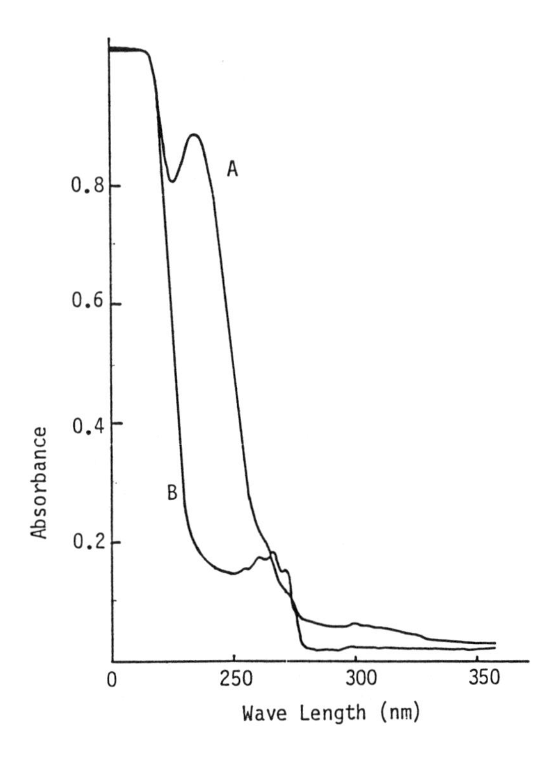

Figure 1. UV spectra of a thin liquid film of 1: A, before irradiation; B, after irradiation. (Reproduced with permission from Ref. 6. Copyright 1986 John Wiley & Sons, Inc.)

to that observed in solution (see Scheme 2). In the air, however, the silyl radicals produced on the surface of the films react with oxygen to give silyl peroxy radicals which would result in formation of solid films. Irradiation of thick films of 2 and 3 also led to crosslinking, however, the inside of the films could not be completely solidified. The photochemical behavior of liquid films of 2 and 3 is not very much different from that of 1, because vinyl- and allyl-substituted silicon compounds are rather inactive toward radical polymerization. In contrast to the photochemical behavior of 2 and 3, polymer 4 gave thick solid films. In this case, the radicals formed in the films in which crosslinking occurred on the surface would induce the radical reaction of butenyl groups inside of the films. Consequently, photocrosslinking occurred no matter what the thickness of the films to give insoluble solid films.

p-Disilanylenephenylene Polymers

Next, we have synthesized p-(disilanylene)phenylene polymers and examined their photochemical behavior in solutions and in solid films (3) (K. Nate, M. Ishikawa, H. Ni, H. Watanabe, Y. Saheki, Organometallics, in press). To our knowledge, silicon polymers, in which the alternate arrangement of a disilanylene unit and a phenylene group is found regurary in the polymer backbone have not been reported so far (10). First, we attempted to prepare poly[p-(tetramethyldisilanylene)phenylene] by the condensation of 1,4-bis(chlorodimethylsilyl)benzene with sodium metal in toluene. However, the polymers thus obtained were mainly insoluble and intractable white solid materials. This may be ascribed to the high crystallinity of the polymers. In order to reduce the crystallinity, we used 1,4-bis(chloroethylmethylsilyl)benzene and 1,4-bis(chloromethylphenylsilyl)benzene as the starting monomers.

The reaction of these monomers with sodium dispersion at 70-90°C under a nitrogen atmosphere afforded poly[p-diethyldimethyldisilanylene)phenylene] (10) and poly[p-(1,2-dimethyldiphenyldisilanylene)phenylene] (11), respectively, in good yields.

$$\text{Cl-Si(R)(Me)-C}_6\text{H}_4\text{-Si(R)(Me)-Cl} \xrightarrow[\text{in toluene}]{\text{2Na}} \text{-[Si(R)(Me)-Si(R)(Me)-C}_6\text{H}_4\text{]}_n\text{-} + \text{2NaCl}$$

10, R = Et

11, R = Ph

Both polymers 10 and 11 are soluble in common organic solvents, melt without decomposition, and can be drawn into the fibers. Molecular weights of the polymers 10 and 11, determined by gel permeation chromatography with tetrahydrofuran as the eluant after purification by reprecipitation from benzene-ethanol, showed a broad monomodal molecular weight distribution. The degree of polymerization depends on particle size of sodium metal. Polymers with molecular weights of 23,000-34,000 are always obtained, if fine sodium particles are used.

The ^{1}H NMR chemical shifts of the polymers 10 and 11 are only slightly dependent on the molecular weight. The ^{1}H NMR spectrum of the polymer 10 shows three shrap resonances at δ 0.34, 0.94, and 7.28 ppm, due to MeSi, EtSi, and phenylene ring protons, while the polymer 11 reveals shrap resonances at δ 0.62, 7.28, and 7.35 ppm, attributed to MeSi and phenyl ring protons. The polymers 10 and 11 show characteristic strong absorption bands at 262 and 254 nm, respectively, significantly red-shifted relative to pentamethylphenyldisilane which exhibits an absorption band at 231 nm.

When a benzene solution of the polymer 10 was photolyzed upon irradiation with a 6-W low-pressure mercury lamp for 2 h, soluble photodegradation products were produced. In order to learn the influence of irradiation time in the molecular weight of photodegradation products, benzene solutions containing 10 with molecular weight of 23,000 were photolyzed for a definite time under the same conditions, and the molecular weights of the resulting photoproducts were determined by GPC. The molecular weight of the photoproducts decreased in the early stages of the photochemical reaction, and passed through a minimum value of 14,500 after 1h, and then increased gradually with increasing reaction time (29,000 after 6 h, and 37,000 after 12 h), indicating that some crosslinking reactions occurred during the photolysis (Figure 2). ^{1}H NMR spectra of the resulting photoproducts show very broad resonances for MeSi, EtSi, and phenyl ring protons. In addition, a multiple resonances with low intensity due to an Si-H proton is observed at δ 4.2 ppm in all the photoproducts. UV spectra of the products show disappearance of the original strong band at 262 nm, and IR spectra reveal a band with medium intensity at 2120 cm^{-1} attributed to the stretching frequencies an Si-H bond.

We carried out the photolysis of 10 in the presence of an excess of methanol in benzene. In contrast to the photolyses in the absence of metanol, the molecular weight of the photoproducts decreased continuosly with increasing reaction time, and finally reached a constant value ($\bar{M}w$ = 1,800), after 4 h-irradiation. The ^{1}H NMR spectrum of the photoproducts shows signals at δ 3.4 and 4.2 ppm, due to CH_3-O and Si-H protons, in addition to the broad signals corresponding to MeSi, EtSi, and phenyl ring protons. The resonances with low intensities at δ 2.2 and 6.2 ppm, assigned tentatively as cyclohexadienyl ring protons are also observed.

Assuming that our assignment for the cyclohexadienyl ring protons is correct, the result indicates that the rearranged silicon-carbon unsaturated compounds are formed in the photodegradative processes. The photorearranged silene intermediates, however, are involved only as minor products for the present polymeric systems. In fact, the ratio of the cyclohexadienyl group to the disilanylenephenylene unit in the photoproducts which show a maximum value of the cyclohexadienyl ring protons, is calculated to be approximately 1:19 by ^{1}H NMR spectroscopic analysis.

When a benzene solution of 10 in the presence of methanol-d_1 was photolyzed under the same conditions, the products containing both CH_3-O and Si-H groups were again obtained. The IR spectrum of the product shows no absorption due to an Si-D stretching frequencies, indicating that no direct reaction of the photo-excited 10 with methanol is involved. The relative ratio of an Si-H to CH_3-O

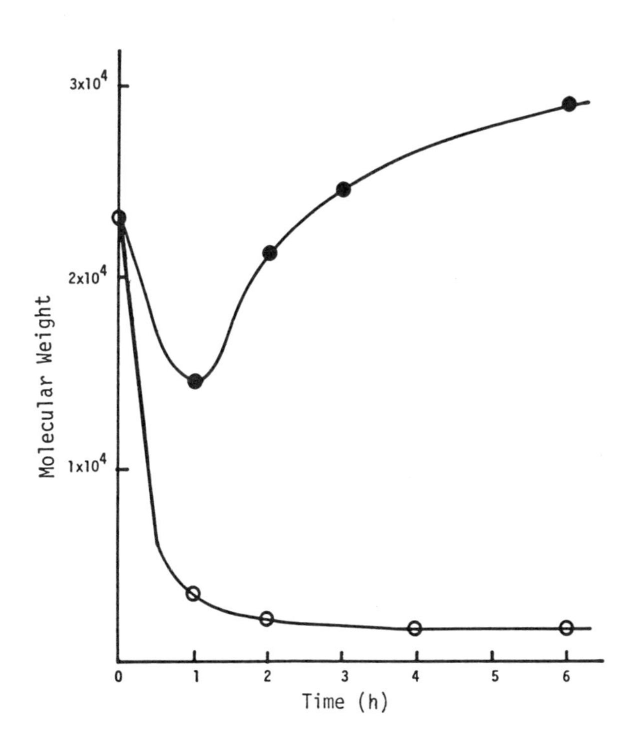

Figure 2. Plot of molecular weights of products vs. irradiation time for the photolysis of 10: A, in benzene; B, in the presence of methanol in benzene.

to $[Me(Et)SiSi(Me)(Et)C_6H_4]_n$ in the photoproduct obtained from 4 h-irradiation of 10 is determined to be 1:1.5-2.0:3.2-3.8(n). These results clearly indicate the formation of a different type of silicon-carbon double bonded intermediates. This type of silene may be produced from homolytic scission of silicon-silicon bonds, followed by hydrogen abstraction of the resulting radical from either a methyl group or methylene of an ethyl group on the others (Scheme 3).

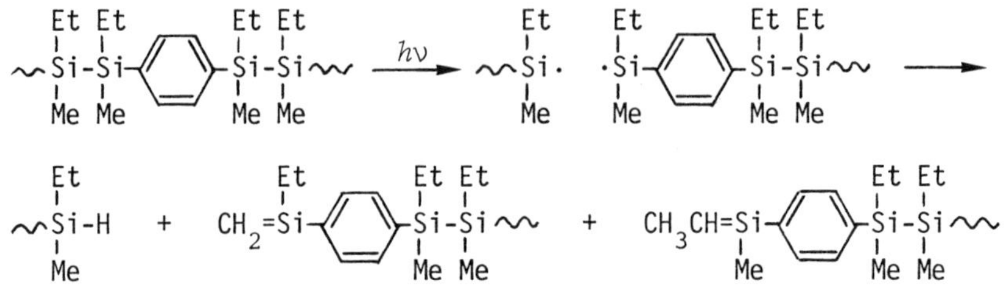

Scheme 3.

On the basis of the methoxy contents in the photoproducts, these silenes seem to be formed as main products in the polymeric systems, although this type of silene is produced only as a minor product in the photolysis of aryldisilanes.

The photolysis of polymer 11 in the absence of methanol proceeded with considerably different fashion from that of 10. Irradiation of a benzene solution of 11 resulted in the decrease of the molecular weight of the photoproducts (Figure 3). No increase of the molecular weight of the products with increasing reaction time was observed. Moreover, the photodegradation of 11 proceeded faster than that of 10. Thus, the photolysis of a benzene solution of 11 for 1 h with a low-pressure mercury lamp led to the formation of the photoproduct whose molecular weight was determined to be 5,900, which remained unchanged after 4 h-irradiation. ^1H NMR spectra of the products display broad signals at δ 0.6, 4.9, and 6.8-7.8, attributed to MeSi, HSi, and phenyl ring protons, respectively.

Irradiation of 11 in the presence of methanol for 1 h afforded the product with molecular weight of 2,000 which remained constant during further irradiation. IR and ^1H NMR spectra show that the resulting products involve both CH_3-O and Si-H groups. The relative ratio of an Si-H and CH_3-O group to $[Ph(Me)SiSi(Me)(Ph)C_6H_4]_n$ is calculated as being 1:1.0-1.7:2.5-3.0(n).

When similar photolysis of 11 in the presence of MeOD was carried out, again the product whose ^1H NMR reveals the resonance due to the Si-H proton was observed. The relative ratio of the Si-H and CH_3-O protons was identical with those of the products obtained in the presence of non-deuterated methanol. The formation of the methoxysilyl group can be understood by the addition of methanol across the silicon-carbon double bonds. ^1H NMR spectra of all photoproducts obtained from the photolyses of 11 in the presence of methanol reveal no resonances attributed to the cyclohexadienyl ring protons. This indicates that the photochemical degradation of the polymer 11 gives no rearranged silene intermediates, but produces

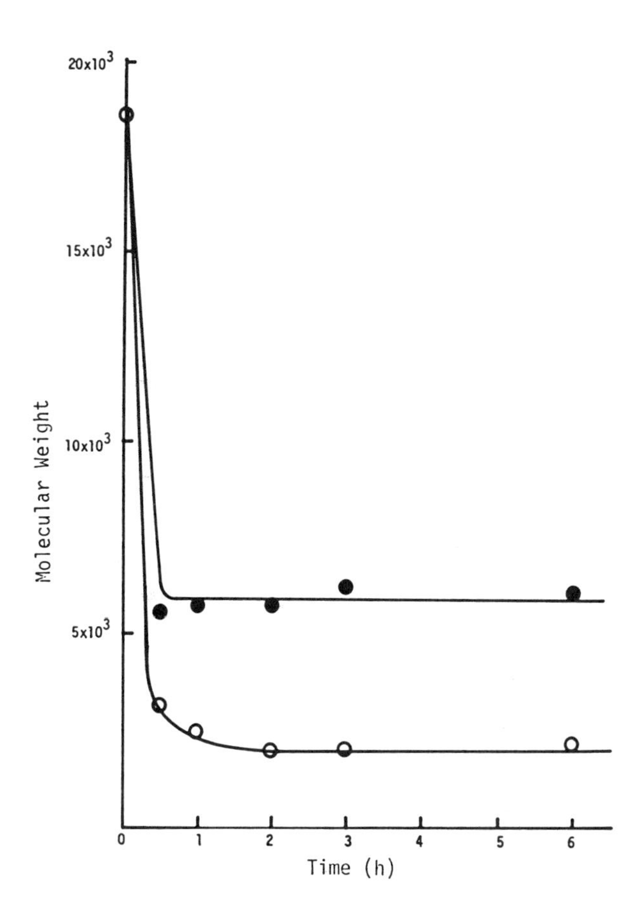

Figure 3. Plot of molecular weights of products vs. irradiation time for the photolysis of 11: A, in benzene; B, in the presence of methanol in benzene.

exclusively silaethene intermediates and hydrosilanes (Scheme 4). It seems likely that the increase of the molecular weight observed in the photolysis of 10 in the absence of methanol may be ascribed to the polymerization of the photorearranged silene intermediates.

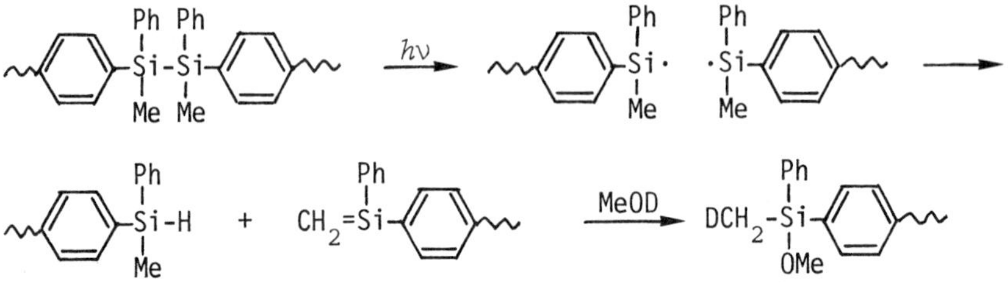

Scheme 4.

The absorption spectra of the polymers 10 and 11 in the solid films exhibit a broad structureless band at 262 and 254 nm, respectively. The absorption maxima of thses bands are not different from those in a THF solution.

We have examined the photochemistry of the thin films of the polymers 10 and 11, in order to learn whether the polymers can be used as positive deep UV resists. Thus, irradiation of thin films with a thickness of approximately 0.1 μm prepared by spin coating of a 5% toluene solution of the polymer 10 on a quartz plate in air with a 500-W Xe-Hg lamp for 30 s gave photodegradation products, which are soluble in 2-ethoxyethanol. The absorption at 262 nm observed in UV spectra of the starting films vanished after UV-irradiation, showing that the scission of the silicon-silicon bonds occurred. IR spectra of the resulting films showed that intensity of the band at 500 cm^{-1} assigned as the Si-Si stretching frequencies decreases remarkably, and strong absorption bands at 3350 and 1060 cm^{-1}, and a broad band at 1720 cm^{-1} appeared after UV-irradiation. The fact that only very weak absorption band at 2140 cm^{-1} due to the Si-H stretching frequencies was observed, indicates that the silyl radical generated photochemically from the scission of the silicon-silicon bonds were scavenged by oxygen in air. Indeed, IR spectra of the products obtained from the similar photolysis under reduced pressure obviously show the absorption due to the Si-H stretching frequencies, but the absence of the band at 3350 and 1720 cm^{-1}.

Irradiation of a solid film prepared from the polymer 11 in air afforded photodegradation products which are soluble in 2-ethoxyethanol. IR spectra of the products show strong absorption bands attributed to Si-OH and Si-O-Si stretching frequencies. In contrast to the products from 10, these products show absorption due to the Si-H bond. This result indicates that some of the silenes would be formed in this system. The intense absorption at 254 nm in the UV spectrum again disappeared after UV-irradiation.

The main route for the photodegradation of the polymers 10 and 11 in the solid films in air consists of radical scission of the silicon-silicon bonds, analogous to that observed in solution, but in the films, the resulting silyl radicals react with oxygen to give siloxanes and silanols as the final products. For the polymer 11, the formation of silenes which would be trapped by oxygen or moisture in the air are also involved as a minor degradative pathway.

Lithographic Applications

It has been reported that organosilicon polymers show high etching resistance against an oxygen plasma, and can be used as the top imaging layer in the double-layer resist system. The typical organosilicon polymers reported to date for the lithographic applications are the siloxane polymers having vinyl groups (11) or chloromethyl groups (12), polystyrenes having trimethylsilyl groups (13), and silyl substituted novolacs (14,15). Polysilanes have also been found to serve as excellent O_2 reactive ion etching barriers for double-layer resist applications (16).

The O_2 reactive ion etching rates of poly(disilanylenephenylene) 10 and 11, and polyimide (PIQ) were measured. As can be seen in Figure 4, the polymers 10 and 11 show very high etching resistance against the oxygen plasma, compared with the PIQ.

The lithographic applications of a double-layer resist system in which the poly[p-(dimethyldiphenyldisilanylene)phenylene] film (0.2 μm thick) was used as the top imaging layer have been examined (K. Nate, T. Inoue, H. Sugiyama and M. Ishikawa, J. Appl. Polym. Sci. in press). In these studies, the PIQ (2.0 μm thick) was used as an underlayer. Thus, the film consisting of the polymer 11 and PIQ prepared on a silicon wafer was exposed to deep UV-light with the use of Canon contact aligner PLA-521 through a photomask for 5 to 6 s (UV intensity: 72 mV/cm^2 at 254 nm). The resulting film was then developed with a 1:5 mixture of toluene and isopropyl alcohol for 15 s and rinsed with isopropyl alcohol for 15 s. A positive resist pattern was obtained after treatment of the film pattern with O_2 RIE under the condition of 0.64 W/cm^2 (RF power:7 MHz, O_2 pressure:3 mtorr).

The SEM photograph of the double-layer resist pattern thus obtained is shown in Figure 5. In this way, submicron level line patterns where the minimum line width is 0.5 μm and the aspect ratio is above 3.0 could readily be obtained.

Summary

Irradiation of thin liquid films of polysiloxanes 1-4, containing phenyldisilanyl units in air led to homolytic scission of silicon-silicon bonds which underwent crosslinking to give solid materials. Irradiation of thick films of 1-3 also led to crosslinking, however, the inside of the films could not be solidified, while polymer 4 afforded thick solid films. Similar irradiation of solid films prepared from p-(disilanylene)phenylene polymers 10 and 11 gave photodegradation products. The main route for photodegradation consists of radical scission of silicon-silicon bonds in the polymer backbone. The resulting silyl radicals react with oxygen to give siloxanes and silanols as the final products. It has been found that polymers 10 and 11 show very high etching resistance against the oxygen plasma, and can be used as the top imaging layer in the double-layer resist system.

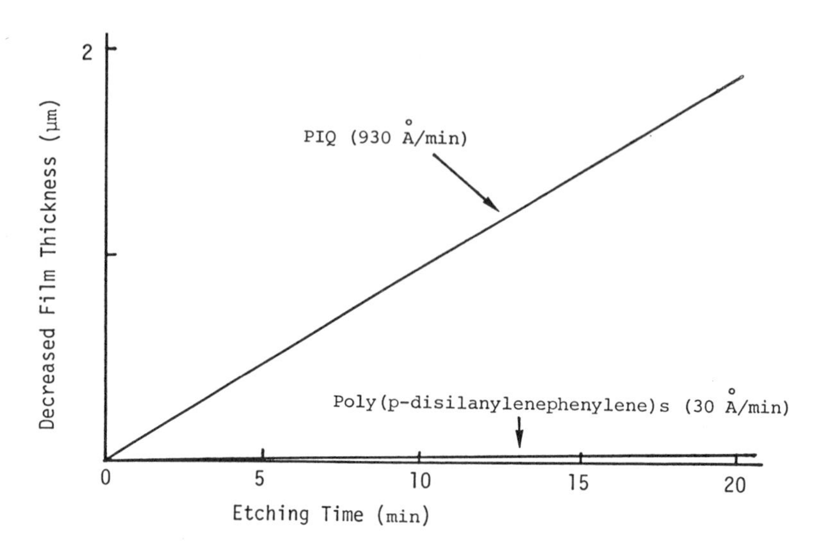

Figure 4. Etching rate of 10 and 11, and polyimide.

Figure 5. Double-layer resist pattern prepared by using 11 as the top imaging layer.

Acknowledgments

The cost of this work was defrayed in part from Grant-in-Aid for Special Project Research, Organic Chemical Resources, 61111002, to which the authors' thanks are due. We also express our appreciation to Shin-etsu Chemical Co., Ltd., and Toshiba Silicone Co., Ltd., for a gift of organochlorosilanes.

Literature Cited

1. (a) Ishikawa, M. Pure Appl. Chem. 1978, 50, 11. (b) Ishikawa, M.; Kumada, M. Adv. Organomet. Chem. 1981, 19, 51.
2. Ishiakwa, M.; Imamura, N.; Miyoshi, N.; Kumada, M. J. Polym. Sci. Polym. Lett. Ed. 1983, 21, 657.
3. Ishikawa, M.; Ni, H.; Matsuzaki, K.; Nate, K.; Inoue, T.; Yokono, H. J. Polym. Sci. Polym. Lett. Ed. 1984, 22, 669.
4. (a) West, R.; David, L. D.; Djurovich, P. I.; Steraley, K. L.; Srinivasan, K. S. V.; Yu, H. J. Am. Chem. Soc. 1981, 103, 735. (b) Zhang, X. H.; West, R. J. Polym. Sci. Polym. Chem. Ed. 1984, 22, 159. (c) Trefonas, P.; West, R.; Miller, R. D. J. Am. Chem. Soc. 1985, 107, 2737. (d) Trefonas, P.; Damewood, Jr.; West, R.; Miller, R. D. Organometallics 1985, 4, 1318.
5. (a) Ishikawa, M.; Fuchikami, T.; Kumada, M. J. Organomet. Chem. 1976, 118, 155. (b) Ishikawa, M.; Oda, M.; Nishimura, K.; Kumada, M. Bull. Chem. Soc. Japan 1983, 56, 2795.
6. Nate, K.; Ishikawa, M.; Imamura, N.; Murakami, Y. J. Polym. Sci. Polym. Chem. Ed. 1986, 24, 1551.
7. Gilman, H.; Atwell, W. H.; Schwebke, G. L. J. Organomet. Chem. 1964, 2, 369.
8. Sakurai, H.; Kumada, M. Bull. Chem. Soc., Japan 1964, 37, 1894.
9. Hague, D. N.; Prince, R. H. J. Chem. Soc. 1965, 4690.
10. Nate, K.; Sugiyama, H.; Inoue, T.; Ishikawa, M. Abstr. Meeting Electrochem. Soc. Oct. 7-12, 1984, New Orleans, L. A. Abstract No. 530, p 778, 1984.
11. Shaw, J. M.; Hatzakis, M; Paraszczak, J.; Liutkus, J.; Babich, E. Regional Technical Conference, "Photopolymers Principles-Process and Materials" p 285, N. Y. November 1982.
12. Morita, M.; Tanaka, K.; Imamura, S.; Tamamura, T.; Kogure, I. The 44th Conference of Japanese Applied Phisics, No. 28a-T-1, Abstract p 243, September 1983.
13. Suzuki, N.; Saigo, K.; Gokan, H.; Ohnishi, Y. The 44th Conference of Japanese Applied Physics, No. 26a-U-7, Abstract p 258, September 1983.
14. Saotome, Y.; Gokan, H.; Suzuki, M; Ohnishi, Y. J. Electrochem. Soc. 1985, 132, 909.
15. Wilkins Jr., C. W.; Reichmanise, E.; Wolf, T. M.; Smith, B. C. J. Vac. Sci. Technol. 1985, B 3(1), 306.
16. Hofer, D. C.; Miller, R. D.; Willson, G. C. Proceedings of SPIE, Advances in Resist Technology, 469, 16, Santa Clara, California, March 1984.

RECEIVED September 1, 1987

Chapter 17

Routes to Molecular Metals with Widely Variable Counterions and Band-Filling

Electrochemistry of a Conductive Organic Polymer with an Inorganic Backbone

Tobin J. Marks, John G. Gaudiello [1], Glen E. Kellogg, and Stephen M. Tetrick

Department of Chemistry and Materials Research Center, Northwestern University, Evanston, IL 60201

This contribution reviews recent results on $[Si(Pc)O]_n$ (Pc = phthalocyaninato) solid state electrochemistry and the structural interconversions that accompany electrochemical doping/undoping processes. In acetonitrile/$(\underline{n}\text{-Bu})_4N^+BF_4^-$, it is found that a significant overpotential accompanies initial oxidation of as-polymerized $[Si(Pc)O]_n$. This can be associated with an orthorhombic→tetragonal structural transformation. Once in the tetragonal crystal structure, cycling between tetragonal doped $\{[Si(Pc)O](BF_4)_y\}_n$ and tetragonal undoped $[Si(Pc)O]_n$ is relatively facile. Evidence is presented for continuous tuning of the band-filling between y = 0.00 and 0.50. In comparison, electrochemical oxidation of monoclinic β-Ni(Pc) under the same conditions is also accompanied by a significant overpotential in forming tetragonal Ni(Pc)-$(BF_4)_{0.48}$. However, electrochemical undoping produces the monoclinic γ-Ni(Pc) phase with far less band structure tunability than in the silicon polymer. Experiments with tosylate as the anion indicate that tetragonal $\{[Si(Pc)O](tosylate)_y\}_n$ can be tuned continuously between y = 0.00 and 0.67. For the anions PF_6^-, SbF_6^-, $CF_3(CF_2)_nSO_3^-$ (n = 0,3,7), it appears that the maximum accessible oxidation levels are largely dictated by packing limitations (anion size). Evidence is also presented for reversible reductive doping of $[Si(Pc)O]_n$ in THF/$(\underline{n}\text{-Bu})_4N^+BF_4^-$.

A major barrier to understanding fundamental relationships between molecular architecture, electronic structure, and charge transport in molecular metals derives from our inability to introduce poten-

[1]Current address: Department of Chemistry, Michigan State University, East Lansing, MI 48824

0097-6156/88/0360-0224$06.00/0

tially informative perturbations without major, deleterious reorganizations in crystal structure. Over the past several years we have shown that robust, highly crystalline, structurally well-defined polymers of the type $[M(Pc)O]_n$, M = Si,Ge; Pc = phthalocyaninato (I,II) (1-5) offer a unique opportunity to experiment with the properties of a (phthalocyanine) molecular metal (III) (6-

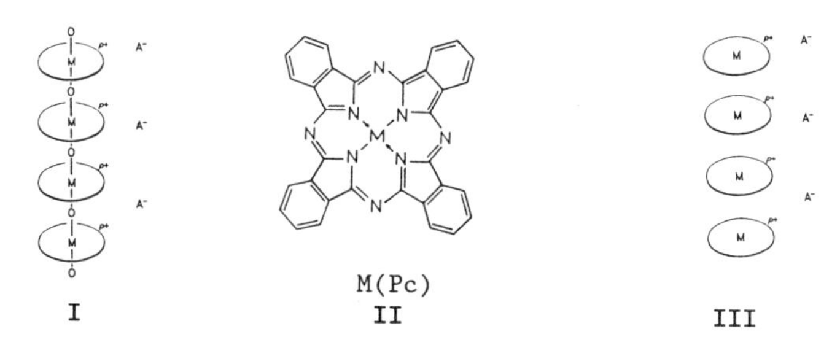

11) under conditions in which the charge-carrying, π radical cation $Pc^{\rho+}$ subunits are rigidly held in a cofacial orientation. In earlier contributions, we discussed in detail the effects on electrical, optical, and magnetic properties, hence band structure, of sequentially varying Pc-Pc interplanar spacing (hence bandwidth) (1-2) and chemically introduced off-axis, charge-compensating counterions (X in $\{[M(Pc)O]X_y\}_n$) (1-5,12-14). However, the scope of chemistry possible increases dramatically when electrochemical approaches (15-18) are used for doping. Thus, we recently communicated (7,19,20) that a far greater range of counterions can be introduced, substantial tuning of the band-filling can be effected, phase boundary/ free energy information can be obtained, and reductive (n-type) doping can be carried out. In addition, the $[M(Pc)O]_n$ system has provided valuable insight into several poorly understood, general aspects of conductive polymer electrochemistry. In view of the rapidly expanding nature of this area, we felt it timely to review a number of recent developments. Thus we summarize here recent results on $[Si(Pc)O]_n$ solid state electrochemistry and closely related structural interconversions.

Electrochemical Methodology

The film-like nature of most conductive polymers is readily adaptable to electrochemical experimentation by simply attaching pressure contacts and immersing the specimen in an electrolyte or by studying films which have been electropolymerized directly on the working electrode (15-18). However, the basic morphological unit of $[Si(Pc)O]_n$ consists of small crystallites (1-5,21-23). We thus utilized two different approaches for electrochemical experiments. Gram-scale reactions can be carried out by doping the polymer as a finely-ground slurry in a standard three-compartment cell (Figure 1). This approach is particularly useful for preparing macroscopic amounts of doped polymer sufficient for physical characterization by a battery of techniques. Alternatively, smaller scale experiments can be carried out by pressing a small amount of the finely ground polymer against a platinum disk electrode (Figure 2). This method has the advantage of more rapid

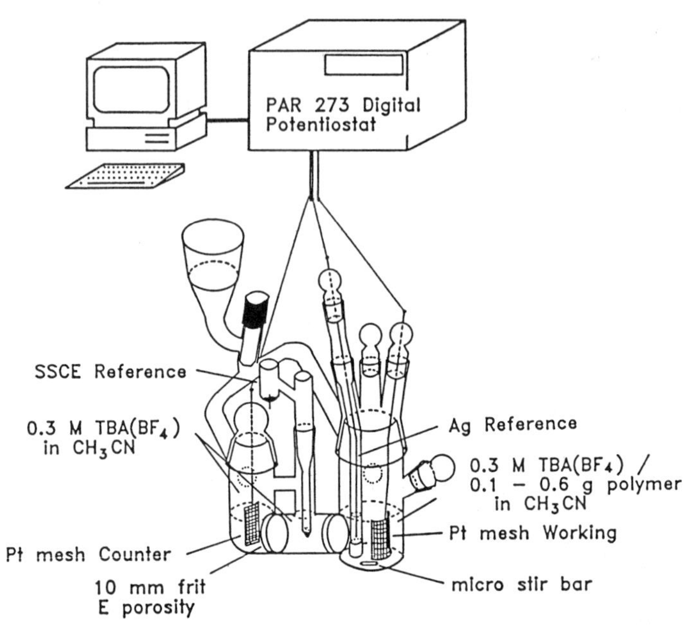

Figure 1. Apparatus for slurry-scale electrochemical experiments with $\{[Si(Pc)O]X_y\}_n$ materials. In the case shown, the equipment is configured for studies in acetonitrile/($\underline{n}$-Bu)$_4$-$N^+BF_4^-$.

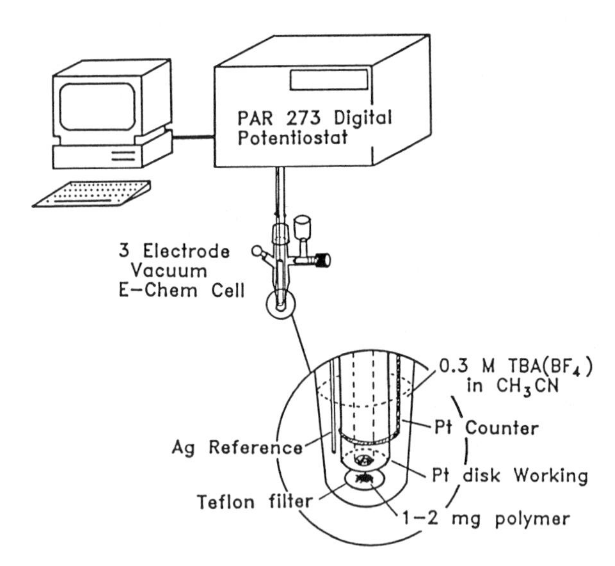

Figure 2. Apparatus for microscale electrochemical experiments with $\{[Si(Pc)O]X_y\}_n$ materials. In the case shown, the equipment is configured for studies in acetonitrile/($\underline{n}$-Bu)$_4N^+BF_4^-$.

kinetics and is analogous to techniques employed for characterizing inorganic electrode materials (24-27). It has not been suitable, in this configuration, for large-scale syntheses. Both types of experiments are controlled by a PAR Model 273 potentiostat/galvanostat interfaced to a Zenith microcomputer using locally developed software. Experiments were carried out under rigorously anhydrous and anaerobic conditions using purified electrolytes and solvents. Both types of experiments utilized silver wire quasi-reference electrodes that were referenced directly to a sodium saturated calomel electrode at the completion of the experiment. In both cases, the key electrochemical experiment is one in which the potential is incrementally stepped and the resulting current flow monitored after each step. When the current decays to a predetermined background level, the next potential step is automatically executed. This experiment has been variously called sequential controlled potential coulometry (CPC), electrochemical potential spectroscopy (ECPS), or electrochemical voltage spectroscopy (EVS) (18,25,28,29), and yields a detailed picture of polymer doping level as a function of applied potential. In the present systems, it has been shown that coulometrically-determined doping stoichiometries are in good agreement with those determined by elemental analysis (19,20,30).

$[Si(Pc)O]_n$ Oxidative Doping in Acetonitrile/$(n\text{-Bu})_4N^+BF_4^-$

Slurry-scale electrochemical doping of $[Si(Pc)O]_n$ in acetonitrile/ $(n\text{-Bu})_4N^+BF_4^-$ is illustrated in Figure 3. Interestingly, the initial oxidation exhibits a substantial overpotential which is not observed on subsequent reoxidation following electrochemical undoping. X-ray diffraction data argue that the initial overpotential (commonly referred to as the "break-in" process (15–18, 32,33) in other conductive polymer systems but poorly understood) is largely structural in origin. Thus, as-polymerized $[Si(Pc)O]_n$ has been assigned a closely-packed orthorhombic crystal structure (3), whereas $\{[Si(Pc)O](BF_4)_y\}_n$ has a tetragonal crystal structure (Figure 4) (12-14). The orthorhombic→tetragonal reorganization necessarily involves 6-8 Å displacements of the tightly-packed $[Si(Pc)O]_n$ chains transverse to the backbone direction. Subsequent electrochemical undoping of tetragonal $\{[Si(Pc)O](BF_4)_y\}_n$ yields a more open $[Si(Pc)O]_n$ phase which can be indexed as an undoped tetragonal crystal structure (Figure 4) (30). This polymer is spectroscopically, diffractometrically, and electrochemically identical to material which can be prepared by thermally undoping tetragonal $\{[Si(Pc)O](I_3)_{0.37}\}_n$ in vacuo.

From a structural standpoint, it is not surprising that tetragonal $[Si(Pc)O]_n$ readily undergoes oxidative doping with a smaller overpotential than the orthorhombic polytype nor that cycling between tetragonal doped and undoped structures is relatively facile. Furthermore, the diffraction data evidence a smooth, monotonic increase in $\{[Si(Pc)O](BF_4)_y\}_n$ lattice parameters (−0.03(2) Å in c and +0.27(6) Å in a) as y = 0→0.50. This is in accord with a simple Veigard's law/homogeneous doping picture. Further support for a continuous tuning of the band-filling is derived from the smooth character of the EVS curves (Figure 3). Transitions between phases of greatly differing structures and free energies would be indicated by steps and plateaus (15-18,24-27). Infrared spectroscopy, X-ray diffraction, and coulometry indicate that repeated electrochemical cycling between doped and undoped states occurs with only minor irreversible decomposition of the polymer. The microdoping experiments are in good agreement with the above electrochemistry (Figure 5), the principal difference being a diminution in all hystereses which are apparently electrode contact related. Although less pronounced than in the slurry experiments, the "break-in" overpotential is clearly evident, while the hysteresis in doping and undoping the tetragonal phase is now very small.

Physical data on the $\{[Si(Pc)O](BF_4)_y\}_n$ system provide an intriguing picture of how the collective properties of a molecular metal vary as band-filling is tuned over a wider range than has ever before been possible (19,20,30,31). As discussed in detail elsewhere both thermoelectric power and optical reflectivity data evidence an evolution from an insulator to a molecular metal. Interestingly, both measurements show an insulator→metal transition at $y \approx 0.15$-0.20, followed by a smooth excursion through a metallic state with progressively less band-filling as $y \rightarrow 0.50$. The close adherence of the behavior (e.g., the plasma frequency, the thermoelectric power) to that predicted by simple tight-binding band theory is gratifying (19,20,31). Magnetic susceptibility measurements evidence a large Curie-like component for $y \lesssim 0.20$, indicating, as also implied by the thermoelectric power and optical results, that there is substantial localization of the electronic structure at low doping levels. This presumably arises from disorder. At higher doping levels, the susceptibility is substantially more Pauli-like as is the case for most phthalocyanine molecular metals (1,2,7-14). Interestingly, four-probe conductivity measurements show a sharp rise in conductivity between $y=0$ and $y \approx 0.20$, followed by a leveling-off beyond $y \approx 0.20$.

What If The Phthalocyanine Rings Are Not Connected By A Polymer Backbone?

To probe the importance of the $[Si(Pc)O]_n$ enforced-stacking superstructure in the electrochemical doping/undoping process, experiments were also carried out with sublimed Ni(Pc), which has a monoclinic, slipped-stack crystal structure (β–Ni(Pc), Figure 6) (34). It is found that the onset potential for oxidative doping in acetonitrile/(n-Bu)$_4N^+BF_4^-$ is roughly comparable to that of orthorhombic $[Si(Pc)O]_n$ (Figure 7). The abrupt rise in y vs. E indicates that few intermediate stoichiometries are accessed before $Ni(Pc)(BF_4)_y$, $y \approx 0.48$, is reached; i.e., there is a major change in crystal structure. X-ray diffraction data indicate that the doped material has a tetragonal crystal structure similar to Ni(Pc)-$(BF_4)_{0.33}$ (9) and $Ni(Pc)(ClO_4)_{0.42}$ (6) (Figure 6). In contrast to $\{[Si(Pc)O](BF_4)_{0.50}\}_n$, electrochemical undoping of $Ni(Pc)(BF_4)_{0.48}$ does not take place over a particularly broad potential range. Rather, the EVS curve is rather steep (Figure 7) suggesting a major structural reorganization upon undoping. Upon complete undoping, powder X-ray diffraction reveals that the electrochemically undoped Ni(Pc) is not similar in structure to tetragonal $[Si(Pc)O]_n$ nor does it exhibit the starting β-Ni(Pc) structure. Rather, the tetragonal crystal structure has collapsed to another known, mono-

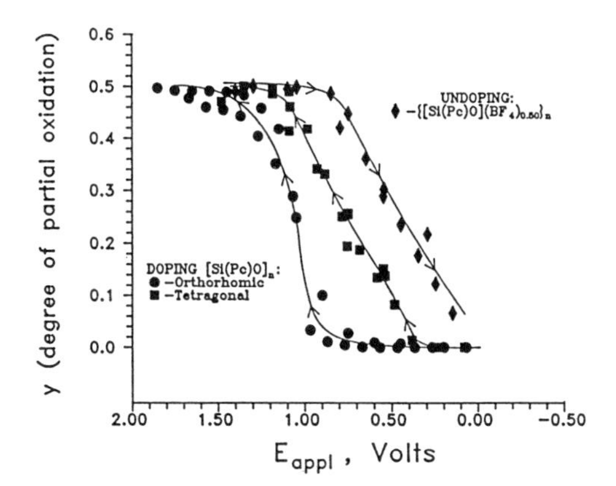

Figure 3. Slurry electrochemical voltage spectroscopy experiments with $\{[Si(Pc)O](BF_4)_y\}_n$ (y = 0.00-0.50) materials in acetonitrile/(n-Bu)$_4N^+BF_4^-$.

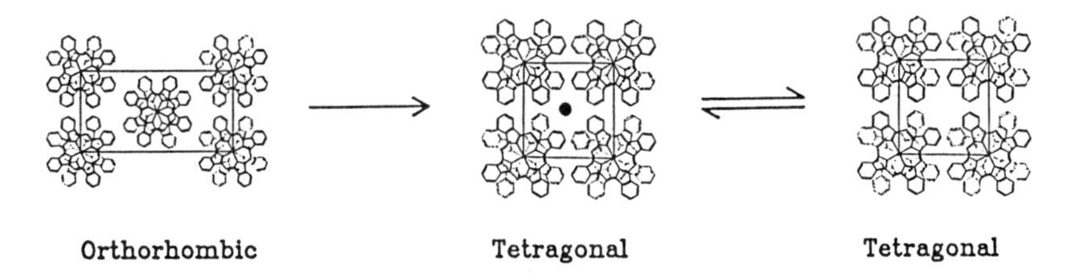

Figure 4. Structural transformations accompanying electrochemical doping and undoping of $\{[Si(Pc)O](BF_4)_y\}_n$ materials, y = 0.00-0.50

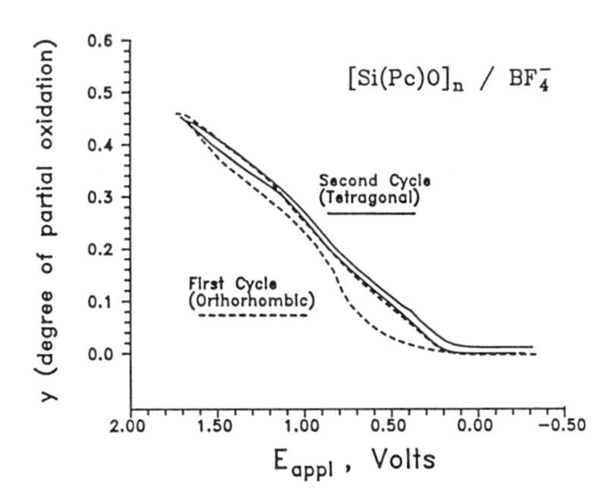

Figure 5. Microscale electrochemical voltage spectroscopy experiments with $\{[Si(Pc)O](BF_4)_y\}_n$ materials in acetonitrile/(n-Bu)$_4N^+BF_4^-$. Data points were acquired at 50 mV intervals.

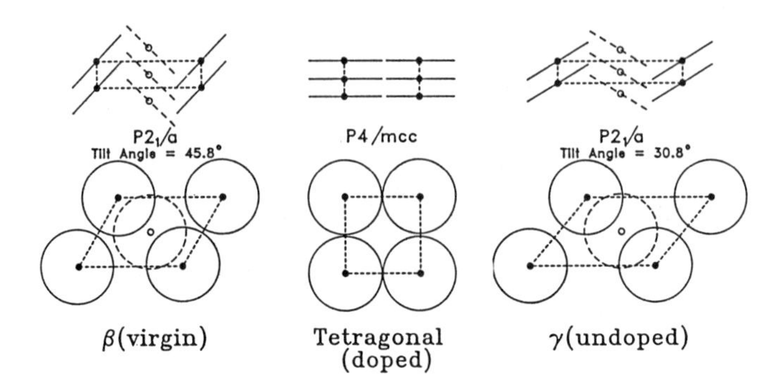

Figure 6. Structural relationships in the electrochemistry of $Ni(Pc)(BF_4)_y$ materials, y = 0.00-0.50.

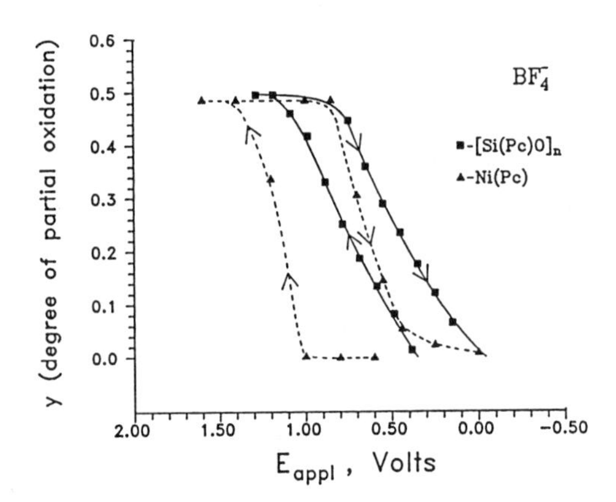

Figure 7. Comparison of slurry electrochemical voltage spectroscopy experiments with Ni(Pc) and tetragonal $[Si(Pc)O]_n$.

clinic slipped-stack γ-Ni(Pc) phase (Figure 6) usually produced by precipitation from acid (35,36). This structure differs principally from the initial β phase in having a smaller tilting angle of the perpendicular to the Ni(Pc) plane with respect to the crystallographic b (stacking) axis. Hence, it is structurally somewhat more closely related to the tetragonal packing arrangement than is the β phase. These results clearly indicate that the enforced M(Pc) stacking is a major structural factor facilitating the electrochemical tuning of the $[Si(Pc)O]_n$ band structure.

$[Si(Pc)O]_n$ Electrochemistry With Other Counterions

The electrochemical techniques described above can be employed with a wide variety of electrolytes to probe the nature of the doping process, factors controlling the ultimate degree of band-filling achievable, and band structure-counterion interactions. In Figure 8 is shown a doping/undoping experiment with tetragonal $[Si(Pc)O]_n$ in acetonitrile/Et_4N^+tosylate$^-$ (p-toluenesulfonate). It can be seen that the band-filling is smoothly tunable over a very broad range: $\{[Si(Pc)O](tos)_y\}_n$, y=0 → ≈0.67. The shape of the EVS curve implies a homogeneous doping process. X-ray diffraction data support this picture and reveal a monotonic expansion in tetragonal lattice parameters (+0.70(5) Å in a and −0.03(2) Å in c) (20,30). Broader linewidths in the $\{[Si(Pc)O](tos)_y\}_n$ diffraction patterns imply a less ordered crystal structure than in $\{[Si(Pc)O](BF_4)_y\}_n$.

The charge transport and optical properties of the $\{[Si(Pc)O]$-$(tos)_y\}_n$ materials as y=0 → 0.67 are reminiscent of the $\{[Si(Pc)O]$-$(BF_4)_y\}_n$ system, but with some noteworthy differences. Again there is an insulator-to-metal transition in the thermoelectric power near y≈0.15-0.20. Beyond this doping stoichiometry, the tosylates also show a continuous evolution through a metallic phase with decreasing band-filling. However, the transition seems somewhat smoother than in the BF_4^- system for y≳0.40, possibly a consequence of a more disordered tosylate crystal structure. Both $\{[Si(Pc)O]$-$(tos)_y\}_n$ optical reflectance spectra and four-probe conductivities are also consistent with a transition to a metal at y≈0.15-0.20. Repeated electrochemical cycling leads to considerably more decomposition than in the tetrafluoroborate system.

Electrochemical experiments with tetragonal $\{Si(Pc)O]_n$ and acetonitrile/$(n\text{-}Bu)_4N^+PF_6^-$ or acetonitrile/$(n\text{-}Bu)_4N^+SbF_6^-$ reveal doping behavior similar to the tetrafluoroborate system with one exception. Limiting stoichiometries obtained are $\{[Si(Pc)O]$-$(PF_6)_{0.47}\}_n$ and $\{[Si(Pc)O](SbF_6)_{0.41}\}_n$. The lower stoichiometries achieved are explicable in terms of the greater sizes of the latter two anions and the space available in the counterion "tunnels" of the $\{[Si(Pc)O]X_y\}_n$ crystal structures. These relationships can be put on a more quantitative footing by considering anion van der Waals diameters as shown in Figure 9. Further verification that packing considerations are important in dictating achievable doping stoichiometries is provided by electrochemical experiments with a series of perfluoroalkylsulfonates as the tetrabutylammonium salts in acetonitrile. As can be seen in Figure 10, EVS curves are similar to other anions, however the "leveling-off" point of limiting Si(Pc) oxidation state falls with increasing anion bulk. In all cases, the X-ray diffraction patterns are consistent with

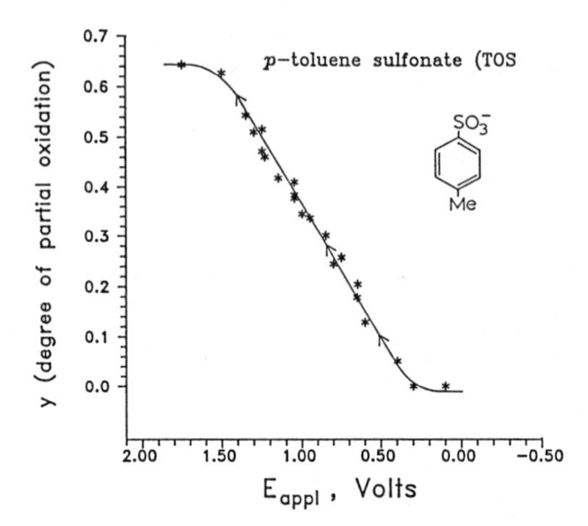

Figure 8. Slurry electrochemical voltage spectroscopy experiments with $\{[Si(Pc)O](tosylate)_y\}_n$ materials, y = 0.00-0.67 in acetonitrile/Et_4N^+(tosylate$^-$).

Counterion	Van der Waals diameter (Å)	Maximum y $\bar{y}_{calc}$	$\bar{y}_{obs}$
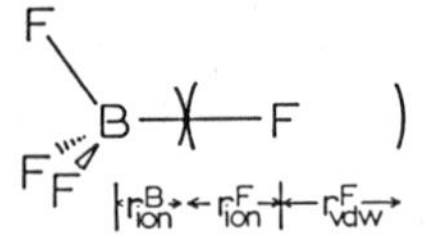 F–B(F)(F)–F (r^B_{ion}, r^F_{ion}, r^F_{vdw})	5.9	0.56	0.50
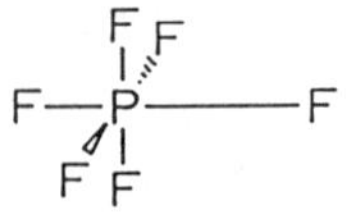 F–P(F)(F)(F)(F)–F	6.4	0.52	0.47
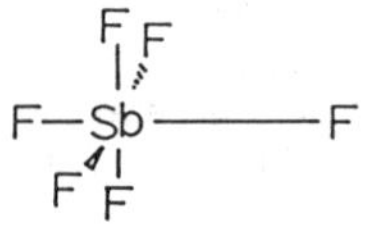 F–Sb(F)(F)(F)(F)–F	6.9	0.48	0.41

Figure 9. Spherical anion size effects on maximum possible $\{[Si(Pc)O]X_y\}_n$ doping levels as seen from a simple packing model.

the now ubiquitous tetragonal $\{[Si(Pc)O]X_y\}_n$ crystal structure. Although van der Waals diameters are more difficult to estimate for these more flexible counterions, assuming elongated $CF_3(CF_2)_nSO_3^-$ geometries yields limiting stoichiometries in reasonable agreement with experiment (Figure 11).

In other work (19,20), we have also shown that sulfate can be electrochemically introduced as a counterion using $[(\underline{n}\text{-Bu})_4N^+]_2SO_4^=$ in acetonitrile. In this case, the final product stoichiometry, $\{[Si(Pc)O](SO_4)_{0.09}\}_n$, is limited by the oxidative stability of the sulfate anion. Thermoelectric power, optical reflectivity, magnetic susceptibility, and four-probe electrical conductivity measurements evidence behavior typical of an $[Si(Pc^{\rho+})O]_n$ compound where $\rho \approx 0.20$. That is, there is no evidence that the more concentrated counterion charge has induced significant localization of the band structure.

Reductive $[Si(Pc)O]_n$ Doping

In principal, it should also be possible to prepare an $[Si(Pc)O]_n$-based metal by introducing electrons into the Si(Pc) lowest unoccupied molecular orbital (LUMO) rather than removing electrons from the highest occupied molecular orbital (HOMO). Molecular orbital calculations at the DVM-Xα level (37-39) indicate that the LUMO is of appropriate Si(Pc) π spatial character to form a conduction band. In contrast to the HOMO, however, it also contains an appreciable amount of Si-O character, so that interesting metal effects may be observed in the physical properties. Initial experiments indicate that tetragonal $[Si(Pc)O]_n$ can indeed be reductively doped to form anionic, extremely air-sensitive $[Si(Pc^{\rho-})O]_n$ products. An example of a microscale doping experiment, which illustrates cycling between p- and n-doping, is shown in Figure 12. Further characterization of this interesting system is in progress.

Conclusions

These results illustrate that electrochemical techniques can be employed to synthesize a vast range of $[Si(Pc)O]_n$-based molecular metals/conductive polymers with wide tunability in optical, magnetic, and electrical properties. Moreover, the structurally well-defined and well-ordered character of the polymer crystal structure offers the opportunity to explore structure/electrochemical/collective properties and relationships to a depth not possible for most other conductive polymer systems. On a practical note, the present study helps to define those parameters crucial to the fabrication, from cheap, robust phthalocyanines, of efficient energy storage devices.

Acknowledgments

This research was supported by the NSF-MFL program through the Materials Research Center of Northwestern University (Grant DMR85-20280) and by the Office of Naval Research. We thank Dr. M. G. Kanatzidis for helpful discussions.

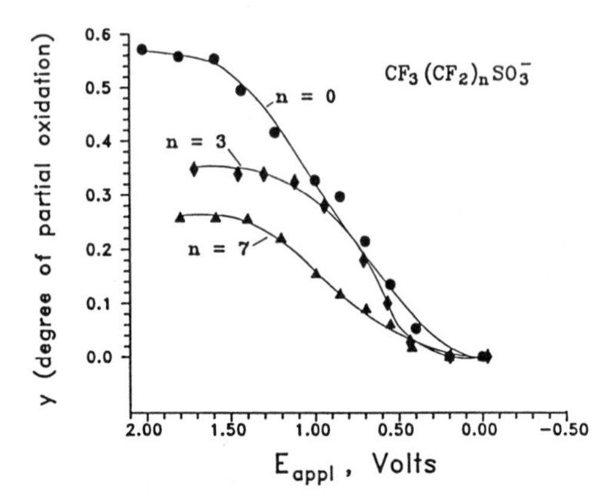

Figure 10. Slurry electrochemical voltage spectroscopy experiments with tetragonal $[Si(Pc)O]_n$ and various perfluoroalkylsulfonates (as the tetrabutylammonium salts in acetonitrile).

Counterion	Van der Waals length (Å)	Maximum y $\bar{y}_{calc}$	$\bar{y}_{obs}$
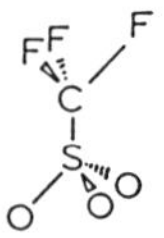	6.4	0.52	0.55
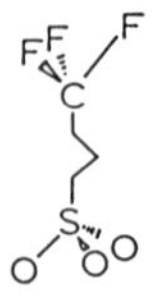	9.4	0.35	0.38
	15.7	0.21	0.26

Figure 11. Perfluoroalkylsulfonate anion size effects on maximum achievable $\{[Si(Pc)O]X_y\}$ doping levels as seen from a simple packing model.

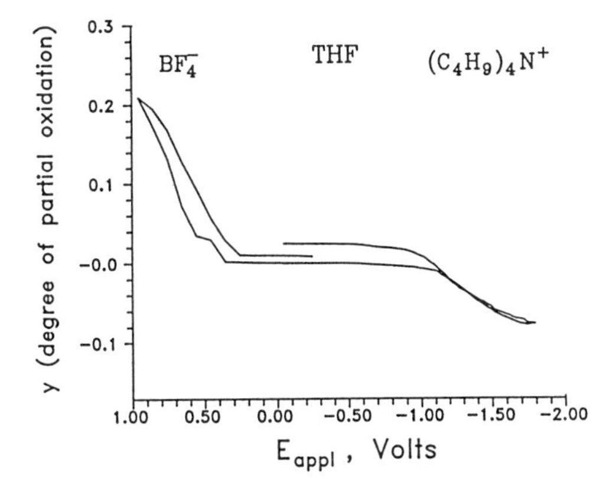

Figure 12. Micro-scale electrochemical voltage spectroscopy showing oxidative as well as reductive doping of tetragonal $[Si(Pc)O]_n$ in THF/($\underline{n}$-Bu)$_4N^+BF_4^-$.

Literature Cited

1. Marks, T. J. Science 1985, 227, 881-889.
2. Diel, B. N.; Inabe, T.; Lyding, J. W.; Schoch, K. F., Jr.; Kannewurf, C. R.; Marks, T. J. J. Am. Chem. Soc. 1983, 105, 1551-1567.
3. Dirk, C. W.; Inabe, T.; Schoch, K. F., Jr.; Marks, T. J. J. Am. Chem. Soc. 1983, 105, 1539-1550.
4. Dirk, C. W.; Inabe, T.; Lyding, J. W.; Schoch, K. F., Jr.; Kannewurf, C. R.; Marks, T. J. J. Polym. Sci., Polym. Symp. 1983, 70, 1-29.
5. Dirk, C. W.; Mintz, E. A.; Schoch, K. F., Jr.; Marks, T. J. J. Macromol. Sci.-Chem. 1981, A16, 275-298.
6. Almeida, M.; Kanatzidis, M. G.; Tonge, L. M.; Marks, T. J.; Marcy, H. O.; McCarthy, W. J.; Kannewurf, C. R. Solid State Commun., in press.
7. Gaudiello, J. G.; Marcy, H. O.; McCarthy, W. J.; Moguel, M. K.; Kannewurf, C. R.; Marks, T. J. Synth. Met. 1986, 15, 115-128.
8. Inabe, T.; Liang, W.-B.; Lomax, J. F.; Nakamura, S.; Lyding, J. W.; McCarthy, W. J.; Carr, S. H.; Kannewurf, C. R.; Marks, T. J. Synth. Met. 1986, 13, 219-229.
9. Inabe, T.; Nakamura, S.; Liang, W.-B.; Marks, T. J.; Burton, R. L.; Kannewurf, C. R.; Imaeda, K.-I. J. Am. Chem. Soc. 1985, 107, 7224-7226.
10. Inabe, T.; Marks, T. J.; Burton, R. L.; Lyding, J. W.; McCarthy, W. J.; Kannewurf, C. R.; Reisner, G. M.; Herbstein, F. H. Solid State Commun. 1985, 54, 501-503.
11. Schramm, C. J.; Scaringe, R. P.; Stojakovic, D. R.; Hoffman, B. M.; Ibers, J. A.; Marks, T. J. J. Am. Chem. Soc. 1980, 102, 6702-6713.
12. Inabe, T.; Gaudiello, J. G.; Moguel, M. K.; Lyding, J. W.; Burton, R. L.; McCarthy, W. J.; Kannewurf, C. R.; Marks, T. J. J. Am. Chem. Soc. 1986, 108, 7595-7608.
13. Inabe, T.; Moguel, M. K.; Marks, T. J.; Burton, R.; Lyding, J. W.; Kannewurf, C. R. Mol. Cryst. Liq. Cryst. 1985, 118, 349-352.
14. Inabe, T.; Kannewurf, C. R.; Lyding, J. W.; Moguel, M. K.; Marks, T. J. Mol. Cryst. Liq. Cryst. 1983, 93, 355-367.
15. Diaz, A. F.; Bargon, J. In Handbook of Conducting Polymers; Skotheim, T. A., Ed.; Marcel Dekker: New York, 1986; Vol. 1, pp 81-115.
16. Street, G. B., ibid. pp. 265-291.
17. Burgmayer, P.; Murray, R. W., ibid., pp. 507-523.
18. Shacklette, L. W.; Toth, J. E.; Murthy, N. S.; Baughman, R. H. J. Electrochem. Soc. 1985, 132, 1529-1535.
19. Gaudiello, J. G.; Almeida, M.; Marks, T. J.; McCarthy, W. J.; Butler, J. C.; Kannewurf, C. R. J. Phys. Chem. 1986, 90, 4917-4920.
20. Almeida, M.; Gaudiello, J. G.; Butler, J. C.; Marcy, H. O.; Kannewurf, C. R.; Marks, T. J. Synth. Met., in press.
21. Zhou, X.; Inabe, T.; Marks, T. J.; Carr, S. H., submitted for publication.
22. Zhou, X.; Marks, T. J.; Carr, S. H. J. Polym. Sci., Phys. Ed. 1985, 23, 305-313.

23. Zhou, X.; Marks, T. J.; Carr, S. H. Polymeric Mats. Sci. Eng. 1984, 51, 651-654.
24. Rouxel, J.; Brec, R. Ann. Rev. Mat. Sci. 1986, 16, 137-162.
25. Thompson, A. H. Rev. Sci. Instrum. 1983, 54, 229-237.
26. Whittingham, M. S. Ann. Chim. Fr. 1982, 7, 204-214, and references therein.
27. Thompson, A. H. J. Electrochem. Soc. 1979, 126, 608-616.
28. Kaufman, J. H.; Mele, J. E.; Heeger, A. J.; Kaner, R.; MacDiarmid, A. G. J. Electrochem. Soc. 1983, 130, 571-574.
29. Kaufman, J. H.; Kaufer, J. W.; Heeger, A. J.; Kaner, R.; MacDiarmid, A. G. Phys. Rev. B. 1982, 26. 2327-2330.
30. Gaudiello, J. G.; Kellogg, G. E.; Tetrick, S. M.; Marks, T. J., submitted for publication.
31. Gaudiello, J. G.; Kellogg, G. E.; Tetrick, S. M.; Almeida, M.; Marks, T. J.; Marcy, H. O.; McCarthy, W. J.; Butler, J. C.; Kannewurf, C. R., submitted for publication.
32. Pickup, P. G.; Osteryoung, R. A. J. Am. Chem. Soc. 1984, 106, 2294-2299, and references therein.
33. Kaufman, F. B.; Schroeder, A. H.; Engler, E. M.; Kramer, S. R.; Chambers, J. Q. J. Am. Chem. Soc. 1980, 102, 483-488.
34. Brown, C. J. J. Chem. Soc. A 1968, 2488-2493.
35. Brown, C. J. J. Chem. Soc. A 1968, 2494-2498, and references therein.
36. Assour, J. M. J. Phys. Chem. 1965, 69, 2295-2299.
37. Pietro, W. J.; Marks, T. J.; Ratner, M. A. J. Am. Chem. Soc. 1985, 107, 5387-5391.
38. Pietro, W. J.; Ellis, D. E.; Marks, T. J.; Ratner, M. A. Mol. Cryst. Liq. Cryst. 1984, 105, 273-287.
39. Ciliberto, E.; Doris, K. A.; Pietro, W. J.; Reisner, G. M.; Ellis, D. E.; Fragalà, I.; Herbstein, F. H.; Ratner, M. A.; Marks, T. J. J. Am. Chem. Soc. 1984, 106, 7748-7761.

RECEIVED September 1, 1987

Chapter 18

A New Approach to the Synthesis of Alkyl Silicates and Organosiloxanes

George B. Goodwin[1] and Malcolm E. Kenney

Department of Chemistry, Case Western Reserve University, Cleveland, OH 44106

A new route to alkyl silicates and organosiloxanes is described. This route has three steps. These are (1) the protonation of a silicate (obtained by collection, mining, or laboratory or commercial synthesis), (2) the esterification of the silicic acid formed by the protonation, and (3) the organodealkoxylation of the alkyl silicate resulting from the esterification. An important feature of this route is that it does not depend on elemental silicon. The route is illustrated with the synthesis of $(EtO)_4Si$ from γ-Ca_2SiO_4, Ca_3SiO_4O, and portland cement, $(EtO)_3SiOSi(OEt)_3$ from $Ca_2ZnSi_2O_7$, $(n\text{-}PrO)_3SiO(n\text{-}PrO)_2SiOSi(On\text{-}Pr)_3$ from $Ca_3Si_3O_9$, the [5.5.1] and [5.3.3] isomers of $(EtO)_{10}Si_6O_7$ from $Cu_6Si_6O_{18}\cdot 6H_2O$ and from $Na_4Ca_4Si_6O_{18}$, and the [5.5.1] and [5.3.3] isomers of $Me_{10}Si_6O_7$ from the [5.5.1] and [5.3.3] isomers of $(EtO)_{10}Si_6O_7$.

As is well known, a number of different kinds of silicate ions are found in silicates. Thus, among gem silicates benitoite, $BaTiSi_3O_9$, contains the cyclotrisilicate ion and aquamarine (beryl), $Be_3Al_2Si_6O_{18}$, contains the cyclohexasilicate ion (1,2). Similarly, among common silicate minerals hemimorphite, $Zn_4Si_2O_7(OH)_2\cdot H_2O$, contains the disilicate ion and enstatite, $MgSiO_3$, contains the infinite chain silicate ion (1). In the case of the common synthetic silicates, portland cement contains phases which can be approximated as Ca_2SiO_4 and Ca_3SiO_4O and which contain the orthosilicate ion (3,4), Figure 1.

Some of the many silicates that are readily available have silicate ions with frameworks that are similar to or the same as those present in common alkyl silicates and common organosiloxanes (5,7). In light of this, a synthetic approach to alkyl silicates and organosiloxanes based on substitution reactions becomes conceivable.

[1]Current address: Glass R&D Center, PPG Industries, Inc., Pittsburgh, PA 15238

0097-6156/88/0360-0238$06.00/0

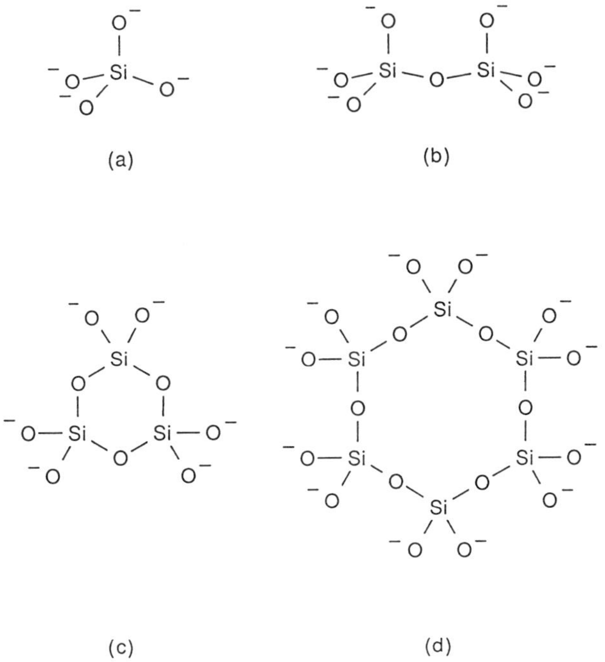

(a) (b)

(c) (d)

Figure 1. Structures of (a) orthosilicate, (b) disilicate, (c) cyclotrisilicate, and (d) cyclohexasilicate ions.

The complete process for synthesizing such species using this approach would entail the acquisition of an appropriate natural silicate or the preparation of an appropriate synthetic silicate and then the conversion of this silicate into the alkyl silicate or organosiloxane by suitable substitution reactions. In terms of bond cleavage, this process could entail no destruction and reformation of framework silicon-oxygen bonds, and, in terms of oxidation number, it would entail no reduction and reoxidation of the silicon.

This process contrasts with the elemental-silicon processes sometimes used for alkyl silicates (8) and the elemental-silicon processes generally used for oligomeric and polymeric organosiloxanes (6,7). Since the silicon in these processes is obtained from quartz, these processes entail, in terms of bond cleavage, the destruction of four silicon-oxygen bonds per silicon and the subsequent reformation of the required number of such bonds. In terms of oxidation number, they entail the reduction of the silicon from four to zero and then its reoxidation back to four, Figures 2 and 3.

Because these processes require reduction and reoxidation of the silicon, they require large amounts of energy per unit of product. This makes them inherently unattractive and makes a search for replacements for them worthwhile. This naturally leads to a consideration of the silicate-based substitution approach to these compounds.

A number of pieces of work indicating that this approach can be developed have been reported. In one, a commercial 3.25:1 SiO_2:Na_2O sodium silicate was successively protonated and butoxylated to a mixture of polymeric butyl silicates (9). In a second, chrysotile asbestos, $Mg_3Si_2O_5(OH)_4$, was successively protonated and allyloxylated to a polymeric allyl silicate, apparently with siloxane framework preservation (10). Also, in a very minor byproduct reaction pseudowollastonite, $Ca_3Si_3O_9$, was protonated and propoxylated to a propyl silicate with framework preservation (11). In other efforts, $(EtO)_3SiOSi(OEt)_3$ was alkyldealkoxylated to organodisiloxanes (12), and oligomeric and polymeric organoalkoxysiloxanes were alkyldealkoxylated to organosiloxanes with or apparently with framework preservation (13,14). Also, monomeric and oligomeric metal silicates were protonated and silylated to organosilylsiloxanes with full or substantial framework preservation (15).

In the present paper, a route to alkyl silicates and organosiloxanes which has elements in common with these pieces of work and which is based on the substitution approach is illustrated and discussed.

THE ROUTE

In the route used in this work, a monomeric or oligomeric silicate is protonated, the resulting silicic acid is esterified, and the alkyl silicate is organodealkoxylated. In each step, the siloxane framework is fully or substantially preserved. For the case starting with a metal disilicate the reactions can be represented as:

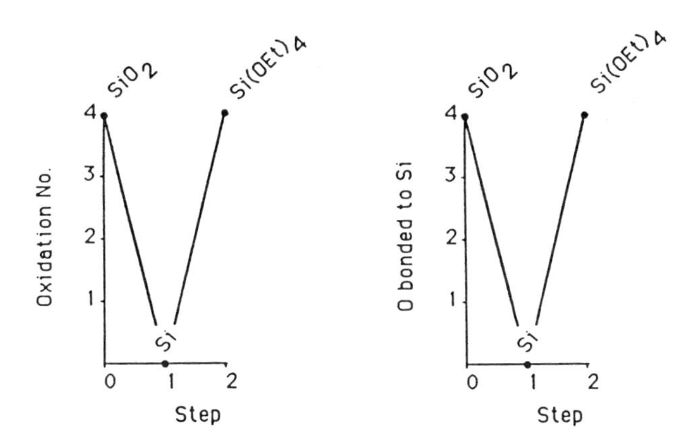

Figure 2. Variation of oxidation number of silicon with step and variation of number of oxygens bonded to silicon with step in direct or elemental-silicon process for $(EtO)_4Si$.

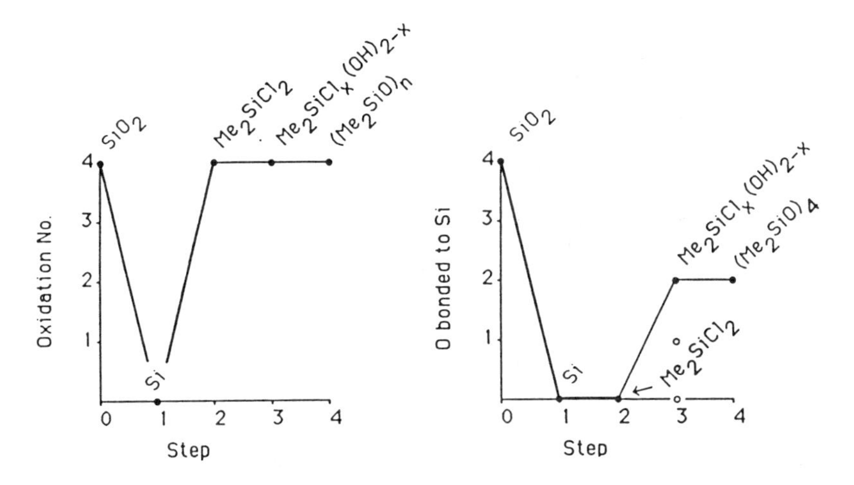

Figure 3. Variation of oxidation number of silicon with step and variation of number of oxygens bonded to silicon with step in usual process for $(Me_2SiO)_4$.

$$O_3SiOSiO_3^{6-} + H^+ + ROH \longrightarrow (HO)_3SiOSi(OH)_3 \cdot xROH \qquad (1)$$

$$(HO)_3SiOSi(OH)_3 \cdot xROH + ROH \longrightarrow (RO)_3SiOSi(OR)_3 \qquad (2)$$

$$(RO)_3SiOSi(OR)_3 + R'MgX \longrightarrow R'_3SiOSiR'_3 \qquad (3)$$

It is possible that some esterification of the silicate ion occurs in this route before the ion is fully protonated and thus that the first two steps of the route overlap.

A novel feature of the route is that it leads to monomeric and oligomeric alkyl silicates from metal silicates in good yield and with full or substantial siloxane framework preservation. Further, it leads to oligomeric organosiloxanes of intermediate molecular weight from alkyl silicates in good yield and with siloxane framework preservation.

EXAMPLES

$(EtO)_4Si$ from γ-Ca_2SiO_4. A dilute HCl-ethanol solution was added slowly to a a cooled suspension of γ-Ca_2SiO_4 (Cerac Inc., Milwaukee, WI) in ethanol. The resulting suspension was filtered and the solid was washed. The filtrate and the washings were combined and slowly added to an ethanol-toluene solution that was being distilled at a moderate rate. After the resultant had been distilled until a substantial amount of distillate had been collected, the suspension obtained was filtered and the solid was washed. The filtrate and washings were combined, concentrated, bulb-to-bulb distilled, and fractionally distilled. With the aid of gas chromatography, gas chromatography-mass spectrometry, and infrared spectroscopy, the product was compared to authentic $(EtO)_4Si$. The results showed that it was quite pure (99.8%) $(EtO)_4Si$ (contained yield 33%).

The $(EtO)_4Si$ could have been converted to Me_4Si with a methyl Grignard reagent if desired (16).

$(EtO)_4Si$ from MONOCLINIC Ca_3SiO_4O. Monoclinic Ca_3SiO_4O (Construction Technology Laboratories, Inc., Skokie, IL) was treated in a manner similar to that used with γ-Ca_2SiO_4. The product was relatively pure (95%) $(EtO)_4Si$ (contained yield 42%).

$(EtO)_4Si$ from PORTLAND CEMENT. Portland cement (Maryneal zero C_3A Type III, Lone Star Industries, Inc., Houston, TX) was also treated in a similar fashion. The product again was relatively pure (94%) $(EtO)_4Si$ (contained yield 41%).

$(EtO)_3SiOSi(OEt)_3$ from $Ca_2ZnSi_2O_7$ (HARDYSTONITE). An HCl-ethanol solution was added slowly to a suspension of $Ca_2ZnSi_2O_7$ (prepared by sintering ZnO and $CaSiO_3$ (wollastonite)) in ethanol. The suspension formed was stirred at ambient temperature and filtered. The filtrate was cooled and added slowly to an ethanol-toluene solution that was being distilled at a moderate rate. After the resultant had been distilled until a substantial amount of distillate had been collected, the mixture produced was decanted. The decantate was concentrated and fractionally distilled. The product was compared to authentic $(EtO)_3SiOSi(OEt)_3$ with the aid of gas chromatography, gas chromatography-mass spectrometry, and infrared spectroscopy. The results showed that it was relatively pure (97%) $(EtO)_3SiOSi(OEt)_3$ (contained yield 29%).

The $(EtO)_3SiOSi(OEt)_3$ could have been converted to $Et_3SiOSiEt_3$ with an ethyl Grignard reagent if desired (12).

$(n\text{-}PrO)_3SiO(n\text{-}PrO)_2SiOSi(On\text{-}Pr)_3$ from $Ca_3Si_3O_9$ (PSEUDOWOLLASTONITE). To a suspension of $Ca_3Si_3O_9$ (prepared by heating $CaSiO_3$ (wollastonite)) in *n*-propanol was added an HCl-*n*-propanol solution (and the addition funnel washings). The resulting mixture was distilled until a substantial amount of distillate had been collected and the suspension produced was filtered. The solid was washed, the filtrate and washings were combined, and the resultant was concentrated and mixed with an HCl-*n*-propanol solution. After the solution formed had been distilled until a substantial amount of distillate had been collected, the remainder was twice diluted with pentane, filtered, and concentrated. The final concentrate was bulb-to-bulb distilled and fractionally distilled. With the aid of gas chromatography, gas chromatography-mass spectrometry, and infrared spectroscopy, the product was shown to be relatively pure (94%) $(n\text{-}PrO)_3SiO(n\text{-}PrO)_2SiOSi(On\text{-}Pr)_3$ (contained yield 18%): IR (neat) 2964 (s), 2938 (s), 2880 (s), 1465 (m) 1088 (vs) cm^{-1}; GCMS *m/z* [rel intensity] 529 [$(M\text{-}OPr)^+$, 7], 235 [$((HO)_7Si_3O_2)^+$, 100].

THE [5.5.1] and [5.3.3] ISOMERS of $(EtO)_{10}Si_6O_7$ from $Cu_6Si_6O_{18}\cdot 6H_2O$ (DIOPTASE). An HCl-ethanol solution was added slowly to a suspension of $Cu_6Si_6O_{18}\cdot 6H_2O$ (Ward's Natural Science Establishment, Inc., Rochester, NY) in an ethanol-toluene solution. The resulting mixture was distilled until a substantial amount of distillate had been collected and the suspension formed was cooled and filtered. The solid was washed and the filtrate and washings were combined. After the solution produced had been concentrated, it was mixed with an HCl-ethanol solution, ethanol, and toluene. The mixture formed was distilled until a substantial amount of distillate had been collected, and the remainder was distilled twice. By gas chromatography, gas chromatography-mass spectrometry, infrared spectroscopy, and ^{29}Si nuclear magnetic resonance spectroscopy, the product was shown to contain a substantial amount (43%) of the [5.5.1] isomer of $(EtO)_{10}Si_6O_7$ (contained yield 20%): GCMS [rel intensity] 685 [$(M\text{-}OEt)^+$, 6], 415 [$((HO)_7Si_6O_8)^+$, 100]; ^{29}Si NMR (39.7 MHz, $CDCl_3$) δ -95.56 (middle type Si), -101.55 (branch type Si). It was also shown to contain a significant amount (29%) of the [5.3.3] isomer of $(EtO)_{10}Si_6O_7$ (contained yield 14%): GCMS *m/z* [rel intensity] 685 [$(M\text{-}OEt)^+$, 12], 415 [$((HO)_7Si_6O_8)^+$, 100]; ^{29}Si NMR (39.7 MHz, $CDCl_3$) δ -94.69 (middle type Si), -96.65 (middle type Si), -101.27 (branch type Si).

THE [5.5.1] and [5.3.3] ISOMERS of $(EtO)_{10}Si_6O_7$ from $Na_4Ca_4Si_6O_{18}$. $Na_4Ca_4Si_6O_{18}$ (prepared by sintering Na_2CO_3, $CaCO_3$, and SiO_2) was treated in a manner similar to that used with $Cu_6Si_6O_{18}\cdot 6H_2O$. A substantial part (46%) of the product was the [5.5.1] isomer of $(EtO)_{10}Si_6O_7$ (contained yield 31%): GCIR (Ar matrix) 2984 (m), 2936 (m), 2905 (m), 1161 (s), 1105 (vs) cm^{-1}. A significant part

(28%) of the product was the [5.3.3] isomer of $(EtO)_{10}Si_6O_7$ (contained yield 20%): GCIR (Ar matrix) 2983 (m), 2935 (m), 2904 (m), 1162 (s), 1106 (vs) cm^{-1}.

THE [5.5.1] and [5.3.3] ISOMERS of $Me_{10}Si_6O_7$ from the [5.5.1] and [5.3.3] ISOMERS of $(EtO)_{10}Si_6O_7$. A solution of CH_3MgCl in tetrahydrofuran was slowly added to a cooled solution of the [5.5.1] and [5.3.3] isomers of $(EtO)_{10}Si_6O_7$ (prepared as above) in tetrahydrofuran. The resulting mixture was stirred for a considerable period of time while being kept cool and then was concentrated. The concentrate was stirred with a mixture of dilute HCl and pentane, and the resulting organic phase was separated and washed with an aqueous NaCl solution. It was then distilled until a substantial amount of distillate had been collected. The oil formed was flash chromatographed and the eluate was twice fractionally distilled. The product was shown by gas chromatography, gas chromatography-mass spectrometry, gas chromatography-infrared spectroscopy, high pressure liquid chromatography, and ^{29}Si nuclear magnetic resonance spectroscopy to contain a substantial amount (60%) of the [5.5.1] isomer of $Me_{10}Si_6O_7$ (contained yield on the basis of contained [5.5.1] isomer of $(EtO)_{10}Si_6O_7$ 38%): IR (neat) 2966 (m), 1260 (s), 1079 (vs) cm^{-1}; GCMS *m/z* [rel intensity] 415 [$(M-Me)^+$, 68]; ^{29}Si NMR (39.7 MHz, $CDCl_3$) δ -19.07 (D type Si); -63.55 (T type Si). In addition, it was shown to contain a significant amount (29%) of the [5.3.3] isomer of $Me_{10}Si_6O_7$ (contained yield on the basis of contained [5.3.3] isomer of $(EtO)_{10}Si_6O_7$ 29%): IR (neat) 2966 (m), 1263 (s), 1083 (vs) cm^{-1}; GCMS *m/z* [rel intensity] 415 [$(M-Me)^+$, 83]; ^{29}Si NMR (39.7 MHz, $CDCl_3$) δ -17.98 (D type Si), -20.93 (D type Si), -62.73 (T type Si). The [5.5.1] and [5.3.3] isomers of $Me_{10}Si_6O_7$ are known compounds (17,18).

DISCUSSION

REACTANTS and CHEMISTRY, STEP 1. The silicates used in the examples given in the first or protonation step of the route have silicate ions with various structures. Thus, γ-Ca_2SiO_4, monoclinic Ca_3SiO_4O, and the principal silicate phases in portland cement, approximately Ca_3SiO_4O and Ca_2SiO_4 contain the orthosilicate ion (3,4), $Ca_2ZnSi_2O_7$ contains the disilicate ion (19), $Ca_3Si_3O_9$ contains the cyclotrisilicate ion (1), and $Cu_6Si_6O_{18}\cdot 6H_2O$ and $Na_4Ca_4Si_6O_{18}$ contain the cyclohexasilicate ion (1,20). The structural and chemical diversity of these silicates show that a wide variety of silicates can be employed in this step.

Various methods can be used to obtain the silicates necessary for the step including small scale collection (e.g., dioptase), mining, synthesis from silica (e.g., $Na_4Ca_4Si_6O_{18}$), and synthesis from other silicates (e.g., $Ca_3Si_3O_9$ and $Ca_2ZnSi_2O_7$). The results obtained so far indicate that the silicates best suited for use are often calcium silicates.

No acids other than HCl have been tried in the step. However, it is likely that it can be carried out with other strong acids.

From the results of several experiments it has been learned that isopropanol and *n*-butanol can be used to make alkyl silicates by the route and thus that they can be used in the first step. This suggests that a variety of alcohols can be used in it (however, as is obvious, the choice of the alcohol is governed by the nature of the alkoxy groups needed in the second step).

The fact that the alkyl silicates produced by the route retain the siloxane framework of the parent silicates fully or substantially, Figure 4, shows that the framework is at least largely retained in the first step. A number of experiments indicate one factor aiding structure retention in this step is the use of an amount of HCl that is just slightly above that which is stoichiometrically required (105-110%). Other experiments indicate that in some instances the use of a low temperature (~-10 °C) likewise helps with structure retention.

Also pertinent to structure retention in this step are results from experiments showing that the route yields alkyl silicates having poor structure retention when the intermediate silicic acid concentration is high and good retention when it is low. These results suggest that a low silicic acid concentration (~0.04M) aids structure retention in the first step. If so, this is easily understandable in terms of the ease with which silicic acids react with themselves.

Since it is well known that alcohols form strong hydrogen bonds, it is very probable that the alcohol in this step hydrogen bonds to the silicic acid (or acids) produced in it, and as a result creates a protective sheath around each silicic acid molecule. This sheath is clearly very important in framework retention in view of the reactivity of silicic acids.

STEP 2. The experiments with ethanol, *n*-propanol, isopropanol, and *n*-butanol already described or mentioned indicate that a variety of alcohols can be used in the second step. It appears that unhindered primary and secondary alcohols generally will be suitable.

The acid in this step clearly functions as a catalyst (acids are known to catalyze the esterification of silanols) (21). The toluene, when employed, serves to drive the esterification to completion by forming a water-toluene-ethanol azeotrope (12% water) (22). It also renders the reaction solution a poor solvent for the byproduct salts and thus facilitates the separation of these salts (the pentane, when used, serves this same function).

The fact that the alkyl silicates produced by the route fully or substantially retain the original siloxane framework also shows that the framework is at least largely retained in the second step. The results pertaining to silicic acid concentration already mentioned lead to the conclusion that a low silicic acid concentration aids this structure retention. Also aiding it, no doubt, is the ability of the alcohol to sheath and protect the silicic acid.

(The ring opening occuring in the conversion of the $Si_3O_9{}^{6-}$ ion to $(n\text{-PrO})_3SiO(n\text{-PrO})_2SiOSi(On\text{-Pr})_3$ could occur in either the first or second step of the synthesis. It is attributable to a proton-assisted cleavage of the siloxane framework enhanced by the strain inherent in the ring. The rearrangement which occurs in the conversion of the $Si_6O_{18}{}^{12-}$ ion to the [5.3.3] isomer of $(EtO)_{10}Si_6O_7$ also could occur in either the first or second step

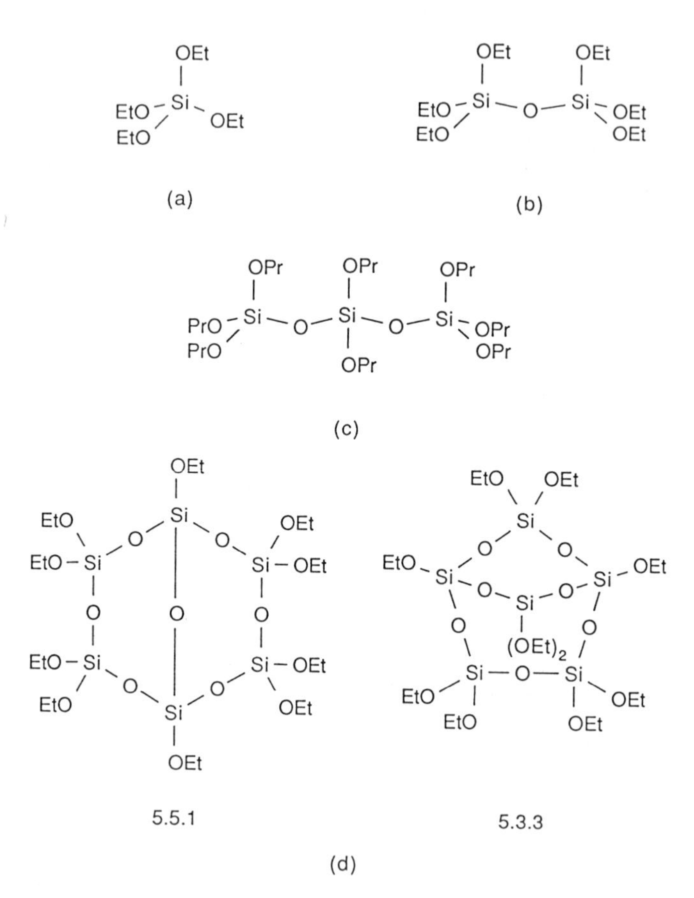

Figure 4. Structures of alkyl silicates produced from (a) γ-Ca_2SiO_4, Ca_3SiO_4O, and portland cement, (b) $Ca_2ZnSi_2O_7$, (c) $Ca_3Si_3O_9$, and (d) $Cu_6Si_6O_{18}\cdot 6H_2O$ and $Na_4Ca_4Si_6O_{18}$.

of the synthesis. It can be ascribed to a proton-assisted framework cleavage followed by ring-contracting ring closures.)

STEP 3. The last or organodealkoxylation step can no doubt be carried out with a variety of Grignard reagents, and thus it is clearly flexible too. From the results of various experiments, it is apparent that framework retention in this step is aided by a low temperature (~0 °C) and a moderate Grignard reagent to alkoxy group ratio (~2:1).

UTILITY. This route appears to offer a path to alkyl silicates that is simple and has the potential of having low energy requirements. It does not offer a corresponding low-energy path to organosiloxanes because a Grignard reagent is used in the organodealkoxylation step and thus elemental magnesium is ultimately required for this step. However, it should be pointed out that elemental magnesium is needed only for the formation of the Si-C bonds in this route, whereas elemental silicon is required for the formation of all bonds in the common route. In this way the new route is advantageous. Further, its existence reemphasizes the fact that possible paths to organosiloxanes that are not based on elemental silicon cannot be summarily dismissed.

The route also provides a satisfactory path to some alkyl silicates for which no other satisfactory routes are available. Some of these alkyl silicates may be of interest in the synthesis of ceramics by the sol-gel technique.

ACKNOWLEDGMENT. We thank Ralph E. Temple, Dale R. Pulver, and Gordon Fearon for helpful discussions. We also thank Diamond Shamrock and Dow Corning for financial support.

LITERATURE CITED

1. Liebau, F. Structural Chemistry of Silicates; Springer-Verlag: Berlin, 1985.
2. Fleischer, M. Glossary of Mineral Species 1980; Mineralogical Record: Tucson, AZ, 1980.
3. Helmuth, R. A.; Miller, F. M.; O'Connor, T. R.; Greening, N. R. In Kirk-Othmer Encyclopedia of Chemical Technology, 3rd ed.; Grayson, M., Ed.; John Wiley: New York, 1979; Vol. 5, p 163.
4. Wells, A. F. Structural Inorganic Chemistry, 5th ed.; Oxford: Oxford, 1984; p 1017.
5. Arkles, B. In Kirk-Othmer Encyclopedia of Chemical Technology, 3rd ed.; Grayson, M., Ed.; John Wiley: New York, 1982; Vol. 20, p 912.
6. Stark, F. O.; Fallender, J. R.; Wright, A. P. In Comprehensive Organometallic Chemistry; Wilkinson, G., Ed.; Pergamon: Oxford, 1982; Vol. 2, p 305.
7. Hardman, B. B.; Torkelson, A. In Kirk-Othmer Encyclopedia of Chemical Technology, 3rd ed.; Grayson, M., Ed.; John Wiley: New York, 1982; Vol. 20, p 922.
8. Ayen, R. J.; Burk, J. H. Mater. Res. Soc. Symp. Proc. 1986, 73 (Better Ceram. Chem. 2), 801.
9. Iler, R. K.; Pinkney, P. S. Ind. Eng. Chem. 1947, 39, 1379.

10. Bleiman, C.; Mercier, J. P. Inorg. Chem. 1975, 14, 2853.
11. Calhoun, H. P.; Masson, C. R. J. Chem. Soc., Dalton Trans. 1980, 1282.
12. Smith, B. Ph.D. Thesis, Chalmers Technical High School, Gothenburg, Sweden, 1951 as quoted in Eaborn, C. Organosilicon Compounds; Academic Press: New York, 1960; p 14.
13. Wacker-Chemie G.m.b.H. Br. Patent 732 533, 1955.
14. Compton, R. A.; Petraitis, D. J. U. S. Patent 4 309 557, 1982.
15. Lentz, C. W. Inorg. Chem. 1964, 3, 574.
16. George, P. D.; Sommer, L. H.; Whitmore, F. C. J. Am. Chem. Soc. 1955, 77, 6647.
17. Jancke, H.; Engelhardt, G.; Magi, M.; Lippman, E. Z. Chem. 1973, 13, 392.
18. Menczel, G. Acta Chim. Acad. Sci. Hung. 1977, 92, 9.
19. Deer, W. A.; Howie, R. A.; Zussman, J. Rock-Forming Minerals; John Wiley: New York, 1962; Vol. 1, p 240.
20. Ohsato, H.; Takéuchi, Y.; Maki, I. Acta Crystallogr., Sect. C: Cryst. Struct. Commun. 1986, 42, 934.
21. Eaborn, C. Organosilicon Compounds; Academic Press: New York, 1960; p 295.
22. Horsley, L. H. Azeotropic Data II; Advances in Chemistry 35; American Chemical Society: Washington, DC, 1962; p 61.

RECEIVED September 1, 1987

POLYPHOSPHAZENES

Chapter 19

Current Status of Polyphosphazene Chemistry

Harry R. Allcock

Department of Chemistry, Pennsylvania State University, University Park, PA 16802

Inorganic polymer chemistry is an area of research that links the classical fields of ceramics, metals, and organic polymers, and provides opportunities for the synthesis of new substances that combine the properties of all three. Polyphosphazenes are inorganic macromolecules that illustrate the possibilities available for a wide range of other inorganic systems. In this article, it will be demonstrated that, depending on the synthesis method and molecular structure, it is possible to bias the properties toward those of ceramics, metals, or organic high polymers, and also to develop polymers of biomedical interest.

Polyphosphazenes comprise some of the most intensively studied inorganic macromolecules. They include one of the oldest known synthetic polymers and many of the newest. In molecular structural versatility, they surpass all other inorganic polymer systems (with over 300 different species now known), and their uses and developing applications are as broad as in many areas of organic polymer chemistry.

Moreover, the molecular structural, synthetic, and property nuances of these polymers illustrate many of the attributes, problems, and peculiarities of other inorganic macromolecular systems. Thus, they provide a "case study" for an understanding of what may lie ahead for other systems now being probed at the exploratory level. In short, an understanding of polyphosphazene chemistry forms the basis for an appreciation of a wide variety of related, inorganic-based macromolecular systems and of the relationship between inorganic polymer chemistry and the related fields of organic polymers, ceramic science, and metals.

This relationship is illustrated in Figure 1. The science of solids is the science of supramolecular systems in which the three-dimensional solid structure is held together by covalent bonds

0097-6156/88/0360-0250$06.00/0

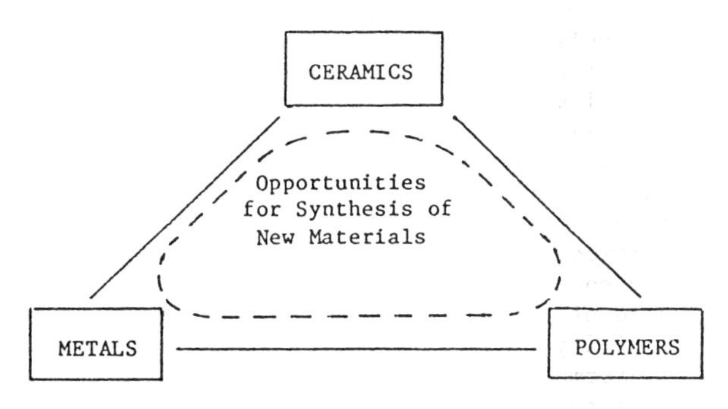

Figure 1. The relationship between the three classical materials areas of ceramics, polymers, and metals.

(ceramics), metallic bonds (metals), ionic attractions (salts), or a combination of covalent and van der Waals forces (linear or branched polymers). In practice, it is convenient to consider solid systems in terms of the three-cornered arrangement shown in Figure 1, in which the hitherto separate and insulated fields of organic polymers, ceramics, and metals form the cornerstones of traditional knowledge and practice for materials research. A crucial point to be emphasized is that future developments in the materials field will almost certainly occur on the connecting lines that join these disciplines or in the central area that lies between all three.

This is one of the main purposes of inorganic polymer research --the search for new and useful compounds and materials that combine the properties of polymers with those of ceramics and/or metals. Such hybrid molecules and supramolecular solids offer the promise of systems with the flexibility, strength, toughness, and ease of fabrication of polymers, with the high temperature oxidative stability of ceramics, and the electrical or catalytic properties of metals. Polyphosphazene chemistry provides an illustration of what is possible in one representative hybrid system.

Macromolecules differ from small molecules in a number of critical properties. First, the linear chain structure confers elasticity, toughness, and strength on the solid state system. This is a consequence of the reorientational freedom of the skeletal bonds and of their ability to absorb impact or undergo elastic deformation by means of conformational changes rather than bond cleavage. However, the nature of the skeletal bonds and the elements involved can have a powerful influence on the torsional barrier for individual skeletal bonds.

Second, linear chain polymers are thermodynamically unstable at elevated temperatures. Entropic influences favor a breakdown to small molecules either by random fragmentation or by depolymerization. The latter process involves a reversion of the polymer to monomer or small molecule rings. Depolymerization to small rings is a feature common to many inorganic polymers at temperatures above 200-250°C.

Third, the introduction of crosslinks between chains confers insolubility and increased solid state rigidity, often accompanied by improved thermal stability. High degrees of crosslinking confer ceramic-type properties on the solid, whether the backbone atoms are carbon atoms or inorganic species.

And finally, irrespective of the types of elements in the backbone, the properties of a linear polymer will depend on the side groups attached to that backbone. This principle underlies all polyolefin and polyvinyl macromolecular science and technology. It applies equally well to inorganic polymer systems.

An important developing area that lies in the region between polymer chemistry, ceramic science, and metals, involves the search for new electrically-conducting solids. Linear polymers may conduct electricity by electronic or ionic mechanisms. As will be discussed, polyphosphazenes have been synthesized that, depending on the side group structure, conduct by either of these two processes.

The historical development of polyphosphazene chemistry is compared in Figure 2 with those of other inorganic polymer systems. Its origins can be traced to the late 1800's, (1) although the first

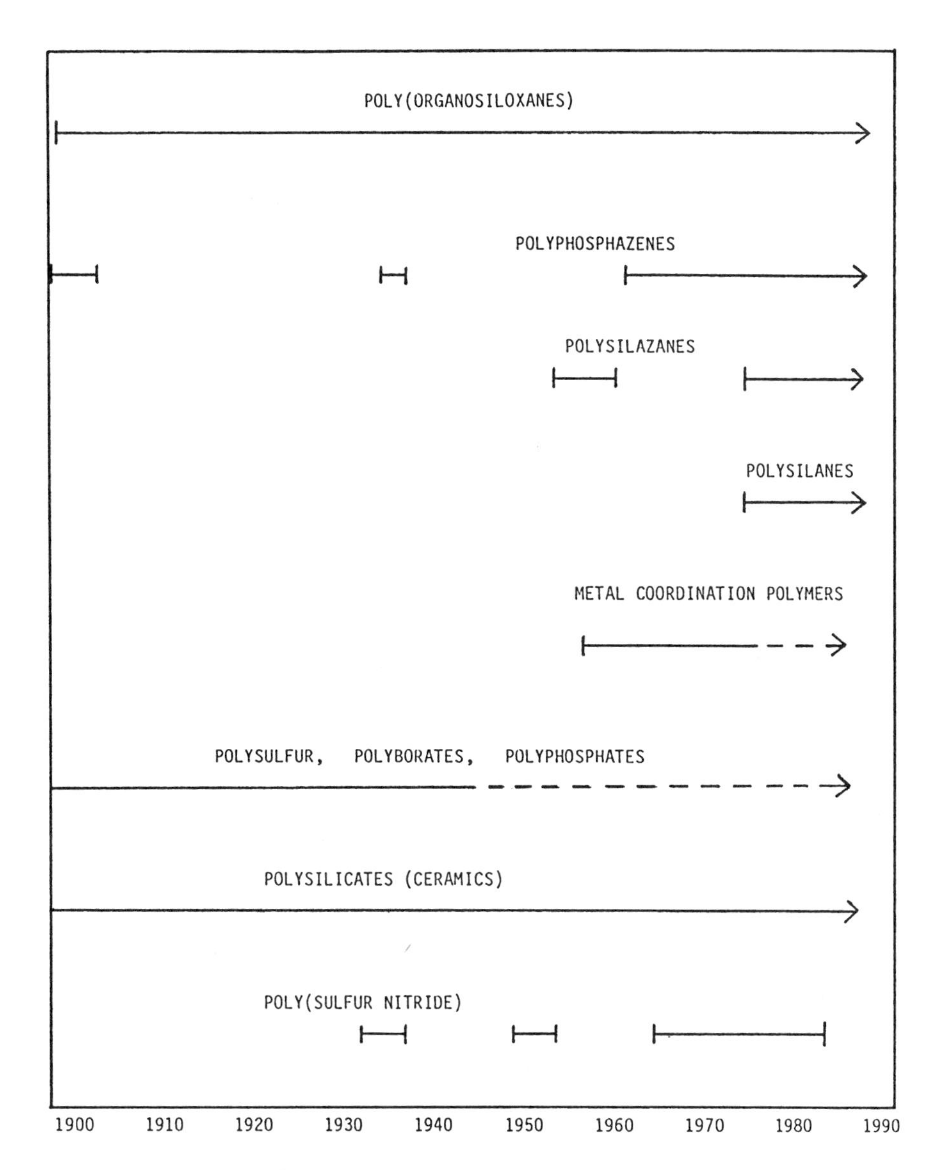

Figure 2. Approximate sequence of development of different areas of inorganic polymer chemistry from 1900 to the present. The broken lines represent a continuing but low level of activity. Vertical bars indicate abrupt beginnings and ends of transient research efforts.

stable poly(organophosphazenes) were not synthesized until the mid-1960's.

Synthesis of Polyphosphazenes

Three approaches have been developed for the synthesis of polyphosphazenes. These are: (1) The macromolecular substitution route; (2) The cyclic trimer or tetramer substitution/polymerization route, and (3) Direct synthesis from organosilylphosphazene monomers. This last method is described in detail in another Chapter and will not be considered further in this review.

Macromolecular Substitution Route. The current surge in polyphosphazene research is mainly a result of the development in the mid 1960's (2-4) of a substitutive route to the synthesis of organo phosphazene high polymers. Before that time, only a sporadic interest in the subject existed because the known polymers, cross linked poly(dihalophosphazenes), (1,5) were insoluble and hydrolytically unstable.

The substitutive method of synthesis is illustrated in Scheme I. Thermal polymerization of hexachlorocyclotriphosphazene (I) yields an uncrosslinked, organic solvent-soluble high polymer (II) (2-4). This reaction must be controlled carefully to avoid the formation of the crosslinked, insoluble polymer known as "inorganic rubber" (1). The cyclic trimer (I), obtained from phosphorus pentachloride and ammonium chloride, is available commercially and is now used on a large scale in the polymerization reaction. Uncrosslinked poly(dichlorophosphazene) (II) is a highly reactive macromolecular intermediate. The chlorine atoms can be replaced by a wide variety of organic or organometallic side groups. Replacement of the chlorine atoms converts the hydrolytically-sensitive intermediate to one of many water-stable polymeric derivatives, usually without cleavage of the backbone bonds.

This reaction is the basis of nearly all the polyphosphazene chemistry and technology developed during the last 20 years. The important consequence of this reaction pathway is that molecular diversity can be achieved by macromolecular substitution rather than by the more conventional method of polymerization or copolymerization of different monomers. It allows polymers with different side groups or combinations of side groups to be prepared with an identical backbone and molecular weight distribution. Thus, direct comparisons can be made of the effects of different side groups on physical properties. It also allows gross or subtle property changes to be designed into a macromolecule, a feature that assists both laboratory studies and technological innovations. Because of these advantages, the substitution route has been a major reason for both the recent expansion of research in polyphosphazenes and the use of these polymers in advanced technology.

Four examples of the effects of changes in side group structure are illustrated in the polymers depicted as II and VI-VIII. This method of synthesis can be used as a prelude to the generation of a wider structural variety if organic reaction chemistry is carried out on the organic side groups. For example, the glucosylphosphazene polymer shown as X has been prepared by the chemistry shown in Scheme II.

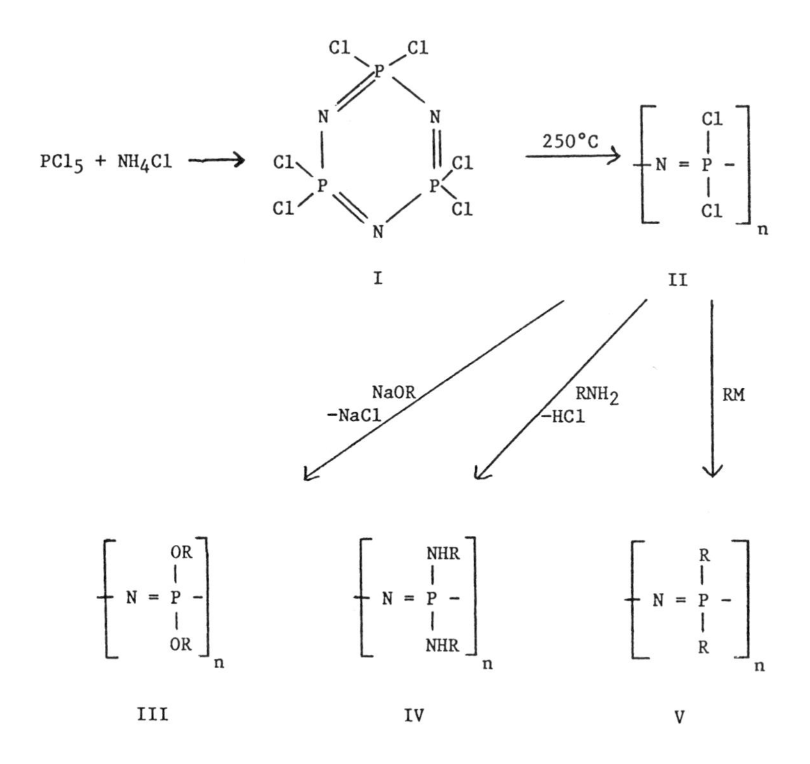

PCl_5 + NH_4Cl
Cl
P
N
I
250°C
II
NaOR
-NaCl
RNH_2
-HCl
RM
OR
NHR
R
n
III
IV
V

Scheme 1.

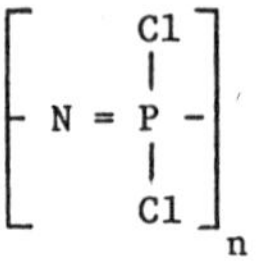

$$\left[N = P(Cl)(Cl) - \right]_n$$

II

(Elastomer: sensitive to water)

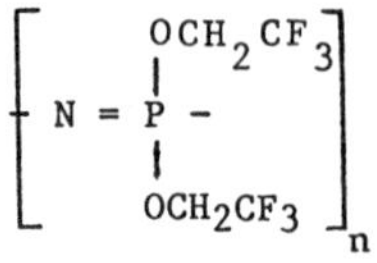

$$\left[N = P(OCH_2CF_3)(OCH_2CF_3) - \right]_n$$

VI

(Microcrystalline, film- and fiber-forming polymer: stable to water, hydrophobic)

$$\left[N = P(OCH_2CF_3)(OCH_2(CF_2)CF_2H) - \right]_n$$

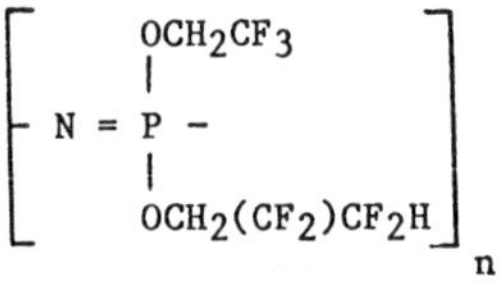

VII

(Elastomer: stable to water, hydrophobic)

$$\left[N = P(NHCH_3)(NHCH_3) - \right]_n$$

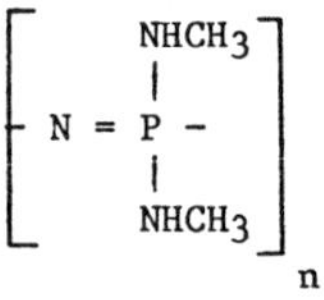

VIII

(Water-soluble and water-stable)

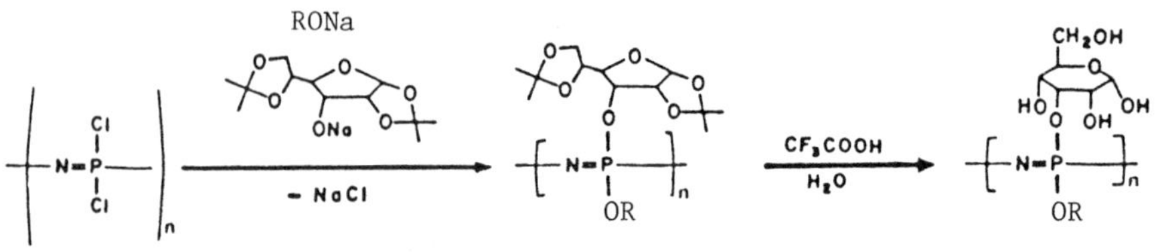

RONa

- NaCl

CF_3COOH

H_2O

CH_2OH

HO OH OH

OR

II IX X

Scheme 2.

Deprotection of X, and subsequent oxidation, reduction, and acetylation reactions can, with care, be carried out without decomposition of the inorganic backbone. Reactions of this type are of particular interest for the synthesis of bioactive or biocompatible polyphosphazenes.

It should be noted that most of the substitution-based synthesis work with poly(organophosphazenes) has been preceded by exploratory studies at the small molecule, model compound level, often with the use of cyclic trimer I as a model for polymer II (6).

Cyclic Trimer Substitution/Polymerization Route. It is often easier to replace chlorine atoms in phosphazenes by organic side groups at the cyclic trimer or tetramer level than at the high polymer level. This is particularly true when organometallic reagents are employed as halogen replacement nucleophiles. Thus, an alternative route to poly(organophosphazenes) involves the synthesis of organo-substituted cyclic trimeric or tetrameric phosphazenes, followed by polymerization of these to the analogous high polymer (Scheme III).

So far, it has proved difficult to polymerize cyclic oligomers that bear organic side groups only, but some species that bear both halogeno and organic side groups can be induced to polymerize. Several cyclic trimers that fall into this category are shown in equations 1-3.

The ring strain inherent in species XV increases its propensity for polymerization. A substantial number of organo-halogenocyclophosphazenes have now been polymerized (7-17). After polymerization, the halogen atoms can be replaced by treatment with alkoxides, aryloxides, primary or secondary amines, or organometallic reagents.

Different Classes of Poly(organophosphazenes)

With this synthetic and molecular structural diversity, polyphosphazene chemistry has developed into a field that rivals many areas of organic polymer chemistry with respect to the tailored synthesis of polymers for specific experimental or technological uses. Indeed, hybrid systems are also available in which organic polymers bear phosphazene units as side groups. This is discussed in another Chapter.

An understanding of modern polyphosphazene chemistry can be obtained by considering the following aspects of the field. (1) Phosphazene polymers with side group systems that allow the high inherent flexibility of the backbone to become manifest. Such polymers are rubbery elastomers (18-20) several of which have been developed in government and industry as high technology materials. Other examples are of interest as solid ionic conduction media (21). (2) Polyphosphazenes with biocompatible or biologically active groups--including several examples that are water-soluble, bio-degradable, or are bioactive as solids. (3) Membrane materials. (4) Polyphosphazenes with transition metals in the side groups, several of which have interesting catalytic or electroactive properties. (5) Polymers with rigid, stackable side groups that can impart either semiconductivity or liquid crystalline character.

The elastomer chemistry and technology are discussed in another Chapter. Aspects 2-5 are illustrated by the examples given in the following sections.

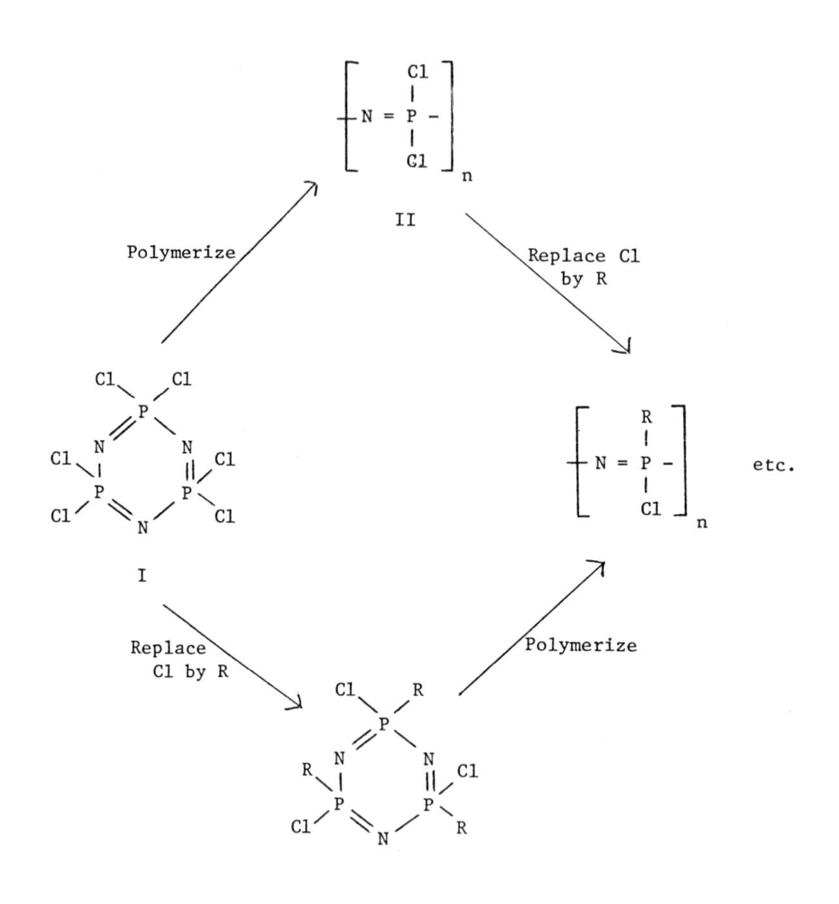
Cl
N = P -
Cl
n
II
Polymerize
Replace Cl
by R
Cl
Cl
P
N
N
Cl
Cl
P
P
Cl
Cl
N
Cl
I
R
N = P -
Cl
n
etc.
Replace
Cl by R
Polymerize
Cl
R
P
N
N
R
Cl
P
P
Cl
R
N

Scheme 3.

Biocompatible or Bioactive Polyphosphazenes

This subject can be considered in terms of five different types of molecules or materials: (a) biologically inert, water-insoluble polymers; (b) water-insoluble polymers that bear biologically active surface groups; (c) water-swellable polymeric gels, or amphiphilic polymers that function as membranes; (d) water-insoluble but bioerodable polymers that erode in aqueous media with concurrent release of a linked or entrapped bioactive molecule; and (e) water-soluble polymers that bear bioactive agents as side groups.

(a) Biologically-Inert, Water-Insoluble Polymers. Polymers of this type are of biomedical interest as materials for the construction of artificial heart valves, blood vessels, soft tissue prostheses, or as coatings for devices such as pacemakers. A number of different polyphosphazenes are being considered for such uses, but special attention has been focused on two classes--those that bear fluoroalkoxy side groups, and those with aryloxy side units (22). Both types are hydrophobic materials that, depending on the side group arrangements, can exist as elastomers or as microcrystalline fiber- or film-forming materials. Preliminary studies have suggested that these two classes of polyphosphazenes are inert and biocompatible in subcutaneous tissue implantation experiments.

(b) Water-Insoluble Polymers that Bear Biologically-Active Surface Groups. Three examples will be given of polyphosphazenes that have these characteristics. First, the anticoagulent heparin has been attached to the surface of a poly(aryloxyphosphazene) via a quaternized aryloxy side group system (23).

Second, dopamine has been linked covalently to a poly(aryloxyphosphazene) via a diazo-coupling technique (24). Experiments showed that rat pituitary cells in culture responded to the surface-bound dopamine in a similar manner to that found when the dopamine was free in solution.

Third, a poly[bis(phenoxy)phosphazene] has been coated on porous alumina particles, surface nitrated, reduced to the amino-derivative, and then coupled to the enzyme glucose-6-phosphate dehydrogenase or trypsin by means of glutaric dialdehyde. The immobilized enzymes were more stable than their counterparts in solution, and they could be used in continuous flow enzyme reactor equipment (25).

(c) Water-Swellable Polymeric Gels or Amphiphilic Polymers that Function as Membranes. Two polyphosphazene systems have been studied in detail (26)--mixed substituent polymers that bear both methylamino and trifluoroethoxy groups (XVI), with the polymer chains lightly crosslinked via the methylamino units by gamma-irradiation, and poly[bis(methoxyethoxyethoxy)phosphazene (XVII), again lightly crosslinked by gamma rays. The latter polymer is water-soluble in the uncrosslinked state (27). It also functions as a solid electrolyte medium for salts such as lithium triflate (21).

(d) Water-Insoluble but Bioerodable Polymers that Erode in Aqueous Media with Concurrent Release of a Linked or Entrapped Bioactive Molecule. The first examples of polyphosphazenes that function in this way were species that contained ethyl glycinate or other amino

$$\underset{\text{XI}}{N_3P_3Cl_5(C_6H_5)} \xrightarrow{\text{heat}} \underset{\text{XII}}{\left[\left(N{=}P(Cl)(Cl) \right)_2 - N{=}P(C_6H_5)(Cl) - \right]_n} \tag{1}$$

$$\underset{\text{XIII}}{N_3P_3Cl_4(CH_3)(CH_2Si(CH_3)_3)} \xrightarrow{\text{heat}} \underset{\text{XIV}}{\left[\left(N{=}P(Cl)(Cl) \right)_2 - N{=}P(CH_2Si(CH_3)_3)(CH_3) - \right]_n} \tag{2}$$

$$\underset{\text{XV}}{N_3P_3F_4[(C_5H_4)_2Fe]} \xrightarrow{\text{heat}} \underset{\text{XVI}}{\left[N{=}P(F)(F) - N{=}P(F)(C_5H_4) - N{=}P(F)(C_5H_4) - \right]_n \ (\text{Fe})} \tag{3}$$

$$\underset{\text{XVI}}{\left[N{=}P(NHCH_3)(OCH_2CF_3) - \right]_n}$$

$$\underset{\text{XVII}}{\left[N{=}P(OCH_2CH_2OCH_2CH_2OCH_3)(OCH_2CH_2OCH_2CH_2OCH_3) - \right]_n}$$

acid ester side groups (28). Hydrolysis results in a breakdown of the polymer to amino acid, ethanol, and phosphate, all of which can be metabolized, and ammonia which can be excreted. Subsequent application of this chemistry has confirmed the biocompatibility of the system and has led to the development of a controlled drug release system (29).

A different approach reported recently (30) involves the synthesis and evaluation of polyphosphazenes that contain both aryloxy and imidazolyl side groups. Hydrolytic removal of the imidazolyl units takes place with a concurrent erosion of the solid polymer in a manner that is appropriate for the controlled release of entrapped drug molecules.

(e) Water-Soluble Polymers that Bear Bioactive Side Groups. Polyphosphazenes that bear methylamino (31), glucosyl, or alkyl ether (27) side groups are soluble in water. Poly[bis(methylamino)-phosphazene] was used in early work as a coordination carrier for platinum antitumor drugs (32). The chemistry now exists for the linkage of a wide range of bioactive agents, such as local anesthetics (33), steroids (34), antibacterial agents (35), or proteins to polyphosphazenes which, in conjunction with the water-solubilizing groups, offers a promise for a variety of pharmacological applications.

Polyphosphazenes with Transition Metals in the Side Groups

This aspect of polyphosphazene chemistry has recently been reviewed elsewhere (36) and only a brief account will be given here.

Metallophosphazenes are a new type of macromolecule designed to bridge the gap between polymers and metals. Although still at an exploratory stage of laboratory development, they may provide access to electronically-conducting polymers, magnetically-active polymers, macromolecular catalysts, electrode mediator systems, or polymers crosslinked by metal atoms.

Transition metals have been linked to cyclo- and polyphosphazenes by four different methods. First, and most obvious, the linkage makes use of the coordinating power of the backbone nitrogen atoms. The platinum dichloride adduct (referred to earlier) falls into this category.

Second, as a logical development of the first approach, polyphosphazenes have been synthesized that bear phosphine units connected to aryloxy side groups (37). The phosphine units bind organometallic compounds, such as those of iron, cobalt, osmium, or ruthenium (38). In several cases, the catalytic activity of the metal is retained in the macromolecular system (39). A similar binding of transition metals has been accomplished through nido carboranyl units linked to a polyphosphazene chain (40).

Third, metallocene units, such as ferrocene or ruthenocene, have been linked to phosphazene cyclic trimers or tetramers and these were polymerized and substituted to give polymers of the type mentioned previously (41). Polyphosphazenes with ferrocenyl groups can be doped with iodine to form weak semiconductors. Polymer chains that bear both ruthenocenyl and ferrocenyl side groups are prospective electrode mediator systems.

Finally, at the cyclic trimer and tetramer levels, compounds of types XVIII and XIX have been synthesized, and metal-substituted high polymers have also been prepared. The development of organometallic reaction chemistry at the small molecule level, as exemplified by compounds XVIII and XIX is an illustration of the key role played by small molecule model compound work in the development of high polymeric phosphazene chemistry.

Polyphosphazenes with Rigid, Stackable Side Groups

The connection between polymer chemistry and ceramic science is found in the ways in which linear macromolecules can be converted into giant "ultrastructure" systems, in which the whole solid material comprises one giant molecule. This transformation can be accomplished in two ways--first by the formation of covalent, ionic, or coordinate crosslinks between polymer chains, and second, by the introduction of crystalline order. In the second approach, strong van der Waals forces within the crystalline domains confer rigidity and strength not unlike that found when covalent crosslinks are present.

A difference between microcrystallite-based ultrastructure and covalently-crosslinked systems is that microcrystallites melt at specific temperatures, allowing the polymer to be fabricated by heating at modest temperatures. Subsequent cooling of the system below the crystallization temperature allows the physical property advantages of the solid state to become manifest. Liquid crystallinity is also possible if some order is retained in the molten state. Crystalline order not only adds mechanical strength, it also provides opportunities for the appearance of other properties that depend on solid state order--such as electronic conductivity. For these reasons, several exploratory studies have been made of the feasibility of attachment of crystallizable side groups to polyphosphazenes. Three examples are given below.

TCNQ-Polyphosphazene Systems. Tetracyanoquinodimethane (XX) salts crystallize in the form of stacked arrays that allow electrical semiconductivity (42). Although this phenomenon has been studied in many laboratories, it has not been possible to fabricate conductive films or wires from these substances because of the brittleness that is characteristic of organic single crystals. However, it seemed possible that, if the flexibility and ease of fabrication of many polyphosphazenes could be combined with the electrical properties of TCNQ, conducting polymers might be accessible.

In recent studies, it has been found that TCNQ units can be attached to quaternized polyphosphazene molecules (43). The quaternization sites are either backbone nitrogen atoms or amino or phosphino units on the side chains. Although the loading of TCNQ was not high, the polymeric system possessed sufficient electrical semiconductivity to suggest that the TCNQ units were forming stacked arrays within the solid matrix. This suggested that other side groups might behave in a similar manner.

Polyphosphazene-Phthalocyanine Structures. Thus, a related study was carried out with copper phthalocyanine units linked covalently to a poly(aryloxyphosphazene) (44). Non-polymeric copper phthalocyanine forms ordered stacked structures in the crystalline state. When

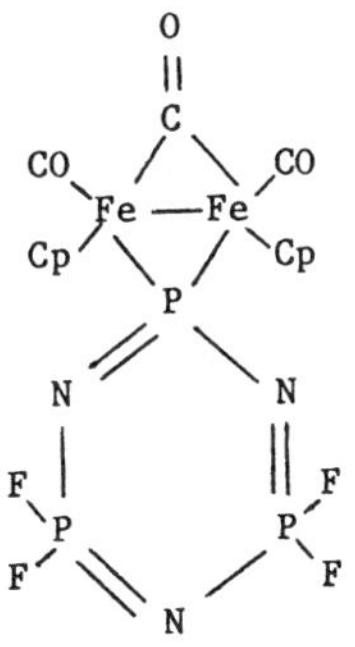

XVIII

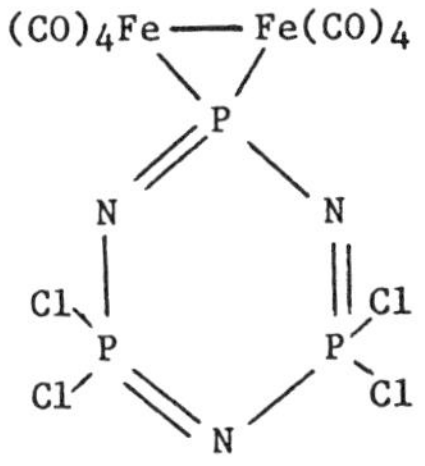

XIX

XX

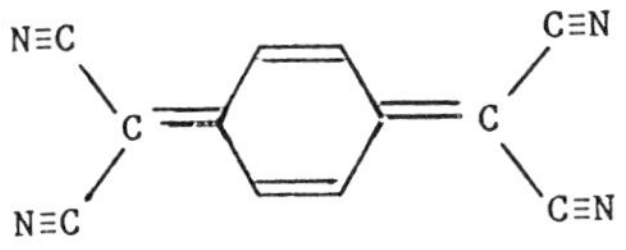

$$\left[N = P\left(O(CH_2CH_2O)_3\text{-}C_6H_4\text{-}N = N\text{-}C_6H_4\text{-}OCH_3 \right)_2 \right]_n$$

XXI

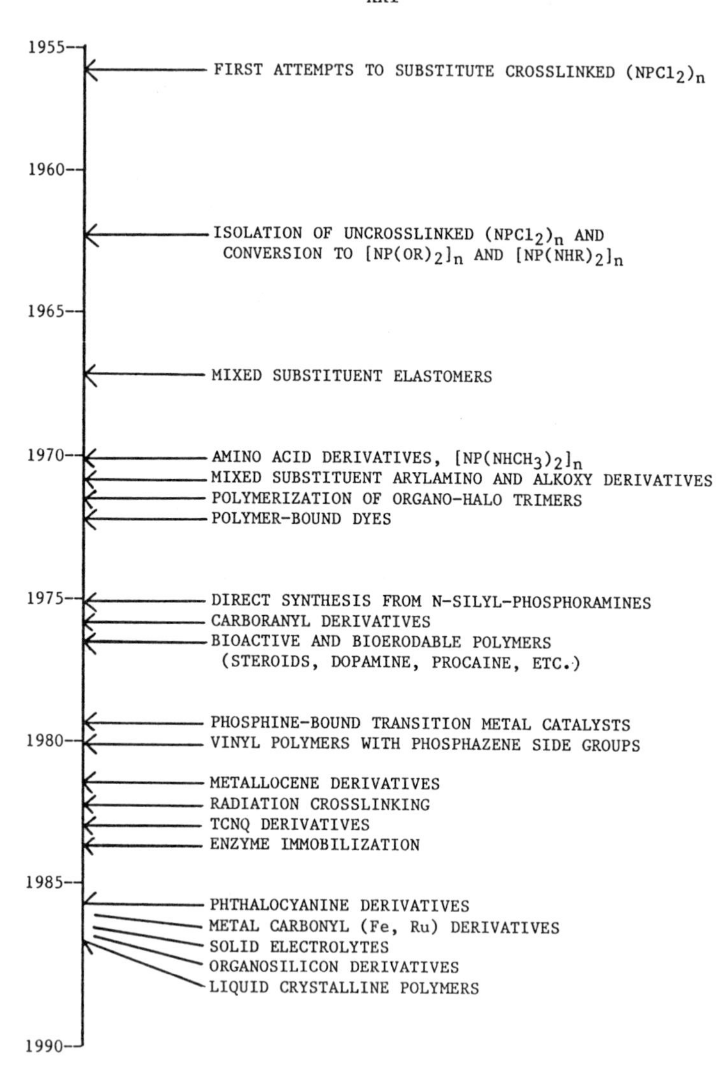

Figure 3. The approximate sequence of developments in the synthesis of poly(organophosphazenes) from the early 1950's to the present.

treated with iodine or other dopants, these are transformed into electrical semiconductors (45-46). In the phosphazene polymeric system, some phthalocyanine group stacking also appears to occur, even though the phthalocyanine concentration is low. Doping with iodine generated weak electrical semiconductivity.

Liquid Crystalline Polymers. The third example is the recent synthesis in our laboratory of a liquid crystalline polyphosphazene (47). The structure of this macromolecule is shown in XXI. This pale yellow polymer shows typical microcrystalline properties at temperatures below 118°C. Heating above 118°C leads to the formation of a mesophase, in which some molecular order is retained. The mesophase melts above 127°C to give an isotropic melt. This is a striking illustration of the way in which a highly flexible polymer chain, coupled to rigid units via flexible spacer groups, allows liquid crystalline stacking of the aromatic azide units to occur. It appears from recent parallel work that the same phenomenon exists even when a non-mesogenic cosubstituent group is present (48).

Historical Perspective

The development of synthetic routes to new polyphosphazene structures began in the mid 1960's (2-4). The initial exploratory development of this field has now been followed by a rapid expansion of synthesis research, characterization, and applications-oriented work. The information shown in Figure 3 illustrates the sequence of development of synthetic pathways to polyphosphazenes. It seems clear that this field has grown into a major area of polymer chemistry and that polyphosphazenes, as well as other inorganic macromolecules, will be used increasingly in practical applications where their unique properties allow the solution of difficult engineering and biomedical problems.

Acknowledgments

Our work has been supported mainly by the Army Research Office, Air Force Office of Scientific Research, Office of Naval Research, and the Public Health Service through the National Heart, Lung, and Blood Institute.

Literature Cited

1. Stokes, H. N. Am. Chem. J. 1879, 19, 782.
2. Allcock, H. R.; Kugel, R. L. Inorg. Chem. 1966, 5, 1016.
3. Allcock, H. R.; Kugel, R. L.; Valan, K. J. Inorg. Chem. 1966, 5, 1709.
4. Allcock, H. R.; Kugel, R. L. Inorg. Chem. 1966, 5, 1716.
5. Seel, F.; Langer, J. Angew Chem. 1956, 68, 461; Z. Anorg. Allg. Chem. 1958, 295, 316.
6. Allcock, H. R. Accts. Chem. Res. 1979, 12, 351.
7. Allcock, H. R.; Moore, G. Y. Macromolecules 1975, 8, 377.
8. Allcock, H. R.; Patterson, D. B. Inorg. Chem. 1977, 16, 197.

9. Allcock, H. R.; Schmutz, J. L.; Kosydar, K. M. Macromolecules 1978, 11, 179.
10. Ritchie, R. L.; Harris, P. J.; Allcock, H. R. Macromolecules 1979, 12, 1014.
11. Scopelianos, A. G.; O'Brien, J. P.; Allcock, H. R. J. Chem. Soc., 1980, 198.
12. Allcock, H. R. Polymer 1980, 21, 673.
13. Allcock, H. R.; Ritchie, R. J.; Harris, P. J. Macromolecules 1980, 13, 1332.
14. Allcock, H. R.; Connolly, M. S. Macromolecules 1985, 18, 1330.
15. Allcock, H. R.; Lavin, K. D.; Riding, G. H. Macromolecules 1985, 18, 1340.
16. Whittle, R. R.; Desorcie, J. L.; Allcock, H. R. Acta. Cryst. 1985, C41, 546.
17. Allcock, H. R.; Brennan, D. J.; Graaskamp J. M. Macromolecules (submitted).
18. Rose, S. H. J. Polym. Sci (B) 1968, 6, 837.
19. Singler, R. E.; Hagnauer, G. L.; Sicka, R. W. ACS. Symp. Ser. 1982, 11.
20. Tate, D. P. J. Polym. Sci., Polym. Symp. 1974, 48, 33.
21. Blonsky, P. M.; Shriver, D. F.; Austin, P. E.; Allcock, H. R. Polym. Mater. Sci. Eng. 1985, 53, 118.
22. Wade, C. W. R.; Gourlay, S.; Rice, R.; Hegyli, A.; Singler, R. E.; White, J. in Organometallic Polymers, Carraher, C. E.; Sheats, J. E.; Pittman, C. U. Eds.; Academic: New York, 1978; 283-286.
23. Neenan, T. X.; Allcock, H. R. Biomaterials 1982, 3, 2, 78.
24. Allcock, H. R.; Hymer, W. C.; Austin, P. E. Macromolecules 1983, 16, 1401.
25. Allcock, H. R.; Kwon, S. Macromolecules 1986, 19, 1502.
26. Allcock, H. R.; Gebura, M.; Kwon, S. (unpublished work).
27. Allcock, H. R.; Austin, P. E.; Neenan, T. X.; Sisko, J. T.; Blonsky, P. M.; Shriver, D. F. Macromolecules 1986, 19, 1508.
28. Allcock, H. R.; Fuller, T. J.; Mack, D. P.; Matsumura, K.; Smeltz, K. M. Macromolecules 1977, 10, 824.
29. Grolleman, C. W. J.; deVisser, A. C.; Wolke, J. G. C.; van der Goot, H.; Timmerman, H. J. Controlled Release 1986, 3, 143.
30. Laurencin, C.; Koh, H. J.; Neenan, T. X.; Allcock, H. R.; Langer, R. S. J. Biomed. Mater. Res. (in press).
31. Allcock, H. R.; Cook, W. J.; Mack. D. P. Inorg. Chem. 1972, 11, 2584.
32. Allcock, H. R.; Allen, R. W.; O'Brien, J. P. J. Am. Chem. Soc. 1977, 99, 3984.
33. Allcock, H. R.; Austin, P. E.; Neenan, T. X. Macromolecules 1982, 15, 689.
34. Allcock, H. R.; Fuller, T. J. Macromolecules 1980, 13, 1338.
35. Allcock, H. R.; Austin, P. E. Macromolecules 1981, 14, 1616.
36. Allcock, H. R.; Desorcie, J. L.; Riding, G. H. Polyhedron, 1987, 6, 119.
37. Allcock, H. R.; Fuller, T. J.; Evans, T. L. Macromolecules 1980, 13, 1325.
38. Allcock, H. R.; Lavin, K. D.; Tollefson, N. M.; Evan, T. L. Organometallics 1983, 2, 267.
39. Dubois, R. A.; Garrou, P. E.; Lavin, K. D.; Allcock, H. R. Organometallics 1986, 5, 460.

40. Allcock, H. R.; Scopelianos, A. G.; Whittle, R. R.; Tollefson, N. M. J. Am. Chem. Soc. 1983, 105, 1316.
41. Allcock, H. R.; Riding, G. H.; Lavin, K. D. Macromolecules 1987, 20, 6.
42. Melby, L. R.; Harder, P. J.; Hertler, W. R.; Mahler, W.; Benson, R.; Mochel, W. E. J. Am. Chem. Soc. 1962, 84, 3374.
43. Allcock, H. R.; Levin, M. L.; Austin, P. E. Inorg. Chem. 1986, 25, 2281.
44. Allcock, H. R.; Neenan, T. X. Macromolecules 1986, 19, 1495.
45. Nohr, R. S.; Kuznesof, P. M.; Wynne, K. J.; Kenney, M. E.; Srebenmann, P. G. J. Am. Chem. Soc. 1981, 103, 4371.
46. Marks, T. J. Science (Washington, D.C.) 1985, 227, 881.
47. Kim, C.; Allcock, H. R. Macromolecules (in press); Polym. Prepr. 1987, 28(1), 446.
48. Singler, R. E.; Willingham, R. A.; Lenz, R. W.; Furukawa, A.; Finkelmann, H. Macromolecules (in press); Polym. Prepr. 1987, 28(1), 488.

RECEIVED September 30, 1987

Chapter 20

Phosphazene Polymers: Synthesis, Structure, and Properties

Robert E. Singler, Michael S. Sennett, and Reginald A. Willingham

Army Materials Technology Laboratory, Watertown, MA 02172–0001

An overview of the synthesis and characterization of a unique class of polymers with a phosphorus-nitrogen backbone is presented, with a focus on poly(dichlorophosphazene) as a common intermediate for a wide variety of poly(organophosphazenes). Melt and solution polymerization techniques are illustrated, including the role of catalysts. The elucidation of chain structure and molecular weight by various dilute solution techniques is considered. Factors which determine the properties of polymers derived from poly(dichlorophosphazene) are discussed, with an emphasis on the role that the organic substituent can play in determining the final properties.

The study of open-chain polyphosphazenes has attracted increasing attention in recent years, both from the standpoint of fundamental research and technological development. The polyphosphazenes are long chains of alternating phosphorus-nitrogen atoms with two substituents attached to phosphorus. These polymers have been the subject of several recent reviews (1-3). Interest has stemmed from the continuing search for polymers with improved properties for existing applications as well as for new polymers with novel properties.

Figure 1 provides an overview of the two step synthesis process, pioneered by Allcock (4) and in use today by a number of workers and laboratories: formation of a soluble reactive polymer intermediate (II) from which is derived a large number of polymers via substitution reactions.

Since the initial disclosure by Allcock, workers have sought to answer various questions: 1) What is the nature of the polymerization process (mechanism)? 2) What is the structure of poly(dichlorophosphazene) that distinguishes it from the insoluble "inorganic rubber" (III)? 3) The substitution process gives a seemingly endless variety of products. What are the limitations or

Published 1988, American Chemical Society

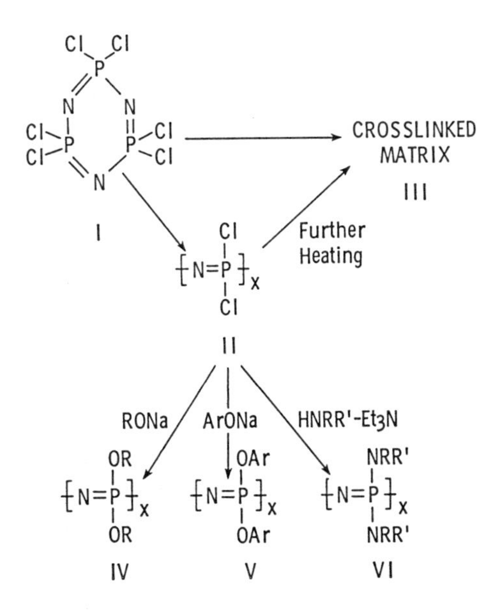

Figure 1. Synthesis of poly(dichlorophosphazene) and poly(organophosphazenes).

controlling factors in the substitution process? 4) How do the above factors control the properties of the poly(organophosphazenes) (eg. IV, V, VI)? 5) Are any of these polymers technologically useful or of commercial interest?

This paper will provide an overview of the polymerization processes and the properties of poly(dichlorophosphazene). This paper will also discuss the various factors which influence the properties of the poly(organophosphazenes) and show how these factors have resulted in a class of polymers with a wide range of properties, including several examples of current commercial importance.

Poly(dichlorophosphazene)

The polymerization of hexachlorocyclotriphosphazene (I) has been the subject of numerous investigations (5). The reaction (I $\longrightarrow$ II, III) is markedly influenced by the presence of trace impurities. The conventional route to II is a melt polymerization at 250 °C of highly purified trimer $(NPCl_2)_3$, sealed under vacuum in glass ampoules. Proper selection of reaction time and temperature is necessary to obtain II and avoid the formation of III. For large scale industrial processes, various acids and organometallic compounds can be utilized as catalysts to prepare soluble polymer, both in bulk and in solution (2). The advantages of catalyzed polymerizations include lower reaction temperatures, higher yields, and the use of conventional large scale equipment.

Size exclusion chromatography (GPC) and other dilute solution techniques have been applied to the characterization of II (6,7). Polymers obtained from the bulk polymerization typically have high molecular weights and broad molecular weight distributions (MWD's). Catalyzed processes generally give narrower MWD's but lower molecular weight polymer. Although questions still remain as to the nature of the polymerization mechanism (7), it is generally thought to be a cationic, chain growth, ring opening polymerization process (Figure 2). Evidence for this includes the effectiveness of Lewis acid catalysts, especially BCl_3, formation of high molecular weight polymer early in the polymerization, and dilute solution parameters obtained on II which point to randomly coiled polymer chains relatively free of long-chain branching for low to moderate conversions to high polymer.

One way to overcome the molecular weight limitations in a solution catalyzed process is by taking advantage of the "living" nature of the polymerization (7). For the BCl_3 catalyzed polymerization, one can add monomer (trimer) to the existing polymer to increase the molecular weight in a stepwise fashion (Figure 3). Trimer is polymerized in the presence of BCl_3 in a trichlorobenzene solution in a sealed ampoule at 210 °C for 48 hours. For the second and third stages, trimer is added in solution equal to the amount in stage 1. The BCl_3 concentration is held constant. Each stage is carried to greater than 95 % conversion. Light scattering measurements on the polymer obtained from stage 3 show $MW > 10^6$, thus confirming that high molecular weight II can be obtained in high conversion in a catalyzed solution process (8).

$[NPCl_2]_3 \longrightarrow [NPCl_2]_n$

BULK - UNCATALYZED

HIGH PURITY TRIMER NECESSARY - OTHERWISE GEL FORMATION

HIGH POLYMER (MW ~ 10^6)

AT LOW CONVERSION (<30%), GEL FREE, 250°C, 40-100 hr

BULK - CATALYZED

TRIMER PURITY LESS CRITICAL

LOWER TEMPERATURES (170°C - 220°C)

WITH HIGHER CONVERSIONS (>50%) OF GEL-FREE POLYMER AT SHORTER TIMES

LOWER MW POLYMER (~10^5)

SOLUTION - CATALYZED

SAME COMMENTS AS IN BULK - CATALYZED

INERT SOLVENT

GENERAL MECHANISM

CATIONIC - CHAIN GROWTH - RING OPENING

Figure 2. General comments on the polymerization process.

$$[NPCl_2]_3 \xrightarrow[\text{TCB, 210°C, SEALED TUBE}]{BCl_3} [NPCl_2]_n$$

STEPWISE PROCESS

FIRST STAGE: 15 wt% TRIMER IN $C_6H_3Cl_3$ (3g/16g). BCl_3 = 0.66g. 48 hr. 210°C. 95% CONVERSION. SOLUBLE POLYMER.

SECOND STAGE: NEW TRIMER SOLUTION ADDED TO POLYMER. $[BCl_3]$ ~ CONSTANT. SAME t, T, % CONVERSION.

THIRD STAGE: REPEAT

STAGE	M_n*	M_w*
1	13,000	37,000
2	100,000	118,000
3	322,000	536,000 (M_W ~ 6 x 10^6)†

*GPC MW DETERMINATION. POLYSTYRENE STANDARDS.
†LIGHT SCATTERING.

SENNETT (1986)

Figure 3. Solution polymerization with BCl_3. Stepwise process.

Poly(organophosphazenes)

The synthesis of poly(organophosphazenes) represents probably the best example of a central theme of inorganic macromolecules: Preparation of a reactive polymeric intermediate, poly(dichlorophosphazene), and subsequent use in a wide variety of side group replacement reactions (Figure 1). This concept has been demonstrated in a number of laboratories (3) and has provided a wide variety of polymers with different properties.

Table I serves to illustrate how the nature and size of the substituent attached to the P-N backbone can influence the properties of the poly(organophosphazenes). The glass transition temperatures range from -84 °C for $(NP(OCH_2CH_3)_2)_n$ to around 100 °C for the poly(anilinophosphazenes). Polymers range from elastomers to flexible film forming thermoplastics or glasses at room temperature.

In the case of poly(alkoxyphosphazenes) (IV) or poly(aryloxyphosphazenes) (V) a dramatic change in properties can arise by employing combinations of substituents. Polymers such as $(NP(OCH_2CF_3)_2)_n$ and $(NP(OC_6H_5)_2)_n$ are semicrystalline thermoplastics (Table I). With the introduction of two or more substituents of sufficiently different size, elastomers are obtained (Figure 4). Another requirement for elastomeric behavior is that the substituents be randomly distributed along the P-N backbone. This principle was first demonstrated by Rose (9), and subsequent work in several industrial laboratories has led to the development of phosphazene elastomers of commercial interest. A phosphazene fluoroelastomer and a phosphazene elastomer with mixed aryloxy side chains are showing promise for military and commercial applications. These elastomers are the subject of another paper in this symposium (10).

Studies have shown that not all phosphazene copolymers are necessarily elastomers (11,12). Figure 5 contrasts semicrystalline homopolymers with an elastomeric copolymer, and then with some semicrystalline aryloxy copolymers. Note in Figure 5 that there is a decreasing order of crystallinity from top to bottom. The intermediate cases represent two classes of crystalline copolymers which are distinguishable by their thermal transition behavior and X-ray crystal structure patterns. Increasing the differences in the size and nature of the substituents on the phenoxy ring will produce amorphous copolymers, but the polyphosphazene unit cell appears to be unusually tolerant of perturbations on a more limited scale (12). As evidenced by the structures in Figure 5, some care must be taken in selecting substituents to achieve desired properties, especially if the goal is to prepare amorphous polymers.

The side chain substituents can affect the properties of the polyphosphazenes in yet another way. Whereas $(NP(OCH_2CH_3)_2)_n$ is amorphous, increasing the side chain length by using long chain alcohols can result in polymers which are semicrystalline (13). Presumably these polymers assume more of the character of poly(ethylene oxide), as the side chain length increases.

The morphology of the semicrystalline polyphosphazenes is complex. Table I provides examples of phosphazenes with two first order transitions denoted by T(1) and Tm. The T(1) is an intermediate transition to a partially ordered state. Between T(1)

Table I. Summary of Transition and Decomposition Temperatures (°C) for Various Polyphosphazenes

POLYMER	T_g*	T(1)*	T_m†	T_d‡
$[Cl_2PN]_n$	-66		33*	
$[(CH_3CH_2O)_2PN]_n$	-84			
$[(CF_3CH_2O)_2PN]_n$	-66	90	240*	360
$[(C_6H_5O)_2PN]_n$	6	160	390	380
$[(3\text{-}ClC_6H_4O)_2PN]_n$	-24	66	370	380
$[(4\text{-}ClC_6H_4O)_2PN]_n$	4	167	365	410
$[([CH_3]_2N)_2PN]_n$	-4			
$[(C_6H_5NH)_2PN]_n$	105			
$[(4\text{-}CH_3OC_6H_4NH)_2PN]_n$	92			266
$[(CF_3CH_2O)(C_3F_7CH_2O)PN]_n$	-77			
$[(CF_3CH_2O)(HCF_2C_3F_6CH_2)PN]_n$	-68			
$[(C_6H_5O)(4\text{-}C_2H_5C_6H_4O)PN]_n$	-27			
$[(C_6H_5)(4\text{-}ClC_6H_4O)PN]_n$	5	77,94		

*BY DIFFERENTIAL THERMAL ANALYSIS OR DIFFERENTIAL SCANNING CALORIMETRY.
†BY THERMAL MECHANICAL ANALYSIS EXCEPT WHERE NOTED.
‡THERMAL DECOMPOSITION TEMPERATURES BY THERMAL GRAVIMETRIC ANALYSIS. MOLECULAR WEIGHT CHANGES HAVE BEEN REPORTED BELOW 200°C.

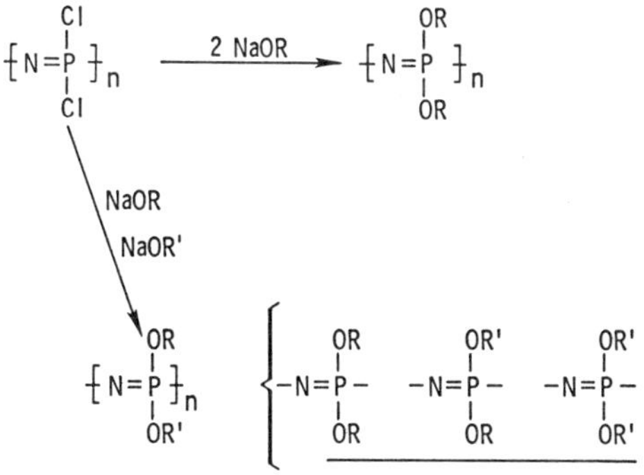

$$[N{=}P(Cl)_2]_n \xrightarrow{2\ NaOR} [N{=}P(OR)_2]_n$$

$$[N{=}P(Cl)_2]_n \xrightarrow{NaOR,\ NaOR'} [N{=}P(OR)(OR')]_n \quad \left\{ -N{=}P(OR)(OR)- \quad -N{=}P(OR')(OR)- \quad -N{=}P(OR')(OR')- \right.$$

Figure 4. Contrasting synthesis of homopolymers and copolymers showing possible copolymer structures which are randomly distributed along the polymer backbone.

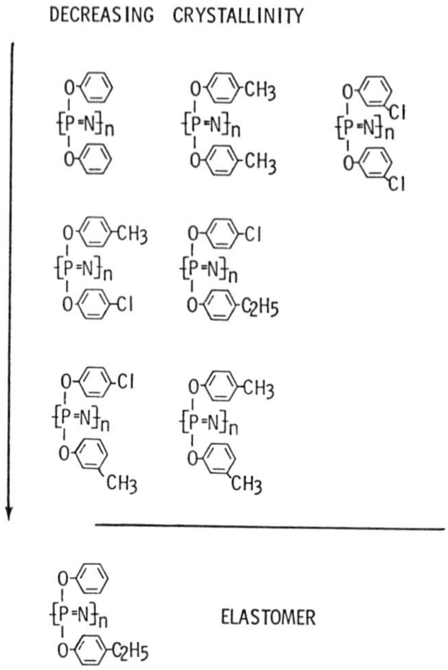

Figure 5. Effect of side chain substituents on polymer crystallinity. Polymers with two para substituents (second row) are more crystalline than polymers with mixed para and meta substituents.

and Tm is a mesomorphic state. However, detailed analysis (14-16) shows that the polymers with a T(1) transition are not nematic or smectic in nature, but rather have a pseudohexagonal phase exhibiting dynamic disorder when characterized by X-ray diffraction experiments. Polyphosphazenes such as $(NP(OCH_2CF_3)2)_n$ have been termed "condis" or conformationally disordered crystals by Wunderlich (17).

To show another example of the effect of side chain structure on polymer properties, it has been recently demonstrated that liquid crystalline side chain phosphazenes can be prepared by attaching a mesogenic group through a flexible spacer to the phosphazene polymer chain (18). Copolymer VII (Figure 6) exhibits a strong reversible liquid crystalline phase between 123 and 175 °C. Microscopic analysis in the liquid crystalline region is shown in Figure 7. A similar polyphosphazene with a different substituted phenylazo mesogen side chain has also been prepared which shows liquid crystalline order (19). Further work is underway to elucidate the exact nature of this state and to prepare additional liquid crystalline side chain polyphosphazenes.

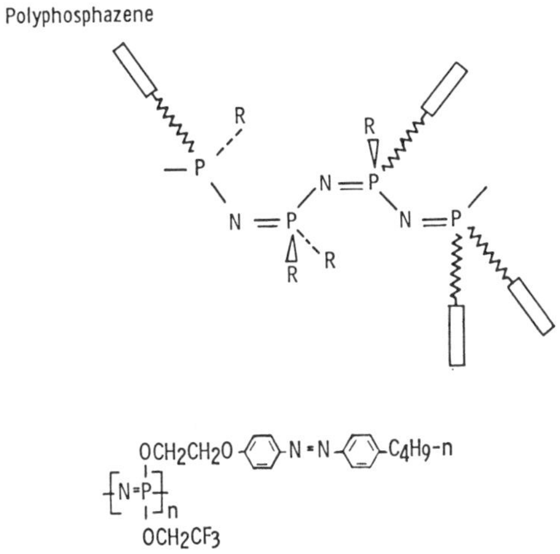

Polyphosphazene

$$\left[N = \underset{OCH_2CF_3}{\overset{OCH_2CH_2O-C_6H_4-N=N-C_6H_4-C_4H_9\text{-}n}{P}} \right]_n$$

Figure 6. General structure for phosphazenes with mesogenic side groups. Example is a mixed substituent polymer (VII) where R represents the trifluoroethoxy group and the mesogen with flexible spacer is represented by the curlicue and rectangular box.

Figure 7. Optical micrograph of VII showing texture of the mesophase at 182 °C. Magnification 320 X.

Conclusion

The polyphosphazenes are high molecular weight polymers with a wide range of novel and potentially useful properties. The large number of different pendant groups with widely varied functionality which can be attached to the P-N backbone demonstrate the unusual molecular design potential of this class of polymers. Undoubtedly, some of these will hold promise for future research and development.

Literature Cited

1. Tate, D.P. and Antowiak, T.A. Kirk-Othmer Encycl. Chem. Technol. 3rd. Ed. 1980, 10, 939.
2. Singler, R.E.; Hagnauer, G.L.; Sicka, R.W. In Polymers for Fibers and Elastomers; Arthur, J.C., Ed.; ACS Symposium Series, No. 260. American Chemical Society, Washington, D.C., 1984, p 143.
3. Allcock, H.R. Chem. & Eng. News, March 18, 1985, p 22.
4. Allcock, H.R. Phosphorus-Nitrogen Compounds; Academic Press, New York, 1972.
5. Hagnauer, G.L. J. Macromol. Sci.- Chem. 1981, A16, 385.
6. Hagnauer, G.L.; Koulouris, T.N. In Liquid Chromatography of Polymers and Related Materials-III; Jack Cazes, Ed.; Marcel Dekker, Inc.; New York, 1981; p 99.
7. Sennett, M.S.; Hagnauer, G.L.; Singler,R.E.; Davies,G. Macromolecules 1986, 19, 959.
8. Sennett, M.S. unpublished results.
9. Rose, S.H. J. Polym. Sci. B 1968, 6, 837.
10. Penton, H.R. In Inorganic and Organometallic Polymers; Zeldin, M.; Allcock, H. and Wynne, K. Eds., ACS Symposium Series, No. xx, American Chemical Society, Washington, D.C., 1987.
11. Dieck, R.L. and Goldfarb, L. J. Polym. Sci. Poly Chem. Ed. 1977, 15, 361.
12. Beres, J.J.; Schneider, N.S.; Desper, C.R.; Singler, R.E. Macromolecules 1979, 12, 566.
13. Allcock, H.R.; Austin, P.E.; Neenan, T.X.; Sisko, J.T.; Blonsky, P.M.; Shriver, D.F. Macromolecules 1986, 19, 1508.
14. Schneider, N.S.; Desper, C.R.; Beres, J.J. In Liquid Crystalline Order in Polymers; Blumstein, A., Ed., Academic Press, N.Y., 1978, p 299.
15. Kojima, M.; Magill, J.H. Makromol. Chem. 1985, 186, 649.
16. Yeung, A.S.; Frank, C.W.; Singler, R.E. Polym. Prepr. ACS Div. Polym. Chem. 1986, 27(2), 214.
17. Wunderlich, B. and Grebowicz, J. In Polymeric Liquid Crystals; Blumstein, A., Ed.; Plenum Press, New York, 1985, 28, 145.
18. Singler, R.E.; Willingham, R.A.; Lenz, R.W.; Furukawa, A.; Finkelmann, H. Macromolecules 1987, in press.
19. Allcock, H.R. and Kim, C. Macromolecules 1987, in press.

RECEIVED September 1, 1987

Chapter 21

Polyphosphazenes: Performance Polymers for Specialty Applications

H. R. Penton

Ethyl Technical Center, Ethyl Corporation, Baton Rouge, LA 70898

Although polyphosphazenes were first synthesized in the laboratory nearly one hundred years ago, their technological importance is only now being realized. Today commercial applications for polyphosphazene specialty elastomers exist in aerospace, marine, oil exploration and industrial fields. This paper describes the properties and applications for this unique class of elastomers.

Nearly one hundred years ago, Stokes reported the first synthesis of a polyphosphazene, which he described as "inorganic rubber" (1). Stokes had synthesized poly(dichlorophosphazene) by the thermal polymerization of hexachlorocyclotriphosphazene. Although the poly(dichlorophosphazene), or chloropolymer as it is often called, had all the characteristics of a strong elastomer, the polymer rapidly hydrolyzed upon exposure to atmospheric moisture; losing its rubber qualities due to excessive crosslinking and ultimate breakdown to form phosphoric acid, ammonia and hydrochloric acid. In addition to the hydrolytic instability of the chloropolymer, thermal polymerization of hexachlorocyclotriphosphazene generally yielded only gelled polymer due to excessive branching and cross-linking as the conversion to polymer exceeded fifty percent (2).

The problems associated with the synthesis and handling of chloropolymer were a major barrier to the development of these polymers until Allcock and Kugel found that chloropolymer could be obtained as a soluble, gel-free polymer if conversions were limited to less than fifty percent (3). Subsequent replacement of the chlorines with metal alkoxides or aryloxides yielded organo-substituted polyphosphazenes which were both thermally and hydrolytically stable (4).

Allcock's discovery that stable polyphosphazenes could be prepared under controlled conditions opened the door for the commercial development of this class of polymers, and in 1970, the first polyphosphazene elastomers were synthesized and the technology subsequently developed by Firestone Tire and Rubber Company (5). Today, polyphosphazenes are commercially available, and represent

0097-6156/88/0360-0277$06.00/0

the first new class of semi-inorganic elastomers to be commercially developed since silicone rubber (6).

Polyphosphazenes are unique in the world of polymers in that the various end products can be derived from a single precursor polymer, polydichlorophosphazene (I). Different polymers are prepared by modifying the organic groups used to displace the chlorines of I. For instance, if I is reacted with the sodium salts of trifluoroethanol and a mixed fluorotelomer alcohol, a poly(fluoroalkoxyphosphazene) elastomer (II) is obtained having a unique set of properties including a wide service temperature range, fuel and oil resistance, low temperature flexability, high temperature stability, and excellent flex fatigue and damping characteristics (7). Replacement of the chlorines with phenoxy and p-ethylphenoxy substituents yields a poly(aryloxyphosphazene) (III); a non-halogenated, flame resistant, low smoke producing elastomer (7).

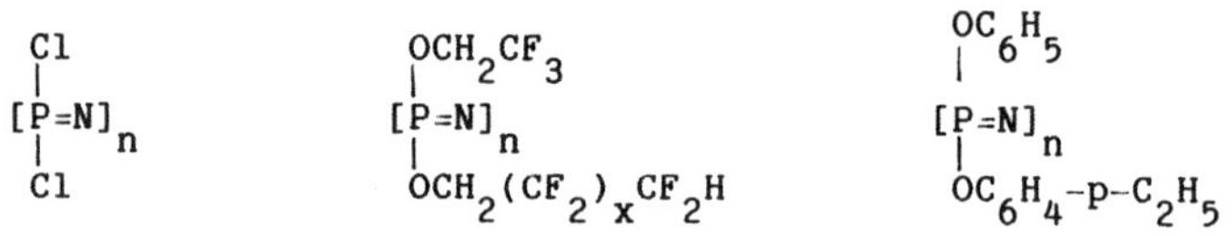

$$\left[\begin{array}{c} Cl \\ | \\ P{=}N \\ | \\ Cl \end{array}\right]_n \qquad \left[\begin{array}{c} OCH_2CF_3 \\ | \\ P{=}N \\ | \\ OCH_2(CF_2)_xCF_2H \end{array}\right]_n \qquad \left[\begin{array}{c} OC_6H_5 \\ | \\ P{=}N \\ | \\ OC_6H_4\text{-p-}C_2H_5 \end{array}\right]_n$$

$x = 1, 3, 5, 7 \ldots$

I II III

Chemistry

The conventional route to prepare I generally involves a high temperature melt polymerization of hexachlorocyclotriphosphazene, or trimer (IV). Recent studies have demonstrated the effectiveness of various acids and organometallics as catalysts for the polymerization of IV (8). Alternate routes for the preparation of chloropolymer which do not involve the ring opening polymerization of trimer have been reported in the patent literature (9, 10). These routes involve a condensation polymerization process and may prove to be of technological importance for the preparation of low to moderate molecular weight polyphosphazenes.

The true value of the chloropolymer (I) lies in its use as an intermediate for the synthesis of a wide variety of poly(organophosphazenes) as shown in Figure 1. The nature and size of the substituent attached to the phosphorus plays a dominant roll in determining the properties of the polyphosphazene. Homopolymers prepared from I, in which the R groups are the same or, if different, similar in molecular size, tend to be semi-crystalline thermoplastics. If two or more different substituents are introduced, the resulting polymers are generally amorphous elastomers. (See Figure 1.)

Applications of Polyphosphazenes

Poly(Fluoroalkoxyphosphazene) Elastomers. When I is substituted with a mixture of trifluoroethoxide and telomer fluoroalkoxides, an elastomer II is obtained having a fluorine content of approximately 55 percent. A small amount of an unsaturated cure site may also be incorporated into the polymer to promote vulcanization.

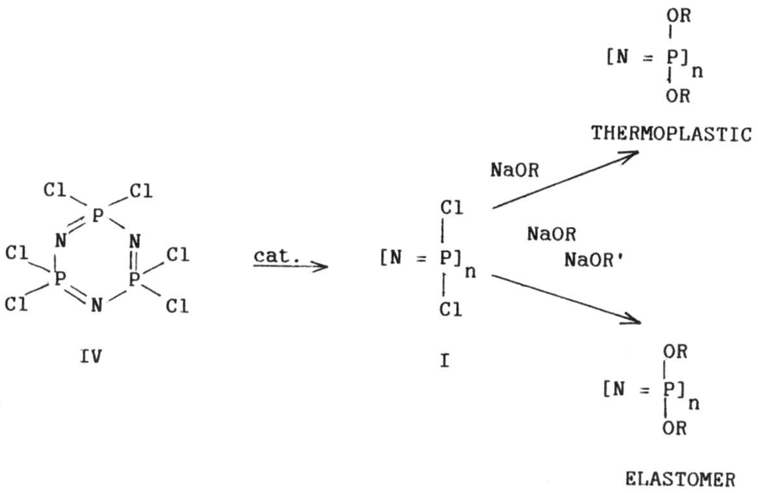

$$\text{IV (hexachlorocyclotriphosphazene)} \xrightarrow{\text{cat.}} \underset{\text{I}}{[N = P(Cl)_2]_n} \begin{cases} \xrightarrow{\text{NaOR}} [N = P(OR)_2]_n \ \text{THERMOPLASTIC} \\ \xrightarrow{\text{NaOR, NaOR'}} [N = P(OR)_2]_n \ \text{ELASTOMER} \end{cases}$$

Figure 1. Synthesis of poly(organophosphazenes).

The fluorine content of II gives it excellent resistance to fuels, oils, most hydraulic fluids and chemicals. Since there are no C-C and C-H bonds along the polymer backbone, II displays excellent resistance to degradation by atmospheric oxygen and ozone. In addition, the inherently flexible nature of the P=N backbone allows this elastomer to be used at temperatures down to -65°C, and gives the polymer excellent flex fatigue resistance over a broad temperature range (-65 to 175°C).

The phosphazene fluoroelastomers can be compounded and processed using conventional rubber equipment. Typical filler systems consist of silicas, clays or carbon blacks either alone or in combination. The resulting compound can be cured with peroxides, sulfur and radiation. Depending on the type and amount of filler used, vulcanizates having an excellent balance of physical properties are obtained (Table I). Elastomers of II offer significant property advantages over fluoropolymer elastomers and fluorosilicones, particularly where a combination of low temperature flexibility as well as high temperature stability, good flex-fatigue resistance and fuel, oil or chemical resistance are required. (See Table I.)

Because of its excellent range of properties and reliability, poly(fluoroalkoxyphosphazene) elastomers are used as seals, gaskets, and shock mounts in demanding military, aerospace, petroleum and industrial applications. In addition, applications under development for this elastomer include fuel hoses for artic use, coated fabrics for protective clothing, sealants, coatings and medical devices.

Poly(Aryloxyphosphazene) Elastomers. Poly(aryloxyphosphazene) elastomers, III, prepared by the reaction of chloropolymer with mixed phenoxides, offer excellent fire resistance without the incorporation of halogens in the polymer or as an additive. In addition, in a fire situation smoke evolution from these polymers is minimal and the resulting off-gases have low toxicity and are non-acidic (11).

Table I. Properties of Poly(fluoroalkoxyphosphazene) Elastomers

PHYSICAL PROPERTY	UNITS	VALUE
Density	g/ml	1.75-1.85
Tensile Strength	MPa (psi)	6.9-13.8 (1000-2000)
100% Modulus	MPa(psi)	2.8-13.8 (400-2000)
Elongation	%	75-250
Compression Set (70 hr @ 150°C)	%	15-25
Hardness, Durometer	Shore A	35-90
TR-10	°C	-56
Brittle Point	°C	-68
Temperature Range	°C	-65 to 175
Tear Resistance, Die B	kN/m (ppi)	to 26 (to 150)

Excellent Bonding to Metals and Fabrics
Excellent Vibration Damping Characteristics
Excellent Flex Life in Dynamic Applications
Excellent Weatherability
Fungus Resistant
Flame Resistant
Resistant to Broad Range of Fluids
Liquid Oxygen Compatible

Poly(aryloxyphosphazene) elastomers can be cured with peroxides, sulfur and radiation. The resulting vulcanizates are resistant to attack by moisture and oils and have been found to have desirable characteristics for electrical insulation applications where fire safety is a concern (Table II) (12). Fire resistant, low smoke, closed cell foams with excellent properties (Table III) have also been developed from poly(aryloxyphosphazene) elastomers (13). Applications for these foams, which can be produced as either slabstock or tube stock, are being developed for military, aerospace and commercial uses. (See Table II and III.)

Summary

Although polyphosphazene chemistry is nearly 100 years old, the technological potential of this class of polymers is only now being realized.

The poly(fluoroalkoxyphosphazene) elastomers offer a unique combination of properties including a wide operating temperature range, excellent fuel and oil resistance and low temperature properties superior to those of the fluorosilicones and fluoropolymer elastomers.

Poly(aryloxyphosphazene) elastomers offer outstanding fire resistance without the presence of halogen. In a fire these elastomers produce low levels of smoke as well as low toxicity and non-corrosive off-gases.

Table II. Properties of Poly(aryloxyphosphazene) Wire and Cable Insulation/Jacketing

PHYSICAL PROPERTY	UNITS	VALUE
Tensile Strength	MPa (psi)	9.0-11.0 (1300-1600)
Elongation	%	200-400
Tear Resistance, Die B	kN/m (ppi)	13 (75)
Halogen	%	0.05
Water Absorption	mg/cm	3.25
Hardness, Durometer	Shore A	83
Heat Aging (168 hr @ 158°C)		
Tensile Retention	%	110
Elongation Retention	%	45
LOI	%	44
NBS Smoke Density		
Flaming	DM	200-225
Non-Flaming	DM	65-75
Dielectric Const., 10 Hz		3.9
Volume Resistivity	ohm-cm	$3.2 \cdot 10^{14}$

Table III. Properties of Poly(aryloxyphosphazene) Foams

PROPERTY	UNITS	VALUE
Density	lb/ft^3	4.5-5.5
Compressive Resistance	psi	2-8
Tensile Strength	psi	25-35
Compression Set	%	25
Thermal Conductivity	$BTU \cdot in/hr \cdot ft^2 {}^\circ C$	0.26
Water Vapor Permeability	perm•in	0.19
LOI		44
NBS Smoke Density		
Non-Flaming		40
Flaming		70
Flame Speed Index		11
Acid Gas Generation	mg HCl/g	0

References

1. H. N. Stokes, Amer. Chem. J., 19, 782 (1897).
2. H. R. Allcock, Phosphorus-Nitrogen Compounds, Academic Press, New York (1972).
3. H. R. Allcock and R. L. Kugel, J. Amer. Chem. Soc., 87, 4216 (1965).
4. H. R. Allcock and R. L. Kugel, Inorg. Chem., 5, 1709 (1966).
5. D. F. Lohr and J. A. Beckman, Rubber and Plastics News, 16 (1982).
6. H. R. Penton, International Rubber Conference Summaries, Stuttgart, Fed. Reb. of Germany, 55 (1985).

7. H. R. Penton, Kautschuk Gummi Kunst., 39, 301 (1986).
8. G. L. Hagnauer, J. Macromol. Sci., Chem. Ed., A16, 385 (1981).
9. E. D. Hornbaker and H. M. Li, U.S. Patent: 4,198,381 (1980).
10. R. De Jaeger, M. Helioui and E. Puskaric, U.S. Patent: 4,377,558 (1983)
11. P. J. Lieu, J. H. Magill and Y. C. Alarie, J. Fire and Flammability, 11, 167 (1980).
12. J. T. Books, D. M. Indyke, W. O. Muenchinger, International Wire and Cable Symposium Proceedings, Cherry Hill, New Jersey (USA), 1 (1984).
13. W. B. Mueller, J. Cellular Plastics, 22, 53 (1986).

RECEIVED September 1, 1987

Chapter 22

Poly(alkyl/arylphosphazenes)

Robert H. Neilson [1], R. Hani [1], G. M. Scheide [1], U. G. Wettermark [1], P. Wisian-Neilson [2], R. R. Ford [2], and A. K. Roy [2]

[1]Department of Chemistry, Texas Christian University, Fort Worth, TX 76129

[2]Department of Chemistry, Southern Methodist University, Dallas, TX 75275

A representative series of the title polymers and copolymers were prepared by the thermally induced condensation polymerization of the corresponding N-silylphosphoranimines. The polymers were characterized by standard dilute solution techniques, thermal analysis (DSC and TGA), multinuclear NMR spectroscopy, and X-ray diffraction measurements. The polymers have molecular weights (M_w) ranging from 25,000 to 200,000 with symmetrical molecular weight distributions ($M_w/M_n \simeq 2$) and, in general, exhibit solution behavior typical of random coil polymers in "good" solvents (e.g., THF or $CHCl_3$). Deprotonation/substitution reactions of the N-silylphosphoranimine "monomers" were used to prepare a variety of functionalized phosphazene precursors containing, for example, silyl, vinyl, and phosphinyl substituents. A chain growth process is suggested as the polymerization mechanism on the basis of studies of polymer molecular weight vs. extent of reaction.

In recent years, many poly(phosphazenes), $[R_2PN]_n$, with a variety of substituents at phosphorus have been prepared and they often exhibit useful properties including low temperature flexibility, resistance to chemical attack, flame retardancy, stability to UV radiation, and reasonably high thermal stability. (1, 2) Compounds containing biologically, catalytically, or electrically active side groups are also being investigated. (3, 4)

The most commonly used synthetic route to poly(phosphazenes) is the ring-opening/substitution method developed by Allcock and coworkers. (1) This procedure involves the initial preparation of poly(dichlorophosphazene), $[Cl_2PN]_n$, by the ring-opening polymerization of the cyclic trimer and subsequent nucleophilic displacement of the chlorine atoms along the chain. In each case, the substituents at phosphorus must be introduced after polymerization since the fully substituted cyclic phosphazenes do not polymerize. (5, 6)

A common feature of poly(phosphazenes) prepared in this manner is that the organic substituents are bonded to phosphorus through oxygen or nitrogen links, thereby providing pathways for decomposition or depolymerization on heating above about 200°C. It has been postulated that directly P-C bonded alkyl or aryl side groups might enhance the thermal or chemical stability of the polymers and give rise to interesting physical properties. (6 - 8) Furthermore, the alkyl substituted poly(phosphazenes), $[R_2PN]_n$, are isoelectronic with the well-known siloxane polymers, $[R_2SiO]_n$, and the extent of this analogy would be interesting to pursue. The published attempts to prepare the directly P-C bonded polymers by the substitution method have generally not been successful. Treatment of poly(dihalophosphazenes) with organometallic reagents (e.g., RMgX or RLi) results in either incomplete substitution under mild conditions or undesired reactions such as chain cleavage and/or crosslinking under more vigorous conditions. (7 - 10)

0097–6156/88/0360–0283$06.00/0

During the last few years, a new general method for the synthesis of poly(phosphazenes) has been under investigation in our laboratory. The approach is based on the premise that suitably constructed N-silylphosphoranimines can eliminate substituted silanes to form cyclic and/or linear phosphazenes (eq 1). This type of *condensation polymerization* reaction has several potential advantages, the most important of which is the ability to incorporate the desired phosphorus substituents directly into the starting Si-N-P compound. This procedure has resulted in the successful preparation of the first fully P-C bonded poly(phosphazene), $[Me_2PN]_n$ (**1**), as described in preliminary reports. (11-13)

$$Me_3SiN{=}\overset{R}{\underset{R'}{P}}{-}X \longrightarrow Me_3SiX + [-N{=}\overset{R}{\underset{R'}{P}}-]_n \qquad (1)$$

We report here the results of our recent studies of poly(alkyl/arylphosphazenes) with particular emphasis on the following areas: (1) the overall scope of, and recent improvements in, the condensation polymerization method; (2) the characterization of a representative series of these polymers by dilute solution techniques (viscosity, membrane osmometry, light scattering, and size exclusion chromatography), thermal analysis (TGA and DSC), NMR spectroscopy, and X-ray diffraction; (3) the preparation and preliminary thermolysis reactions of new, functionalized phosphoranimine monomers; and (4) the mechanism of the polymerization reaction.

Results and Discussion

Synthesis. The N-silyl-P-trifluoroethoxyphosphoranimine reagents, used as "monomers" in this polymer synthesis, are readily prepared from either PCl_3 or $PhPCl_2$ in a straightforward, 3-step reaction sequence as described elsewhere. (14-16) These compounds are obtained as colorless, distillable, air-sensitive liquids in yields of 50-75% based on starting PCl_3 or $PhPCl_2$.

On heating at ca. 160-200°C for 2-12 days, these phosphoranimines readily eliminate $Me_3SiOCH_2CF_3$ (eq 2) to form the polymeric phosphazenes, e.g., **1**-**4**. Copolymers

$$Me_3SiN{=}\overset{R}{\underset{R'}{P}}{-}OCH_2CF_3 \xrightarrow{-\ Me_3SiOCH_2CF_3} [-N{=}\overset{R}{\underset{R'}{P}}-]_n \qquad (2)$$

$$[-N{=}\overset{R}{\underset{R'}{P}}-]_n \qquad\qquad [(-N{=}\overset{R}{\underset{Ph}{P}}-)_x(-N{=}\overset{R}{\underset{R}{P}}-)_y]_n$$

1: R = R′ = Me
2: R = R′ = Et
3: R = Me, R′ = Ph
4: R = Et, R′ = Ph

5: R = Me
6: R = Et
x = y

such as **5** and **6** are prepared by heating equimolar mixtures of the dialkyl- and phenyl/alkyl substituted phosphoranimines.

These thermolysis reactions normally produce polymeric products, free of the cyclic analogs, in essentially quantitative yield and in sufficient purity to give satisfactory elemental analysis upon removal of the silyl ether byproduct under vacuum. Final purification is generally achieved by precipitation of the polymer into a non-solvent such as hexane. With the exception of poly(diethylphosphazene) (**2**), which is insoluble in all common solvents (see below), the new polymers are readily soluble in CH_2Cl_2 and $CHCl_3$. In addition, the phenyl substituted compounds (**3**-**6**) are soluble in THF and various aromatic solvents. None of the polymers are water-soluble; however, $[Me_2PN]_n$ (**1**) is soluble in a 50:50 water/THF mixture.

All of the soluble polymers (**1** and **3**-**6**) give high resolution NMR spectra (^1H, ^{13}C, and ^{31}P) that are completely consistent with their proposed structures. As observed for other types of poly(phosphazenes), the ^{31}P chemical shifts of these alkyl/aryl substituted polymers are consistently ca. 15-30 ppm upfield from those of the analogous cyclic trimers and tetramers. Some important structural information is provided by ^{13}C NMR spectroscopy, particularly for the phenyl/alkyl derivatives **3** and **4**. These polymers are rare examples of phosphazenes that *contain two different substituents at each phosphorus atom in the chain*. Thus, they have the possibility of being stereoregular. The fact that the structures are completely *atactic*, however, is confirmed by the observation of three doublets in the P-Me region of the ^{13}C NMR spectrum (ca. 22 ppm) in a 1:2:1 intensity ratio.

More recently, we have extended the use of the condensation polymerization method to include many other combinations of alkyl and aryl groups in both homopolymers as well as copolymers. Future systematic studies of these materials should provide valuable information on the structure/property relationships which are applicable to this class of polymers.

The possibility of incorporating unsaturated groups into the polymers, as potential sites for crosslinking, is also being explored. For example, the vinyl (**7**, **8**), allyl (**9**), and butenyl (**10**) substituted phosphoranimines have been prepared and subjected to the usual thermolysis and cothermolysis conditions (eq 3). In these cases, all of which have a terminal

$$x\ Me_3SiN{=}P(R)((CH_2)_nCH{=}CH_2)OCH_2CF_3 \quad + \quad y\ Me_3SiN{=}P(R')(Me)OCH_2CF_3 \longrightarrow \text{Swellable, crosslinked polymers} \quad (3)$$

R′ = Ph, Me

7: R = Me, n = 0
8: R = Ph, n = 0
9: R = Ph, n = 1
10: R = Me, n = 2

Swellable, crosslinked polymers (x:y = 1:2, 1:5, 1:10, 1:20)

vinyl group, crosslinking occurs during the thermal polymerization process, leading to the formation of tough, insoluble rubber-like materials. Only when very high proportions of comonomers (e.g., x:y ≃ 1:50) are employed, do we obtain soluble, characterizable polymers. In contrast, when *substituted vinyl* groups (see below) are involved, the thermal crosslinking is suppressed so that soluble phosphazenes bearing intact -CH=CHR substituents are obtained.

Characterization. The series of poly(alkyl/arylphosphazenes) (**1**-**6**) was studied by a variety of standard dilute solution techniques including viscosity measurements, membrane osmometry, size exclusion chromatography, and light scattering. Taken together, these studies demonstrate (17) that **the poly(alkyl/arylphosphazenes) exist as extended, flexible chains in good solvents such as THF or $CHCl_3$, with average chain lengths of several hundred to a thousand repeat units and symmetrical molecular weight distributions** ($M_w/M_n \simeq 2$).

The intrinsic viscosities (ca. 0.15-0.50 dL/g) were determined at 30°C in $CHCl_3$ for the dimethyl polymer **1** and in THF for the phenyl substituted polymers **3-6**. In all cases, the plots of reduced viscosity vs. concentration were quite linear at low concentration (ca. 0.1-1.0%) with Huggins constants in the range (ca. 0.30-0.45) characteristic of good polymer/solvent interactions. Absolute number average molecular weights (M_n), determined by membrane osmometry, fall in the general range of 20,000-100,000, with those of the dimethyl (**1**) and phenyl/methyl (**3**) polymers and the corresponding copolymer (**5**) typically being greater than 50,000.

Initially, the routine characterization of these poly(phosphazenes) was hindered by their anomalous behavior in size exclusion chromatography (SEC) experiments. The analyses were attempted on commercial columns, with either μStyragel or glass bead packings, at a variety of temperatures and concentrations. In addition, several different solvents including mixed solvent systems were investigated. In almost all cases, severely tailing, non-reproducible chromatograms were obtained, suggestive of an adsorption type of interaction between the polymer and the column materials. Recently, however, we found that *these SEC problems are completely circumvented by the addition of a small amount (ca. 0.1 weight percent) of an ionic species such as $(n\text{-}Bu)_4NBr$ to the THF mobile phase*. Under these conditions, classic SEC behavior is observed and consistent, non-tailing chromatograms are obtained. Moreover, the molecular weights (M_n) measured by SEC, relative to narrow molecular weight polystyrene standards, agree very well (within ca. 20-30%) with the values determined by membrane osmometry.

The absolute weight average molecular weights (M_w) of several samples of $[Ph(Me)PN]_n$ from different preparations were determined by light scattering measurements in dilute THF solutions. The M_w values, which range from 73,000 to 202,000, are consistently (ca. 30%) higher than those obtained by SEC determinations. These results, when combined with the membrane osmometry data, confirm that the SEC experiments provide valid representations of the molecular weight distributions of these poly(alkyl/arylphosphazenes). The radius of gyration $\langle S^2 \rangle_z^{1/2}$ for the highest molecular weight sample was found to be 249 Å which is consistent with the polymer having a randomly coiled configuration in solution. Indeed, the calculated ratio $\langle S^2 \rangle_z^{1/2}/M_w$ (ca. 0.31 $Å^2$ mol/g) is essentially identical to the value obtained for high molecular weight poly(dichlorophosphazene). (18)

Also, a good correlation between M_w (light scattering) and intrinsic viscosity is observed for $[Ph(Me)PN]_n$ over the molecular weight range of the samples studied. Thus, the Mark-Houwink relationship, $[\eta] = K(M_w)^a$, yields values of $K = 1.44 \times 10^{-4}$ (with $[\eta]$ in dL/g) and a = 0.66. These data again indicate a well solvated, extended-chain structure of the polymer in THF.

In DSC experiments, the symmetrically substituted poly(dialkylphosphazenes) **1** and **2** show sharp endothermic melt transitions [4.0 and 7.7 kJ/(mole of repeat unit), respectively], indicating a fairly high degree of crystallinity. On the other hand, none of the phenyl substituted polymers (**3** and **4**) and copolymers (**5** and **6**) show a melt transition; therefore, the side group asymmetry disrupts the crystalline order in the solid state. Poly(dimethylphosphazene) (**1**) shows a glass transition of -46°C which increases to -3°C and 37°C for the 25% (**5**) and 50% (**3**) phenyl substituted analogues. A similar trend is noted for the phenyl/ethyl derivatives (**4** and **6**), although poly(diethylphosphazene) itself does not exhibit a discernible glass transition.

Thermogravimetric analysis (TGA) of these poly(phosphazenes) shows their decomposition onset temperatures in an inert atmosphere to be ca. 350 to 400°C, depending on the side group. These temperatures are ca. 25-75°C higher than that reported for commercial materials based on the fluoroalkoxy substituted polymer, $[(CF_3CH_2O)_2PN]_n$. (19) Interestingly, methyl rather than phenyl side groups yield the more stable materials, as shown by the results for polymers **1**, **3**, and **5**. We have also recently reported that silylmethyl groups (i.e., Me_3SiCH_2) have a significant stabilizing influence (20).

Samples of the poly(dialkylphosphazenes) **1** and **2** displayed X-ray powder diffraction patterns characteristic of crystalline regions in the materials. The peaks in the diffraction pattern of **1** were of lower amplitude and greater angular breadth than those of **2**. These data indicate that poly(diethylphosphazene) (**2**) is highly crystalline while poly(dimethylphosphazene) (**1**) is more amorphous with smaller crystalline zones. This high degree of crystallinity is probably responsible for the insolubility of **2** as noted above. All of the phenyl substituted polymers **3**-**6** were found to be quite amorphous in the X-ray diffraction studies, a result that is further evidence for an atactic structure of the poly(alkylphenylphosphazenes) **3** and **4** and for a random substitution pattern in the copolymers **5** and **6**.

Derivatization reactions. In addition to changing the substituents in the primary precursor phosphines, $(Me_3Si)_2NPRR'$, two other approaches have been used to prepare polymers with more complex substituents attached to the backbone by direct P-C linkages. These are (a) alteration of R and R' in the immediate N-silylphosphoranimine precursors, $Me_3SiN{=}P(OCH_2CF_3)RR'$, and (b) alteration of R and R' in the preformed poly(alkyl/arylphosphazenes), $[RR'P{=}N]_n$. Some examples of the former method are described below and in other papers (21 - 23), while the use of the latter approach to prepare several silylated poly(alkyl/arylphosphazenes) has been recently reported. (20)

The basic reaction which is employed in the derivatization of the phosphoranimine monomers is the deprotonation of **11** with n-BuLi followed by quenching with various electrophiles (eq 4).

$$Me_3SiN{=}\overset{\displaystyle R}{\underset{\displaystyle Me}{P}}OCH_2CF_3 \xrightarrow[(2)\ EX]{(1)\ n\text{-}BuLi} Me_3SiN{=}\overset{\displaystyle R}{\underset{\displaystyle CH_2E}{P}}OCH_2CF_3 \qquad (4)$$

11: R = Me or Ph

12: E = alkyl, allyl, $SiMe_2R'$, PPh_2

This process has also been extended to include the Peterson olefination reaction (eq 5) of **12** (where E = Me_3Si) to give a series of substituted vinyl derivatives **13**. Both **12** and **13** have typically been obtained in high yield and have been fully characterized by NMR spectroscopy and elemental analysis.

$$Me_3SiN{=}\overset{\displaystyle Me}{\underset{\displaystyle CH_2SiMe_3}{P}}OCH_2CF_3 \xrightarrow[(2)\ R_2C{=}O\ \ (3)\ Me_3SiCl]{(1)\ n\text{-}BuLi} Me_3SiN{=}\overset{\displaystyle Me}{\underset{\displaystyle HC{=}CR_2}{P}}OCH_2CF_3 \qquad (5)$$

13

$R_2C(O) = Me_2C(O)$, $PhC(O)Me$, $PhC(O)H$, $PhC(O)CF_3$, $PhC(O)CH{=}CH_2$, etc.

A number of copolymers **14** have been prepared by cothermolysis of the new derivatives, **12** and **13**, with the simplest phosphoranimine precursors, $Me_3SiN{=}P(OCH_2CF_3)(Me)R$ (**11**). The copolymers **14** derived from the Peterson olefination products **13** are soluble, non-crosslinked materials with molecular weights in the 30,000 - 100,000 range. This implies

$$[N{=}\overset{R}{\underset{Me}{P}}]_x[N{=}\overset{R}{\underset{E'}{P}}]_y$$

14

R = Me or Ph

E' = CH_2PPh_2, CH=CHR' (R' = Me, Ph)

that these systems might allow for subsequent crosslinking via a "curing" process. Finally, the series of copolymers bearing pendant diphenylphosphine ligands have reasonably highmolecular weights when the ratio x:y is ca. 5:1 and the phosphines readily coordinate to $Fe(CO)_4$ upon treatment with $Fe_2(CO)_9$.

In addition to providing many new precursors to functionalized poly(alkyl/arylphosphazenes), the deprotonation/substitution reactions of these N-silylphosphoranimines serve as useful models for similar chemistry that can be carried out on the preformed polymers. New reactions and experimentation with reaction conditions can first be tried with monomers before being applied to the more difficult to prepare polymeric substrates. A considerable amount of preliminary work [e.g., with the silylated monomers (22) and polymers (20)] has demonstrated the feasibility of this model system approach.

The polymerization process. The successful synthesis of poly(alkyl/arylphosphazenes) by the condensation polymerization process leads to a number of questions concerning the mechanism of this reaction (eq 1). It was initially assumed that a step growth mechanism, as is usually the case for condensation reactions, was operative since the polydispersity (M_w/M_n) values of the polymers are close to the theoretical limit of 2.0 expected for a step growth process. The possibility of a step growth mechanism is precluded, however, by the results of experiments in which the polymerization is stopped prior to completion. Even when the reaction is ca. 20% complete, the contents of the polymerization ampoule are found to be fairly high molecular weight ($M_w \simeq 62{,}000$) polymer and unreacted monomer. No indication of the presence of linear or cyclic oligomers is found in the NMR spectroscopic and SEC studies of these reaction mixtures. Such data clearly indicate that some type of a **chain growth mechanism** must be occurring.

While our studies of the nature of the polymerization process have not yet fully elucidated the mechanism, they have produced a very significant practical improvement. In a study of the effect of varying the leaving group on phosphorus, a series of alkoxy and aryloxy substituted phosphoranimines were prepared in high yield by treatment of the P-bromophosphoranimine, $Me_3SiN{=}PR(R')Br$, with the appropriate alcohol or phenol in the presence of Et_3N. The alkoxy derivatives are considerably more stable than the trifluoroethoxy analogs and do not yield poly(phosphazenes) upon thermolysis. On the other hand, **the thermal decomposition of the phenoxy substituted monomers is an efficient, high yield, and inexpensive new preparative route to the poly(alkyl/arylphosphazenes)**.

Conclusion

The condensation polymerization of N-silyl-P-trifluoroethoxyphosphoranimines is a useful synthetic route to a new class of polymers, i.e., the poly(alkyl/arylphosphazenes). In our efforts to expand the scope and potential applications of this methodology, studies are continuing in a variety of areas including: (1) synthesis of new monomers containing reactive functional groups [e.g., $(CH_2)_nCH{=}CH_2$, CH_2PPh_2, CH_2Br, and $SiMe_2R$]; (2) thermolysis reactions of many of these functionalized monomers; (3) derivatization reactions of preformed polymers by (a) deprotonation/substitution of P-Me groups, (b) electrophilic substitution on the P-Ph groups, (c) acid coordination to the backbone nitrogens; and (4) studies of the effects of leaving group variation in the alkoxy and aryloxy substituted phosphoranimines.

Acknowledgments

The authors thank the U.S. Army Research Office for financial support of this research at TCU and SMU. The light scattering and the thermal analysis experiments were conducted, respectively, by Dr. G.L. Hagnauer (Army Materials Technology Laboratory, Watertown, MA) and Dr. J.J. Meister (SMU).

Literature Cited

1. (a) Allcock, H.R. *Chem. Eng. News* **1985**, *63(11)*, 22.
   (b) Allcock, H.R. *Angew. Chem., Int. Ed. Engl.* **1977**, *16*, 147.
2. Singler, R.E.; Hagnauer, G.L.; Sicka, R.W. *ACS Symposium Series* **1984**, *260*, 143.
3. Allcock, H.R. *ACS Symposium Series* **1983**, *232*, 49.
4. Blonsky, P.M.; Shriver, D.F.; Austin, P.; Allcock, H.R. *J. Am. Chem. Soc.* **1984**, *106*, 6854.
5. Allcock, H.R. *Acc. Chem. Res.* **1979**, *12*, 351.
6. Allcock, H.R.; Patterson, D.B. *Inorg. Chem.* **1977**, *16*, 197.
7. Allcock, H.R.; Evans, T.L.; Patterson, D.B. *Macromolecules* **1980**, *13*, 201.
8. Allcock, H.R.; Desorcie, J.L.; Harris, P.J. *J. Am. Chem. Soc.* **1983**, *105*, 2814.
9. Allcock, H.R.; Chu, C.T.-U. *Macromolecules* **1979**, *12*, 551.
10. Evans, T.L.; Patterson, D.B.; Suszko, P.R.; Allcock, H.R. *Macromolecules* **1981**, *14*, 218.
11. Wisian-Neilson, P.; Neilson, R.H. *J. Am. Chem. Soc.* **1980**, *102*, 2848.
12. Neilson, R.H.; Wisian-Neilson, P. *J. Macromol. Sci.-Chem.* **1981**, *A16*, 425.
13. Wisian-Neilson, P.; Roy, A.K.; Xie, Z.-M.; Neilson, R.H. *ACS Symposium Series* **1983**, *232*, 167.
14. Neilson, R.H.; Wisian-Neilson, P. *Inorg. Chem.* **1982**, *21*, 3568.
15. Wisian-Neilson, P.; Neilson, R.H. *Inorg. Chem.* **1980**, *19*, 1875.
16. Wisian-Neilson, P.; Neilson, R.H. *Inorg. Synth.*, in press.
17. Neilson, R.H.; Hani, R.; Wisian-Neilson, P.; Meister, J.J.; Roy, A.K.; Hagnauer, G.L. *Macromolecules* **1987**, *20*, 910.
18. Hagnauer, G.L. *ACS Symposium Series* **1980**, *138*, 239.
19. Critchley, J.P.; Knight, G.J.; Wright, W.W. "Heat Resistant Polymers"; Plenum Press: New York, 1983; Chapter 8, and references cited therein.
20. Wisian-Neilson, P.; Ford, R.R.; Neilson, R.H.; Roy, A.K. *Macromolecules* **1986**, *19*, 2089.
21. Roy, A.K.; Hani, R.; Neilson, R.H.; Wisian-Neilson, P. *Organometallics* **1987**, *6*, 378.
22. Wettermark, U.G.; Wisian-Neilson, P.; Scheide, G.M.; Neilson, R.H. *Organometallics* **1987**, *6*, 959.
23. Roy, A.K.; Wettermark, U.G.; Wisian-Neilson, P.; Neilson, R.H. *Phosphorus and Sulfur*, in press.

RECEIVED September 1, 1987

Chapter 23

Hybrid Inorganic–Organic Polymers Derived from Organofunctional Phosphazenes

Christopher W. Allen

Department of Chemistry, University of Vermont, Burlington, VT 05405

Carbon chain polymers with pendant inorganic ring systems can be prepared by homo- or copolymerization reactions of cyclophosphazenes with olefinic exocyclic groups. The resulting polymers and copolymers exhibit flame retardancy and have a large number of reactive sites for further synthetic transformations. In alkenylpentafluorocyclotriphosphazenes the reactivity of the olefinic center may be mediated by placement of an electron donating group on the olefin or by placing an insulating function between the olefin and the phosphazene. A broad range of homo- and copolymers have been prepared from vinyloxyphosphazenes. These materials undergo facile thermal decomposition. The volatile products of these processes have been identified and some insight has been gained into the kinetics and mechanism of the polymer thermolysis reactions.

The primary approach to the development of main group inorganic polymer chemistry has been in the preparation, characterization and utilization of polymers with an inorganic backbone which may, or may not, be protected by organic substituents (**1a**). Important members of this class of materials which are discussed in this volume and elsewhere include:

$$\begin{array}{c} R \\ | \\ (AB)_n \\ | \\ R \end{array} \quad \textbf{1a.} \qquad\qquad \begin{array}{c} R' \\ | \\ (CHCH_2)_n \\ | \\ (AB)_x \\ R_2 \end{array} \quad \textbf{1b.}$$

R_2AB: R_2SiR_2Si; SS; R_2SiO; R_2PN; SN $\qquad$ x = 2-4

0097–6156/88/0360–0290$06.00/0

poly(siloxanes), poly(phosphazenes) and more recently poly-(silanes), poly(carbosilanes) and poly(silazanes). Our approach has been to invert the roles of the organic and inorganic entities thereby producing polymers with carbon chain backbones and inorganic substituents (**1b**). Vinyl polymers with transition metal substituents have been studied extensively with most interest being shown in metallocene derivatives.[1,2] Inorganic main group entities as substituents on carbon chain polymers have received sporatic study with the most interest being shown in acyclic substituents involving organosilanes.[3] Vinyl carboranes can also be polymerized and copolymerized to polymeric species having the boron cluster as a substituent.[4] Our primary focus has been on the use of organofunctional cyclophosphazenes as precursors to the desired polymers.[5] We have conducted a systematic study of the effect of variation of both the phosphazene and olefinic components within the general structure of an organofunctional phosphazene (**2**). These studies have allowed for both an understanding of the polymerization behavior of this novel class of monomers and the incorporation of certain of the useful properties, such as flame retardency, of the phosphazenes into traditional organic polymers.

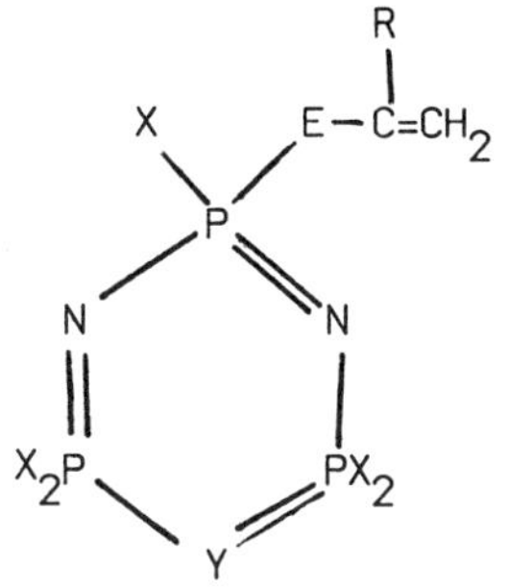

X=Cl,F
R=Me,OEt,OMe
Y=N,NPX_2N
E=O,C_6H_4

2

ALKENYL PHOSPHAZENES

We intially demonstrated that our basic approach was feasible by preparing copolymers (**3**; R=Me) of 2-(propenyl)pentafluorocyclotriphosphazene (**4**) and styrene or related monomers.[6,7] The copolymers, which have good solubility and thermal stability, do not burn or produce smoke when exposed to a flame. Nucleophilic substitution reactions of the types which are well known in phosphazene chemistry[8] can be carried out on the phosphazene moiety in these copolymers thus allowing for the preparation of a broad spectrum of related materials.[7] However, the strong electron withdrawing nature of the $N_3P_3F_5$ unit[9] transforms the propenyl group into a highly polar olefin which undergoes competitive anionic addition in the course of its preparation[10] and serves as a termination site in the copolymerization reaction.[6]

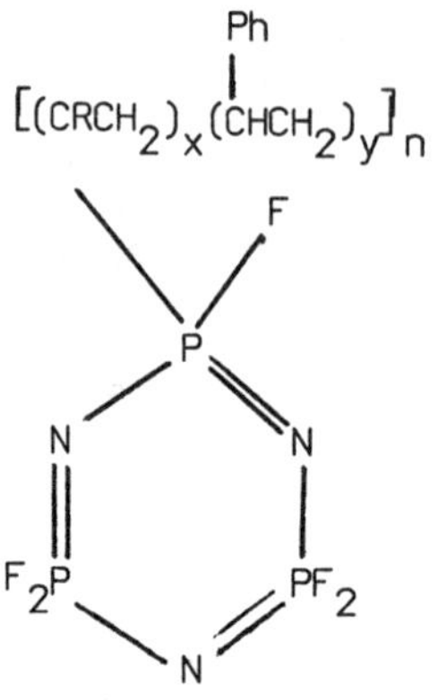

Ph

$[(CRCH_2)_x(CHCH_2)_y]_n$

F

P

N N

F_2P PF_2

N

3

The electron deficient nature of the olefin can be mediated in two ways either by placement of an electron donating substituent on the olefin[11] or by insulating the olefin from the phosphazene ring by other groups. The introduction of an alkoxy function as the electron donating group on the olefin leads to 2-(1-ethoxy vinyl)pentafluorocyclotriphosphazene, $N_3P_3F_5C(OEt){=}CH_2$ (**5**).[11] A comparison of the ^{13}C nmr chemical shifts of the olefinic β-carbon in **4** and **5** gives evidence for a significant reduction in the olefin polarity by addition of the electron donating ethoxy function. The change in olefin polarity is also demonstrasted by the absence of anionic addition to the olefinic center in **5**. Facile copolymerization of **5** with styrene or methylmethacrylate occurs to yield materials (e.g. **3**; R=OEt) with higher degrees of phosphazene incorporation but properties similar to the copolymers derived from **4**. The use of an aryl group as an insulating function[13,14] is exemplified in α-methylstyrylphosphazenes such as the meta and para isomers of $N_3P_3F_5C_6H_4C(CH_3){=}CH_2$ (**6**) which also have been induced to undergo copolymerization with styrene and methylmethacrylate.[14] These later compounds have the highest degree of phosphazene incorporation into copolymers which we have yet observed. Reactivity ratio and Q,e calculations show that, in general, (i.e., in **4**,**5**,**6**) the phosphazene moiety exhibits a very strong electron withdrawing effect without any significant mesomeric interaction. Thus, the model of the electronic structure of the olefinic center in alkenylphosphazenes which arises from quantitative studies of reactivity in copolymerization reactions is very similar to that deduced from spectroscopic studies of arylphosphazenes.[9]

Further examination of these reactivity and Q,e data allow for a clear understanding of the polymerization behavior and how it is modified by the phosphazene. Although most systems are best described by the terminal model, the reactivity patterns exhibited in the copolymerization of α-methylstyryl pentafluorocyclophosphazene (**6**) with methylmethacrylate can only by quantitatively fit by a pennultimate model.[14]

In order to obtain homopolymers from the organofunctional phosphazenes, monosubstituted rather than 1,1-disubstituted olefins of the type described above, are required. The styrene derivative, $N_3P_3F_5C_6H_4CH{=}CH_2$, was prepared from the phenylmethyl ether derivative, $N_3P_3C_6H_4C(OMe){=}CH_2$, via hydrostannation and subsequent β-elimination of the trialkyltin methoxide.[14] Polymerization of the styryl phosphazene lead to a high molecular weight polymer with pentafluorocyclotriphosphazene units at the para position of each phenyl group (**7**; $E{=}C_6H_4$, X=F). The TGA of this material shows a significant retention of involatile material to over 1000°, suggesting the intriguing possibility

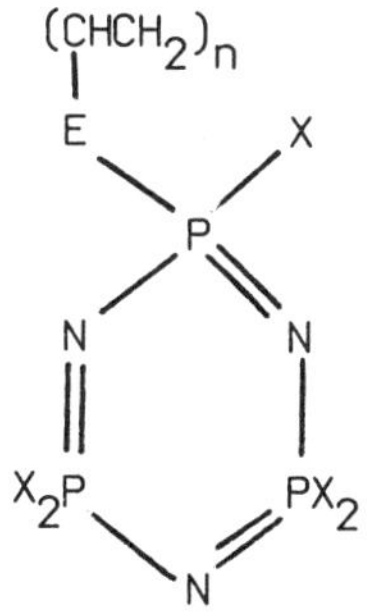

$(CHCH_2)_n$–E, X on P; cyclic N_3P_3 ring with PX_2 groups — **7**

of using the pyrolysis of polymers of type **1b** as a new route to ceramic solids.

VINYLOXYPHOSPHAZENES

In an attempt to generate convenient routes to organofunctional phosphazenes which could undergo homopolymerization reactions, we explored the reactions of lithium enolates, particularly that of acetaldehyde, with halophosphazenes which leads to the series of organofunctional monomers $N_3P_3Cl_{6-n}(OCH{=}CH_2)_n$[15,16] (n=1-6), $N_3P_3F_{6-n}(OCH{=}CH_2)_n$(n=1-5)[17] and $N_4P_4Cl_{8-n}(OCH{=}CH_2)_n$ (n=1,2).[18] In these systems, an oxygen atom acts as the insulating function between the phosphazene and the olefin. Other, related monomers, can be prepared by nucleophilic substitution reactions with, for example alkoxide ions, of the chlorine atoms in $N_3P_3Cl_5OCH{=}CH_2$ (**8**). The appropriate monosubstituted derivative (n=1) in many cases has been polymerized by radical initiation to yield the linear high polymer with a cyclophosphazene moiety as part of each repeating unit, e.g. **7**; E=O, X=Cl.[19] However, not all of the vinyloxyphosphazene monomers will undergo radical polymerization; those with amino substituents are unreactive. The ^{13}C nmr data indicate that these species electronically resemble vinyl ethers (which do not undergo radical polymerization) whereas the reactive derivatives resemble vinyl acetate. These data demonstrate an excellent example of electronic effect transmission in cyclophosphazene systems.

The large number of reactive sites per monomer unit in these highly functionalized polymers has allowed for further synthetic transformations based on the chemistry of the phosphazene unit thereby allowing preparation of the amino substituted polymers by reaction of 7 (E=O, X=Cl) with the appropriate amine. The broad range of vinyloxyphosphazene monomers which are available allow for the formation of copolymers of two different organofunctional phosphazene monomers. Consequently, we can apply polymerization reactivity studies to directly explore differences in chemical behavior based on changes in substituent, ring size, etc. of the inorganic monomer. We have prepared a series of copolymers of 8 and its tetrameric analog, $N_4P_4Cl_7OCH=CH_2$ and shown that 8 has the greater reactivity in this system. In a joint study with Dr. van de Grampel of the University of Groningen, we have explored the copolymerization of 8 with a vinyloxycyclophospha-(thia)zene.[20]

As opposed to the alkenylphosphazene polymers, the vinyloxychlorophosphazene polymers are thermally labile undergoing an exothermic decomposition via HCl elimination as low as 100° followed by a more complex endothermic process at higher temperatures. The first step represents a mild thermal cross-linking reaction with low weight loss. The solid state kinetics of this first step have been measured. The activation energy for the first step is greater for the tetramer than for the trimer derivative. Since this is the reverse of the behavior observed for displacement reactions of the $(NPCl_2)_3$ and $(NPCl_2)_4$[8], it implies that the barrier is primarily associated with the chain reorganization energy needed to bring the phosphorus-chlorine bond in proximity with a carbon-hydrogen group. After the exothermic cross-linking step, there is a second process which appears to result in the elimination of oxobridged phosphazene dimers, e.g. $(N_3P_3Cl_5)_2O$ and, by implication, to build an oxo cross-link between the polymer chains. Copolymerization with organic monomers suggest a similarity of behavior between the vinyloxyphosphazene and vinylacetate. The thermal lability of the polymers containing the vinyloxyphosphazene is even more pronounced in the copolymers. In order to overcome some of these problems a larger insulating function was sought. The reactions of 2-hydroxyethyl methylmethacrylate with $N_3P_3Cl_6$ leads to $N_3P_3Cl_5OCH_2CH_2OC(O)C(CH_3)=CH_2$ which undergoes radical or thermal homo and copolymerization which if not carefully controlled is accompanied by extensive cross-linking.

EXTENSIONS TO OTHER SYSTEMS

Our studies have convinced us that this approach to hybrid organic-inorganic polymers is a valuable one leading to a wide variety of new, interesting, materials. In addition to new phosphazenes, we are now expanding our studies to the preparation of pentamethylvinylcyclotriborazene polymers and we are investigating various silicon containing systems.

ACKNOWLEDGMENTS

This work was supported in part by the Office of Naval Research; collaborative efforts with Dr. van de Grampel are supported, in part, by a NATO travel grant.

LITERATURE CITED

1. Pittmann, Jr., C.U.; Rausch, M.D., Pure Appl. Chem., **Pure Appl. Chem.**, 1986, **58**, 617.
2. Dzhardimalieva, G.I.; Zhorin, V.A.; Ivkva, I.N.; Pomogailo, A.D.; Enikopolyan, N.S., **Dokl. Akad. Nauk SSSR, 1986, 287,** 654.
3. Billingham, N.C.; Jenkins, A.D.; Kronfli, E.B.; Walton, D.R.M. **J. Polym. Sci., Polym. Chem. Ed.**, 1977, **15**, 683; Lakshmana Rao, V.; Babu, G.N., **J. Macromol. Sci., Chem.**, 1983, **A20**, 527; Ledwith, A.; Chiellini, E.; Solaro, R., **Macromolecules**, 1979, **12**, 240; Carbonaro, A.; Greco, A.; Vassi, I.W., **Eur. Polym. J.**, 1968, **4**, 445; Tumanova, I.A.; Semenov, O.B.; Yanovsky, Yu. G., **Inter. J. Polym. Mater.**, 1980, **8**, 225.
4. Wright, J.R.; Kingen, T.J., **J. Inorg. Nucl. Chem.** 1974, **36**, 3667; Klingen, T.J., **ibid**, 1980, **42**, 1109.
5. Allen, C.W., **J. Polym. Sci., Polym. Symp.**, 1983, **70**, 79.
6. Dupont, J.G., Allen, C.W., **Macromolecules.**, 1979, **12**, 169.
7. Allen, C.W.; Dupont, J.G., **Ind. Eng. Chem. Prod. Res. Dev.**, 1979, No. 18, 80.
8. Allcock, H.R., **Phosphorus-Nitrogen Compounds**; Academic Press: New York, 1972; Allen, C.W., "Cyclophosphazenes and Related Compounds" in **The Chemistry of Inorganic Rings**, Ed. I. Haiduc and D.B. Sowerby, Academic Press, in press.
9. Allen. C.W.; White, A.J., **Inorg. Chem.**, 1974, **13**, 1220; Allen, C.W., **J. Organometal. Chem.**, 1977, **125**, 215; Allen, C.W.; Green, J.C., **Inorg. Chem.**, 1980, **19**, 1719.
10. Allen, C.W.; Bright, R.P.; Ramachandran, K., **ACS Symp. Ser.**, Phosphorus Chemistry, 1981, **171**, 321.
11. Allen, C.W.; Bright, R.P., **Inorg. Chem.** 1983, **22**, 1291.
12. Allen. C.W.; Birght, R.P., **Macromolecules**, 1986, **19**, 571.
13. Shaw, J.C.; Allen, C.W., **Inorg. Chem.**, 1986, **25**, 4632.
14. Allen, C.W.; Shaw, J.C., **Phosphorus and Sulfur**, 1987, **30**, 97.
15. Allen, C.W.; Ramachandran, K.; Bright, R.P.; Shaw, J.C., Inorg. Chim. Acta, 1982 **64**, L109.
16. Ramachandran, K.; Allen, C.W., **Inorg. Chem.** 1983, **22**, 1445.
17. Allen, C.W.; Bright, R.P., **Inorg. Chim. Acta** 1985, **99**, 107.
18. Brown, D.E.; Allen, C.W., **Inorg. Chem.**, 1987, **26**, 934.
19. Allen, C.W.; Ramachandran, K.; Brown, D.E., **Inorg. Syn.**, in press.
20. van de Grampel, J.C.; Jekel, A.P.; Dhathathreyan, K.; Allen, C.W.; Brown, D.E., Abstract 319, 193rd American Chemical Society Meeting, Denver, April, 1987.

RECEIVED September 1, 1987

Chapter 24

Polybis(pyrrolyl)phosphazene

R. C. Haddon [1], S. V. Chichester [1], and T. N. Bowmer [2]

[1]AT&T Bell Laboratories, Murray Hill, NJ 07974–2070
[2]Bell Communications Research, Inc., Red Bank, NJ 07701–7020

Current progress in the synthesis and properties of pyrrolylphosphazenes is summarized. The differences in reactivity of the cyclic trimer $(NPCl_2)_3$, and high polymer $(NPCl_2)_x$, toward the pyrrolide nucleophile are discussed. Efforts to induce electronic conductivity in the polyphosphazenes are reviewed with particular emphasis on polybis(pyrrolyl)phosphazene.

It is clear from the contents of this volume that polyphosphazene chemistry is currently receiving a great deal of attention. Commercialization of these materials is still in its infancy but appears to hold significant potential.[1] Much of the current level of interest in this field stems from the synthetic versatility of the phosphazene polymers, as developed by Allcock and coworkers.[2–4] As a result of this, the preparative chemistry of the polyphosphazenes is the most highly developed of all the inorganic polymer systems.[4] This has lead to the synthesis of a wide variety of substituted polyphosphazenes,[2–5] many of which were specifically designed for a given chemical, mechanical or biological property. Noticeably absent from the properties exhibited by the polyphosphazenes has been electronic conductivity and in general the materials are insulators.[4] It should be noted, however, that suitably functionalized polyphosphazenes have recently been shown to serve as electrolytes for *ionic* conductivity.[6]

The question of electronic conductivity in the polyphosphazenes inevitably raises questions regarding the electronic structure of the phosphazene linkage.[7–12] This matter has been the subject of controversy in the literature, but experimentally the situation is now well known.[4,13] In spite of the fact that the phosphazene backbone is fully conjugated, bond equalized and possesses bond lengths which are indicative of partial double bond character, the evidence suggests that these are localized systems.

0097–6156/88/0360–0296$06.00/0

The possibility of inducing electronic conductivity within the P-N linkage is being actively pursued within a number of research groups,[4,11,14,15] but the phosphazenes themselves present a daunting challenge (in spite of their proximity to polysulfurnitride in the Periodic Table).

An alternative approach to electronic conductivity in the polyphosphazenes, utilizes the 'outrigger approach' in which the appropriate functionality for electronic mobility is provided by the substituents.[4] A successful implementation of this methology which made use of polyphosphazene bound copper phthalocyanine units was recently reported by Allcock and Neenan.[16] Iodine doping of these materials led to conductivities in the semiconductor range.[16] It is also possible to prepare semiconducting tetracyanoquinodimethane (TCNQ) salts, in which quaternized polyphosphazenes serve as counterions.[17] In a variant of the outrigger approach we have synthesized polybis(pyrrolyl)phosphazene (PBPP);[18] this new polymer undergoes electrochemical oxidation to produce conducting films at the anode.[18]

Apart from the ability of PBPP to undergo electroxidation, this polymer is significantly different from previously reported polyphosphazenes, both in preparation and in properties. Although a complete picture of the chemistry of this polymer has yet to emerge, our current progress will be reviewed in the present article.

Preparation of Pyrrolylphosphazenes

The first pyrrolylphosphazenes were apparently prepared by McBee and coworkers in 1960, and although the work was never documented in the chemical literature, hexakis-(pyrrolyl)cyclotriphosphazene (**1**) and octakis-(pyrrolyl)cyclotetraphosphazene were described in a Technical Report of the Defense Technical Information Center.[19] Compound **1** was reported to be produced in 26% yield from the interaction of hexachlorocyclotriphosphazene [$(NPC\ell_2)_3$] with excess potassium pyrrolide in refluxing benzene over a 24 hour period. Lithium pyrrolide and pyrrolyl magnesium bromide were found to be unsatisfactory reagents for the preparation of **1**.

The yield of the potassium pyrrolide reaction is considerably improved by the use of a phase transfer catalyst.[20] However, in exploratory work on this substitution reaction we have found that sodium pyrrolide reacts with $(NPC\ell_2)_3$ in tetrahydrofuran at room temperature to produce **1** in quantitative yield in less than 3 hours. The structure of the compound[20] is reproduced in Figure 1.

In general, the small molecule substitution chemistry provides an excellent model for the analogous high-polymer synthesis.[4,13] However, in attempting to extrapolate from the reaction of pyrrolyl salts with $(NPC\ell_2)_3$ to the corresponding reaction with the high polymer $(NPC\ell_2)_x$, we discovered significant discrepancies in the two reaction pathways.[21] High yields of **1** may be isolated from the reaction of $(NPC\ell_2)_3$ with all of the alkali metal pyrrolide salts which were tested, except lithium. In the case of the polymer $(NPC\ell_2)_x$, however, we found that some of the alkali metal pyrrolides brought about substantial chain cleavage and occasionally produced the cyclic trimer (**1**) in high yield. Furthermore there is apparently a facile cross-linking reaction available to PBPP, as a substantial fraction of the isolated polymer is often insoluble (although chemical analyses indicate complete replacement of chloride).

There is little mention in the literature of the use of amide salts in substitution reactions on chlorophosphazene precursors. The anilide anion was shown to be a powerful nucleophile in substitution reactions on various trimer derivatives, but investigations of such reactions with the high polymer have not been reported.[22] Where strong nucleophiles (such as amide salts) with low steric requirements are employed, the usual pentacoordinate transition state (Scheme 1), may be a viable reaction intermediate which can undergo alternative modes of decomposition, perhaps involving chain cleavage and/or cross-linking.

$$[N{=}PCl_2]_x + 2x\ C_4H_4N^-K^+ \xrightarrow{\text{THF}} [N{=}P(NC_4H_4)_2]_x + 2x\ KCl \quad (1)$$

(PBPP)

Nevertheless, through scrupulous purification of the reaction components and rigorous control of the reaction conditions it is possible to isolate the polymer in a state of good purity, by the reaction of potassium pyrrolide with $(NPC\ell_2)_x$ in tetrahydrofuran at room temperature (Equation 1). Addition of water to the reaction mixture precipitates the polymer as a white rubbery solid which hardens on drying. A ^{31}P NMR spectrum of a typical reaction product is given in Figure 2.

Properties of Polybis(pyrrolyl)phosphazene (PBPP)

The solubility properties of the polymer are very sensitive to the mode of isolation.[21] PBPP is partially soluble in tetrahydrofuran immediately after precipitation from the reaction mixture, but when the polymer has been dried the solubility is less predictable.

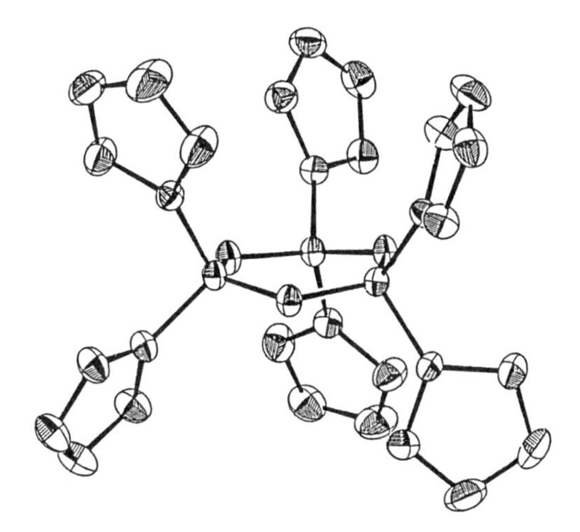

Figure 1. Structure of crystalline $[NP(NC_4H_4)_2]_3$ (**1**).[20]

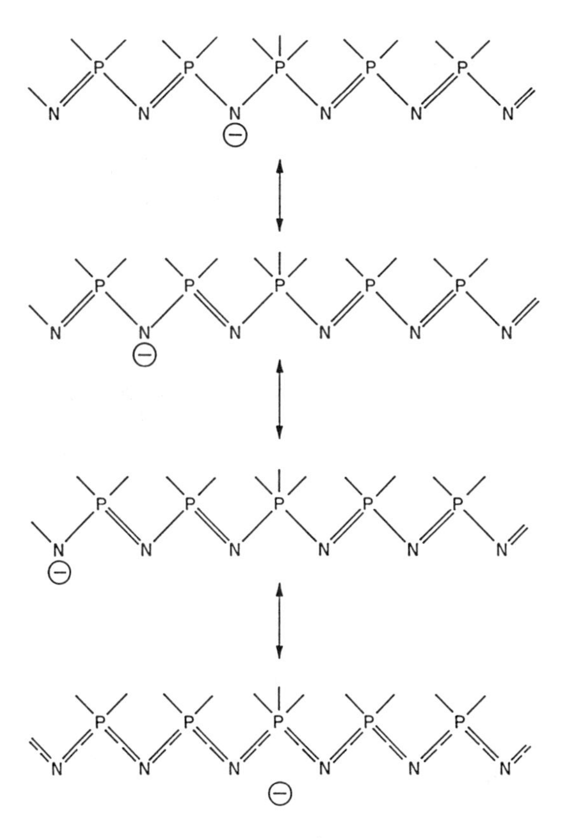

Scheme 1.

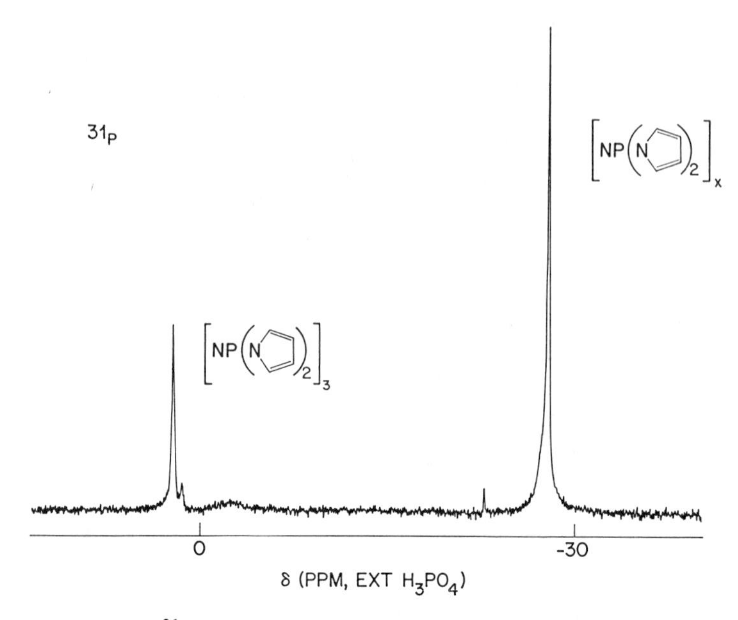

Figure 2. ^{31}P NMR spectrum of reaction to produce PBPP.

Brittle colorless films of PBPP may be cast from tetrahydrofuran solution. The insoluble portion of PBPP is swelled by the tetrahydrofuran and gives rise to free-standing films on solvent evaporation. Differential scanning calorimetry experiments on PBPP show a glass transition temperature at 40°C, and some indication of a melting transition at 170°C.

PBPP undergoes electrochemical oxidation to produce black films with conductivities in the semiconductor range.[17] More highly conducting materials are produced by heating PBPP in an inert atmosphere (without doping).[21] As may be seen from the thermogravimetric analysis in Fig. 3, PBPP exhibits some resistance toward the thermal depolymerization process which is characteristic of many polyphosphazenes.[13] The conductivity and thermal involatility probably arise from the tendency of PBPP to undergo cross-linking as exemplified in Scheme 2. The structure found for the trimer (Fig. 1),[20] suggests that these should be viable processes in the polymer.[20]

Future Prospects

Although the picture is far from complete, the available evidence suggests that PBPP is rather different from most polyphosphazenes. The polymer may be induced to be an electronic conductor, but perhaps as a result of this tendency to cross-link, the material is more sensitive and difficult to handle than most polyphosphazenes and the thermal depolymerization reaction is inhibited.

Nevertheless the synthetic versatility offered by the polyphosphazenes

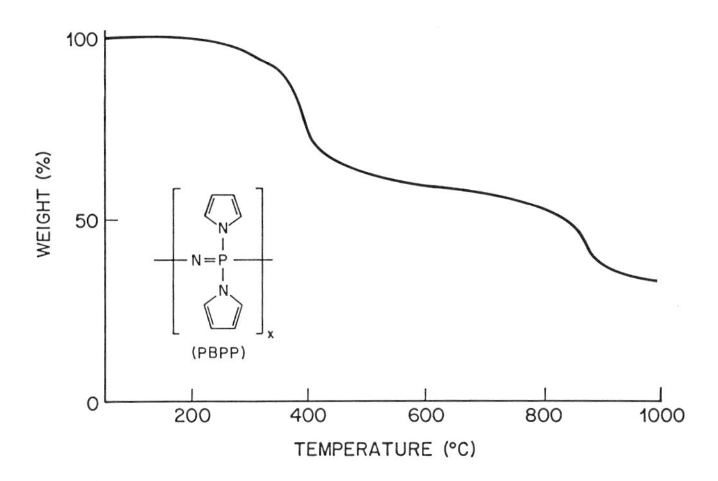

Figure 3. Thermogravimetric analysis of PBPP under nitrogen.

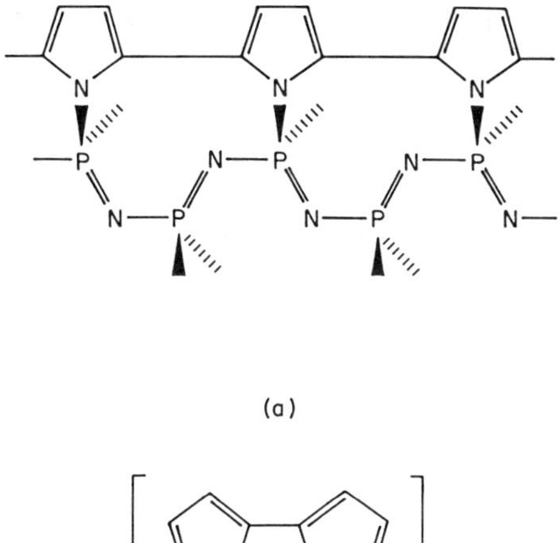

(a)

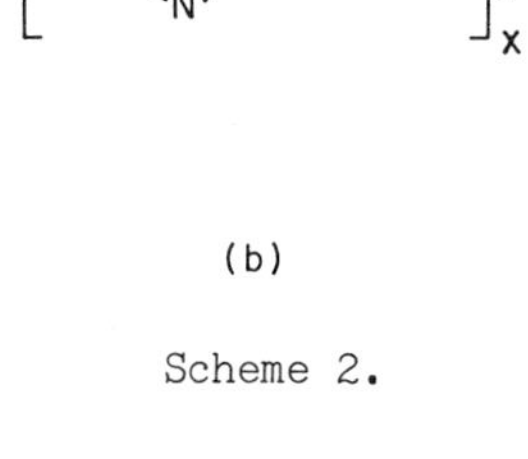

(b)

Scheme 2.

suggests that PBPP[18] and the phthalocyanine derivatives[16] represent the forerunners of a new class of polymers with interesting electronic solid state properties.

LITERATURE CITED

1. Penton, H. R. *Polym. Prepr.* **1987**, *28*, 439.
2. Allcock, H. R.; Kugel, R. L.; Valan, K. J. *Inorg. Chem.* **1966**, *5*, 1709.
3. Allcock, H. R.; Kugel, R. L. *Inorg. Chem.* **1966**, *5*, 1716.
4. Allcock, H. R. *Chem. Eng. News* **1985**, *63* (11), 22.
5. Neilson, R. H.; Ford, R. R.; Hani, R.; Roy, A. K.; Scheide, G. M.; Wettermark, U. G.; Wisian-Neilson, P. *Polym. Prepr.* **1987**, *28*, 442.
6. Blonsky, P. M.; Shriver, D. F.; Austin, P.; Allcock, H. R. *J. Am. Chem. Soc.* **1984**, *106*, 6854.
7. Craig, D. P.; Paddock, N. L. In "*Nonbenzonoid Aromatics*"; Snyder, J. P., Ed.; Academic Press, New York, 1971, Vol. 2.
8. Dewar, M. J. S.; Lucken, E. A. C.; Whitehead, M. A. *J. Chem. Soc.* **1960**, 2423.
9. Salem, L. "*Molecular Orbital Theory of Conjugated Systems*"; Benjamin: New York, 1966; p. 158.
10. Haddon, R. C. *Chem. Phys. Lett.* **1985**, *120*, 310.
11. Haddon, R. C.; Mayo, S. L.; Chichester, S. V.; Marshall, J. H. *J. Am. Chem. Soc.* **1985**, *107*, 7585.
12. Ferris, K. F.; Friedman, P.; Friedrich, D. M., ACS Meeting, Denver, 1987.
13. Allcock, H. R. "*Phosphorus-Nitrogen Compounds*", Academic Press, New York, 1972.
14. Wisian-Neilson, P.; Roy, A. K.; Xie, Z.-M.; Neilson, R. H. *ACS Symposium Series* **1983**, *232*, 167.
15. Barendt, J. M.; Haltiwanger, R. C.; Norman, A. D. *Inorg. Chem.* **1986**, *25*, 4323: Thompson, M. L.; Tarassoli, A.; Haltiwanger, R. C.; Norman, A. D. *Inorg. Chem.* **1987**, *26*, 684.
16. Allcock, H. R.; Neenan, T. X. *Macromol.* **1986**, *19*, 1495.
17. Allcock, H. R.; Levin, M. L.; Austin, P. E. *Inorg. Chem.* **1986**, *25*, 2281.
18. Haddon, R. C.; Stein, S. M.; Chichester-Hicks, S. V.; Marshall, J. H.; Kaplan, M. L.; Hellman, M. Y. *Mater. Res. Bull.* **1987**, *22*, 117.
19. McBee, E. T.; Johncock, P.; French, S. E.; Braendlin, H. P. *U.S. Govt. Res. Rep. AD* **1960**, 254985.
20. Craig, S. L.; Cordes, A. W.; Stein, S. M.; Chichester-Hicks, S. V.; Haddon, R. C. *Acta Crystallogr.*, in press.
21. Haddon, R. C.; Chichester-Hicks, S. V.; Bowmer, T. N. to be published.
22. Allcock, H. R.; Smeltz, L. A. *J. Am. Chem. Soc.* **1976**, *98*, 4143.

RECEIVED September 22, 1987

Chapter 25

Skeletal Stabilization as the Basis for Synthesis of Novel Phosph(III)azane Oligomers and Polymers

Elizabeth G. Bent, Joseph M. Barendt, R. Curtis Haltiwanger, and Arlan D. Norman

Department of Chemistry and Biochemistry, University of Colorado, Boulder, CO 80309

Reactions of $PhPCl_2$ or $P(NR_2)_3$ (R = alkyl) with orthoaminated benzenes yield new phosph(III)azanes which contain extended P(III)-N skeletons stabilized by bridging o-C_6H_4 groups. The novel $[(PhP)(C_6H_4N_2PPh)]_n$ is formed from the $PhPCl_2$ - 1,2-$(NH_2)_2C_6H_4$ reaction; the dimer (n = 2) is fully characterized. From 1,2-$(NH_2)_2$-C_6H_4 or 1,2,3-$(NH_2)_3C_6H_3$ reactions with $P(NR_2)_3$ (R = Me, Et), phosphazanes are obtained which contain P_3N_2 or P_4N_3 skeletal units stabilized around the arene rings. The intermediates $C_6H_4(NH)_2(PNEt_2)$ and $C_6H_4(NH)$-$[N(PNEt_2)_2](PNEt_2)$ in the 1,2-$(NH_2)_2C_6H_4$ - $P(NEt_2)_3$ reaction have been characterized. The two-phosphorus compound displays 1,4- A-B condensation monomer reactivity. Preliminary studies of its oligomer/polymer chemistry are reported.

Inorganic polymers based on alternating main group element-nitrogen skeletons (e.g. I - IV) are of interest for their potential as elastomers, high-temperature oils, electrical conductors, biological molecule carriers, and precursors to ceramic materials (1 - 6).

$$-\!\!\left(\,\overset{\backslash\,/}{P}{=}\overset{|}{N}\,\right)\!\!-_n \qquad -\!\!\left(\,\overset{|}{P}{-}\overset{|}{N}\,\right)\!\!-_n \qquad -\!\!\left(\,\overset{|}{B}{-}\overset{|}{N}\,\right)\!\!-_n \qquad -\!\!\left(\,\overset{\backslash\,/}{Si}{-}\overset{|}{N}\,\right)\!\!-_n$$

I II III IV

Significant recent advances have occured with phosphazenes (1) and to a lesser extent silizanes (6); in contrast, polymers based on phosphorus(III) and nitrogen are virtually unknown. Because such systems offer opportunities for metal coordination, and in their oxidized forms could be valuable polymer precursors to PON ceramics (2), we have begun systematic studies of the structural and reactivity factors necessary for their formation.

Reactions that in principle could yield linear phosph(III)azanes generally do not (7, 8); instead they produce low molecular weight oligomers. Among these, the four-membered ring

0097-6156/88/0360-0303$06.00/0

1,3,2,4-diazadiphosphetidines (V) are most common. Only with small substituents on phosphorus and nitrogen (R and R' = Me or Et) have higher oligomers (e.g. VI) been obtained (9).

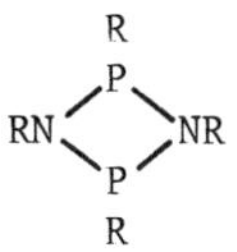

V 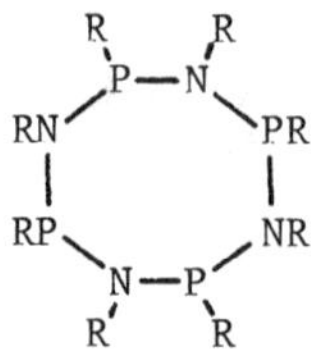 VI 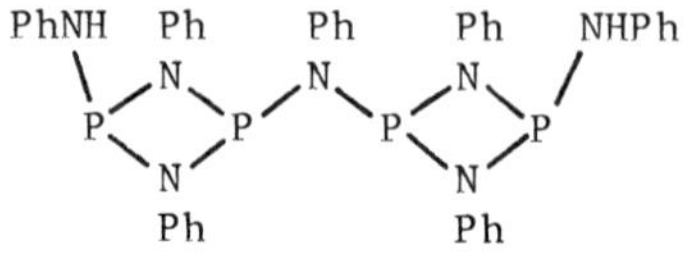 VII

Reactions of PX_3 with RNH_2 could form cyclic-linear polymers based on connected P_2N_2 rings, e.g. $[P_2(NR)_3]_n$; however, so far only the dinuclear oligomer VII has been obtained (8). Apparently, a combination of small bond angles around phosphorus and strong intergroup repulsions between substituents on phosphorus and nitrogen (1, 7, 8, 10) strongly favor small ring oligomers and disfavor polymers. In no case has P(III)-N oligomer-polymer interconversion, analogous to what occurs with phosphazenes and silazanes (1, 6), been seen.

Because low oligomers of linear phosph(III)azanes appear thermodynamically preferred at all reasonable temperatures, we have undertaken a synthetic strategy involving skeletal stabilization as a technique to obtain these polymers. We seek to construct phosphorus-nitrogen skeletons which contain stabilizing units between adjacent nitrogen atoms, for example as in VIII, creating rigid extended units which will not

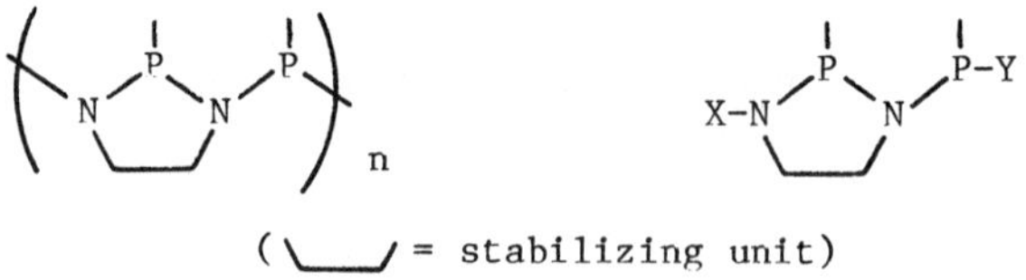

(= stabilizing unit)

VIII IX

allow small ring formation but will favor molecular extension based on rigid 1,4- reactive units. Thus we have sought to synthesize reactive A-B type intermediate monomers (IX) or oligomers/polymers (VIII) directly based on such systems. Our studies so far are prototypical and have concentrated mainly on reaction strategies; however, they indicate the approach has considerable potential.

Amine-Phosphorus(III) Halide Condensation Reactions

Since amine-phosphorus(III) halide condensation reactions are well-established low-energy routes to phosphorus-nitrogen bonds, they provide a logical starting point for examination of skeletally stabilized molecule synthesis. For our initial studies, reactants that are phenyl substituted as opposed to alkyl substituted on phosphorus were chosen, so as to maximize the possibility of obtaining crystalline products.

Reaction of $PhPCl_2$ with $1,2-(NH_2)_2C_6H_4$ in toluene in the presence of Et_3N to remove hydrogen chloride proceeds smoothly at reflux as:

$$2n\ PhPCl_2 + n\ 1,2-(NH_2)_2C_6H_4 \xrightarrow[-Et_3NHCl]{Et_3N} 1/n\ [(PhP)(C_6H_4N_2PPh)]_n \quad (1)$$

X

The character of product X depends on reaction dilution. Oligomers ($n \geq 2$)/polymers form; dimer formation is minimized at high reactant concentrations.

The structure of dimer XI (X, where n = 2), obtained by x-ray single crystal analysis (Figure 1), shows that the desired skeletal stabilization based upon 1,4- condensation units has been achieved. XI is a cyclophane type molecule which contains an eight membered ring of alternating phosphorus and nitrogen atoms. Adjacent nitrogen atoms are bridged by $o-C_6H_4$ units into diazaphosphole rings. The C_{2v} symmetry of the molecule is maintained in solution also, since the P(31) NMR spectrum exhibits two temperature independent triplets (J = 17.2 Hz) at 85.0 ppm and 112.0 ppm.

In addition to being a skeletally stabilized molecule, XI contains an unusual ring structure. The ring has four nitrogen and two phosphorus atoms in a plane with the two remaining phosphorus atoms above the ring, which along with the oriented phenyl groups forms a well-defined cavity in a boat shaped eight membered ring. This geometry is undoubtedly imposed by the skeletal stabilizing unit. Other eight membered phosphorus(III)-nitrogen rings, $(MePNMe)_4$ (9) and $(PrNCH_2CH_2NP)_4$ (11) are crown shaped containing all chemically equivalent phosphorus atoms. In XI, because the phosphorus atoms are of two sets and one set is in a highly protected environment, the possibility of selective, "cavitand," coordination at phosphorus sites exists.

Detailed characterization of the oligomer/polymer products (X)

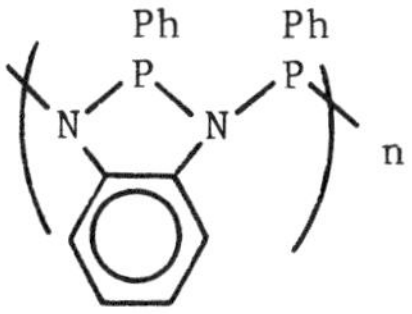

X

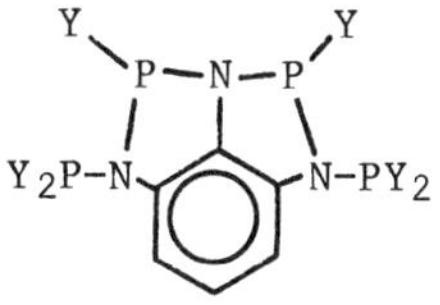

XIII

from Equation 1 are in progress. The products are viscous oils, soluble in hydrocarbon solvents, consistent with that expected of linear oligomer/polymers. They exhibit broad approximately equal area P(31) resonances at 95 - 102 ppm and 114 - 120 ppm, consistent with phosphorus environments of bridging and diazaphosphole types. Given the physical properties and the correlation of P(31) and H(1) NMR spectral parameters between dimer XI and the higher order products, we infer X has the cyclic-linear structural properties characteristic of XI. Studies of the oligomer/polymer products to

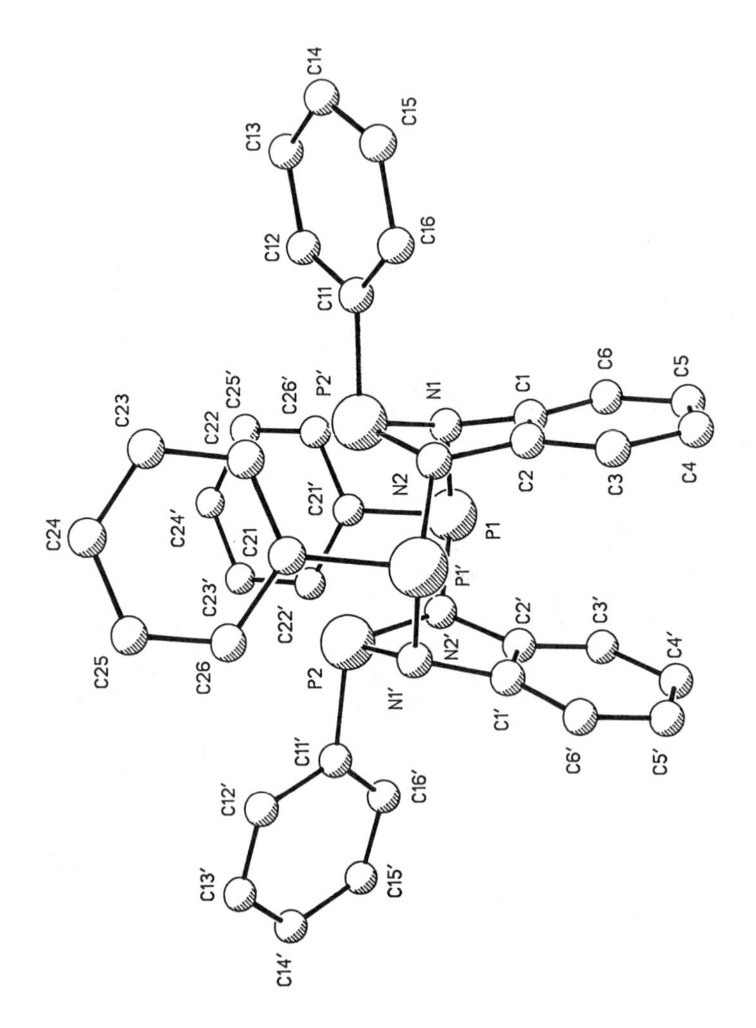

Figure 1. The structure of $[(PhP)(C_6H_4N_2PPh)]_2$ (XI).

maximize molecular weights and to determine reaction generality with other types of phosphorus substitution are in progress.

Transamination Condensations

Amine-phosphorus halide reactions involve hydrogen halide (or amine hydrohalide) formation. They can be complicated through acid catalysis or side reactions. Thus, the possible formation of skeletally stabilized products through transamination reactions in which no hydrohalide products are obtained is of interest. Several reactions have been examined, again all using stabilization by 1,2-aminated benzene units. Reactions involving $RP(NR_2)_2$ and $P(NR_2)_3$ are being studied, however, the latter have initially been easier to understand because they involve less complex stereochemistry at the phosphorus centers in the products.

Tris(dialkylamino)phosphines react with ortho-amino benzenes to form extended phosphorus-nitrogen skeletons (i) around the arene rings or (ii) of the 1,4-stabilized type, depending upon conditions. Reactions of $1,2\text{-}(NH_2)_2C_6H_4$ with excess $P(NEt_2)_3$ result in formation of the triphosphazane XII (Y = NEt_2) and tetraphosphazane XIII (Y = NMe_2), respectively (12, 13). The structures of XII (as its disulfide) and XIII (as its tetrasulfide) are shown in Figures 2 and 3. Surprisingly, these are the most extended linear phosphorus(III)-nitrogen skeletons fully characterized to date.

Under conditions of approximately equimolar reactant ratios [$1,2\text{-}(NH_2)_2C_6H_4$: [$P(NEt_2)_3$], intermediates with considerable potential in subsequent polymer synthesis can be identified. At a 1.0 : 0.8 ratio during 24 hr at 60°C, reaction occurs to form the labile monophosphorus XIV as the principle product (Scheme 1). At

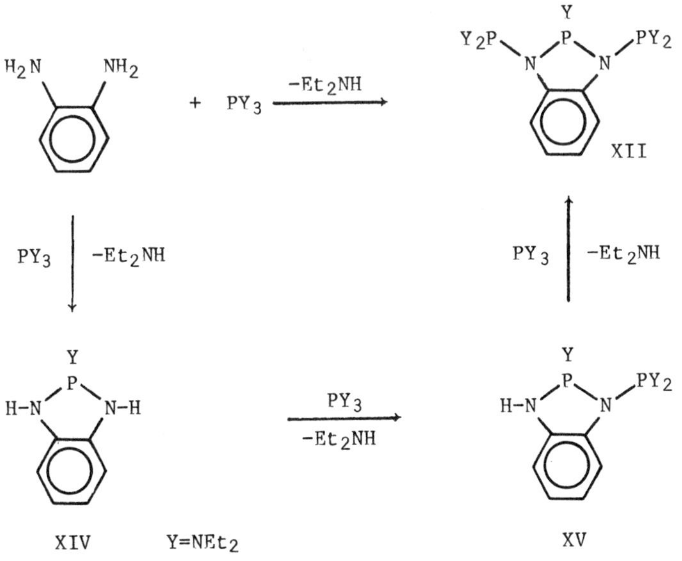

Scheme 1.

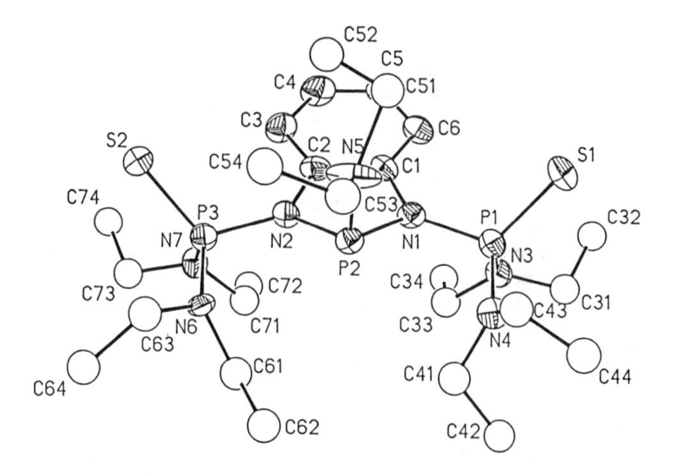

Figure 2. The structure of $C_6H_4N_2P(NEt_2)[P(NEt_2)_2]_2$, the disulfide of XII.

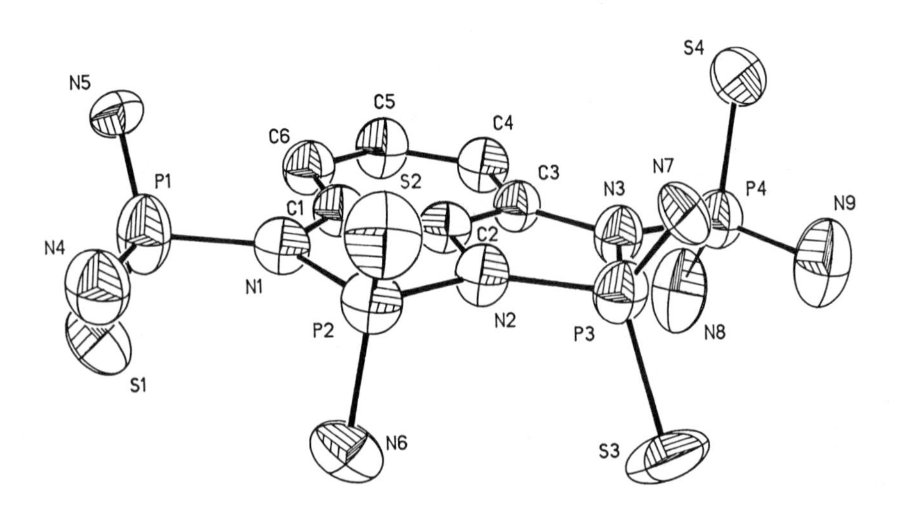

Figure 3. The structure of $C_6H_3N_3[P(S)NMe_2]_2[P(S)(NMe_2)_2]_2$, the tetrasulfide of XIII.

lower ratios (1.0 : 2.0), reaction proceeds in 2 hr at 80°C mainly to the two-phosphorus product XV. XIV and XV exhibit P(31) NMR spectra consisting of a singlet (89.0 ppm) and two doublets (105.8, 94.0 ppm; $^2J_{pp}$ = 41.5 Hz), respectively. In both cases the resonances are in chemical shift regions expected and characteristic of aminophosphines.

Both XIV and XV are highly reactive and neither can be isolated from the reaction mixture in the absence of Et_2NH; however, upon reaction with elemental sulfur, conversion to the phosphorus sulfide analogs (XVI and XVII) occurs. The latter are crystalline solids stable indefinitely at 25°C in the absence of Et_2NH. Characterization of XIV and XV depends on spectral data of the species in solution and upon correlation of these data with structural data obtained on the stable analogs XVI and XVII. The structure of XVI has been established by x-ray analysis (Figure 4).

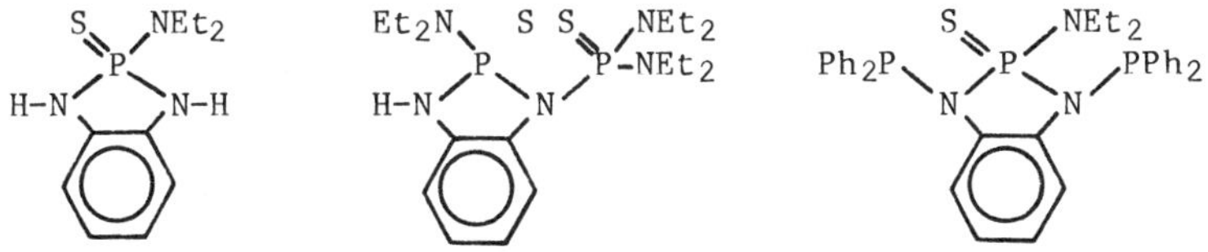

XVI XVII XVIII

Compounds XIV - XVII have potential for use in subsequent phosphazane oligomer/polymer syntheses because of their available functionality. In fact, treatment of XVI with two equivalents of Ph_2PCl in the presence of Et_3N yields the P(III)-P(V)-P(III) triphosphazane XVIII quantitatively. Compounds XV and XVII are especially interesting because they are A-B type 1,4- stabilized reactive monomers (prepolymers) (14). Because A-B monomers contain both elements of functionality necessary for condensation, precise control of the reactant ratio can be maintained and products with the highest possible molecular weights can be expected (14).

The successful use of XV or XVII in polymer synthesis requires that elimination of Et_2NH via 1,4- processes competes effectively with 1,2- processes. It is known that partially alkylated diamines

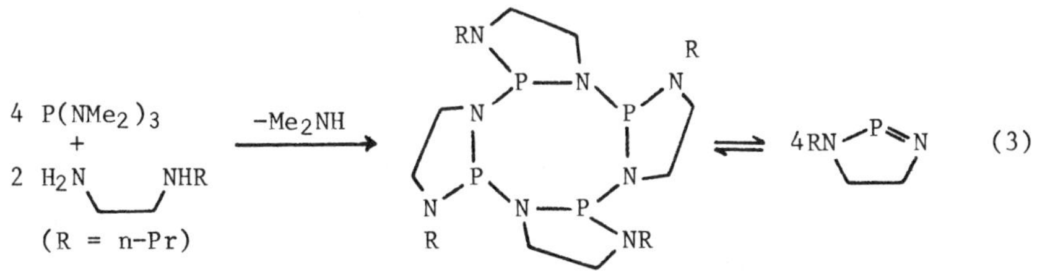

$$4\ P(NMe_2)_3 + 2\ H_2N\text{—}CH_2CH_2\text{—}NHR \xrightarrow{-Me_2NH} \text{XIX} \rightleftharpoons 4\ \text{XX} \qquad (R = n\text{-}Pr) \qquad (3)$$

react with $P(NMe_2)_3$ to form cyclic tetramers (XIX) which exist in equilibrium with their monomers (XX), apparently as a result of 1,2-

elimination of Me_2NH from an initial product analogous to XIV (11, 15). Similarly, we observe 1,2- Et_2NH elimination from XIV in the monomer-dimer equilibrium:

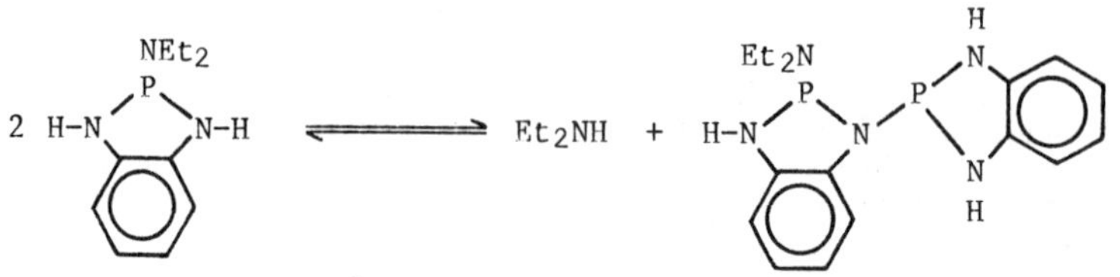

$$2\ \mathrm{XIV} \rightleftharpoons Et_2NH + \mathrm{XXI} \quad (4)$$

However, preliminary studies of XV polymerization indicate that 1,4-elimination of Et_2NH occurs according as:

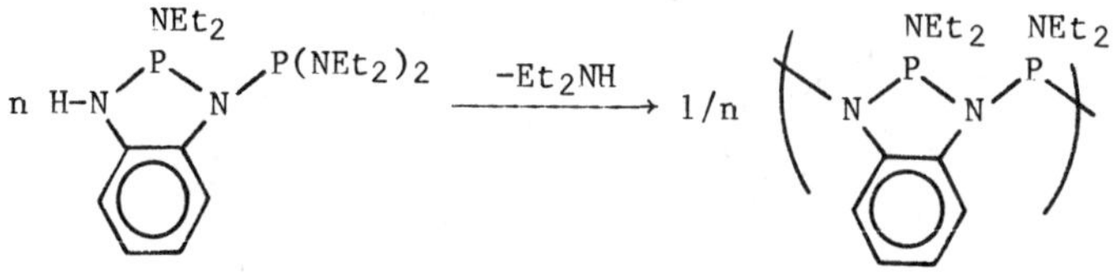

$$n\ \mathrm{XV} \xrightarrow{-Et_2NH} 1/n\ (\mathrm{XXII})_n \quad (5)$$

In vacuo at 25°C as Et_2NH is removed, product exhibiting broad P(31) NMR resonances in the 80 - 86 ppm and 100 - 112 ppm regions form. This occurs reversibly; readdition of Et_2NH reforms XV. Some 1,2-elimination is evident, also. Treatment of the 1.00 : 2.00 reaction mixture with sulfur yields traces of oxidized condensation dimer XXIII which we have characterized by single crystal x-ray analysis (Figure 5). The degree of oligomerization of product oil (XXII), which is soluble in hydrocarbon solvents and thermally stable for periods of weeks at ambient temperatures, remains to be established and improved.

Reactions involving alkoxy- or aryloxyphosphines with amines might also be expected to yield phosphazane oligomers/polymers. However, $PhP(OPh)_2$ with 1,2-$(NH_2)_2C_6H_4$ produced only the Arbuzov type rearrangement product $C_6H_4(NH)_2P(O)Ph$.

Conclusions

Although our studies of skeletally stabilized phosphazanes are in early stages of development, the results so far indicate that through use of carefully selected stabilizing units, systems based on 1,4-building blocks can be obtained. The extent to which other stabilizing groups affect oligomer/polymer formation reactions and how to control and optimize molecular weights are vital questions to be examined next.

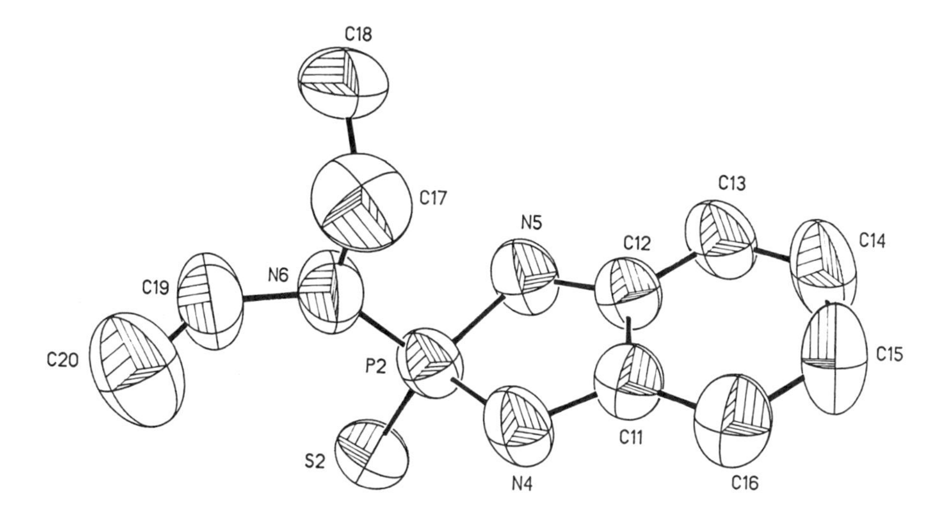

Figure 4. The molecular structure of $C_6H_4(NH)_2[P(NEt_2)]$ (XVI).

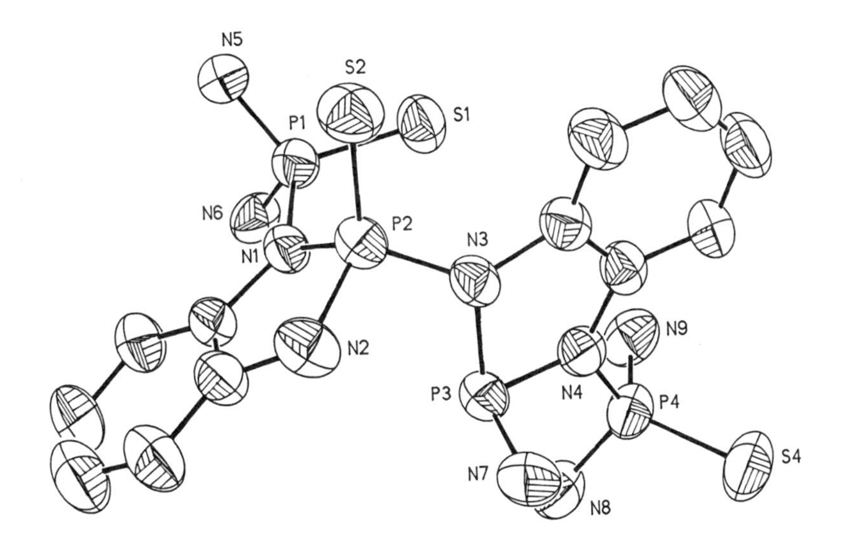

Figure 5. The molecular structure of the 1,2-condensation dimer of $C_6H_4(NH)[NP(NEt_2)_2](PNEt_2)$ (XV), as the trisulfide derivative. Ethyl groups on N5, N6, N7, N8 and N9 omitted for clarity.

Acknowledgments

The authors thank the National Science Foundation and the Colorado Advanced Materials Institute Polymer Research Center for generous financial support.

Literature Cited

1. Allcock, H. R. Chem. and Eng. News 1985, 63(5), 22.
2. Ray, N. H. In Inorganic Polymers, Academic Press, London, 1978.
3. Allcock, H. R. In Heteroatom Ring Systems and Polymers, Academic Press, New York, 1967.
4. Wynne, K. J.; Rice, R. W. Ann. Rev. Mater. Sci. 1984, 14, 297.
5. Westwood, A. R. C.: Skalny, J. P. In Cutting Edge Technologies, National Academic Press, Washington, 1984, pg. 117.
6. Seyferth, D.; Wiseman, G. H.: Prud'home, C. J. Am. Ceram. Soc. 1984, 66, C13.
7. Keat, R. Topics in Current Chemistry 1982, 88, 1677.
8. Thompson, M. L.; Tarassoli, A.; Haltiwanger, R. C.; Norman, A. D. Inorg. Chem. 1987, 26, 684.
9. Zeiss, W.; Schwarz, W.; Hess, H. Angew Chem. Int. Ed. Engl. 1977, 16, 407.
10. Keat, R.; Murray, L.; Rycroft, D. C.; J. Chem. Soc. Dalt. Trans. 1982, 1503.
11. Malavaud, C. A.; N'Gando M'Pondo, T.; Lopez, L.; Barrans, J.; Legros, J. P. Can. J. Chem. 1984, 62, 43.
12. Barendt, J. M.; Haltiwanger, R. C.; Norman, A. D. Inorg. Chem. 1986, 25, 4323.
13. Barendt, J. M.; Haltiwanger, R. C.; Norman, A. D. J. Am. Chem. Soc. 1986, 108, 3127.
14. Allcock, H. R.; Lampe, F. W. In Contemporary Polymer Chemistry, Prentice-Hall, Englewood Cliffs, N. J., 1981.
15. Lehousse, C.; Haddad, M.; Barrans, J. Tetrahedron Lettr. 1982,

RECEIVED September 1, 1987

ORGANOELEMENT-OXO POLYMERS DERIVED FROM SOL-GEL PROCESSES

Chapter 26

Structure of Sol–Gel-Derived Inorganic Polymers: Silicates and Borates

C. J. Brinker [1], B. C. Bunker [1], D. R. Tallant [1], K. J. Ward[1], and R. J. Kirkpatrick [2]

[1]Sandia National Laboratories, Albuquerque, NM 87185
[2]Department of Geology, University of Illinois, Urbana, IL 61801

Many of the structural features of sol-gel-derived inorganic polymers are rationalized on the basis of the stability of the M-O-M condensation products in their synthesis environments. Structures which emerge in solution reflect a successive series of hydrolysis, condensation and, depending on the acid or base concentration, restructuring reactions. M-O-M bonds which are unstable with respect to hydrolysis or alcoholysis are generally absent. During consolidation, the condensation process continues within a very stiff matrix in a relatively inert environment. Under these conditions, metastable species may form which are temporarily stabilized by the high matrix viscosity and the absence of water and alcohol.

The sol-gel process for making glasses and ceramics is now an intensely studied topic in materials science. Within just the last six years, at least eight symposia have been devoted largely to this subject (1-8). This excitement results from the fact that sol-gel processing provides a means to "grow" ceramic polymers in solution at room temperature, shape them by casting, film formation, or fiber drawing, and consolidate them to dense glasses or polycrystalline ceramics at temperatures often less than half the conventional processing temperature (Figures 1 and 2). Low temperature processing has several advantages. There are the obvious reductions in energy consumption and impurities (due to containerless processing and the use of distillable precursors), but more importantly, it allows the synthesis of materials which are thermodynamically unstable and therefore which are generally impossible to obtain by conventional high temperature processing. Examples include avoidance of phase separation (9) and crystallization (10) in binary silicate glasses, synthesis of extremely high surface area substrates and films (11), and formation of thermodynamically unstable oxynitride glasses (12).

The most common sol-gel process employs metal alkoxides of network forming elements ($M(OR)_x$ where M is Si, B, Ti, Al, etc. and R is often an alkyl group) as monomeric precursors. In alcohol/water solutions the alkoxide groups are removed stepwise by hydrolysis reactions, generally employing acid or base catalysts, and are

NOTE: This chapter is part II in a series.

0097–6156/88/0360–0314$06.00/0

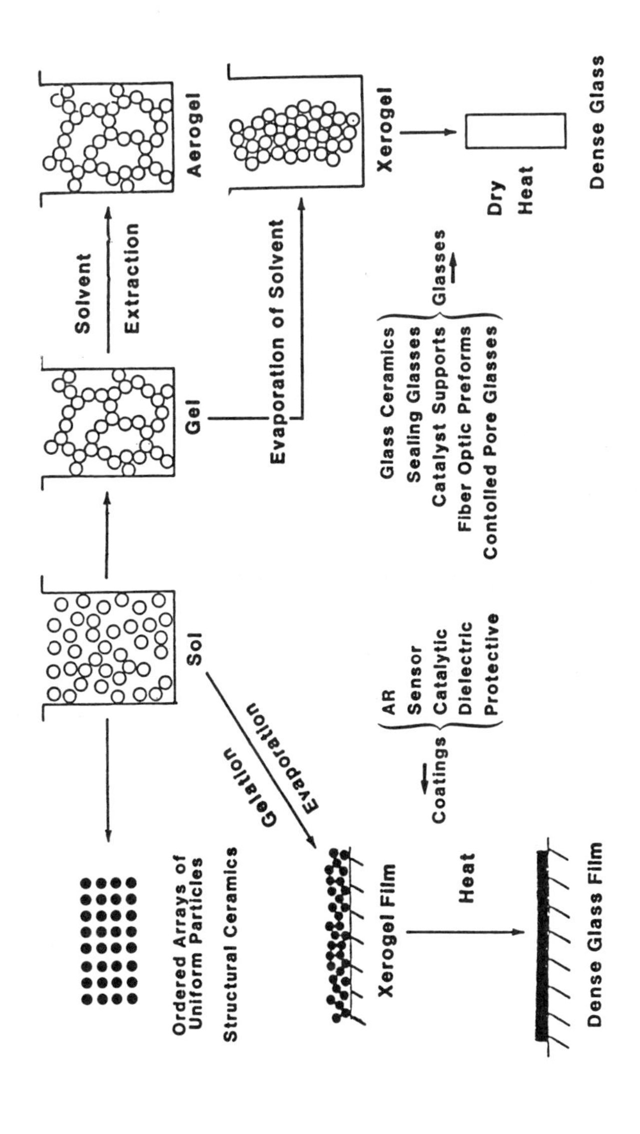

Figure 1. Schematic of the sol-gel process for preparing glasses and ceramics.

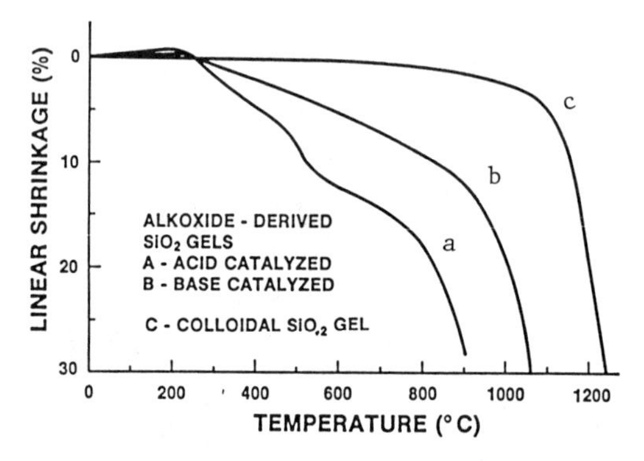

Figure 2. Linear shrinkage versus temperature at 1°C/min heating rate for silica gels prepared by three different methods: a) two-step acid hydrolysis, b) two-step base hydrolysis, c) colloidal process. (Reproduced with permission from Ref. 4. Copyright 1984 Wiley-Interscience.)

replaced by hydroxyl groups (Equation 1). Subsequent condensation reactions involving the hydroxyl groups yield polymeric solution species composed of M-O-M linkages (Equations 2 and 3). It should be

$$Si(OR)_4 + H_2O \rightleftarrows (RO)_3Si - OH + ROH \quad (1)$$

$$2(RO)_3Si - OH \rightleftarrows (RO)_3Si - O - Si(OR)_3 + H_2O \quad (2)$$

$$(RO)_3Si - OH + Si(OR)_4 \rightleftarrows (RO)_3Si - O - Si(OR)_3 + ROH \quad (3)$$

emphasized that although Equations 1-3 in the forward direction are generally used to describe the sol-gel process, these reactions are all reversible. The reverse reaction rates depend on pH, temperature, water and solvent concentrations, as well as the connectivity of M within the polymer from the standpoint of steric and inductive considerations (13-14). Structural consequences of the reversibility of Equations 1-3 constitute the primary topic of this chapter.

Gelation occurs when the growing polymers link together to form a network which spans the entire solution volume. At the gel point, physical properties such as the viscosity and elastic modulus rise abruptly. The gel is best described as a viscoelastic material composed of interpenetrating solid and liquid phases: the solid network prevents the liquid from rapidly escaping and the liquid prevents the network from spontaneously collapsing (15). Gelation rather than precipitation is a benefit in materials technology, because it provides a means for near net shape processing. The sudden increase in viscosity "freezes-in" shapes obtained by casting, film deposition, or fiber drawing.

The gel itself is not an end product. Normally it is dried either by slow evaporation in a convection oven or by supercritical fluid extraction in an autoclave. The dried gels are called xerogels and aerogels, respectively (see Figure 1). Evaporation of solvent causes shrinkage primarily due to surface tension, which exerts an effective hydrostatic compressive stress on the gel proportional to the curvature of the liquid-vapor meniscus. As the drying gel shrinks, condensation reactions continue as reactive terminal groups encounter one another, causing the gel to become progressively stiffer. When the gel is sufficiently stiff to withstand the compressive stresses accompanying drying, further solvent removal creates interconnected porosity within the gel. Supercritical fluid extraction eliminates liquid-vapor interfaces, which greatly reduces the drying shrinkage (16).

Both xerogels and aerogels are characteristically high surface area materials (surface areas normally exceed 500 m^2/g). Unlike wet gels, many uses exist for dried gels due to their high surface areas and small pore sizes (typically, < 20 nm diameters). Examples include catalyst supports (17), ultrafiltration media (18), antireflective coatings (19-20), and ultra-low dielectric constant films. (Lenahan, P. M. and Brinker, C. J., unpublished results.)

Porous xerogels and aerogels are consolidated to form dense glasses and ceramics by thermal treatments, causing a second stage of shrinkage (see Figure 2) (21). In the amorphous systems which we discuss, final densification occurs by viscous sintering in which the shrinkage rate is proportional to the surface energy divided by the product of viscosity and the pore dimension (22). Because of the very high surface areas and very small pore dimensions, gels may be completely densified at such low temperatures that the viscosity is not reduced below about 10^{13} poises (21-22). At such high viscosities the kinetics of phase separation and crystallization are severely retarded (10).

Structural Evolution of Inorganic Polymers and Gels

The structures of sol-gel-derived inorganic polymers evolve continually as products of successive hydrolysis, condensation and restructuring (reverse of Equations 1-3) reactions. Therefore, to understand structural evolution in detail, we must understand the physical and chemical mechanisms which control the sequence and pattern of these reactions during gelation, drying, and consolidation. Although it is known that gel structure is affected by many factors including catalytic conditions, solvent composition and water to alkoxide ratio (13-14), we will show that many of the observed trends can be explained on the basis of the stability of the M-O-M condensation product in its synthesis environment.

This paper summarizes the results of structural studies of silicate and borate gels, both in solution and during drying and consolidation. Using the combined results of NMR, IR and Raman spectroscopies and small angle x-ray scattering (SAXS), we consider the structures of gel-derived inorganic polymers on several length scales. First, there is the local environment of the metal atom: the second nearest neighbors may be either alkyl groups, protons, or other metal atoms. On a larger length scale, oligomeric species (dimers, trimers, tetramers, etc.) may be linear, branched or cyclic. Finally, on the largest length scales, structures may be dense with well-defined interfaces, uniformly porous, or tenuous networks characterized by a fractal dimension, D (for mass fractals, D relates the polymer mass to its radius according to: $M \propto r^D$, where in three dimensions $D < 3$) (23).

Silicate Systems

Silicate polymers and gels have been synthesized by the hydrolysis of tetraethylorthosilicate or tetramethylorthosilicate (TEOS or TMOS) in acidic, neutral, or basic solutions employing H_2O/OR ratios from 2 to over 50. The stability of siloxane bonds toward hydrolysis or alcoholysis in the H_2O/ROH synthesis medium depends strongly on the activity of hydroxyl ions, H_2O/ROH molar ratio, and bond strain imposed by specific bonding configurations. For example, the solubility of amorphous silica (a-SiO_2) increases by over 3 orders of magnitude as the pH is increased from 2 to 9 (24) and decreases by a factor of 28 when water (pH 9.5) is replaced by a 90 wt.% MeOH/10 wt.% H_2O mixture (24). As shown in the following examples, structures on all length scales are profoundly influenced by the relative stability of the siloxane bond in the solution environment.

Availability of Monomer. Under most solution conditions employed for gel formation, monomer is rapidly consumed to form a distribution of oligomeric species (13). Thereafter the only possible source of monomer is that which is produced by hydrolytic (or alcoholic) depolymerization of siloxane bonds (dissolution). According to Iler (24) the dissolution of a-SiO_2 above pH 2 is catalyzed by OH^- ions which are able to increase the coordination of silicon above 4 weakening the surrounding siloxane bonds to the network. The end result is the preferential production of small concentrations of monomeric, soluble silicate species in solution.

The effect of availability of monomer on structural development is evident from the results of computer simulations of two irreversible random growth processes: monomer-cluster aggregation (25), in which monomers react only with a growing cluster, and cluster-cluster aggregation (26), in which monomers react with one another and with clusters in a random manner. Keefer (27) has shown that under reaction limited conditions, monomer-cluster aggregation results in dense, non-fractal "particles." By comparison, diffusion-limited conditions result in fractal clusters (25). In cluster-cluster aggregation, most monomers are consumed to form clusters which in turn react with one another forming larger clusters, etc. In the latter case, i.e., in the relative absence of monomers, the growing clusters are ramified, tenuous networks characterized by a fractal dimension ($D = 1.7$ and 2.0 for diffusion-limited and reaction limited conditions, respectively) (26). Most importantly, these simulations predict that fully dense structures emerge only when a source of monomers is available (and condensation is sufficiently slow with respect to diffusion to be in reaction-limited conditions). This results primarily because branched clusters ($D > 1.5$) are sterically precluded from interpenetrating one another to efficiently fill space (28), whereas, under reaction-limited conditions, monomers can penetrate all regions of the growing polymer and react primarily at the most favored sites.

In situ SAXS investigations of a variety of sol-gel-derived silicates are consistent with the above predictions. For example, silicate species formed by hydrolysis of TEOS at pH 11.5 and $H_2O/Si = 12$, conditions in which we expect monomers to be continually produced by dissolution, are dense, uniform particles with well defined interfaces as determined in SAXS experiments by the Porod slope of -4 (non-fractal) (Brinker, C. J., Hurd, A. J. and Ward, K. D., in press). By comparison, silicate polymers formed by hydrolysis at pH 2 and $H_2O/Si = 5$, conditions in which we expect reaction-limited cluster-cluster aggregation with an absence of monomer due to the hydrolytic stability of siloxane bonds, are fractal structures characterized by $D = 1.9$ (Porod slope = -1.9) (29-30).

Restructuring. In addition to providing a source of monomers, a second consequence of the depolymerization process is that it provides a means for continually restructuring the growing polymers. When the forward and reverse rate constants are comparable, silicate species may undergo repeated depolymerization/repolymerization reactions. Because depolymerization occurs preferentially at least stable sites, restructuring ultimately forms stable configurations at the expense of unstable ones, moving the system toward thermodynamic equilibrium. Conversely, structures formed under conditions in which depolymerization/repolymerization is suppressed, for example near the isoelectric point, are far from equilibrium and highly metastable.

At large length scales restructuring is manifested as ripening (24). In order to minimize interfacial area, small and/or weakly polymerized species depolymerize and preferentially condense with larger, more highly polymerized species. A dramatic example of this phenomenon occurs in the Stöber process (31) for preparing monodisperse, a-SiO_2 colloids, in which TEOS is hydrolyzed near pH 12 in solutions containing H_2O/Si ratios of 7 to 25. Bogush and Zukoski (32) have shown that during this process aggregates composed of many 10-20 nm particles, which emerge initially, completely restructure within 3-4 hours to form considerably fewer 300 nm spheres, greatly reducing the interfacial area. Conversely the fractal structures which form under virtually all acidic or neutral synthesis conditions have enormous interfacial areas: in a mass fractal there is no distinction between surface and interior (33).

Restructuring can also have a pronounced effect after gelation. Aging gels at elevated pH, where dissolution/repolymerization results in interparticle or intercluster "neck" growth (24), reduces the surface area and increases the modulus (34) due to an increase in the contact area between adjacent clusters. Because the shrinkage rate during drying is proportional to the surface energy divided by the modulus, the effect of ripening is to reduce the drying shrinkage, thus increasing the pore volume and pore size in the dried gel (34).

At intermediate length scales restructuring is evident from the ring statistics. According to NMR investigations conducted on aqueous silicate systems at high pH (35), a typical sequence of condensation products is monomer, dimer, linear trimer, cyclic trimer, cyclic tetramer, higher order rings. If depolymerization/ repolymerization occurs freely, providing a source of monomer and a mechanism for ring opening, the distribution of ring sizes in solution should approach that achieved in the thermodynamic equilibrium of the melt (average of six silicons per ring). If monomer is available but depolymerization is suppressed, there exists no mechanism for ring opening, and the most frequent ring size will be the smallest stable ring which emerges during the above sequence of condensation, i.e., the cyclic tetramer. (See discussion of ring strain to follow.) If monomer is quickly consumed and depolymerization is suppressed, we expect a random distribution of stable rings.

An example of this behavior occurs in silicate gels prepared in aqueous solution by leaching 10 Na_2O•30 B_2O_3•60 SiO_2 glasses at 70°C (36). Leaching selectively removes sodium and boron creating a silicate layer which extensively repolymerizes in solution to form a gel. In-situ Raman spectroscopy of the gel layers during leaching indicates that under acidic conditions the most prevalent ring contains 4 silicons (Raman band at 485 cm^{-1} in Figure 3) (36). Under neutral and basic conditions the repolymerized silicates are composed primarily of larger rings (Raman band at 431 cm^{-1} in Figure 3). By comparison, more conventional gel forming conditions, involving the hydrolysis and condensation of TEOS in alcoholic solutions, normally result in a more random distribution of stable rings (37), which suggests that monomers are less available than in aqueous conditions.

Because fully polymerized silicon species are more stable with respect to hydrolysis than weakly polymerized ones (24-36), the effect of restructuring at short length scales is manifested as the maximization of Q^4 species at the expense of Q^1-Q^3 species. (Note: In Q terminology, the superscript denotes the number of bridging oxygens (-O-Si) to which the silicon nucleus is bonded.) Conversely, under conditions where restructuring is inhibited, the pattern of condensation is more random in solution and less fully polymerized species are retained in the final gel.

The effects of restructuring at short length scales are evident in several studies of sol-gel silica polymerization. Based on

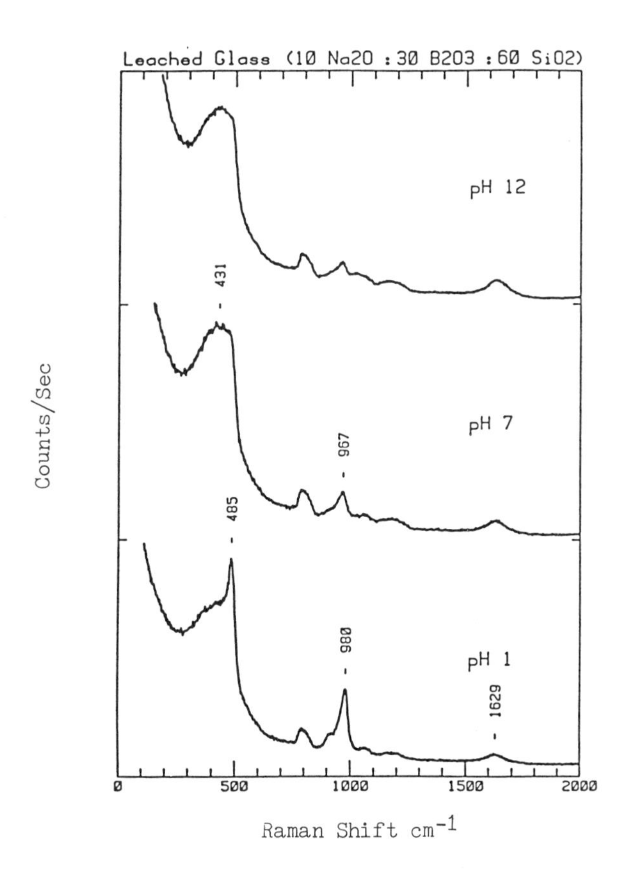

Figure 3. Raman spectra of dried gels prepared by leaching 10-Na_2O•30-B_2O_3•60-SiO_2 glass in aqueous solution at pH 1, 7, or 12 (after reference 40). The broad band centered at 430 cm^{-1} is associated with rings containing 5 or more silicons. The sharp band at 490 cm^{-1} has been assigned to a breathing mode of a ring containing 4 silicons. (Reproduced with permission from Ref. 36. Copyright 1986 Materials Research Society.)

capillary gas chromatography and ^{29}Si NMR results, Klemperer and co-workers (38-39) have shown the predicted trends in the molecular weight distributions and silicon speciation prior to gelation. Under acidic conditions they observe a normal distribution of molecular weight in which intermediate molecular weight species are maximized with respect to high and low molecular weight species. Under basic conditions they observe an inverted molecular weight distribution in which high and low (Q^0) molecular weight species are maximized with respect to intermediate molecular weight species. ^{29}Si NMR of hydrolyzed solutions of TMOS also shows a trend of increasing Q^4 species at the expense of Q^1-Q^3 species with increasing pH (40).

Solid state MASS ^{29}Si NMR spectra of dried gels prepared with 4 equivalents of water show a similar trend. Q^1-Q^4 species are evident in gels prepared under neutral conditions (41), whereas under more basic conditions, Q^1 species are absent and Q^2 species are greatly reduced relative to Q^4 (see NMR results presented below). Thus the effects of restructuring are preserved in the fully dried gel.

Borate Systems

Borate polymers and gels are a second example of how the stability of M-O-M bonding influences solution structure of inorganic polymers. Primary structural units composed of three to eight borons arranged in various configurations of 3-fold rings (connected by borate bonds involving tetrahedral borons) are stable in aqueous solution at intermediate pH; however, gelation is not observed (42). This is due to the hydrolytic instability of B-O-B bonding between primary units. FTIR investigations (Figure 4) of crystalline alkali borates (Figure 5a) indicate that B-O-B bonds linking trigonal borons contained in separate units (identified by the ~1260 cm^{-1} band in Figure 4a) rapidly hydrolyze upon exposure to water vapor (Figures 4a and b) (43). Thus, in order to form infinite polymeric networks in a solution environment, we must either increase the stability of borate bonding toward hydrolysis or alcoholysis or eliminate molecules responsible for depolymerization from the solution medium.

Based on the results of leaching studies of borosilicate glasses (36), the stability of B-O-B bonds toward hydrolysis increases with the sequential replacement of trigonal borons with tetrahedral borons, presumably because tetrahedral boron is coordinately saturated and cannot participate in sp^3d intermediates. This is consistent with our observations that in a mixed alcohol/methoxyethanol solvent there is a minimum fraction of tetrahedral borons which must be exceeded in order to form gels (43). Corresponding FTIR investigations of borate polymers and gels synthesized in alcohol/water solutions reveal that trigonal B-O-B bonding between primary units (both borons trigonally coordinated) is absent (Figure 4c) (43). Figure 5b depicts a polyborate structure based on the tetraborate primary unit which contains the minimum fraction of tetrahedral borons required to form an infinite network excluding all trigonal B-O-B bonding between units (borate bonds linking primary units contain one trigonal and one tetrahedral boron). The fraction of tetrahedral borons required for this tetraborate network (0.25) is close to the minimum fraction observed experimentally viz. 0.19 (gelation may occur without complete incorporation of boron in the network). Similarities between the IR spectra of partially hydrolyzed crystalline lithium tetraborate and

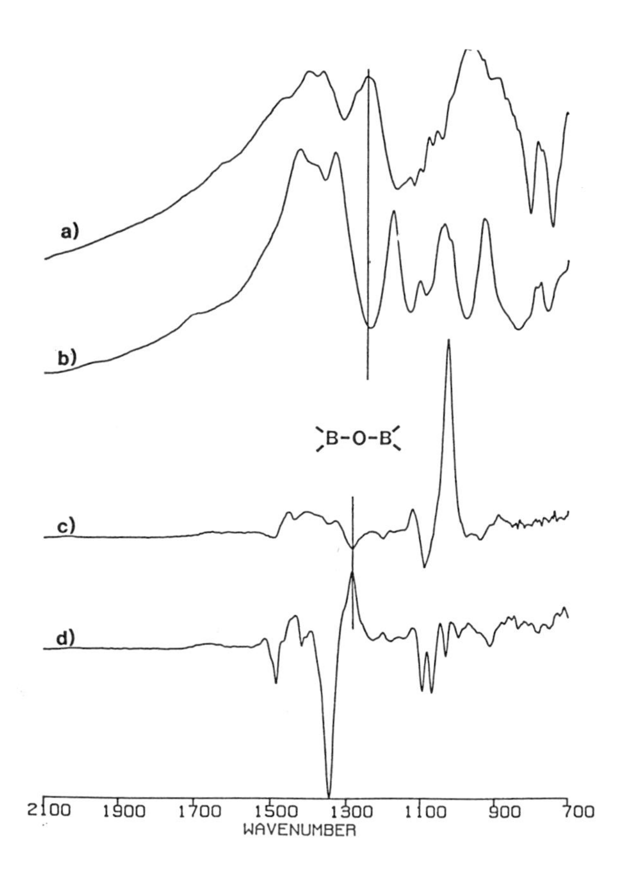

Figure 4. FTIR spectra: a) crystalline, anhydrous lithium tetraborate; b) hydrated, crystalline lithium tetraborate; c) difference spectrum (55 hour spectrum - 27 hour spectrum) during polymerization in methoxyethanol; d) difference spectrum (48 hour spectrum - 24 hour spectrum) during polymerization in THF.

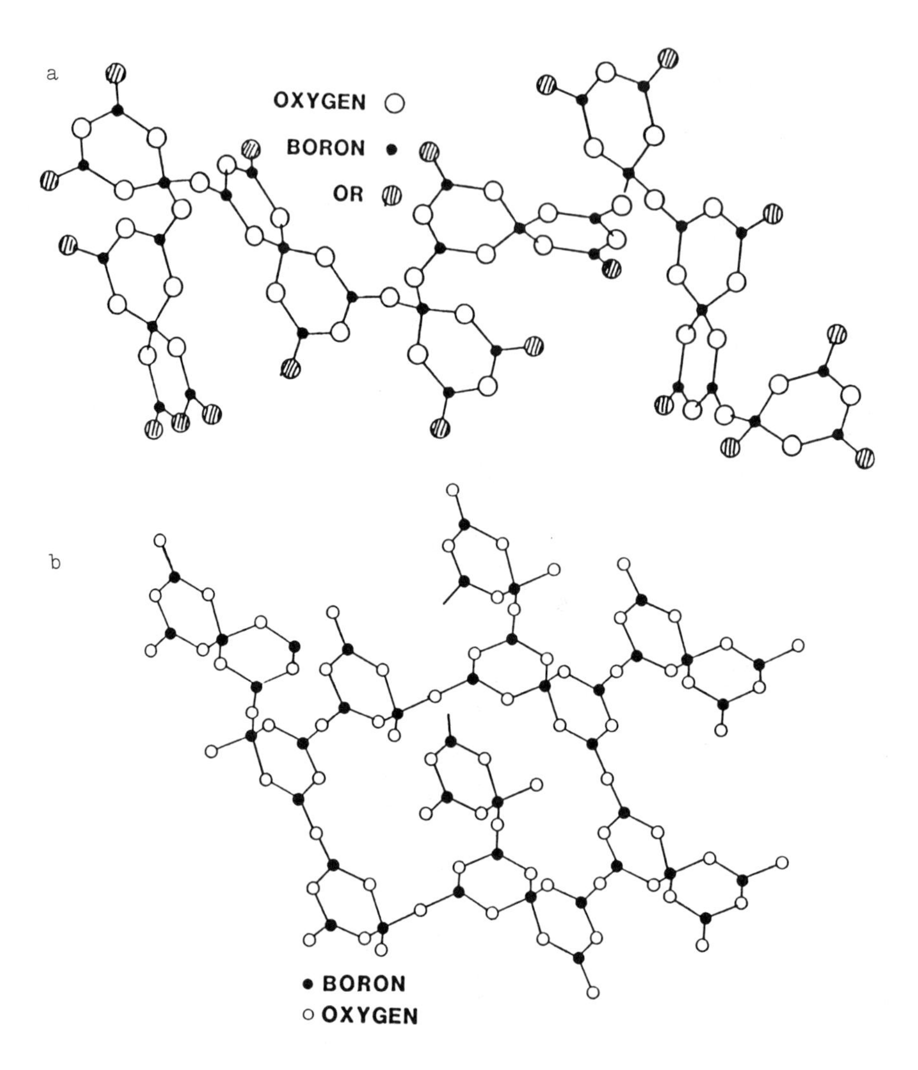

Figure 5. a) a portion of the crystalline tetraborate network showing linkages between primary units: ≡B-O-B= and =B-O-B= (after reference 43); b) tetraborate network which excludes all =B-O-B= bonding between units. (Reproduced with permission from Ref. 43. Copyright 1986 Materials Research Society.)

lithia borate polymers formed near the gel point (Figures 4b and c) lend further support to the structural model portrayed in Figure 5.

Replacement of methoxyethanol with THF reduces the minimum N_4 required for gelation to about 0.17. FTIR investigations indicate that the stability of trigonal =B-O-B= bonding between units is significantly increased in THF (indicated by the ~ 1290 cm^{-1} band in Figure 4d) compared to aqueous or mixed alcohol/water environments (Figures 4b and c, respectively). This emphasizes the sensitivity of gelation to the chemical stability of borate bonding between units. THF without a labile proton is unable to participate in dissociative alcoholysis reactions, resulting in the increased stability of trigonal borate bonding between units.

Gel Structure During Consolidation

Although porous xerogels are hard, stiff materials which appear in many ways to be glass-like, condensation reactions continue to occur during the heat treatments employed for their consolidation. Two factors distinguish structural evolution during consolidation from that which occurs in solution. First, the concentrations of water and alcohol available for depolymerization (Equations 2 and 3) are substantially lower during consolidation. Second, the inorganic network is much more highly condensed and therefore much stiffer. At the moderate temperatures employed for consolidation, the viscosity is often no lower than 10^{13} poises (21-22). The relative absence of reactive molecules combined with the extremely stiff matrix preclude the possibility of restructuring: the condensation products are therefore well preserved during heating at $T \ll T_g$.

The Raman spectra (0-1400 cm^{-1}) shown in Figure 6 illustrate the structural changes which accompany the consolidation of silica gels. The 1100°C sample is fully dense, whereas the 50 and 600°C samples have high surface areas (1050 and 890 m^2/g), respectively. The important features of the Raman spectra attributable to siloxane bond formation are the broad band at about 430 cm^{-1} and the sharp bands at 490 and 608 cm^{-1}(which in the literature have been ascribed to defects denoted as D1 and D2, respectively). The D2 band is absent in the dried gel. It appears at about 200°C and becomes very intense at intermediate temperatures, 600-800°C. Its relative intensity in the fully consolidated gel is low and comparable to that in conventional vitreous silica. By comparison the intensities of the 430 and 490 cm^{-1} bands are much more constant. Both bands are present at each temperature, and the relative intensity of the 430 cm^{-1} band increases only slightly with respect to D1 as the temperature is increased. Figure 7 shows that in addition to elevated temperatures the relative intensity of D2 also decreases upon exposure to water vapor.

Figure 8 shows the ^{29}Si MASS and ^{1}H cross polarization (CP MASS) spectra obtained on heated samples and ^{29}Si MASS spectra collected after exposure of the 600 and 1100°C samples to water vapor. (Brinker, C. J., Kirkpatrick, R. J., Tallant, D. R., Bunker, B. C. and Montez, B., submitted.) The three prominent peaks at chemical shifts (δ) of about -91, -101, and -110 ppm correspond to Q^2, Q^3, and Q^4 silicon sites, respectively (44). The relative intensities of these peaks in the MASS spectra are proportional to the relative concentrations of the different silicon species. The positions of these peaks in both the MASS and CP MASS spectra are correlated with the average Si-O-Si bond angle, ϕ, for bridging oxygens bound to the

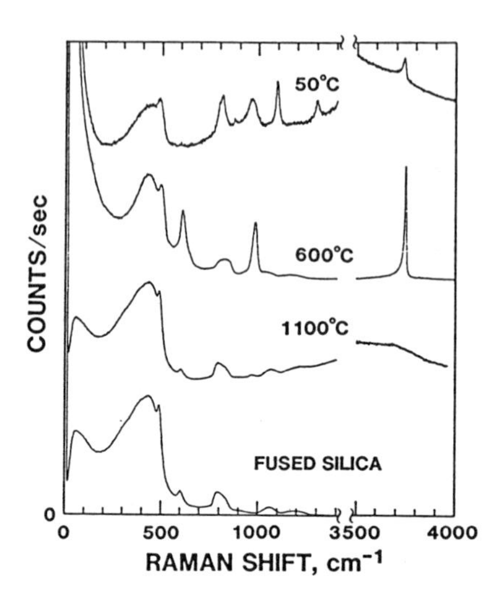

Figure 6. Raman spectra of silicates: silica gel dried at 50°C, heated to 600°C, consolidated at 1100°C, and conventionally prepared fused silica.

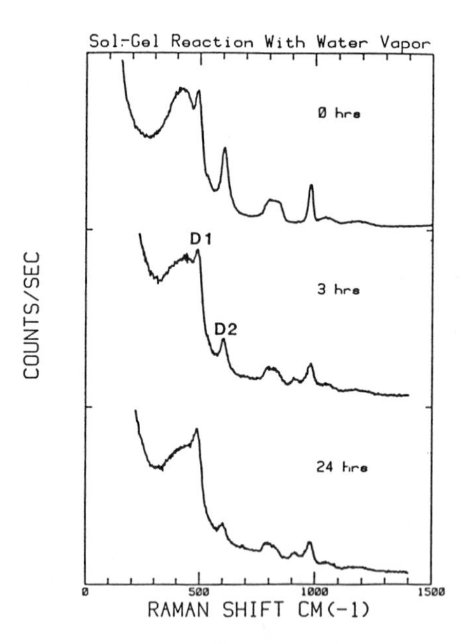

Figure 7. Raman spectra of silica gel heated to 600°C, cooled to room temperature and exposed to water vapor (100% RH) for the indicated times. (Reproduced with permission from Ref. 47. Copyright 1986 Materials Research Society.)

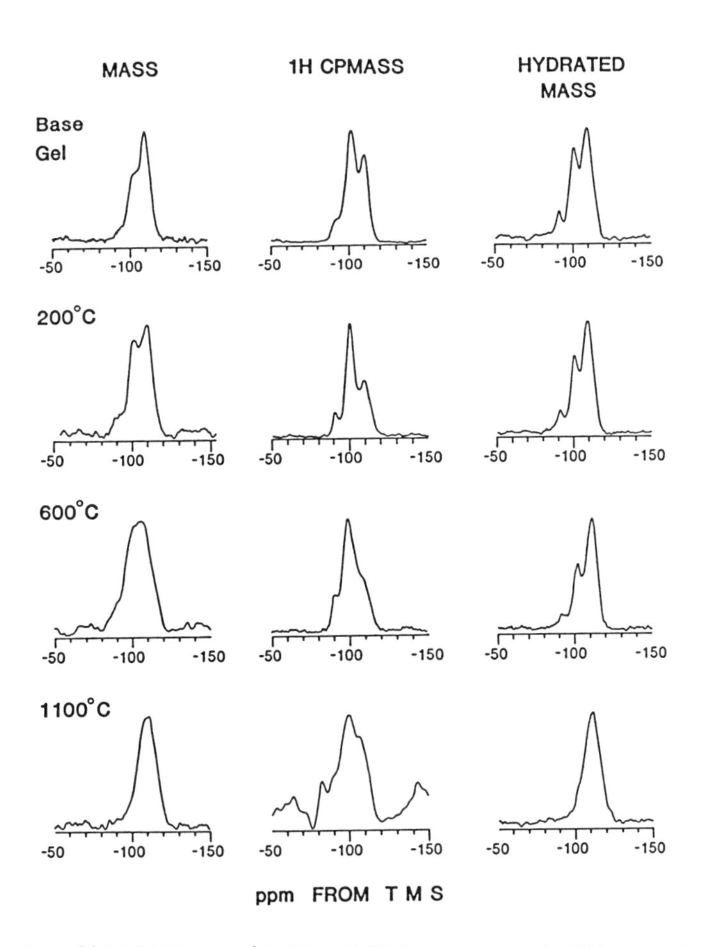

Figure 8. ^{29}Si MASS and ^{1}H CPMASS NMR spectra of silica gels: dried at 50°C, heated to 200 and 600°C, and consolidated at 1100°C. Hydrated samples were exposed to 100% RH at 25°C for 24 hours prior to analysis.

silicon nucleus (45). For example, for Q^4 resonances: $\delta(ppm) = -0.59(\phi) - 23.21$ (correlation coefficient = 0.982).

The behavior of D2 in the Raman experiments is strongly correlated with the Q^4 chemical shift, δ^4, in the NMR spectra. δ^4 equals about -110 to -111 ppm when D2 is absent or when it exhibits low relative intensities comparable to those in conventional vitreous silica, for example the 50 and 1050°C sample spectra and the rehydrated 600°C sample spectrum. From the regression equation cited above -110 to -111 ppm corresponds to ϕ = 147 to 149°, values quite close to the average ϕ in conventional v-SiO_2, 151° (45). The average δ^4 is shifted downfield to about -107 ppm in the 600°C sample in which D2 is observed to be quite intense. Deconvolution of this peak reveals two Q^4 resonances at -110 and -105 ppm. -105 ppm corresponds to ϕ = 138°, which is very near the equilibrium ϕ calculated for the isolated cyclic trisiloxane molecule, $H_6Si_3O_3$, (ϕ = 136.7°) (46). The positions of the Q^2 and Q^3 resonances, however, appear to be totally unaffected by the presence or absence of D2 (as shown in the 600°C CP MASS sample spectrum).

The NMR data clearly associate the formation of the species responsible for D2 with the presence of Q^4 silicon sites with reduced values of ϕ and, conversely, the elimination of this species with an increase in the average ϕ. These observations, combined with our previous results on gels (37-47), identify the D2 species as a cyclic trisiloxane (3-membered ring) as originally proposed by Galeener to explain D2 in conventional v-SiO_2 (48). 3-membered rings are absent in the as-dried gel. They form at intermediate temperatures predominately on the silica surface by the condensation of isolated vicinal silanol groups located on unstrained precursors via the following reaction:

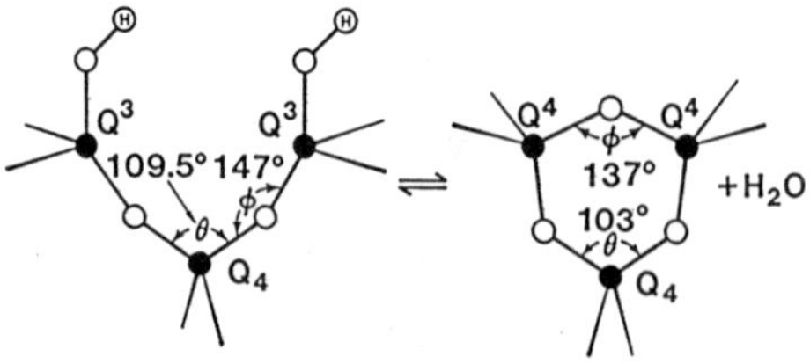

(4)

The heat of formation of 3-membered rings according to Reaction 4 is calculated to be quite endothermic (ΔH_f = 23 kcal/mole) due to the strain energy required to reduce ϕ to 136.7° from its equilibrium value, 151°, and to reduce the tetrahedral (O-Si-O) angle to 103.3° from 109.5° (46).

The altered environments of siloxane bonds contained within 3-membered rings greatly influences their stability. The changes in tetrahedral symmetry around silicon activate empty d-acceptor orbitals (49), making Si more acidic. Decreasing ϕ transfers electron density into the lone pair orbitals of the bridging oxygen, making the oxygen more basic and the silicon more acidic (Gibbs, G. V., private communication, 1987). The increased polarity and enhanced acid/base properties of strained Si-O bonds promotes the adsorption of water on the bond and the subsequent bond hydrolysis reaction (the reverse of Equation 4) according to the following mechanism (50):

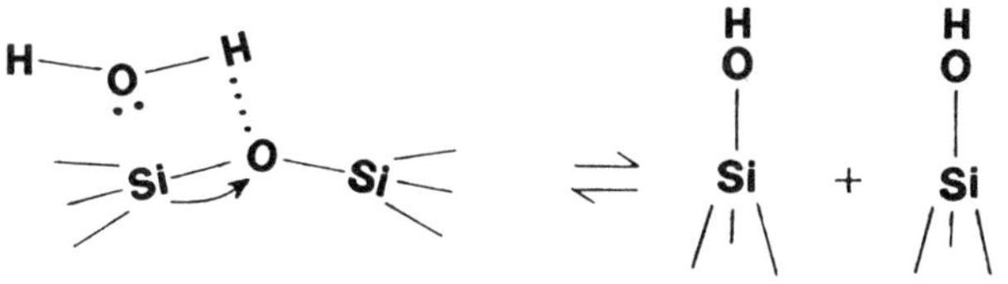

$$\equiv Si\text{-}O\text{-}Si\equiv + H\text{-}O\text{-}H \rightleftharpoons \equiv Si\text{-}OH + HO\text{-}Si\equiv \qquad (5)$$

This explains the enhanced reactivity of D2 species toward water vapor compared to the structures responsible for the D1 and 430 cm^{-1} bands (Figure 7), which are assigned to ring breathing vibrations of unstrained four-membered and higher order rings, respectively.

The identification of the structures responsible for D1 and D2 and their stabilities toward hydrolysis are further supported by investigations of the hydrolysis behavior of the corresponding isolated ring molecules, octamethylcyclotetrasiloxane (OMCT) and hexamethylcyclotrisiloxane (HMCT) (51). OMCT is stable in aqueous environments, whereas HMCT hydrolyzes with a pseudo-first order rate constant 3.8 ($\pm$ 0.4) x 10^{-3} min^{-1} ($t_{1/2}$ = 3.0 hours). This latter value is comparable to the rate constant for D2 hydrolysis, 5.2 ($\pm$ 0.5) x 10^{-3} min^{-1} ($t_{1/2}$ = 2.2 hours) and is 75x greater than the rate constant describing hydrolysis of unstrained, conventional a-SiO_2 (as estimated by extrapolation of the data in reference (52).

Although 3-membered rings are apparently kinetically stabilized by the very high matrix viscosity and the relative absence of water and other reactive molecules during heating at intermediate temperatures, we expect them to be unstable in the solution environment during gel formation and at elevated temperatures where the surface can reconstruct. This is evident from the Raman spectra of the dried and consolidated gels and explains the general absence of three-membered rings in virtually all ^{29}Si NMR studies of silica polymerization during sol-gel processing and the reduced relative intensities of D2 in all fully consolidated gels. (Note: Three membered rings are stable species in aqueous solutions above pH 12 (53). This has been rationalized by considerations of charge separation. When the hydroxyl groups are fully deprotonated the three-membered ring provides an increased separation of charge compared to 4-membered and higher order rings (54).)

Conclusions

Although a number of factors affect structural evolution in sol-gel processing, many of the observed trends can be explained on the basis of the stability of the M-O-M bond in its synthesis environment. Structures which form in solution are the products of a successive series of hydrolysis, condensation and restructuring reactions. M-O-M bonds which are unstable with respect to hydrolysis and alcoholysis are absent. In borates this criterion precludes gelation for systems with insufficient fractions of 4-coordinated borons. In silicates this criterion is manifested as continual restructuring under basic conditions in which siloxane bond hydrolysis is facilitated, whereas the structures which evolve under acidic conditions reflect the products of a more or less random series of hydrolysis and condensation reactions which have undergone little restructuring. During consolidation at intermediate temperatures, the condensation process continues within a very stiff matrix in a relatively inert environment. Under these conditions the stability criteria are mitigated. In silicates we observe the formation of strained 3-membered rings which rapidly hydrolyze upon exposure to water vapor and which are preferentially destroyed at elevated temperatures where reconstruction is possible. Thus we conclude that, during consolidation, metastable species are kinetically stabilized by the high matrix viscosity and the relative absence of reactive molecules.

Literature Cited

1. Proceedings of the International Workshop on Glasses and Glass Ceramics from Gels, Oct. 8-9, 1981, J. Non. Cryst. Solids 1982, 48, 1-230.
2. Proceedings of the Second International Workshop on Glasses and Glass Ceramics from Gels, July 1-2, 1983, J. Non. Cryst. Solids 1984, 63, 1-300.
3. Proceedings of the Third International Workshop on Glasses and Glass Ceramics from Gels, Sept. 12-14, 1985, J. Non. Cryst. Solids 1986, 82, 1-436.
4. Ultrastructure Processing of Ceramics, Glasses and Composites, Hench, L. L. and Ulrich, D. R., Eds.; (Wiley Interscience: New York, 1984.
5. Science of Ceramic Chemical Processing, Hench, L. L. and Ulrich, D. R., Eds.; Wiley Interscience: New York, 1986.
6. Better Ceramics Through Chemistry, Brinker, C. J., Clark, D. E. and Ulrich, D. R., Eds.; Elsevier - North Holland: Amsterdam, 1984.
7. Better Ceramics Through Chemistry II, Brinker, C. J., Clark, D. E. and Ulrich, D. R., Eds.; Mat. Res. Soc.: Pittsburgh, PA, 1986.
8. Proceedings of the IIIrd Int'l Conference on Ultrastructure Processing of Glasses, Ceramics and Composites, Feb. 24-47, 1987 to be published.
9. Yamane, M. and Kojima, T., J. Non. Cryst. Solids 1981, 44, 181-190.
10. Uhlmann, D. R., Zelinski, B. J., Silverman, L., Warner, S. B., Fabes, B. D. and Doyle, W. F. In Science of Ceramic Chemical Processing, Hench, L. L. and Ulrich, D. R., Eds.; Wiley Interscience: New York, 1986; p 173-183.
11. Brinker, C. J. and Mukherjee, S. P., Thin Solid Films 1981, 77, 141-148.
12. Brinker, C. J., Haaland, D. M. and Loehman, R. E. In Better Ceramics Through Chemistry, Brinker, C. J., Clark, D. E. and Ulrich, D. R. Eds.; Elsevier - North Holland: Amsterdam, 1984, 179-184.
13. Brinker, C. J., Keefer, K. D., Schaefer, D. W., Assink, R. D., and Kay, B. D., J. Non. Cryst. Solids 1984, 63, 45-59.
14. Keefer, K. D. In Better Ceramics Through Chemistry, Brinker, C. J., Clark, D. E. and Ulrich, D. R., Eds.; Elsevier North Holland: Amsterdam, 1984; p 15-24.

15. Tanaka, T., Sci. Amer. 1981, 124.
16. Zarzycki, J. Prassas, M. and Phalippou, J., J. Mat. Sci. 1982, 17, 3371-79.
17. Stephens, H. P. Dosch, R. G. and Stohl, F. V., Ind. R and D. Eng. Chem. Prod. Res. and Dev. 1985 24, 15.
18. Larbot, A., Alary, J. A., Fabre, J. P., Guizard, C. and Cot, L. In Better Ceramics Through Chemistry II, Brinker, C. J., Clark, D. E. and Ulrich, D. R., Eds.; Mat. Res. Soc.: Pittsburgh, PA, 1986; p 659.
19. Pettit, R. B., Ashley, C. S., Reed, S. T. and Brinker, C. J. to be published in Sol-Gel Technology, L. C. Klein, Ed.; Noyes: Park Ridge, NJ, 1987.
20. Pettit, R. B. and Brinker, C. J., Solar Energy Materials 1986, 14, 269-287.
21. Brinker, C. J., Drotning, W. D. and Scherer G. W. In Better Ceramics Through Chemistry, Brinker, C. J., Clark, D. E. and Ulrich, D. R., Eds.; Elsevier - North Holland: Amsterdam, 1984; p 25.
22. Scherer, G. W., Brinker C. J. and Roth, E. P., J. Non-Cryst. Solids 1985, 72, 369.
23. Schaefer, D. W. and Keefer K. D. In Better Ceramics Through Chemistry, Brinker, C. J., Clark, D. E. and Ulrich, D. R., Eds.; Elsevier - North Holland: Amsterdam, 1984; p 1.
24. Iler, R. K., The Chemistry of Silica; John Wiley, NY, 1979.
25. Witten, T. A. and Sanders, L. M., Phys. Rev. Lett. 1981, 47, 1400.
26. Meakin, P., Phys. Rev. Lett. 1983, 51, 1119.
27. Keefer, K. D. In Better Ceramics Through Chemistry II, Brinker, C. J., Clark, D. E. and Ulrich, D. R., Eds.; Mat. Res. Soc.: Pittsburgh, PA, 1986; p 295-304.
28. Witten, T. A. and Cates, M. E., Science 1986, 232, 1607.
29. Schaefer, D. W., Keefer, K. D. and Brinker, C. J., Polym. Preprints. Am. Chem. Soc., Div. of Polym. Chem., 1983, 24, 239.
30. Brinker, C. J., Keefer, K. D., Schaefer, D. W. and Ashley, C. S. J. Non-Cryst. Solids 1982, 48, 47.
31. Stöber, W., Fink, A. and Bohn, E., J. Coll. and Int. Sci. 1968, 26, 62.
32. Bogush, G. H. and Zukoski, C. F. Proceedings of the 44th Annual Meeting of the Electron Microscopy Society of America, Bailey, G. W., Ed.; San Francisco Press: San Francisco, CA, 1986; p 846-847.
33. Schaefer, D. W. and Keefer, K. D. In Better Ceramics Through Chemistry II, Brinker, C. J., Clark, D. E., and Ulrich, D. R. Eds., Mat. Res. Soc.: Pittsburgh, PA, 1986; p 277.
34. Brinker, C. J. and Scherer, G. W. In Ultrastructure Processing of Glasses, Ceramics and Composites, Feb. 24-27, 1987; p 33-59.
35. Engelhardt, V. G., Altenburg, W., Hoebbel D. and Wieker, W. Z. Anorg. Allg. Chem. 1977, 418, 43.
36. B. C. Bunker in Better Ceramics Through Chemistry II, Brinker, C. J., Clark, D. E., and Ulrich, D. R., Eds.; Mat. Res. Soc.: Pittsburgh, PA, 1986; p 49.
37. C. J. Brinker, D. R. Tallant, E. P. Roth and C. S. Ashley, J. Non. Cryst. Solids 1986, 82, 117.
38. Klemperer, W. G. and Ramamurthi, S. D., Polymer Preprints, 1987, 28, 432-433.
39. Klemperer, W. G. and Ramamurthi, S. D., this Proceedings.

40. Kelts, L. W., Effinger, N. J. and Melpolder, S. M., J. Non. Cryst. Solids 1986, 83, 353.
41. Klemperer, W. G., Mainz, V. V. and Millar, D. M. In Better Ceramics Through Chemistry II, Brinker, C. J., Clark, D. E. and Ulrich, D. R., Eds.; Mat. Res. Soc.: Pittsburgh, PA, 1986; p 15.
42. Edwards, J. A. and Ross, V., J. Inorg. Nucl. Chem. 1960, 15, 329.
43. Brinker, C. J., Ward, K. J., Keefer, K. D., Holupka, E. and Bray, P. J. In Better Ceramics Through Chemistry II, Brinker, C. J., Clark, D. E., and Ulrich, D. R., Eds., Mat. Res. Soc.: Pittsburgh, PA, 1986; p 57.
44. Sindorf, D. W. and Macill, G. E., J. Am. Chem. Soc. 1983, 105, 1487.
45. Oestrieke, R., Yang, W. H., Kirkpatrick, R. J., Herrig, R. L., Navrotsky, A. and Montez, B., Geochim. Cosmochim. Acta, submitted.
46. O'Keefe, M. and Gibbs, G. V., J. Chem. Phys. 1984, 81, 876.
47. Brinker, C. J., Tallant, D. R., Roth, E. P. and Ashley, C. S. In Defects in Glasses, Galeener, F. L., Griscom, D. L. and Weber, M. J., Eds.; Mat. Res. Soc.: Pittsburgh, PA, 1986; p 387.
48. Galeener, F. L. In The Structure of Non-Crystalline Materials, Gaskell, P. H., Parker, J. M. and Davis, F. A., Eds.; Taylor & Francis: London, 1982; p 337.
49. Boudjonk, P., Kapfer, C. A. and Cunico, R. F., Organometallics 1983, 2, 336.
50. Michalske, T. A. and Bunker, B. C., J. Appl. Phys. 1984, 56, 2686.
51. Balfe, C. A., Ward, K. J., Tallant, D. R. and Martinez, S. L. In Better Ceramics Through Chemistry II, Brinker, C. J., Clark, D. E. and Ulrich, D. R., Eds.; Mat. Res. Soc. Proc.: Pittsburgh, PA, 1986; p 619.
52. Stöber, W., Adv. Chem. Series 1967, 67, 161.
53. Harris, R. K. and Knight, C. T. G., J. Chem. Soc. 1983, Faraday Trans. 2, 79, 1525.
54. Dent Glasser, L. S. and Lachowski, E. E., J. Chem. Soc. 1980, Dalton Trans., 393; ibid 399.

RECEIVED October 23, 1987

Chapter 27

Organically Modified Silicates as Inorganic–Organic Polymers

H. K. Schmidt

Fraunhofer-Institut für Silicatforschung, Neunerplatz 2, D–8700 Würzburg, Federal Republic of Germany

The combination between inorganic and organic polymeric materials on an atomic scale depends strongly on methods for synthesizing inorganic polymeric networks suitable to the thermal stability of organic materials. The sol-gel process as a "soft-chemistry" method has been proved to be a proper tool for building up inorganic network incorporating organic components. A review over examples for material developments and possibilities of tailoring by chemistry using sol-gel techniques in combination with organic components (organically modified silicates) is given.

Novel materials have always played an important role for the development of new technologies. Since the requirement of modern technologies with respect to material properties became more and more specific it was necessary to develop composites: Properties of different types of basic materials have to be combined in order to fulfil these specific requirements. Moreover, natural raw materials, especially in the field of ceramics, could not meet the requirements for a lot of desired purposes (e.g. purity, homogeneity, reactivity), so novel raw materials were developed by chemical synthesis.

One of the most important chemical routes to ceramics is the sol-gel process (1-5) which was proved to open unique new possibilities to improve material or processing properties (e.g. coatings, powders, fibers and even monolithic materials). One of the key steps on the way from the (mono- or oligomeric) precursor to the solid is the polycondensation step, which defines to a great deal the structure of a prepolymer or polymer to be formed and thus influences processing and final material properties essentially. One of the major advantages of sol-gel techniques is the fact that the network forming step of the inorganic polymer is carried out at rather low temperatures in organic or aqueous solutions (compared to classical glass melting or ceramic firing temperatures). This leads to the possibility of incorporating organic components into inorganic polymers.

The formation of pure inorganic materials by the sol-gel route requires heating: dense glasses can be prepared around Tg; ceramic

0097–6156/88/0360–0333$06.00/0

materials have to be sintered at high temperatures, even if in most cases substantially lower firing temperatures can be applied when sol-gel derived reactive powders are used. The pure inorganic materials are brittle due to their high (three-dimensional) network connectivity. Therefore, sufficient relaxation and consequent densification only can occur at higher temperature. The introduction of organic groupings can reduce the overall network connectivity and reduce Tg and relaxation to substantially lower temperatures. Thus, dense materials can be prepared at temperatures at which organic components are stable. Different materials have been developed during the last decade using this basic reaction principles (6-43). They became known as ormosils (organically modified silicates or ceramers). A better expression, therefore, is ormocer (organically modified ceramics), since the $[SiO_4]^{4-}$ tetrahedron is not a non-dispensable component.

Structures

Inorganic networks, especially glasslike ones can be characterized by the ratio of network forming (e.g. $[SiO_4]^{4-}$) to network modifying (e.g. $\equiv Si{-}O^-Na^+$) units. Network formers (e.g. SiO_2, B_2O_3, Al_2O_3, TiO_2) are (in opposition to the majority of organic units) three-dimensional crosslinking units.

Organic groupings incorporated into this network (e.g. by $\equiv Si{-}R$) act as network modifiers as shown in Figure 1a.

A convenient method to introduce organic functions into inorganic networks is the use of substituted siliceous acid esters and the sol-gel route. The basic reactions, therefore, are given in Equations 1 and 2. Equation 1 describes the sol-gel reaction to a borosilicate glass,

$$NaOR + B(OR)_3 + Si(OR)_4 \xrightarrow[-\ HOR]{+\ H_2O} NaOH + B(OH)_3 + Si(OH)_4$$

$$\xrightarrow{-\ H_2O} \underset{\text{gel}}{(Na_2O{\cdot}B_2O_3{\cdot}SiO_2){\cdot}H_2O} \xrightarrow[\text{heat}]{-\ H_2O} \underset{\text{glass}}{Na_2O{\cdot}B_2O_3{\cdot}SiO_2} \qquad \text{R e.g. methyl,} \tag{1}$$

and Equation 2 shows a generalized sol-gel reaction including different organic groupings

$$u\ Me(OR)_4 + v(HX)_nSi(OR)_{4-n} + w(YX)_m\ Si(OR)_{4-m} \longrightarrow$$

$$\xrightarrow[-\ HOR]{+\ H_2O} -O-\left[\begin{matrix} | \\ O \\ | \\ -Me- \\ | \\ O \\ | \end{matrix}\right]_u -O-\left[\begin{matrix} XH \\ | \\ -Si-O- \\ | \\ XH \end{matrix}\right]_v \left[\begin{matrix} XY \\ | \\ -Si-O- \\ | \\ XY \end{matrix}\right]_w \tag{2}$$

Me = network forming metal
m, n = 2

X e.g. $-CH_2-$, $-C_6H_4-$
Y e.g. $-NH_2$, $-CHO$, $-COOH$, vinyl, epoxy, methacrylate

Figure 1b gives the scheme of another type of structure: Organic polymeric chains are introduced into an inorganic network by polymerization. Numerous types of reactions can be used for synthesis like vinyl polymerization (30-34), epoxide polymerisation (23), or methyl methacrylate polymerization (21).

Figure 1c shows a typical structure of two independent (interpenetrating) networks. A combination of all types of structures is possible, of course.

Material Development and Properties

Introduction of Organic Network Modifiers. The introduction of organic network modifiers into SiO_2 glasses leads to drastic changes of properties. SiO_2 glass, for example, has a thermal expansion coefficient of about $0.5 \cdot 10^{-6} \cdot K^{-1}$, monomethyl-$SiO_2$ glass ($[CH_3SiO_{3/2}]_n$) about $100 \cdot 10^{-6} \cdot K^{-1}$ (40).

In our investigations the adsorption behaviour of CO_2 as a function of organic modification in porous SiO_2 systems was measured. Figure 2 shows the comparison of three adsorbents, synthesized under equal reaction conditions (hydrolysis and condensation of $Si(OC_2H_5)_4$, CH_3-$Si(OC_2H_5)_3$ and $NH_2(CH_2)_3$-$Si(OC_2H_5)_3$ (am) in 50 vol.% CH_3OH with 0.1 N HCl and stoichiometric amount of water).

The influence of the modification on adsorption behavior is obvious. The extreme high load at low pCO_2 values can be attributed to a weak dipole-dipole interaction between the aminogroup and CO_2, the high load of CH_3 group containing adsorbents to a hydrophbic interaction. The example demonstrates the influence of structural changes of inorganic polymers by organic modification on selected properties.

A chemical heat pump was developed by use of two different types of adsorbents with CO_2 (41). Another example for material development is the synthesis of a mild abrasive powder for smoothing the human skin (acne patients) (11,16,42).

Organic Network Formers. As indicated above, an additional organic network can be built up by organic polymer synthesis within an inorganic network. The basic principles are shown in Equations 3 to 5 with a vinyl, methyl methacrylate and epoxide polymerization:

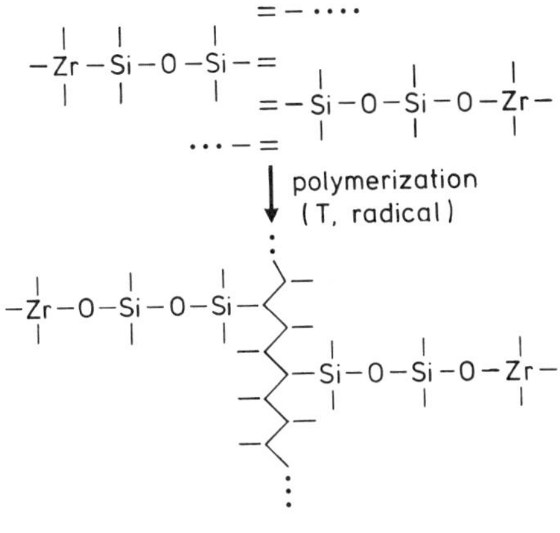

polymerization
(T, radical)

(3)

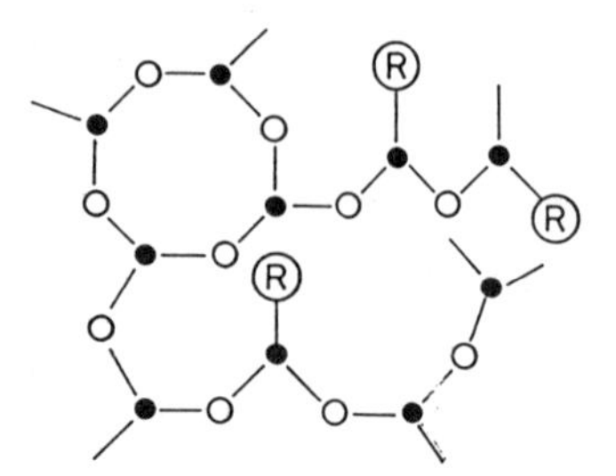

Figure 1a. Scheme of a glasslike structure, modified by organic groupings
● Si
○ O.

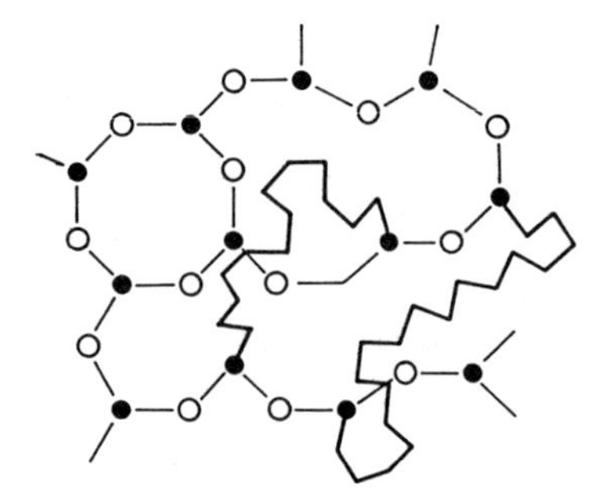

Figure 1b. Scheme of a glasslike structure, modified by additional, covalently bonded organic polymer chains.

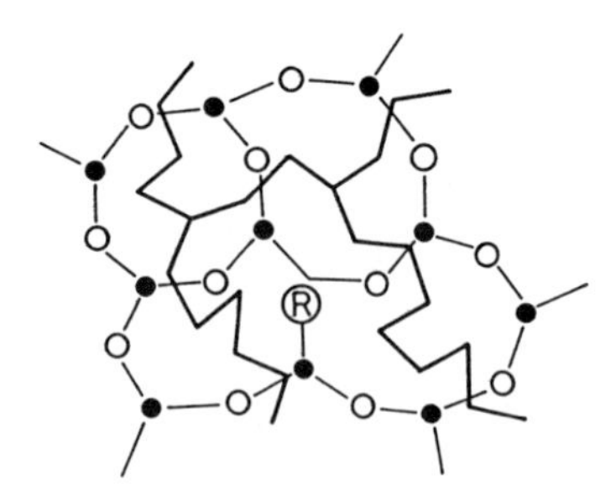

Figure 1c. Scheme of a glasslike structure, modified by an additional, polymeric network (interpenetrating).

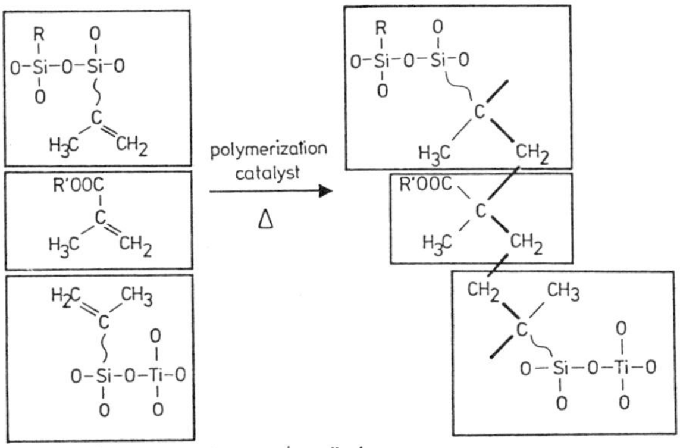

polymerization catalyst Δ (4)

R = glycol group containing organic radical
R' = CH_3 ; C_2H_4OH
$\sim$ = $-C_3H_6OOC-$

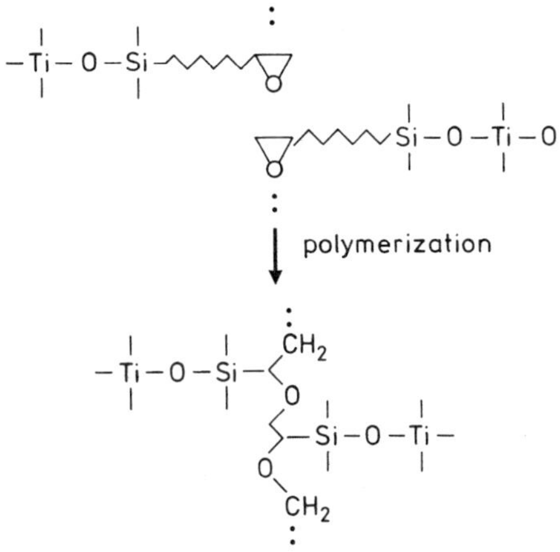

polymerization (5)

Since sol-gel polycondensation and organic polymerization usually take place under different reaction conditions, the sequence of the reactions (organic polymerization prior to sol-gel condensation or vice versa) can be chosen by reaction conditions.

As an example, an adhesive material could be synthesized according to Equation 4 with an excellent adhesion to glass surface under wet conditions without the necessity to apply silane primers. The network of this adhesive has a glasslike structure and the adhesive power can be optimized by optimizing the content of $\equiv$SiOR and $\equiv$SiOH groups (34,30) (Figure 3).

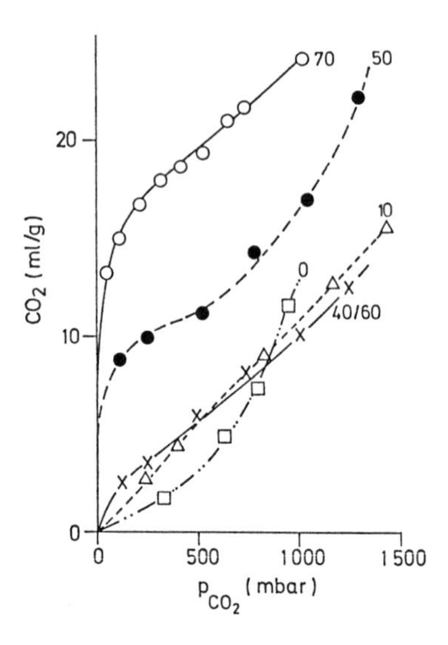

Figure 2. Adsorption isotherms of CO_2 on different network modified adsorbents: 70, network former SiO_2 to network modifier (am) ratio (molar): 30:70; 50, 50:50; 10, 90:10; 0, 100% SiO_2; 40/60, 40 SiO_2 and 60 $CH_3SiO_{3/2}$ (molar); specific surface areas all between 200 and 300 m^2/g.

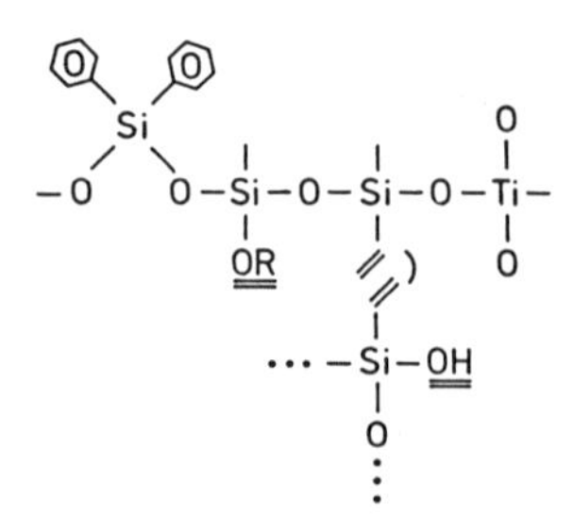

Figure 3. Reactive adhesive.

The adhesive was developed for application to seal food containers with alumina foils and to seal glass plates. In this case, the vinyl polymerization step can be carried out by UV light radiation through the already mounted glass plates with a photoactive radical initiator to cure the viscous adhesive material. Another application of this material is the development of diffusion barrier containing protective coatings on medieval stained glasses (31).

To introduce intrinsic hydrophilicity into silicone like structures (necessary to provide wetting within high oxygen diffusion materials for contact lenses), the ring opening reaction by H_2O addition was proved to be an adequate method (21) (Equation 6).

$$=\overset{OR}{\underset{|}{Si}}\sim\sim\sim\text{(epoxide)} + H_2O \xrightarrow{H^+} =\overset{OR}{\underset{|}{Si}}\sim\sim\sim CHOHCH_2OH \qquad (6)$$

Since sol-gel condensation and ring opening reaction take place under the same reaction conditions (H_2O + H^+) and a reesterification reaction according to Equation 7 occurs,

$$\begin{array}{l} CHOHCH_2OH + RO-\overset{|}{\underset{|}{Si}}\sim\sim\sim \longrightarrow \\ CHOH-CH_2-O-\overset{|}{\underset{|}{Si}}\sim\sim\sim + HOR, \end{array} \qquad (7)$$

a hydrolysis and condensation method was developed using an ester formation reaction as a water generation source, which leads to preferred condensation only. After reaching the desired degree of condensation, the addition of water enhances ring opening. This method (CCC = chemical controlled condensation) could be widely used in order to control different reaction rates of substituted and unsubstituted esters or different alkoxides (e.g. $Si(OR)_4$ and $Ti(OR)_4$). Very homogeneous materials could be synthesized with optical quality. Glycol group containing hydrophilic condensates, however, showed extremely poor mechanical properties. In order to improve the mechanical strength of this material, PMMA polymeric chains were synthesized with photoactivated radical initiators. The results are shown in Figure 4. The strengthening within this system is about a factor of five. This seems to be an extremely important finding with respect to the possibility of strengthening organic polymers by incorporating inorganic networks into organic polymers. The structure of the described contact lense polymer (21) is shown in Figure 5.

The polymerization reaction (Equation 5) leads to materials with extremely hard surfaces. Based on this, a new type of scratch resistant coating for different purposes could be developed. Table I gives some data of the high scratch resistant coating developed for CR 39 plastic eye glass lenses (39).

The data show the unique mechanical properties of this coating. The important item is that this coating can be applied at room temperature and cured at 90 °C and gives a clear film with optical quality. Typical properties of inorganic polymers (hardness) and organic polymers (coating as a lacquer and low temperature processing) could be combined in one and the same polymer.

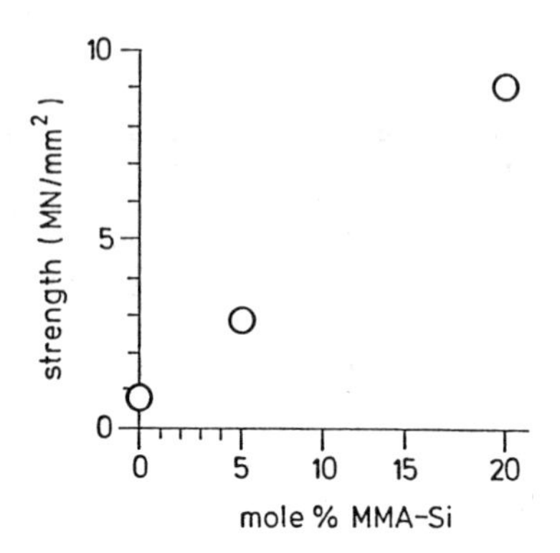

Figure 4. Increase of tensile strength by methacrlyate addition to the polymer system $TiO_2/SiO_2/SiO_{3/2}$-R; R = $-(CH_3)_2O-CH_2-CHOH-CH_2OH$

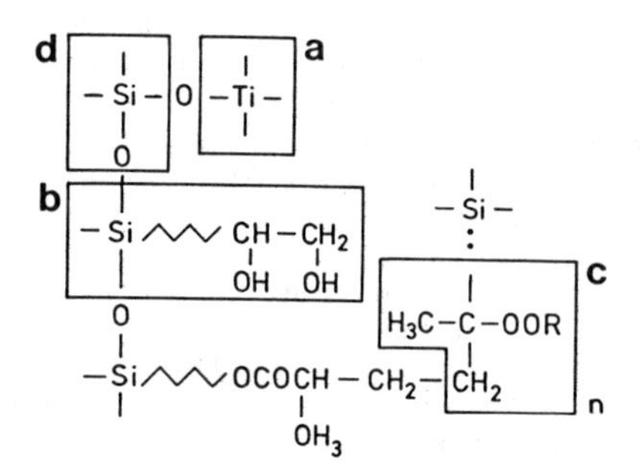

Figure 5. a, d: network formers
b: hydrophilic component
c: strengthening component

Table I. Scratch and abrading tests of different coatings composition (mole-%): MeO_2: 20; epoxysilane: 50; SiO_2: 30

	Coatings and Polymers				
Test	CR 39 (uncoated)	PMMA (uncoated)	Coating 1	2	3
A (load in g)	1 - 2	< 1	10	20 - 30	50
B (haze in %) (200 rev.)	12 - 13	> 20	-	1.5	-
C (haze in %)	4	-	-	1.2	-
	15	-	-	6	-

1 Zr-containing)   diamond powder
2 Ti-containing) coatings   boron carbide powder
3 Al-containing)

A vickers diamond scratch test: diamond load of the first visible scratch (by microscope)
B taber abrader: haze after abrasing
C special abrasion test with abrasive powders: haze after abrading.

Another application led to sensitive layers for coatings: ormocers including organic network formers as well as network modifiers could be developed (35) with reactive groupings for SO_2, NO_x and CO to be used as sensitive layers for new sensors. The reaction of a sensitive ormocer layer with gas molecules leads to a change of electric properties of the material. This change can be either monitored by a capacitance device (Figure 6) or by a field effect transistor (Figure 7).

Interpenetrating Networks. Sol-gel techniques may be used for building up interpenetrating networks by simultaneous or consecutive synthesis of both types of networks. As an example, an amorphous -O-Ti-O-Ti-O- containing interpenetrating network was synthesized by use of a ethyl vinyl acetate copolymer dissolved in toluene as liquid phase. $Ti(OR)_4$ was added slowly, hydrolysis and condensation were carried out through a H_2O vapor pressure controlled atmosphere. The resulting insoluble polymer exhibits a high scratch resistance in the diamond scratch test. Diamond loads of up to 20 g (vickers diamond) do not result in visible scratches (microscope observation). The mechanical strength of the polymer increases remarkably.

Conclusion

The combination of inorganic-organic polymers on a molecular level opens an interesting possibility of synthesizing new materials. Organic polymer synthesis and sol-gel techniques seem to be suitable techniques for this. The field is just at the beginning of its development.

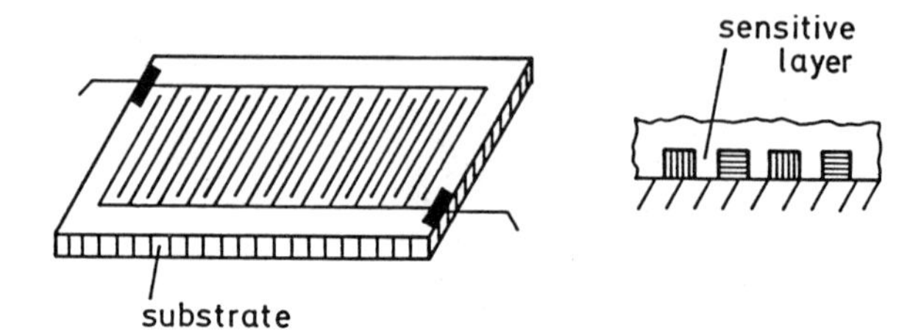

Figure 6. Scheme of a capacitance device sensor with an interdigitated structure.

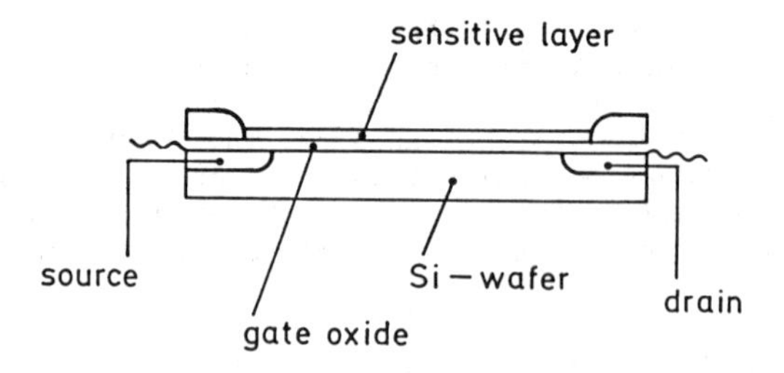

Figure 7. Scheme of a field effect transistor sensor.

Acknowledgments

The author thanks his coworkers Dr. J. Strutz, Dr. B. Seiferling, and Dr. G. Philipp for supporting the experimental work and the Bundesminister für Forschung und Technologie of the Federal Republic of Germany and industrial companies for the financial support.

Literature Cited

1. Dislich, H. Angew. Chem. 1971, 83, 428-35.
2. Roy, R. J. Amer. Cer. Soc. 1969, 52, 344.
3. Mazdiyasni, K. S.; Dolloff, R. T.; Smith, J. S. J. Amer. Cer. Soc. 1969, 52, 523-26.
4. Yoldas, B. E. J. Mater. Sci. 1977, 12, 1203-08.
5. Mackenzie, J. D. J. Non-Cryst. Solids 1981, 48, 1-10.
6. Schmidt, H.; von Stetten, O. Ger. Pat. 27 58 507, 1977.
7. Schmidt, H.; Scholze, H. Ger. Pat. 27 58 414, 1977.
8. Schmidt, H.; Scholze, H. Ger. Offen. 27 58 415, 1977.
9. Scholze, H.; Schmidt, H.; Böttner, H. Ger. Pat. 29 25 969, 1979.
10. Scholze, H.; Schmidt, H.; Tünker, G. Ger. Pat. 30 11 761, 1980.
11. Schmidt, H.; Kaiser, A. Ger. Pat. 3 048 369, 1980.
12. von Stetten, O.; Schmidt, H. Ger. Pat. 32 23 101, 1981.
13. Schmidt, H.; Philipp, G.; Kreiner, Ch. Ger. Pat. 31 43 820, 1981.
14. Schmidt, H.; Kaiser, A.; Scholze, H. Ger. Pat. 32 17 047, 1982.
15. Schmidt, H.; Kaiser, A.; Scholze, H. J. Non-Cryst. Solids 1982, 48, 65-77.
16. Schmidt, H; Kaiser, A.; Patzelt, H.; Scholze, H. J. Phys. 1982, 43 , Coll. C9, Suppl. 12, 275-78.
17. Schmidt, H.; von Stetten, O.; Kellermann, G.; Patzelt, H.; Naegele, W. In Radioimmunoassay and Related Procedures in Medi-cine 1982; International Atomic Energy Agency, Ed.; Vienna, 1982, pp 111-21.
18. Schmidt, H.; Kaiser, A.; Tünker, G. Ger. Pat. 33 40 935, 1983.
19. Schmidt, H.; Philipp, G. Ger. Pat. 34 07 087, 1984.
20. Kaiser, A.; Schmidt, H. J. Non-Cryst. Solids 1984, 63, 261-71.
21. Philipp, G.; Schmidt, H. J. Non-Cryst. Solids 1984, 63, 283-92.
22. Schmidt, H. FhG-Berichte 1984, H. 1, 18-28.
23. Schmidt, H. Mat. Res. Soc. Symp. Proc. 1984, 32, 327-35.
24. Schmidt, H.; Scholze, H.; Kaiser, A. J. Non-Cryst. Solids 1984, 63, 1-11.
25. Kaiser, A.; Schmidt, H.; Böttner, H. J. Membr. Sci. 1985, 22, 257-68.
26. Schmidt, H.; Scholze, H. In Glass ... Current Issues; Wright, A. F.; Dupuy, J., Ed.; Martinus Nijhoff Publ., Dordrecht-Boston-Lancaster 1985, pp 263-80.
27. Schmidt, H.; Philipp, G. In Glass ... Current Issues; Wright, A. F.; Dupuy, J., Ed.; Martinus Nijhoff Publ., Dordrecht-Boston-Lancaster 1985, pp 580-91.
28. Schmidt, H. J. Non-Cryst. Solids 1985, 73, 681-91.
29. Schmidt, H.; Hutter, F.; Haas, K.-H.; Obermeier, E.; Steger,U.; Endres, H.-E.; Drost, St. Ger. Offen. 35 26 348, 1985.
30. Schmidt, H.; Scholze, H.; Tünker, G. J. Non-Cryst. Solids (1986) 80, 557-63.

31. Tünker, G.; Patzelt, H.; Schmidt, H.; Scholze, H. Glastechn. Ber. 1986, 59, 272-78.
32. Schmidt, H.; Scholze, H. In Springer Proceedings in Physics, Vol. 6: Aerogels, 1986, pp 49-56.
33. Philipp, G., Schmidt, H. J. Non-Cryst. Solids 1986, 82, 31-36.
34. Schmidt, H.; Philipp, G.; Patzelt, H.; Scholze, H. Collected papers, XIV. Intl. Congr. on Glass 1986, Vol. II., pp 429-36.
35. Hutter, F.; Haas, K.H.; Schmidt, H. In Proceedings of 2nd International Meeting on Chemical Sensors; Aucouturier, J.-L. et al., Ed.; Bordeaux, 1986; pp 443-46.
36. Schmidt, H. Seiferling, B. Mat. Res. Soc. Symp. Proc. 1986, 73, 739-50.
37. Philipp, G.; Schmidt, H.; Patzelt, H. Ger. Pat. 35 36 716, 1986.
38. Schmidt, H.; Patzelt, H.; Tünker, G.; Scholze, H. Ger. Offen. 34 39 880, 1986.
39. Schmidt, H.; Seiferling, B.; Philipp, G.; Deichmann, K. Third International Conference on Ultrastructure Processing of Ceramics, Glasses and Composites, San Diego, 1987, Publisher: John Wiley (in press)
40. Dislich, H. Ger. Pat. 12 84 067, 1968.
41. Schmidt, H.; Strutz, J.; Gerritsen, H.-G.; Mühlmann, H. Ger. Offen. 35 18 738, 1986.
42. Kompa, H. E.; Franz, H.; Wiedey, K. D.; Schmidt, H.; Kaiser, A.; Patzelt, H. Ärztliche Kosmetologie 1983, 13, 193-200.
43. Wilkes, G. L.; Orler, B.; Huang, H. H. Polym. Prepr. 1985, 26, 300-2.

RECEIVED September 1, 1987

Chapter 28

Sol–Gel Preparation and Properties of Fibers and Coating Films

S. Sakka [1], K. Kamiya [2], and Y. Yoko [2]

[1]Institute for Chemical Research, Kyoto University, Uji, Kyoto-Fu 611, Japan
[2]Faculty of Engineering, Mie University, Tsu, Mie-Ken 514, Japan

The sol-gel process to prepare SiO_2 glass fibers and TiO_2 films has been reviewed. It has been known that the hydrolysis conditions such as molar ratio of water to alkoxide and reaction temperature are critical to the desired forms of the gel product (fiber, film or bulk). Some properties of the resultant products have been examined. Especially, TiO_2 films have been attempted to use as a photoanode for decomposition of water, and their photoelectrochemical properties are described in comparison with the results previously obtained for single crystal and polycrystalline TiO_2, and are discussed in terms of the microstructure of the film.

This paper describes preparation of silica glass fibers and titanium dioxide coating films through the sol-gel process using metal alkoxides as starting materials. The preparation of silica glass fibers described here uses the low temperature process in which gel fibers are drawn from a viscous sol at near room temperature and are converted to glass fibers by heating at 600° to 800°C. In this case, the key point is to find the composition of the silicon alkoxide solutions which become spinnable in the course of hydrolysis and condensation.

It was shown (1-3) that the silicon alkoxide solutions become spinnable when an acid is used as catalyst and the water content of the starting solution is small at less than 4 or 5 in the water to silicon alkoxide molar ratio. Recently, it has been found that this rule for the possibility of drawing fibers is only valid for the solutions reacted in the open system and no spinnability is found in the solutions reacted in the closed system (4). It has also been found that the addition of very large amounts of acid to the starting solution produces relatively large round-shaped particles, preventing the occurrence of spinnability (4). These will be discussed in the first half of this paper.

0097–6156/88/0360–0345$06.00/0

In the second half of the present paper, coating films based on the sol-gel process will be described. Dip-coating using metal alkoxide solutions makes it possible to prepare the coating films which provide the substrate with fresh or improved properties including mechanical, optical, electrical, magnetic, chemical and catalytic functions (2,5,6). It should be noted that this method is suitable for preparing very thin coating films (5). In this paper the optical and photoelectrochemical properties of TiO_2 coating films are discussed in terms of the microstructure of the film.

Preparation of silica glass fibers

Tetraethoxysilane-water-alcohol-hydrochloric acid solutions of appropriate compositions become viscous and spinnable in the course of hydrolysis and condensation of $Si(OC_2H_5)_4$. Fig.1 shows the time change of the viscosity of a $Si(OC_2H_5)_4$ solution with the $[H_2O]/[Si(OC_2H_5)_4]$ ratio of 2 for three different temperatures (1). It is catalyzed by hydrochloric acid. The viscosity increases with time as the reactions proceed. When the viscosity reaches about 10 poises, the solution becomes sticky and spinnable, which makes it possible to draw fibers. Fibers can be drawn by immersing a glass rod in the solution and pulling it up. It is seen from Figure 1 that the time required for the solution to reach the drawable state is shorter for higher temperatures. It should be noted that all starting solutions do not necessarily become drawable. The composition of the solution must be appropriate, in order for the fiber drawing to be possible. Table 1 lists some conditions appropriate for fiber formation. Solutions 1 and 2 have lower water content at the water to tetraethoxysilane ratios 1 and 2, respectively and show spinnability. Solutions 3 and 4, which have higher water contents at the ratios 5 and 20, respectively, do not show spinnability. It has been found that the tetraethoxysilane solutions with the $[H_2O]/[Si(OC_2H_5)_4]$ ratio lower than 5 can be spun into fibers in the course of reactions. Heating of gel fibers to 600°C to 800°C gives silica glass fibers.

Table 1. Composition and properties of tetraethoxysilane solutions

Solution	$Si(OC_2H_5)_4$ (g)	H_2O r	C_2H_5OH (ml)	Conc. of SiO_2(wt%)	Spinnability	Gel. time(h)
1	169.5	1.0	324	33.3	Yes	233
2	178.6	2.0	280	42.3	Yes	240
3	280.0	5.0	79	61.0	No	64
4	169.5	20.0	47	33.5	No	138

The mol ratio $[HCl]/[Si(OC_2H_5)_4]$ is 0.01 for all solutions.
r denotes the ratio $[H_2O]/[Si(OC_2H_5)_4]$.

Spinnability of the alkoxide solution

It was assumed that linear polymeric particles are formed in the low water content solutions which show spinnability on the way of progressing hydrolysis-polycondensation reaction. In order to confirm this, the molecular weights and intrinsic viscosities of the solutions listed in Table 1 have been measured (3). Figure 2 shows the log $\bar{M}_n$ versus log$[\eta]$ plots. The slope of the plot α is larger than 0.5, that is, 0.75 and 0.64 respectively for solutions 1 and 2 of

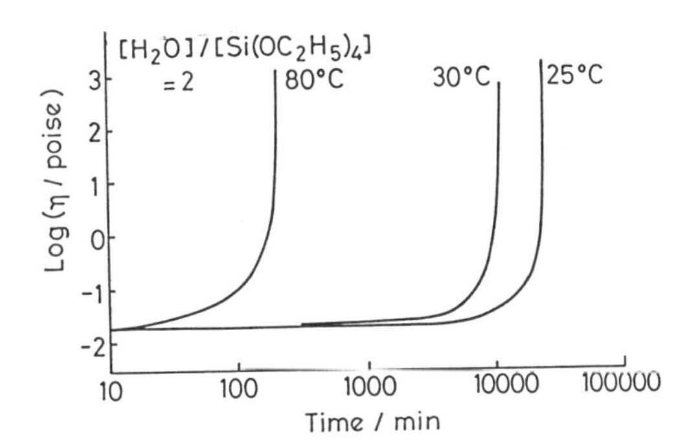

Figure 1. Variations of the viscosity of a tetraethoxysilane solution as a function of time. (Reproduced with permission from Ref. 1. Copyright 1984 Plenum Press.)

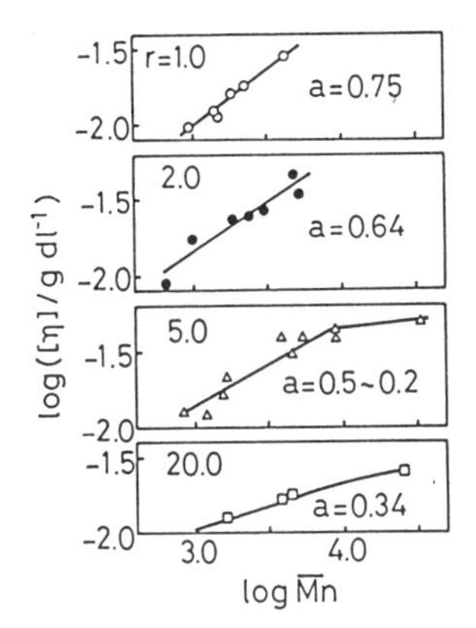

Figure 2. Relation between the intrinsic viscosity and the number-averaged molecular weight of trimethylsilylated alkoxide polymeric particles for $Si(OC_2H_5)_4$ solutions with different r values. (Reproduced with permission from Ref. 3. Copyright 1984 North Holland Physics Publishing.)

which the water content is low. It is smaller than 0.5, that is, 0.2 and 0.34 respectively for solutions 3 and 4 of which the water content is high. It is known that for the polymer solutions [η] is related to $\overline{M}_n$ by the expression $[\eta] = k\overline{M}_n^{\alpha}$, where k is a constant depending on the kind of the polymer, solvent and temperature. The present experimental results on the $[\eta]$-$\overline{M}_n$ relationships can be interpreted to show that in solutions 1 and 2 the siloxane linear polymeric particles are formed in the solution, while in solutions 3 and 4 with higher water contents three-dimensional or spherical growth of alkoxide polymers are predominant.

However, the nature of the polymeric particles is not exactly known yet; whether they are linear polymers with a single chain, those with double or triple chains, or linear aggregates of fine, round SiO_2 beads is not known.

Further examination into the factors affecting the drawability of the alkoxide solution

It has been shown that the spinnability occurs in the silicon alkoxide solution when an acid is used as catalyst and the water content is small at less than 4 in the water to metal alkoxide ratio.

Further examination has shown that the acid content should be small in order for the solution to become spinnable in the course of hydrolysis and polycondensation. It has been found (4) that very large concentrations at more than 0.15 in the [HCl]/[Metal alkoxide] ratio of acid catalyst produce round-shaped particles in the tetraethoxysilane (7) and tetramethoxysilane solutions, and so no spinnability appears.

So far the results have been shown in which the metal alkoxide solutions are reacted in the open system. It has been shown that the metal alkoxide solutions reacted in the closed container never show the spinnability even when the starting solutions are characterized by the l[illegible] acid content and low water content (4). It has been also shown f[illegible] the measurements of viscosity behavior that the solution remai[illegible]tonian in the open system, while the solution exhibits str[illegible] viscosity (shear-thinning) in the closed system.

[illegible]th of gel-derived fibers

[illegible]ensile strength of gel-derived fibers have been measured. It [illegible] been found (8) that gel-derived fibers are a little weaker than [illegible]ilica glass fibers, especially when the diameter is large, as shown in Figure 3. This figure also shows that ZrO_2-SiO_2 fibers exhibit similar tensile strength to fused SiO_2 fibers when the fiber diameter is small.

Color of TiO_2 coating films

TiO_2 films have been prepared on a glass substrate from the solution of $Ti(O\text{-}iC_3H_7)_4:H_2O:C_2H_5OH:HCl=1:1:8:0.08$ in molar ratio (9, 10). The coating TiO_2 films exhibit interference colors which are varied with the film thickness. Figure 4 shows the absorption spectra of gel-derived TiO_2 films with varying thickness, together with the color of the film. The wave-like spectra caused by the interference agree with those calculated on the basis of optics.

Photoelectrochemical properties of TiO_2 films

In order to examine the photoelectrochemical properties of TiO_2

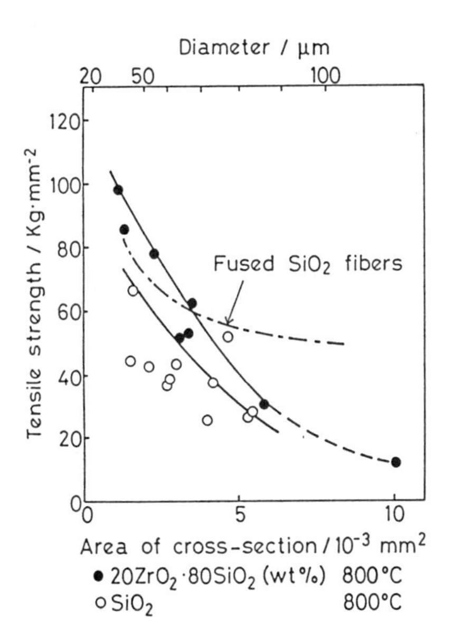

Figure 3. Tensile strength vs. cross-sectional area for sol-gel derived ZrO_2-SiO_2 and SiO_2 glass fibers heated at 800°C. (Reproduced with permission from Ref. 8. Copyright 1985 Uchida Rokakuho.)

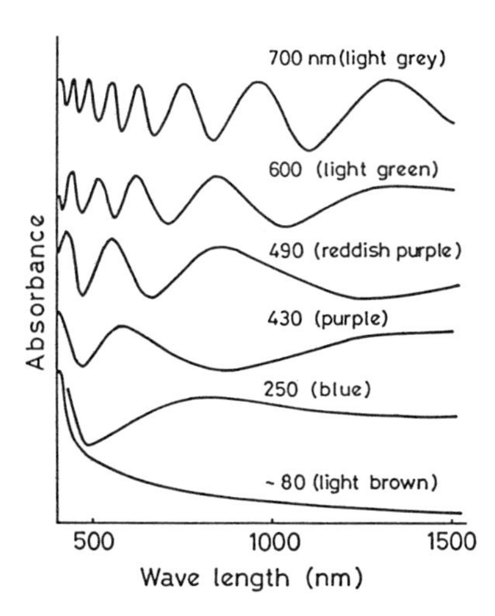

Figure 4. Optical absorption spectra of sol-gel derived TiO_2 coating films with different thicknesses.

films (9,10), the coating films up to more than 2 μm in thickness which are n-type simiconductors have been prepared on a nesa glass substrate by dip-coating technique using $Ti(O\text{-}iC_3H_7)_4\text{-}H_2O\text{-}C_2H_5OH\text{-}HCl$ solution. TiO_2 gel films were obtained by dipping nesa glass substrates in the titanium alkoxide solution and pulling them up at constant speed of 0.15 mm/s. The TiO_2 film thus obtained was subjected to heat treatment at 500°C for 10 min. The film thickness was increased by repeating the above operation. Anatase fine crystals are precipitated in the film.

The apparatus used for studying the photoelectrochemical behavior (11) of the TiO_2 film electrode is shown in Figure 5. Platinum plate of 35 × 25 mm and saturated calomel electrode (S.C.E.) were employed as a counter and a reference electrode, respectively. A 500 W Xenon lamp was used for illuminating the TiO_2 electrode.

Figure 6 shows the current-bias potential curves for TiO_2 film

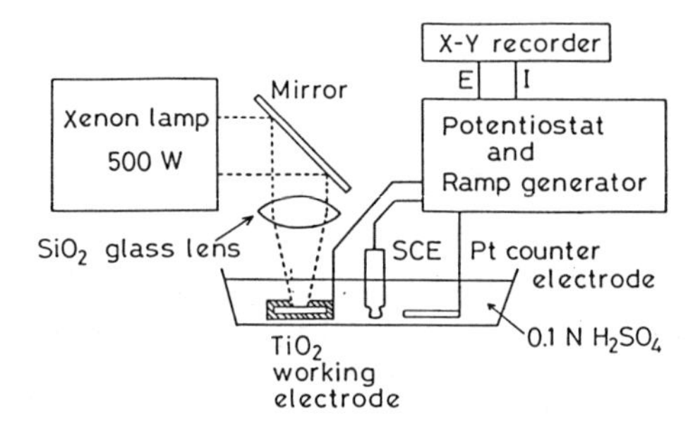

Figure 5. Apparatus for measurement of photoelectrochemical decomposition of water into hydrogen and oxygen. (Reproduced with permission from Ref. 10. Copyright 1987 The Ceramic Society of Japan.)

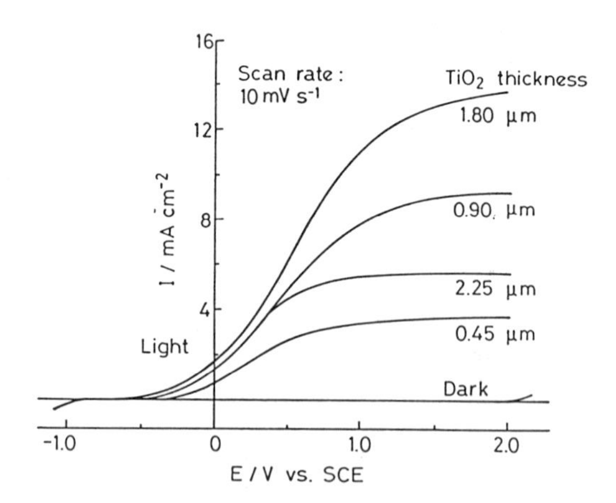

Figure 6. Photocurrent-bias characteristics of TiO_2 coating films with different thicknesses prepared by the sol-gel method. (Reproduced with permission from Ref. 10. Copyright 1987 The Ceramic Society of Japan.)

electrodes with different thicknesses under illumination. It is seen that at any bias potentials the photocurrent increases with increasing film thickness, reaches a maximum at about 1.8 μm and then falls with further increase in the film thickness. It should be noticed that the TiO_2 film electrode of 1.8 μm thick shows a maximum photocurrent as high as 14 $mA \cdot cm^{-2}$ at the bias potential above 1.5 V. This is comparable to or slightly higher than the values reported for single crystal (11) and polycrystalline TiO_2 (12), and much higher than those for the TiO_2 film electrode prepared by other methods such as chemical vapor deposition (13) and oxidation (14) and anodization (15) of Ti metal. The high efficiency of the dip-coated TiO_2 film may be attributed to the porous nature of the film as described below.

As shown in Figure 7, the photocurrent of the as-prepared TiO_2 film electrode of 0.9 μm in thickness increases with heating time, reaches a maximum at a heating time of 20 min and then increases with further heating time. This behavior may be explained on the basis of the changes in the surface structure and carrier concentration of the film with heating time.

The scanning electron microscopy (Figure 8) could not reveal the possible porous structure of the coating film, probably due to the very small size (in the order of several tens A) of pores at the film surface (8). In order to confirm the presence of pores in the TiO_2 coating film, the porosity of the film was estimated from the refractive indices calculated from the interference spectra shown in Figure 4 and the refractive index of the solid anatase crystal. Figure 9 shows the porosity of the film thus obtained as a function of heating temperature. The decrease of porosity with increasing heating temperature, although the method used for estimation is indirect. This suggests that there are pores in the coating film and the porosity decreases as the heat treatment progresses.

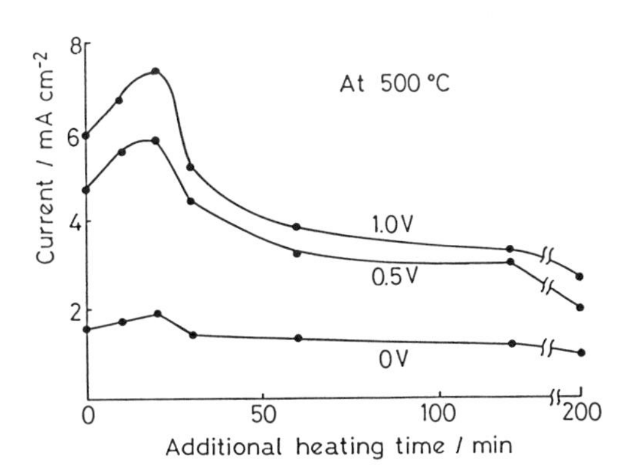

Figure 7. Change of the photocurrent with heating time for a TiO_2 coating film of 0.9 m thick prepared by the sol-gel method. (Reproduced with permission from Ref. 10. Copyright 1987 The Ceramic Society of Japan.)

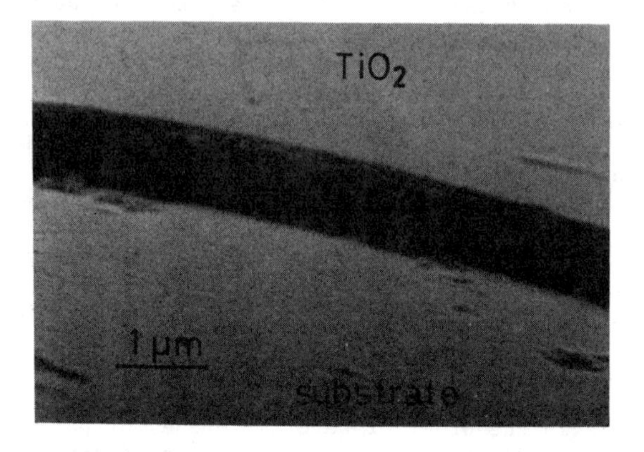

Figure 8. SEM photograph of a TiO_2 coating film of 0.9µm thick heated at 500°C for 20 min.

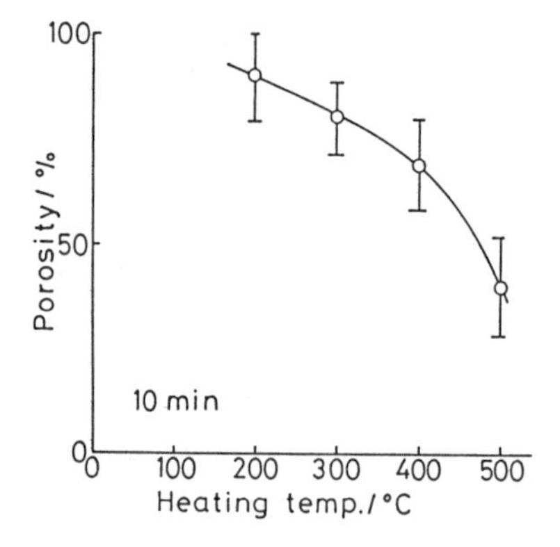

Figure 9. Change of the calculated porosity of a TiO_2 coating film with heating temperature. Each sample was heated at a given temperature for 10 min.

References

1. Sakka, S.; Kamiya, K. Mat. Sci. Res. Vol.17, Emergent Process Methods for High Technology Ceramics; Plenum Press: New York, 1984; p 83-94.
2. Sakka, S. Am. Ceram. Soc. Bull. 1985, 64, 1463-1466.
3. Sakka, S.; Kamiya, K.; Makita, K.; Yamamoto, Y. J. Non-Crystal. Solids 1984, 63, 223-235.
4. Sakka, S.; Kozuka, H.; Kim, S.-H. The Third International Conference on the Ultrastructure Processing of Ceramics, Glasses and Composites; San Diego, Feb. 24-27, 1987.
5. Sakka, S. J. Non-Crystal. Solids 1985, 73, 651-660.
6. Dislich, H. J. Non-Crystal. Solids 1983, 57, 371-388.
7. Orgaz-Orgaz, F.; Rawson, H. J. Non-Crystal. Solids 1986, 82, 57-68.
8. Kamiya, K.; Yoko, T.; Iwanaka, M.; Sakai, A.; Sakka, S. Zirconia Ceramics 1985, 5, 39-52.
9. Yoko, T.; Kamiya, K.; Sakka, S. Denki-Kagaku (Electrochemistry) 1986, 54, 284-285.
10. Yoko, T.; Kamiya, K.; Sakka, S. Yogyo-Kyokai-Shi 1987, 95, 150-155.
11. Fujishima, A.; Honda, K.; Kikuchi, S. Kogyo-Kagaku-Zasshi 1969, 72, 108-113.
12. Mavroides, J. G.; Tchernev, D. I.; Kafalas, J. A.; Kolesar, D.F. Mat. Res. Bull. 1975, 10, 1023-30.
13. Hardee, K. L.; Bard, A. J. J. Electrochem. Soc. 1975, 122, 739-42.
14. Fujishima, A.; Kohayakawa, K.; Honda, K. J. Electrochem. Soc. 1975, 122, 1487-89.
15. Tamura, H.; Yoneyama, H.; Iwakura, C.; Murai, T. Bull. Chem. Soc. Jpn. 1977, 50, 753-54.

RECEIVED October 2, 1987

Chapter 29

New Hybrid Materials Incorporating Poly(tetramethylene oxide) into Tetraethoxysilane-Based Sol–Gel Glasses

Structure–Property Behavior

Hao-Hsin Huang, Raymond H. Glaser, and Garth L. Wilkes

Department of Chemical Engineering, Polymer Materials and Interfaces Laboratory, Virginia Polytechnic Institute and State University, Blacksburg, VA 24061–6496

A novel hybrid system incorporating triethoxysilane endcapped poly(tetramethylene oxide) (PTMO) with tetraethoxysilane (TEOS) has been successfully prepared utilizing a sol–gel process. Compared with the TEOS–PDMS systems reported earlier, these new materials show higher extensibility and tensile strength. Although the high functionality of triethoxysilane endcapped PTMO helps promote good dispersion of this oligomeric component, some degree of local phase separation still results, though not of pure PTMO character. This is supported by the observation of a single peak in the dynamic mechanical tanδ spectrum and a broad maximum in the SAXS profile. The effects of PTMO molecular weight and TEOS content on the structure–property behavior have been studied, and a simplified model for these systems is suggested. Titanium isopropoxide has also been successfully incorporated into the hybrid systems by employing a different reaction procedure; the resulting materials show higher modulus and tensile strength relative to the pure TEOS–PTMO systems. Significant aging at ambient temperatures is observed for all PTMO-modified materials and is attributed to further diffusion-limited curing of the TEOS-based species.

Sol-gel chemistry and associated processes are not new, for over the last twenty-five years this area of science has been studied extensively (1-12). More recently, utilization of sol-gel reactions has become a practical method for producing low temperature glasses - one of the primary advantages being energy savings. The traditional fusion process for manufacturing glasses requires high firing temperatures (> 1400°C) whereas sol-gel glasses require much lower firing temperatures (the order of 900°C) to produce a suitable material. In addition to the energy savings afforded by lower manufacturing temperatures, the sol-gel process makes possible high quality multicomponent glasses with various properties which are difficult to make using the traditional fusion process due to crystallization and phase separation. B, Al, Na, Zn, Zr are but a few of the metal atoms that have been succesfully incorporated into the SiO_2 matrix to form multicomponent glasses (1-4).

0097–6156/88/0360–0354$06.50/0

A major limitation of the sol-gel process, though, is that large crack-free monolithic structures are difficult to make due to the loss of structural integrity (6,7). During the gelation/drying process cracks form as a result of internal drying stresses in thick (> 100 mils) samples. In fact, in pure glass systems, cracking during gelation/drying can reduce the final material to a powder if the drying process occurs to rapidly or if the material has a thickness of many mils. Because of this limitation, the sol-gel technique has been used, primarily, to produce homogeneously mixed multicomponent powders which, when fired at low temperatures, form high purity multicomponent glasses (2).

The basic sol-gel reaction can be viewed as a two-step network-forming polymerization process. Initially a metal alkoxide (usually TEOS, $Si(OCH_2CH_3)_4$) is hydrolyzed generating ethanol and several metal hydroxide species depending on the reaction conditions. These metal hydroxides then undergo a step-wise polycondensation forming a three-dimensional network in the process. The implication here is that the two reactions, hydrolysis and condensation, occur in succession; this is not necessarily true (8,9). Depending on the type of catalyst and the experimental conditions used, these reactions typically occur simultaneously and in fact may show some reversibility.

It occurred to one of the authors (G.L.W.) that oligomeric and polymeric compounds with appropriate functional groups could be incorporated into such inorganic glass networks using the sol-gel process. The incorporation of organic compounds (elastomers in particular) would hopefully impart flexibility to the hybrid sol-gel material. This would make possible the manufacture of crack-free monolithic structures which could be, in principle, heat treated or used as modified or hybrid sol-gel glasses. Indeed, hybrid sol-gel materials incorporating polydimethylsiloxane (PDMS) oligomers of various contents and molecular weight into glass networks have been successfully made and studied under a variety of experimental conditions [13-16]. The potential applications for such materials as coatings or fibers are immediately obvious; even the possibility of manufacturing structural components exists provided these new "hybrid" sol-gel materials have the desired physical properties and are processable.

Historically, hydroxy-terminated PDMS (polydimethylsiloxane) was the first elastomer chosen by our group to be incorporated into a TEOS glass network (13). The primary reasons for this were threefold. First, the similarity between the molecular structure of PDMS and the silicate glass was thought to possibly enhance the incorporation of the elastomer into the network and to reduce the possibility of phase separation prior to gelation. Secondly, hydroxy-terminated PDMS is commercially available and requires no functionalization of the chain ends; the silanol function present at each end of the PDMS chain can undergo step-growth polycondensation with hydrolyzed TEOS species thereby forming a hybrid network with more or less inorganic glass-like behavior depending on the ratio of the reactant species. Thirdly, PDMS has considerable high temperature stability (relative to most organic elastomers) which might be of utility in various applications.

Hydroxy-terminated PDMS, however, has disadvantages. The monofunctional ends limit the number of connections between the polymer (or oligomer) molecule and the glass network to two. This limitation raises the possibility that some PDMS molecules are not tied at both ends to the glass network if the polycondensation does not go to completion i.e. there may be "dangling" or loose PDMS chains in the final sol-gel material. This occurance of free ends would indeed be anticipated since the extent of reaction most likely is not 100%. Hence, the physical properties, specifically the mechanical behavior of the overall material, would be expected to suffer as a result of loose PDMS chains in the system. Disregarding this potential problem, the mechanical behavior of the sol-gel hybrids are, ultimately, influenced by the mechanical behavior of the modifying elastomer;

the intrinsic properties of PDMS are not the most desirable - another disadvantage. PDMS has a relatively low modulus (at ambient temperature) and generally does not exhibit high elongation or tensile strength. The final "major" disadvantage of PDMS is a lack of miscibility in water which jeopardises the homogeneity of materials made using high water concentrations.

The limitations inherent to PDMS prompted the decision to study different polymers as possible "glass-modifying" components. It should be noted though that it was the successful incorporation of PDMS molecules into the TEOS glass network that opened the possibility of other macromolecules being incorporated into glass networks via the sol-gel process. Poly(tetramethylene oxide) (PTMO) chains endcapped with triethoxy silane was one system chosen as a possible "glass-modifier". Other systems are currently being studied and shall be reported in future publications. Reported in this symposium paper are the properties of hybrid organic-glass materials made by incorporating PTMO oligomers and low molecular weight polymers into silicate glass networks as well as silicate-titanate glass networks. For comparative purposes the properties of some PDMS-glass systems will be presented though these systems are discussed in detail elsewhere (14,16).

As a final note to this introduction, it is known that several experimental parameters (type of catalyst, pH, water content, reaction temperature etc.) influence the progression of the overall sol-gel reaction and consequently determine the network structure of the final material (8,17,18). The physical properties of sol-gel materials are expected to be related directly to the type of network formed; therefore, solid state ^{29}Si NMR, Raman spectroscopy and small-angle x-ray scattering (SAXS) studies have been undertaken to characterize the structure formed in sol-gel materials reacted under different conditions. The results of these structural studies conducted on both pure glass and elastomer modified sol-gel materials will be presented in detail in future publications, though preliminary SAXS data will be presented in this paper relating to the PTMO-glass materials. Again, the primary focus of this symposium paper is to present the general physical-mechanical properties of PTMO modified sol-gel glasses.

Experimental

Materials. High purity tetraethoxysilane (TEOS) was obtained from the Fluka company. Endcapped polytetramethyleneoxides (PTMO) with three different molecular weights (before endcapping) - 650, 1000, and 2000 g/mol - were generously supplied by Dr. Carlson. Titanium isopropoxide (Ti-isop) was obtained from Aldrich Chemical company. Tetrahydrofuran and isopropanol were used as the cosolvents for both the TEOS-PTMO and Ti containing systems. Hydrochloric acid (10N) was the catalyst for the TEOS-PTMO system, while for the Ti containing system glacial acetic acid was used to catalyse the reactions.

Reaction Scheme. The acid catalysed reactions for preparing the hybrid materials in the present study can be illustrated by the simplified scheme shown below:

HYDROLYSIS

$$Si(OR)_4 + 4H_2O \xrightarrow{H^+} Si(OH)_4 + 4ROH$$

$$(RO)_3Si\text{–}(PTMO)\text{–}Si(OR)_3 + 6H_2O \xrightarrow{H^+} (HO)_3Si\text{–}(PTMO)\text{–}Si(OH)_3 + 6ROH$$

R is C_2H_5

POLYCONDENSATION (not stoichiometrically balanced)

$$Si(OH)_4 + (HO)_3Si\text{–}(PTMO)\text{–}Si(OH)_3 \xrightarrow{H^+} \text{–}\underset{\underset{|}{O}}{\overset{\overset{|}{O}}{Si}}\text{–O–}\underset{\underset{|}{O}}{\overset{\overset{|}{O}}{Si}}\text{–(PTMO)–}\underset{\underset{|}{O}}{\overset{\overset{|}{O}}{Si}}\text{–O–}\underset{\underset{|}{O}}{\overset{\overset{|}{O}}{Si}}\text{–} + H_2O$$

Although the reaction scheme shows a complete hydrolysis before condensation begins, this is likely not correct as stated earlier. The relative rates and extents of these two reactions will particularly depend on the amount of water added and the acidity of the system (10,11). The high functionality of the triethoxysilane endcapped PTMO oligomer should enhance the incorporation of PTMO molecules into the TEOS network. It was also assumed that the reactivities would be the same between silanol groups from silicic acid and endcapped PTMO. Therefore, no preferential condensation was expected and the deciding factors for which type of condensation (self- or co-) took place would be the diffusivities and local concentrations.

Example Procedures. Due to the fast reaction rate of Ti-isop, two different procedures were employed for TEOS-PTMO and Ti-containing systems.

(a) For the TEOS-PTMO system — eight milliliters of isopropanol and 2 ml of THF were first added to a 100 ml flask, then 10 ml of TEOS and appropriate amount of PTMO were added and mixed thoroughly until the solution appeared to be homogeneous. Then, a mixture of HCl and deionized water were added to the base solution under rapid agitation. The reaction usually proceeded rapidly at ambient temperature, and the system quickly turned into a viscous liquid. After approximately 30 seconds, the liquid was cast into Teflon coated petri dishes and covered with parafilm. After 1-2 days, the parafilm was removed and the gel was dried under ambient conditions for another week prior to testing.

(b) For the Ti containing system — appropriate amounts of TEOS and PTMO were added to a round bottom flask at ambient temperature, then the solution was diluted with a 4:1 mixture of isopropanol:THF so that the final concentration of reactants was 40 % by volume. Glacial acetic acid was measured, with a molar ratio of acid:TEOS = 2:1, into the solution and mixed thoroughly under a nitrogen environment overnight. The desired amount of titaniumisopropoxide was then added and the reaction was carried out under gentle reflux for 1/2 to 2 hours. The final solution was cast into covered petri dishes and allowed to gel under ambient conditions for at least one week prior to testing.

By comparing the present TEOS-PTMO system with the TEOS-PDMS system reported earlier, some differences should be noted:

1. The PTMO is more miscible with water than PDMS and, thus, the homogeneity should be somewhat improved in TEOS-PTMO systems.
2. By endcapping PTMO with trifunctional silanes, the resulting PTMO molecules will have a functionality, f, of 6. This should reduce the gelation time considerably and potentially enhance the dispersion of the PTMO species.
3. Unlike the silanol terminated PDMS, the endcapped PTMO will have to undergo hydrolysis before the condensation can take place.
4. Since reactants were mixed thoroughly with solvents before adding the acid solution, both the TEOS and PTMO should be well dispersed within the system.

All the above changes favor the formation of a more homogeneous system than that of TEOS-PDMS.

Sample Nomenclature. This is illustrated in Table I. The loading of the glassy component(TEOS) is expressed in percentage of the total weight of glassy and oligomeric components used.

Table I. Description of Sample Nomenclature

example: TEOS(50)–PTMO(2000)–50–0.04–RT

item	description
TEOS(50)	silane used (wt%)
PTMO(2000)	oligomer (m.w.)
50	% stoichi. amount of water added
0.04	molar ratio of acid/silane
RT	reaction temperature (°C)

Characterization Methods. Stress-strain experiments were carried out with an Instron model 1122. Dogbone samples of 10mm in length were used, and the initial strain rate was 2 mm/min. Dynamic mechanical data were obtained utilizing a DDV-IIC Rheovibron Dynamic Viscoelastometer. Most samples were tested within the temperature range of -100°C to 220°C with a heating rate of 2-3°C/min. A frequency of 11 Hz was selected for all the dynamic mechanical experiments.

A Siemens Kratky camera system was utilized for small angle x-ray scattering (SAXS) measurements in conjunction with an M. Braun position sensitive detector from Innovative Technology Inc.. Wide angle x-ray diffraction was obtained utilizing a Philips table-top x-ray generator.

Results and Discussion

All the cast films in this study were transparent though the addition of titanium did result in yellowish or orangish samples. Also, the titanium containing materials showed considerably greater toughness than the TEOS-PDMS systems. Although the higher PTMO oligomer, PTMO(2000), can crystallize in pure form under ambient conditions, wide angle x-ray diffraction (WAXD) experiments did not show any sign of crystallization in the final hybrid materials. This observation strongly implies that the incorporation was successful since crosslinking and mixing with TEOS would reduce the mobility of the PTMO molecules and, thus, suppress the crystallization process.

In earlier studies of TEOS-PDMS systems, results indicate that the molecular weight of the oligomer and the TEOS content have a significant effect on the final structure-property behavior. For the TEOS-PTMO systems, these two variables have also been studied and the results will now be addressed.

Table II. Mechanical Properties of TEOS-PTMO-50-0.04-RT

sample no.	wt% of TEOS	m.w. of PTMO	elongation at break (%)	ultimate strength (MPa)	Young's modulus (MPa)
1	50	650	94	4.8	5.2
			88	6.2	6.7
2	50	1000	93	2.7	3.5
					3.8
					3.9
3	50	2000	113	2.1	2.8
			95	1.1	1.9
4	60	2000	119	5.5	5.5
			150	6.2	4.6
			119	3.3	4.0
5	70	2000	57	30.4	105
			68	30.6	86
			80	31.3	81

Effect of PTMO Molecular Weight

Three different molecular weights - 650, 1000, 2000 g/mol - of PTMO (uncapped) were used for reactions in which the initial weight percent of the endcapped oligomeric component was 50%. A representative set of stress-strain curves are shown in Figure 1, and the collective results are listed in Table II. (All the experiments were carried out approximately 15 days after the reactions, therefore, a comparison between samples should be valid. However, the mechanical properties did show some changes with time; the details of this aging behavior will be discussed in the latter portion of this paper.) The TEOS-PTMO materials show considerably higher extensibility than comparable TEOS-PDMS materials (ca 105% for 50% TEOS-PTMO (2000Mw) vs ca 8% for 50% TEOS-PDMS (1700Mw) (16)). However, the ultimate strength of the two types of materials are roughly equal (ca 1.6 MPa for 50% TEOS-PTMO (2000Mw) vs ca 1.1 MPa for 50% TEOS-PDMS (1700Mw)) and the TEOS-PDMS material exhibits a substantially higher Young's modulus (ca 12 MPa for 50% TEOS-PDMS (1700Mw) vs ca 2.4 MPa for 50% TEOS-PTMO (2000Mw)). As shown in Figure 1, the initial modulus decreases with increasing PTMO molecular weight as would be expected. Whereas for the elongation at break, no distinct trend is observed though it does show a minimum for samples made with the lowest molecular weight. Dynamic mechanical spectrum for this series of samples are shown in Figures 2a&b, along with the spectrum of a TEOS-PDMS sample shown in Figure 3 for comparison. The general behavior of the storage modulus is somewhat similar in these two hybrid systems. Specifically, a glassy state with a magnitude of 10^9 Pa. is observed at low temperatures, followed by a glass transition region

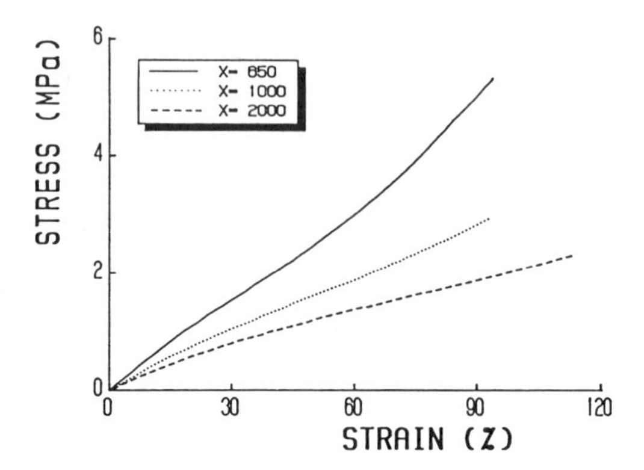

Figure 1. Stress-Strain behavior of TEOS-PTMO materials made with 50 wt% TEOS (PTMO MW = 650, 1000, 2000).

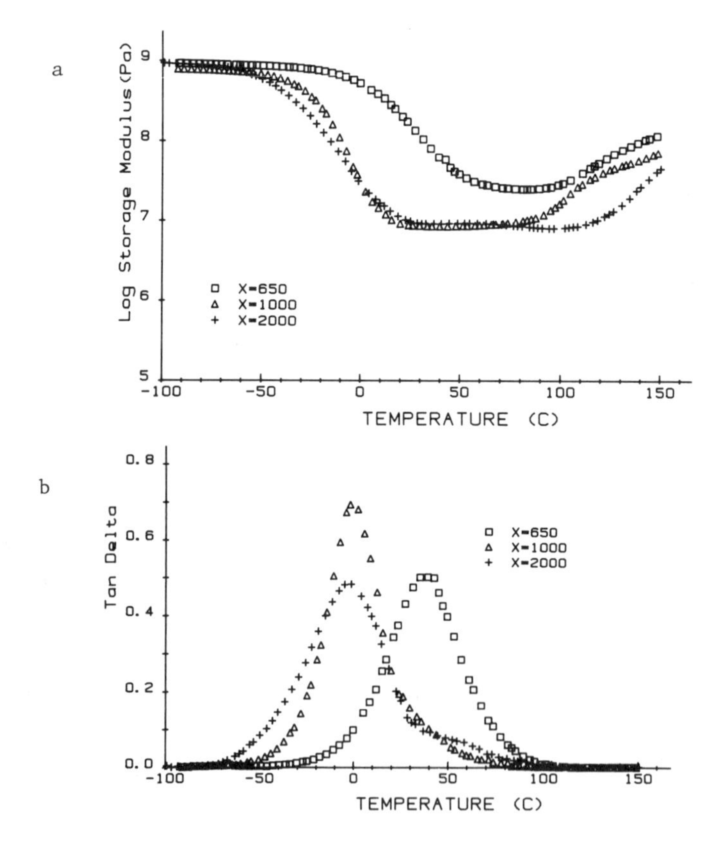

Figure 2. Dynamic mechanical spectra of TEOS-PTMO materials made with 50 wt% TEOS (PTMO MW = 650, 1000, 2000) (a) storage modulus, (b) tanδ.

after which the modulus decreases to ca. 10^7 Pa. corresponding to a rubbery state of the material. Depending on the oligomeric species used, the onset and the broadness of the glass transition is rather different. Finally, an increase of modulus occurs at high temperatures. The onset and magnitude of change, however, depends on the specific system. The PTMO(650) sample always shows a higher modulus than the other two materials, and the onset temperature of the glass transition is also higher. With regard to the tanδ behavior, the difference between Figure 2 and Figure 3 is rather significant. A single peak, instead of a bimodal shape shown by the PDMS sample, is observed for the TEOS-PTMO materials, and the magnitude is always higher than that of the TEOS-PDMS hybrids. While the PTMO(650) sample displays a peak near 38°C, the other two samples show a peak near -2°C.

To further understand the structure of these samples, SAXS experiments were carried out and the resulting profiles are shown in Figure 4. The smeared intensity was plotted against the angular scattering variable, s, which is defined as:

$$s = (2\sin\theta)/\lambda$$

where λ is the wavelength and θ is the internal scattering angle or one half of the radial scattering angle. The important point to note is that a maximum is observed in all three scattering curves, while for the earlier reported TEOS-PDMS systems, the scattering behavior always showed a monotonic decrease with s (16). In addition, the position of the maximum for the TEOS-PTMO system shifts toward smaller angles as the PTMO molecular weight increases. Such results strongly imply that a "correlation length" exists in these samples which is directly related to the molecular weight of the oligomer being incorporated; this indirectly suggests that the correlation length is of "interparticle" origin.

In previous reports on TEOS-PDMS systems (16), a model was suggested to interpret the dynamic mechanical results. Since the PDMS used was silanol terminated, self-condensation could take place at the beginning of the reaction, before significant hydrolysis of TEOS occurs. This results in a siloxane-rich phase which shows a T_g near -106°C, while for the more dispersed siloxane molecules, a higher T_g (near -10°C) was observed. However, for the present system, as mentioned in the experimental procedure, the PTMO molecules have three functional groups at each end that can undergo hydrolysis and condensation reactions. Therefore, both ends of the oligomers are certainly likely to be connected to "TEOS crosslinks" of the final network. The distribution of the PTMO chain length between such "crosslinks" would be close to, if not the same as, the molecular weight distribution of the starting oligomers. Moreover, the homogeneity of the solution before the addition of water should further suppress phase separation, assuming there is no preferential condensation of either component. These reasonable assumptions will be of further use shortly. However, it is of value to first recall some of the general features of the sol-gel behavior of pure TEOS.

According to the model suggested by Brinker et al for the pure sol-gel TEOS glass (17), acid catalysed systems without enough water for complete hydrolysis will first produce chain-like or lightly branched molecules. These molecules then undergo further condensation to form a dense network. Whereas for the base catalysed systems, particulate-like molecules or "Eden cluster" like species are often generated and the aggregation of these particles result in gelation (18). Since intimate incorporation of oligomers into the TEOS network is the goal of this work, it is reasonable to chose the acid catalysed route so that particulate growth of TEOS can be avoided. Hence, one can generalize the model suggested above for the acid catalysed pure TEOS network to interpret the hybrid system at hand. If an oligomeric chain is incorporated into an acid catalysed TEOS network, two types of motional restrictions on the chain are possible:

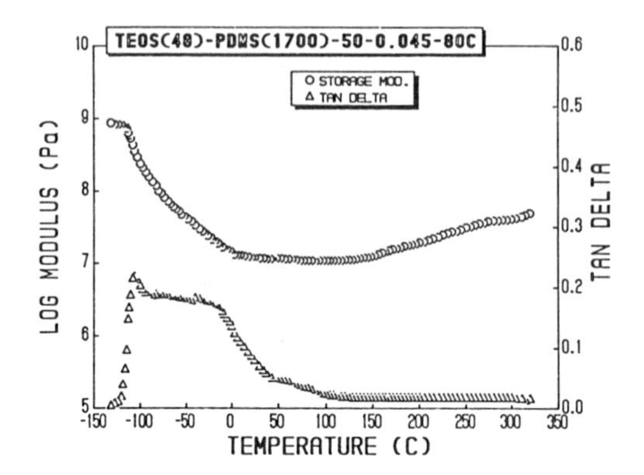

Figure 3. Dynamic mechanical spectra of TEOS–PDMS materials made with 48 wt% TEOS (PDMS MW = 1700), storage modulus and tanδ.

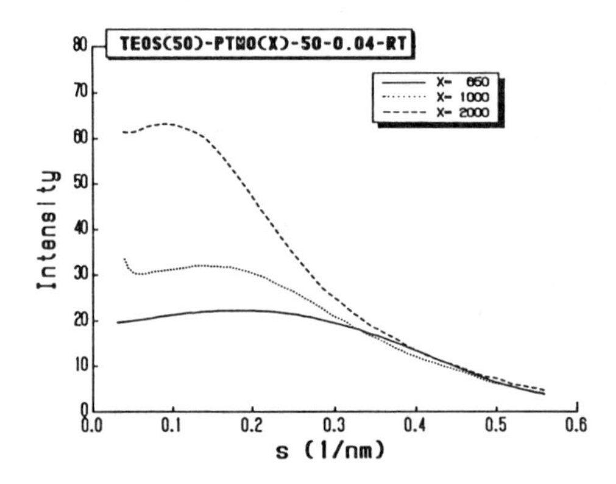

Figure 4. SAXS profiles of TEOS-PTMO materials made with 50 wt% TEOS (PTMO MW = 650, 1000, 2000).

1. That caused by connecting both ends to the network (hereafter referred to as TYPE 1).
2. That caused by interaction with surrounding, lightly condensed, TEOS species, or by possibly being encapsulated in a highly condensed TEOS environment (hereafter referred to as TYPE 2).

Due to these restrictions, more thermal energy will be required to mobilize the incorporated oligomeric chain, which results in an increase of its T_g. This phenomenon has been observed in TEOS-PDMS systems by both dynamic mechanical spectroscopy (16) as well as DSC (19). As to the magnitude of the increase of T_g, it will depend on the extent of restriction being imposed on the oligomeric chain, i.e. a greater restriction on the oligomer should result in a larger increase in its T_g. For relatively short oligomeric chains, restraining the ends by connecting them to the network would impose considerable limitations on the motion of the chain. As the chain length increases, this limiting effect caused by endlinking (TYPE 1) would diminish due to the coil-like nature of the oligomeric chain (20). As for the TYPE 2 restriction, it should be closely related to the dispersion of the oligomeric chain or, in other words, to the mixing of the two components. A better mixing of the system should result in higher interaction between components and thus a greater TYPE 2 restriction. Utilizing these concepts and our earlier assumptions, the experimental results can be rationalized. As the molecular weight of PTMO increases, both types of restrictions decrease due to greater chain length and poorer mixing. Consequently, the increase of T_g should be less significant. This may help explain the shift of T_g toward lower temperatures as PTMO molecular weight increases.

In addition to the change of T_g, the modulus shows a decreasing trend as the molecular weight of PTMO increases. This may be explained as follows:

1. A higher TYPE 2 restriction caused by a better mixing in the system with PTMO(650) would increase the stiffness of the oligomeric chain, and this should result in a higher modulus. As the PTMO molecular weight increases, the encapsulation of oligomeric chains by condensed TEOS species would become more difficult. Such a decrease in the TYPE 2 restrictions would make the PTMO chains less stiff and, as a result, cause the modulus to decrease.
2. Although the compositions of the reactants are the same for all samples, the actual content of PTMO decreases as its molecular weight decreases. This is due to the finite weight of the endcapping groups. Therefore, lower molecular weight of PTMO represents higher glass content and, as expected, higher modulus.

Another important factor is the extent of reaction of the system: a more highly condensed system will surely show a higher modulus. However, this factor for the present system is still undetermined. Further structural characterization is currently being undertaken to understand this aspect. The increase of modulus at high temperatures, which was also observed in the earlier reported TEOS-PDMS systems, is partially attributed to further curing of the TEOS based network. However, the increase in modulus observed at higher temperature is reversible to some degree (as in the TEOS-PDMS systems), therefore, further curing is not an adequate explaination of the phenomenon. Annealing experiments are currently underway to asses the reversibility - if any - of the thermally induced increase in modulus observed for the TEOS-PTMO systems.

The observed maxima in the SAXS scattering profiles result from regular (or partially regular) fluctuations of the electron density within the TEOS-PTMO materials. Such maxima in SAXS profiles are typically observed in systems which exhibit microphase separation. The distance or "correlation length" that characterises this "periodicity" in electron density fluctuation, and hence the microphase separation, is simply the inverse of the scattering parameter s at which the maximum is displayed. It must be understood, however, that this value will only be approximate for our data due to the smearing effect caused by the slit geometry of the kratky camera. (Desmearing will shift the peak, or maximum,to somewhat larger s values.) There is also a dependency on the type of "periodic" morphological texture, i.e. lamellar, dispersed etc.. For a two phase system in which one component forms a domain structure, the correlation distance will then represent an average interdomain spacing. As shown in Figure 4, a maximum is observed in each of the scattering profiles and the correlation distance, calculated by taking the inverse of s, increases as the molecular weight of PTMO systematically increases from 5 nm to 10 nm. As suggested above, both ends of an endcapped PTMO chain are believed to be connected to the final network. However, such "connection points" are not likely to be single condensed TEOS molecules. Instead, they are more likely to be clusters formed by groups of condensed TEOS molecules.

A highly schematic model of such a system is shown in Figure 5. This model attempts to convey, visually, the types of structure present in the TEOS-PTMO hybrid materials. The crowded "x" areas represent the "higher electron density" condensed TEOS species while the coiled interconnecting lines represent the "lower electron density" PTMO species. Note that the separation of these two species is not absolute and that the TEOS species is not completely condensed (Raman data on PDMS systems suggest that the uncondensed TEOS ends are -OEt rather than -OH (15)). Also demonstrated, in terms of the inverse of the scattering parameter s, is the "correlation length" between two TEOS rich clusters. The electron density difference between these clusters and richer PTMO regions (but not pure PTMO - based on the dynamic mechanical results) would certainly cause scattering to take place based on the electron density difference of the components - see Table III. Since these clusters are all separated by the distance of one PTMO chain, a correlation distance shown by SAXS is expected and its magnitude should increase as the PTMO chain becomes longer. As to the broadness of the maximum in SAXS profile, it may be partially attributed to the molecular weight distribution of PTMO oligomers and/or the size distribution of the TEOS clusters. This model does not preclude the existence of linear molecules formed by partially condensed TEOS; also it is consistent with both the mechanical and SAXS results. Investigations are currently underway to provide further support for this simplified model.

Table III. Calculated Electron Densities of all of the Components

material	electron density (mole e^-/cc)
TEOS	0.5119
PDMS	0.5135
PTMO	0.5682
SiO_2	1.10

Effect of TEOS Content

To study the influence of the initial composition of the system on the final structure, three TEOS contents - 50, 60, and 70 wt% - were prepared. The solution became less viscous as TEOS content increased, and the gelation time tended to be longer for samples containing more TEOS as would be expected based on network growth behavior. The dried gels with higher TEOS content appeared to be stiffer but displayed more cracking, which caused difficulty in preparing large uniform materials. However, all samples showed good transparency and some flexibility - the latter decreasing as TEOS content increased.

Figure 6 shows the stress-strain curves of samples made with different TEOS contents; their mechanical properties were listed earlier in Table II. The initial modulus increases slightly as TEOS content changes from 50 to 60 wt%, however, it increases almost twenty times as the TEOS content reaches 70 wt%. Such a large difference implies that a significant change occurs in the structure, which is further supported by the dynamic mechanical results shown in Figures 7a&b. From the storage modulus spectra little difference is observed between the two samples with lower TEOS content, and the general behavior is similar to those discussed earlier in Figure 2. However, the 70% sample shows a large increase in the magnitude and displays a two-step decrease of the modulus.

The corresponding tanδ spectra are shown in Figure 7b, in which a systematic increase of the transition temperature and decrease in magnitude are observed as the TEOS content increases. The T_g transition in the tanδ curves broadens with increasing TEOS content eventually displaying a bimodal type of behavior as the TEOS content reaches 70%. Bimodal glass transition behavior generally indicates two different species within a material undergoing separate glass transitions or two different environments for the same species. In the TEOS-PTMO materials the observed T_g peaks in the tanδ curves are primarily due to the PTMO oligomers (linear TEOS species could also, conceivably, contribute to the T_g peak in the tanδ curves). Therefore, the bimodal peak observed in the tanδ curve of the 70% TEOS-PTMO material strongly implies that two different environments exist for the PTMO chains in this particular system. The differences here being the result of the chemical environment i.e. a "bimodal distribution" in the level of TYPE 2 interactions.

In all of our work, the initial weight percentages of the reactants have been used to express the composition of the hybrid systems. However, these values do not represent the final composition of the dried gel. Since 70% of the reactant mass is lost by converting TEOS to SiO_2, the final inorganic glass content is less than that indicated by the initial TEOS weight percentage. Based on previous studies of two-phase block copolymers (21), the structure of the final material will vary considerably with the composition of the system. As the volume fraction of component 1 is low, it may form isolated domains dispersed in the matrix of component 2. If the two components are of approximately equal volume fraction, then the final structure may be of a co-continuous "lamellar" nature of both components. Finally, as the volume fraction of component 1 becomes high, it may form the matrix with dispersed domains of component 2. For the hybrid two-component system at hand, changes in phase behavior may also be possible although component purity is not expected to be high in this case.

Figure 8 shows the relationship between the initial PTMO wt% and the final PTMO vol% at different degrees of reaction, i.e. the percentage of TEOS that converts to SiO_2. This is theoretically calculated by assuming that all the regions formed by the condensed TEOS have a density of 2.2 g/cm^3 (that of the amorphous silicon dioxide), and the error caused by any possible voids is neglected. Despite the simplified assumptions, such a calculation should give an approximation to the real case. Another parameter used in constructing this figure is the degree of reaction, which represents the percentage of TEOS that has been converted to SiO_2. In light of earlier results from the TEOS-PDMS systems, a value between 70 and 80 for the

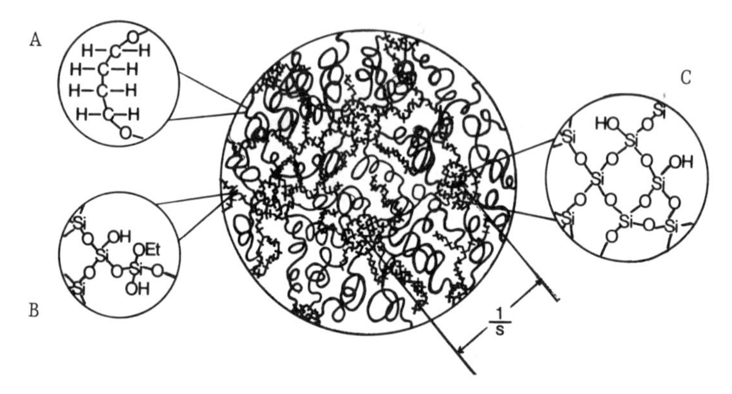

Figure 5. A schematic model for the structure in TEOS-PTMO hybrid systems, (A) PTMO chain, (B) linear species based on partially condensed TEOS, (C) cluster formed by highly condensed TEOS. 1/s corresponds to the correlation length observed in SAXS profiles.

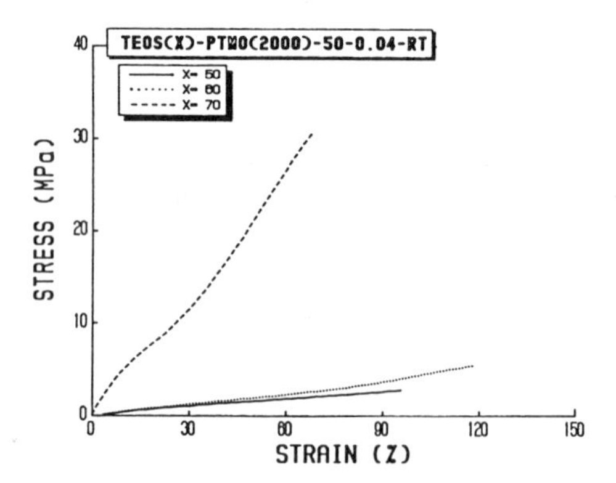

Figure 6. Effect of TEOS content on the stress-strain behavior of TEOS-PTMO materials (PTMO MW = 2000).

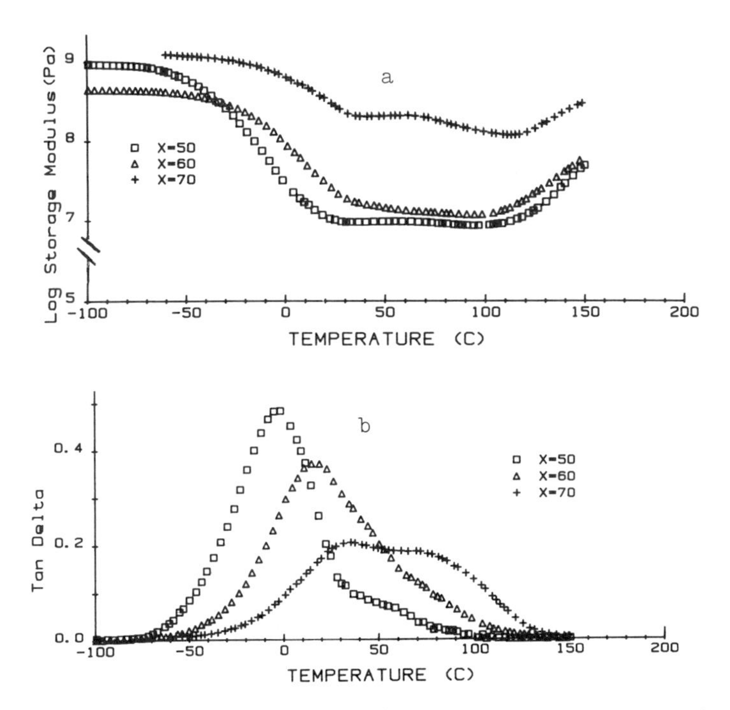

Figure 7. Effect of TEOS content on the dynamic mechanical behavior of TEOS-PTMO materials (PTMO MW=2000), (a) storage modulus and (b) tanδ.

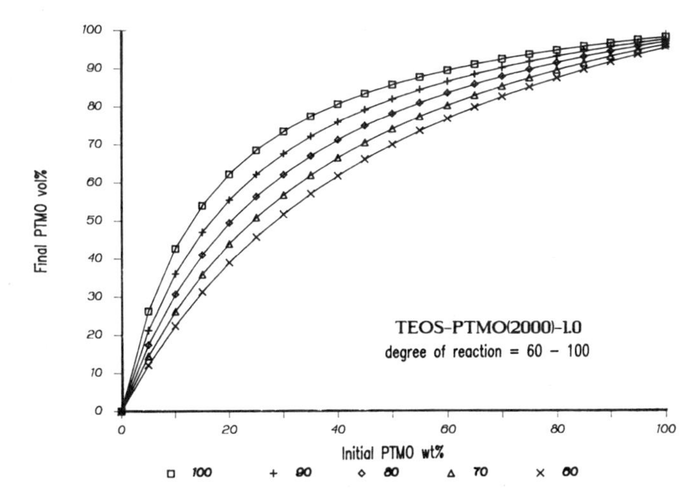

Figure 8. Calculated PTMO volume fraction in the TEOS-PTMO gel with respect to the initial PTMO weight fraction. The degree of reaction is the percentage of TEOS which converts into SiO_2.

degree of reaction would be a good approximation. With this assumed value (say 70%), the final vol% of PTMO for the system with initially 70 wt% TEOS will be close to 50% according to Figure 8. If one considers the error caused by the density assumption of the TEOS based species, the final vol% of PTMO should be even lower. Thus, for such a system the possibility of a structural inversion should be marginal.

In terms of the cluster model suggested previously, it should be possible for PTMO chains to be trapped inside TEOS based clusters as more TEOS is added and phase inversion is approached. Such encapsulation of PTMO chains in TEOS based clusters is expected to be more prevalent in systems with high TEOS content (70%) than in systems with lower TEOS content (50% & 60%) for this reason. The motion of encapsulated PTMO chains (high TYPE 2 restriction) would be greatly restricted and result in a higher glass transition temperature than for PTMO chains with less TYPE 2 restriction (systems with lower TEOS content). It is this explaination that may help account for the two broad transitions observed in the sample with 70% TEOS as well as the drastic increase of the modulus. Restated, the increase of T_g as the TEOS wt% changes from 50 to 60 can be attributed to an increased TYPE 2 restriction of the PTMO chains as more TEOS is added to the system. Note: true continuity of the TEOS network does not, however, seem to exist since yield behavior is not observed in the stress strain curves as might be expected.

The samples produced with 70 wt% of TEOS did show a great improvement in the tensile strength of these hybrid materials (see Table II), however, the occurrence of film cracking again became serious. The possibility of using drying control chemical additives (DCCA) as discussed by Wallace and Hench (22) is currently under investigation; hopefully, by this method, the cracking problem can be solved without a significant change in the properties.

Effect of Titanium Addition

Table IV lists the mechanical stress-strain data for a series of hybrid TEOS-PTMO materials containing different levels of Ti-isop in the starting reaction mixture. These materials, with the exception of the first one, were all made using a modified reaction scheme (see experimental section) in order to incorporate the titanium into the network. The starting reaction mixtures in all cases contained 50% by weight of the glass precursors (TEOS and Ti-isop) and 50% by weight of PTMO(2000) (endcapped with triethoxysilane). One set of samples without titanium was made in order to compare the effects of the reaction scheme on the observed mechanical properties.

Comparing the mechanical properties of the two PTMO-TEOS (0wt% Ti) materials, one first observes a somewhat lower stress at break, a lower elongation at break, and a higher initial modulus for the material made using the modified reaction scheme. The difference between the two materials is believed to be due to the nature of the network formed under the different reaction conditions. The material made using the modified reaction scheme could be argued to have a more completely developed network because of the higher initial modulus and lower elongation at break. It should be noted, though, that the values of ultimate stress and initial modulus for both materials are of the same order of magnitude. Therefore, the differences between the network structures formed by the two reaction schemes are viewed as rather small in significance. The addition of titanium to the system, however, introduces drastic changes in the hybrid network structure as is evidenced primarily by the much larger observed values for stress at break and initial modulus.

Table IV. Mechanical Properties of Ti Containing PTMO(2000) Systems

reaction procedure	wt% of $Ti(OPr)_4$	elongation at break (%)	ultimate strength (MPa)	Young's modulus (MPa)
a	0	113	2.1	2.8
		95	1.1	1.9
b	0	21	1.3	6.9
		11	0.8	7.1
		22	0.9	5.9
		13	1.0	7.2
		19	1.2	7.2
b	15	151	12.4	22.8
		158	11.3	20.0
		98	8.6	23.7
b	30	52	10.7	67.2
		69	11.8	67.0
		51	10.1	66.8
b	30†	63	11.8	64.1
		73	12.1	82.5
		46	12.1	65.6

† opaque samples (different reaction batch)

The addition of 15wt% Ti-isop to the reaction mixture results in a three-fold increase in the ultimate stress, a seven-fold increase in the elongation at break and a three-fold increase in the initial modulus over the system without Ti-isop. The upward trend in the mechanical properties, however, is continued only for the initial modulus when the Ti-isop content is further increased from 15 to 30wt%. The 30wt% Ti-isop material showed a large decrease in elongation at break as compared to the 15wt% Ti-isop material (68% vs 154%) while the ultimate stress remained approximately the same (12 MPa). As mentioned, the initial modulus for the 30wt% Ti-isop material increased three-fold over that of the 15wt% Ti-isop material. The initial modulus value for the 30wt% Ti-isop material was almost one order of magnitude greater than that observed for the TEOS-PTMO material containing no Ti-isop. Obviously the addition of Ti-isop is strongly effecting the final structure in the hybrid material.

It has been proposed that Ti catalyzes the polycondensation reaction in the sol-gel scheme (23). This could explain the large increase in the mechanical properties observed on the initial addition of titanium. In the presence of a condensation catalyst the network forming step-growth polymerization might be "less hindered" by vitrification. In other words, since the energy barrier for condensation is lowered by the catalyst, neighboring silanol functions condense more readily in the presence of a catalyst. Therefore, since vitrification limits the diffusion process - reactive groups come in contact less often - the extent of the condensation reaction for catalyzed systems (materials containing Ti-isop) is higher than that of uncatalyzed systems. These latter statements are, at this stage, of a highly speculative nature and data must be obtained on how the extent of reaction(s) varies with Ti-isop content.

A "tighter" developed glass network would, logically, result in higher modulii and lower elongations. Also, since the titanium acts as a catalyst, the initial addition would have the greatest effect on the reaction; further addition, however, may not be as effective in changing the structure and, hence, the mechanical properties of the overall material. In short, initial addition of titanium would produce a large change in the structure and, therefore, the mechanical properties while further addition of titanium would result in less significant changes. Again, this model, though attractive in explaining the observed mechanical behavior, is still highly speculative and needs to be substantiated or disproven by structural analysis and determination of the extent of reaction of the titanium containing materials.

The SAXS (small-angle x-ray scattering) profiles of the Ti-isop containing samples in Figure 9 show a broadened "peak" the center of which is shifted to somewhat higher values of s as one compares a 15wt% Ti sample with a 0wt% Ti sample made by procedure 'b' (s - ca 0.10 nm^{-1} for 15 wt% Ti vs s - ca 0.06nm^{-1} for 0 wt% Ti). However, the peaks in the SAXS profiles of a 0wt% Ti sample made by procedure 'a' (see Figure 4) and a 15 wt% Ti sample are qualitatively the same. The observed peaks are broad in both cases and are centered at s = 0.1 nm^{-1} which corresponds to an approximate correlation distance of 10 nm (ignoring smearing effects). The 15 wt% Ti sample shows greater than a two-fold increase in the scattering intensity though. This intensity difference likely results from the higher electron density mismatch of the PTMO with the Ti-isop species relative to the TEOS based species. Increasing the titanium content to 30wt% does not change the location or the general shape of the SAXS "peak" but the observed intensity is greater than in the 15wt% Ti sample. These SAXS profiles indicate that a change in structure occurs upon the initial addition of titanium in systems made by reaction procedure 'b' - probably due to the catalytic effect of the titanium on condensation. The lack of Ti is undoubtably why the 0 wt% Ti samples made by procedure 'b' are somewhat different from all other samples. The SAXS profiles hint that 0 wt% Ti samples made by procedure 'a' and titanium containing materials made by procedure 'b' are structurally the same; increasing the titanium content of the material results in only subtler changes in structure. This correlates well with the observed stress-strain data and lends support to the hypothesis that titanium acts as a condensation catalyst. It should be noted, however, that at this time SAXS data for the Ti containing TEOS-PTMO materials can only be analyzed in a qualitative fashion. Electron densities for the samples must be calculated before SAXS data can be analyzed quantitatively; therefore the extent of reaction must be determined. Investigations of this nature are currently being undertaken.

The dynamic-mechanical behavior of a titanium containing TEOS-PTMO hybrid is compared to that of a material without titanium in Figure 10. The storage modulus decreases only one order of magnitude for the titanium containing materials while those without titanium show a decrease of two orders of magnitude at temperatures beyond the glass transition, T_g (Figure 10a). All materials when heated for the first time show a rise in the storage modulus beginning at approximately 150°C. This rise is attributed to further curing or condensation occuring as a result of the elevated temperature. Here again, the titanium containing compounds exhibit a smaller change (rise) when compared to the pure TEOS-PTMO systems. This is strong additional evidence that the condensation reaction in the titanium sol-gels has proceeded to a higher level than in non titanium containing sol-gels (when both are gelled at ambient temperature).

Comparison of the tanδ curves (Figure 10b) results in similar conclusions: the titanium containing TEOS-PTMO hybrids are distinctly different from the pure TEOS-PTMO hybrids. Both the 15 and 30wt% Ti materials exhibit broad tanδ transitions centered at approximately -30°C while the comparable 0wt% Ti material exhibited a "sharper" tanδ transition at approximately -5°C. Furthermore, the magnitude of the transitions decreased (from 0.5 to less than 0.2) upon the addition of titanium. The decrease in the magnitude of the glass transition could

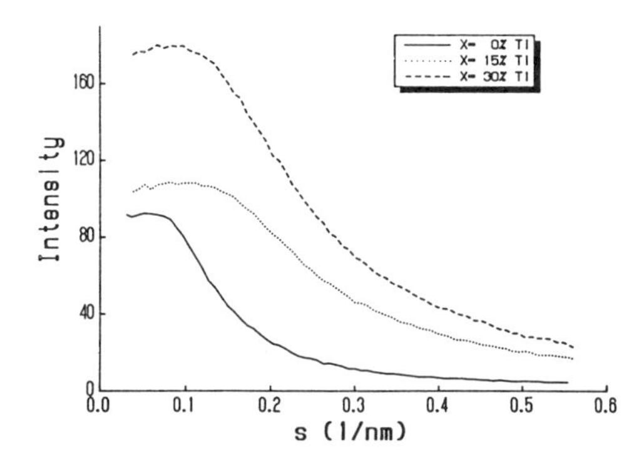

Figure 9. SAXS profiles of titanium containing TEOS-PTMO materials made by procedure 'b' (wt% Ti-isop. = 0, 15, 30).

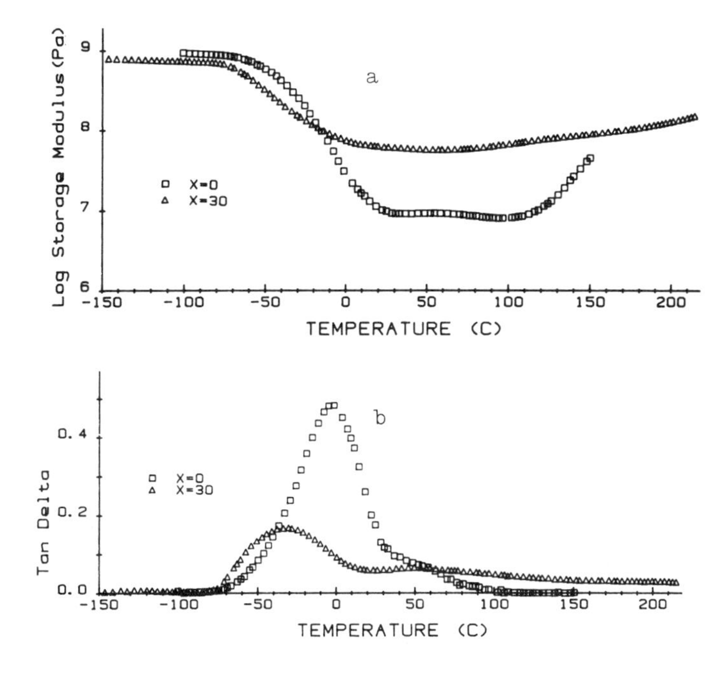

Figure 10. Effect of titanium addition on the dynamic mechanical behavior of TEOS-PTMO materials (PTMO MW = 2000), (a) storage modulus and (b) $\tan\delta$.

simply be the result of the higher modulus of the titanium containing materials. Recall that the "second" (weak) tanδ transition reaching from ca 30°C to ca 120°C in the 30 wt% Ti-isop (Figure 10b) suggests the presence of encapsulated PTMO chains in this system.

The distinctly lower glass transition temperatures observed in the titanium TEOS-PTMO hybrids as opposed to the pure TEOS-PTMO hybrids (comparing the tanδ response) is clearly indicatative of structural differences between these two systems. Since the glass transition is lowered relative to TEOS-PTMO materials - closer to the transition temperature of the pure PTMO material - it could be argued that the titanium containing materials are less homogeneously mixed: i. e. that there are PTMO rich "phases" in the titanium containing systems which would give rise to a T_g nearer to that of the pure PTMO oligomer. However, large scale phase separation would be difficult to accept in light of the similarities of the SAXS curves mentioned earlier. Recall that the SAXS data showed that the peak position does not change as titanium is added to the hybrid systems.

The Aging Effect

In a network forming process, gelation takes place as the molecular weight becomes infinitely large and, as a result, the viscosity of the system increases drastically. The high viscosity after gelation considerably decelerates the crosslinking process through diffusion limitation of the chemical reaction. However, chemical crosslinking does not necessarily stop after gelation; the chemical reaction may continue towards completion, albeit slowly. Hence, given sufficient time, a network forming system will continue to crosslink after gelation until all reactive sites are spent or vitrification occurs quenching mobility. Though vitrification, like gelation, does not necessarily terminate the chemical crosslinking, the diffusion limitation is so great that additional reaction in a vitrified material can be, generally, ignored. Once vitrified, large scale molecular motions are not possible in the system; therefore, reactive sites can no longer come together and the chemical crosslinking reaction is effectively quenched. Therefore, a slow continuation of the crosslinking reaction after gelation is what we will refer to as aging. If vitrification should happen to occur before gelation, aging - as we have defined it - would not be expected to take place.

A time effect on crosslink density - or aging process - has been studied in detail for organic network systems such as the thermoset epoxys and a Time-Temperature-Transformation (TTT) diagram is often used to represent such phenomenon (24). An aging process or time dependent curing has also been noted for the pure sol-gel TEOS system. Consequently, an aging process might also be expected to occur in modified TEOS materials such as the hybrid PTMO-TEOS systems presented in this paper. Also, as a result of further curing or aging of the network structure, material properties should change with time. Aging is, therefore, of great concern in materials that are to be used in the gelled state (i. e. without being fired, further cured, or vitrified).

In principle, aging will take place in the time period between gelation and vitrification (assuming that once vitrified the material ceases to age). Therefore, aging is governed primarily by two factors: the extent of reaction required for gelation (if this is very high little aging can take place) and diffusion/molecular mobility in the gel. The first factor is determined by the average functionality of the system in question and is independent of polymer properties as predicted by Flory (25). The second factor, diffusion in the gel, varies with the chemistry of the system and generally increases as the reaction (aging) temperature is raised beyond the glass transition temperature of the system and the size of the species decreases. The lack of long term aging in the PDMS-TEOS systems presented in earlier publications (16) can be attributed to a combination of these two factors. The difunctional PDMS oligomers lower the average functionality of the system thereby raising the extent of reaction required for gelation; this limits the amount of aging possible since the system is closer to complete reaction when it reaches the gel point. The low glass transition temperature of the PDMS species (see Figure 3b) affords a certain degree of oligomer mobility in materials gelled at ambient temperature which in turn allows the TEOS-PDMS systems to "age" quickly; i. e. the sol-gel reaction reaches its limiting degree of completion in several days instead of several months. In the TEOS-PTMO systems, however, the aging process at ambient temperature should be more substantial than in the TEOS-PDMS system due to the following: a higher average functionality in the system and lower oligomer mobility in gelled materials (as evidenced by glass transition temperatures that approach ambient conditions - see Fig 2b). The higher PTMO functionality results in gelation of the TEOS-PTMO systems at a lower extent of reaction as compared to the TEOS-PDMS systems. Therefore, more reactive sites are available for potential aging in the gelled TEOS-PTMO systems than in the gelled TEOS-PDMS systems. Furthermore, the higher glass transition temperatures observed in gelled PTMO materials (Figure 3) indicates considerable diffusion limitation (low oligomer mobility), which causes the chemical aging process to proceed slowly. Hence, the time period between gelation and vitrification increases and long time aging becomes significant.

Mechanical properties of a TEOS-PTMO material at various aging times are shown in Table V. Indeed, the changes are significant. Modulus increases graduately with time and so does the ultimate strength; the elongation at break decreases relatively rapidly in the first 50 days. Note that the properties observed after aging 315 days can be duplicated by annealing a virgin sample at 110°C for two hours (see Table V) which indicates that aging is accelerated at elevated temperatures and could be used as a "post casting" treatment to stabilize the network.

Table V. Aging effects on Samples of TEOS(50)-PTMO(2000)-50-0.04-RT

days after reaction	elongation at break (%)	ultimate strength (MPa)	Young's modulus (MPa)
15	104	1.6	2.4
24	68	1.3	2.8
34	79	1.8	3.2
53	44	1.6	4.8
315	18	2.9	16.8
8†	18	2.9	16.5

† treated 110°C for 2 hours

For comparison, Table VI shows the effects of room temperature aging on the mechanical properties of a 30wt% Ti 2900 Mw PTMO-glass system. Using the initial modulus as the primary gauge of aging one observes that the most drastic changes occur in the first few days after gelation (as might be expected). The overall effect of aging on the titanium containing materials is to triple the initial modulus (ca 20 MPa to ca 60 MPa) and double the measured stress at break (ca 6 MPa to ca 13 Mpa). The room temperature aging process in the titanium containing materials in terms of the inital modulus is essentially finished after thirty days, however, further aging results in a decrease in the measured elongation at break (ca 110% at 30 days to ca 60% at 105 days).

Table VI. Aging Effect on a 30wt% Ti Containing PTMO(2900) System

days after reaction	elongation at break (%)	ultimate strength (MPa)	Young's modulus (MPa)
3	90	5.2	20.3
	130	6.1	21.6
	97	5.8	19.8
	116	5.9	20.0
7	110	10.3	49.0
	114	10.8	49.6
	109	11.0	52.9
	119	11.0	44.2
33	115	14.4	62.7
	105	13.6	64.4
	106	14.3	67.2
105	65	13.2	55.5
	56	12.8	66.2
	39	10.1	60.0
	55	11.8	61.6
	56	13.0	76.7
4†	70	11.4	52.0
	85	12.7	49.6
	69	12.2	60.4
	89	12.6	51.8

† cured at 80°C under vacuum for 4 days

Annealing gelled samples at elevated temperatures has been found to accelerate the aging process for TEOS-PTMO systems (both with and without Ti) as is expected. Comparing the mechanical spectra of annealed and virgin - unannealed - materials similarities and differences are observed (spectra not shown). Most notably, the storage modulus in annealed samples does not drop as severely after the glass transition as in virgin samples. Also, the storage modulus remains level i. e., no upturn is observed at the end of the spectrum. The annealed storage modulus corresponds to the level observed at the end of the scan of a virgin material (after the upturn). Also, the glass transition regions are almost identical, the only difference being a lower intensity observed for the Tg of the annealed specimen (no doubt a result of the higher storage modulus).

SAXS was employed to study the structural changes in the TEOS-PTMO materials as a function of aging time. The resulting data show an increase in the scattering intensity though the characteristic distance does not change (Figure 11). This can be understood in terms of the structural model suggested earlier. As mentioned, some of the condensed TEOS species form clusters which serve as the connecting sites for PTMO chains. Such clusters are expected to contain unreacted ethoxy and silanol groups that can undergo further condensation with time. The condensation of these groups would result in an increase of the electron density of the TEOS based clusters. Since the observed scattering intensity is related to the electron density difference between the TEOS rich cluster and the PTMO richer

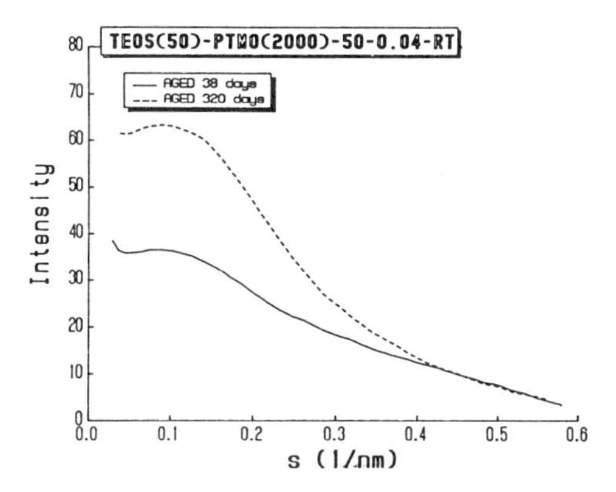

Figure 11. Aging effect on the SAXS profile of a TEOS-PTMO material (PTMO MW=2000)

region, the scattering intensity should increase as the electron density of the cluster increases with aging. On the other hand, the peak position and the tail region of the SAXS curve should remain relatively unaltered since the aging process as described would not change the correlation distance between the TEOS rich clusters (or conversely the PTMO rich regions).

The aging data presented represent only the initial efforts to probe the time dependence observed in the TEOS-PTMO hybrid materials. Comprehensive studies are currently underway which will, it is hoped, provide the structural as well as mechanical data necessary to understand this phenomenon.

Conclusions

Hybrid materials with TEOS and triethoxysilane encapped PTMO have been successfully prepared by the sol-gel process. High transparency is observed for most samples, and many of the mechanical properties are considerably improved relative to the earlier reported TEOS-PDMS systems.

From the dynamic mechanical spectroscopy, an increase of PTMO molecular weight from 650 to 2000 results in a decrease in both the modulus and the glass transition temperature of the final product. The SAXS results indicate that a correlation distance exists in the samples, and this distance increases as PTMO molecular weight increases. A cluster model is thus suggested to account for the experimental results.

The TEOS content is changed from 50 to 70 wt% to study its effect on the final structure of the materials. Besides an increase of the stiffness and glass transition temperature, a speculated change in structure is observed at 70 wt% TEOS. This change is in line with that anticipated from simple theoretical consideration of composition.

Titaniumisopropoxide, which is known to have a fast rate of condensation, has been successfully added to the TEOS-PTMO systems by employing a different reaction procedure. The resulting materials are yellow or orange in color but usually transparent. The modulus increases significantly with the addition of titanium, and the ultimate strength is improved with respect to the pure TEOS-PTMO hybrid systems. This is tentatively attributed to a catalytic effect on condensation caused by the titanium component, which results in a tighter network structure. Glass transition temperatures observed in dynamic mechanical spectra indicate that the phase separation is also enhanced for Ti containing systems.

Aging is significant in both the pure and Ti containing TEOS-PTMO systems and, similar to other reported pure sol-gel glasses, it is due to further curing of the TEOS based species. This results in an increase of the modulus and a decrease of the elongation at break. However, no change is observed in the correlation length shown by SAXS. Therefore, a structural alteration via a change in the dispersion of the PTMO oligomers is not the cause of these observations. Further study on this subject is currently being undertaken.

Acknowledgments

The authors wish to thank Dr. James G. Carlson for providing the endcapped PTMO oligomers.

Literature Cited

1. Dislich, H. ***Angew. Chem. Int. Ed. Engl.*** 1971, ***10(6),*** 363.
2. Yoldas, B. E. ***J. Mat. Sci.*** 1977, ***12,*** 1203.
3. Mukherjee, S. P. ***J. Non-Cryst. Solids*** 1980, ***42,*** 477.
4. Dislich, H. ***J. Non-Cryst. Solids*** 1983, ***57,*** 371.
5. Mackenzie, J. D. ***J. Non-Cryst. Solids*** 1982, ***48,*** 1.
6. Yamane, M.; Aso, S.; Okano, S.; Sakaino, T. ***J. Mat. Sci.*** 1979, ***14,*** 607.
7. Sakka. S; Kamiya, K. ***J. Non-Cryst. Solids*** 1980, ***42,*** 403.
8. Yoldas, B. E. ***J. Mat. Sci.*** 1979, ***14,*** 1843.
9. Yamane, M.; Inoue, S.; Yasumori, A. ***J. Non-Cryst. Solids*** 1984, ***63,*** 13.
10. Zerda, T. W.; Artaki, I.; Jonas, J. ***J. Non-cryst. Solids*** 1986, ***81,*** 365.
11. Strawbridge, I.; Craievich, A. F.; James, P. F. ***J. Non-cryst. Solids*** 1985, ***72,*** 139.
12. Schmidt, H. ***Mat. Res. Soc. Symp. Proc.*** 1984, ***32,*** 327.
13. Wilkes, G. L.; Orler, B.; Huang, H. ***Polym. Prep.*** 1985, ***26(2),*** 300.
14. Huang, H.; Orler, B.; Wilkes, G. L. ***Polym. Bull.*** 1985, ***14(6),*** 557.
15. Huang, H.; Glaser, R. H.; Wilkes, G. L. ***Polym. Prep.*** 1987, ***28(1),*** 434.
16. Huang, H.; Orler, B.; Wilkes, G. L. ***Macromolecules*** 1987, ***20(6),***
17. Brinker, C. J.; Scherer, G. W. ***J. Non-Cryst. Solids*** 1985, ***70,*** 301.
18. Schaefer, D. W.; Keefer, K. D. ***Mat. Res. Soc. Symp. Proc.*** 1986, ***73,*** 277.
19. Parkhurst, C. S.; Doyle, L. A.; Silverman, L. A.; Singh, S.; Anderson, M. P.; McClurg, D.; Wnek, G. E.; Uhlmann, D. R. ***Mat. Res. Soc. Symp. Proc.*** 1986, ***73,*** 769.
20. Critchfield, F. E.; Koleske, J. V.; Magnus, G.; Dodd, J. L. ***J. Elastoplastics*** 1972, ***4,*** 22.
21. Hashimoto, T.; Shibayama, M.; Fujimura, M.; Kawai, H. ***Memoirs of the Faculty of Engineering, Kyoto University,*** 1981, 43.
22. Wallace, S.; Hench, L. L. ***Mat. Res. Soc. Symp. Proc.*** 1984, ***32,*** 47.
23. Philipp, G.; Schmidt, H. ***J. Non-Cryst. Solids*** 1984, ***63,*** 283.
24. Enns, J. B.; Gillham, J. K. ***J. Appl. Polym. Sci.*** 1983, ***28,*** 2567.
25. Flory, P. J. ***Principles of Polymer Chemistry*** Cornell University Press, 1953.

RECEIVED September 1, 1987

BORON-CONTAINING POLYMERS

Chapter 30

Precursors to Nonoxide Macromolecules and Ceramics

C. K. Narula, R. T. Paine, and R. Schaeffer

Department of Chemistry, University of New Mexico, Albuquerque, NM 87131

There is great interest in developing molecular precursors for boron-nitrogen polymers and boron nitride solid state materials, and one general procedure is described in this report. Combinations of B-trichloroborazene and hexamethyldisilazane lead to formation of a gel which, upon thermolysis, gives hexagonal boron nitride. The BN has been characterized by infrared spectroscopy, x-ray powder diffraction and transmission electron microscopy.

Non-oxide ceramic materials are typically prepared from simple starting reagents by using classical energetic reaction conditions (high temperature, high pressure, CVD, etc.). Although these routes provide useful materials for many applications, there is increasing need for alternate low energy procedures which permit more flexible processing of the ceramic end-product. One synthetic route which has attracted recent attention involves the combination of simple reagents into an adduct containing all the elements of interest for the final solid state product. Following thermal or photochemical activated intramolecular elimination-condensation chemistry, an oligomeric or polymeric ceramic precursor is obtained which can be easily handled and converted by pyrolysis to the desired ceramic end-product. This concept has been successfully applied in the production of silicon carbide and silicon nitride (1). Relatively little attention has been given, on the other hand, to the development of molecular polymeric precursors for Group 13-15 binary ceramics. The potential importance of polymeric precursors in this area can be illustrated by a brief overview of selected chemistry known to provide technologically important boron nitride.

Boron nitride may be obtained in three primary crystalline modifications (2): α, β, and γ. The most commonly encountered α form has a graphitic structure (hexagonal cell, a = 2.504 Å, c = 6.661 Å). For many years, this modification has been prepared from combinations of cheap boron and nitrogen containing reagents, e.g. $B(OH)_3$ and $(NH_2)CO$, $B(OH)_3$, C and N_2 or KBH_4 and NH_4Cl (3-5). More

0097-6156/88/0360-0378$06.00/0

recently, chemical vapor deposition (CVD) techniques utilizing mixtures of BCl_3 and NH_3, BF_3 and NH_3 or B_2H_6 and NH_3 have been employed for the synthesis of h-BN (3,6-10). Pertinent to our own studies, several investigators have also used substituted borazenes as precursors in classical thermolysis or CVD preparations of h-BN. For example, Constant and Feurer (11) pyrolyzed hexachloroborazene, $(ClBNCl)_3$ at 900-950°C on silicon substrates and observed the deposition of amorphous BN which was subsequently converted to h-BN by electron bombardment. In thorough, systematic thermolysis studies reported by Rice and coworkers (12) and Paciorek and coworkers (13), it was found that a variety of substituted borazenes $(XBNY)_3$ produced boron rich BN solid state materials.

At the initiation of our studies several years ago, there had been relatively few well documented reports on the formation of polymeric B-N compounds. An explanation for this situation may be found in the historical development of molecular boron-nitrogen chemistry (14-16). In the period 1950-1970, a great deal of attention was given to the evolution of syntheses for monomeric aminoboranes, iminoboranes and borazenes and progress made was impressive. For a variety of reasons, advances in B-N polymer chemistry were frustrated. After that period in the USA, further efforts to explore polymeric materials were few and this area attracted little attention. Nonetheless, several reports in the patent and open literature suggested that borazene compounds could be employed as monomers for modest molecular weight oligomers (13,17-24). Furthermore, a tantalizing patent report in 1976 by Taniguchi and coworkers (25) suggested that spinnable fibers could be obtained from a borazene $(H_2NBHPh)_3$ based polymer which were, in turn, converted to boron nitride. More recently, Paciorek and coworkers have reported on the formation of boron nitride materials from several condensed substituted borazenes (23,24).

Recognizing the need for fundamental synthetic chemistry leading to boron-nitrogen preceramic polymers, we embarked on a program to develop improved condensation routes for aminoboranes, iminoboranes and borazenes. In particular, borrowing from the principles of organic polymer syntheses, chemistry was sought which would result in the crosslinking of borazene monomer building blocks. It was rationalized that crosslinking through amide bridges would produce preceramic oligomers with "excess" nitrogen (e.g. N:B = 1.5) which, upon pyrolysis, might retain more nitrogen than found in pyrolysis products from monomeric borazenes. In addition, it was anticipated that the crosslinking might result in boron nitride materials with a porous macrostructure.

Results

It is well known that some silylamines, in combination with haloboranes, provide a facile route to aminoboranes (Equation 1) (26). Meller and Füllgrabe (19) have previously reported that a

$$R_2BCl + R'_2NSiMe_3 \rightarrow R_2BNR'_2 \quad (1)$$

variation of this reaction will provide coupled substituted

borazenes. Indeed, the following model reactions were examined in our work and bis-borazinyl amines 1 and 2 were isolated in high yield (Equations 2 and 3). Compounds 1 and 2 were fully characterized by elemental analysis and spectroscopic techniques (27).

$$(Cl)(Me)_2B_3N_3Me_3 + (Me_3Si)_2NH \rightarrow \underset{\mathbf{1}}{[Me_2B_3N_3Me_3]_2NH} \quad (2)$$

$$Cl(Me)_2B_3N_3Me_3 + (Me_3Si)_2NMe \rightarrow \underset{\mathbf{2}}{[Me_2B_3N_3Me_3]_2NMe} \quad (3)$$

The ^{11}B NMR spectra for these compounds show resonances at 37.4 and 28.0 (1) and 37.2 and 27.6 (2), respectively.

Extending the model chemistry to the formation of preceramic polymers, we found that oligomeric gels are obtained by related reactions involving B-trichloroborazenes as shown in Equations 4-6. In each case, a transparent gel formed which took the shape of its container. Some specific experimental details for the formation of 3 are provided, and a summary of the processing of the oligomer is given in Figure 1. The gels may be formed in a variety

$$Cl_3B_3N_3H_3 + (Me_3Si)_2NH \xrightarrow{CH_2Cl_2} \underset{\mathbf{3}}{[(HN)_3B_3N_3H_3]_n} + Me_3SiCl \quad (4)$$

$$Cl_3B_3N_3H_3 + (Me_3Si)_2NMe \xrightarrow{CH_2Cl_2} \underset{\mathbf{4}}{[(NMe)_3B_3N_3H_3]_n} + Me_3SiCl \quad (5)$$

$$Cl_3B_3N_3Me_3 + (Me_3Si)_2NH \xrightarrow{CH_2Cl_2} \underset{\mathbf{5}}{[(NH)_3B_3N_3Me_3]_n} + Me_3SiCl \quad (6)$$

of other solvents, and the processing of these materials is described separately (27).

Trichloroborazene, $Cl_3B_3N_3H_3$ (50 mmol) and $(Me_3Si)_2NH$ (75 mmol) were combined at -78°C in CH_2Cl_2 and the mixture stirred for 30 min at -78°C and then slowly warmed to 25°C over two hours. A colorless gel formed which was isolated by vacuum evaporation of the volatiles. The resulting colorless glassy solid was pyrolyzed in vacuo at 900°C for 24 hours in a quartz tube and the evolved volatiles identified as NH_3 and NH_4Cl. The remaining solid was briefly (2 hours) heated in air at 1200°C in order to remove minor carbon impurities and to improve crystallinity. This solid was then treated at room temperature with 40% aqueous HF to remove boric acid and silica formed in small quantities. The solid obtained at 900°C was identified as boron nitride; however, the majority of the material was amorphous. After treatment at 1200°C, white crystalline boron-nitride was obtained in about 55% yield.

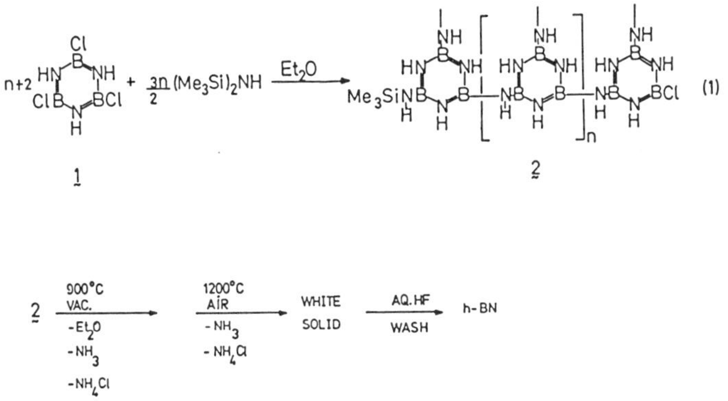

$$n{+}2\ \mathbf{1}\ +\ \frac{3n}{2}(Me_3Si)_2NH \xrightarrow{Et_2O} \mathbf{2} \quad (1)$$

$$\mathbf{2} \xrightarrow[-Et_2O,\ -NH_3,\ -NH_4Cl]{900^{\circ}C\ VAC.} \xrightarrow[-NH_3,\ -NH_4Cl]{1200^{\circ}C\ AIR} \text{WHITE SOLID} \xrightarrow[\text{WASH}]{\text{AQ. HF}} \text{h-BN}$$

Figure 1. Summary of borazene gel processing.

The boron nitride obtained in this study was characterized by infrared spectroscopy, powder x-ray diffractometry and transmission electron microscopy. Trace elemental analyses were also performed by energy dispersive x-ray analysis and carbon arc emission spectroscopy. Representative spectra are displayed in Figures 2-4.

The infrared spectrum of 3 (Figure 2) contained in a KBr pellet shows a broad absorption in the region 1425-1065 cm^{-1}, a sharp absorption at 710 cm^{-1}, and the spectrum closely resembles spectra reported in the literature (6,8,28). The broad band centered at 3300 cm^{-1} suggests the presence of retained N-H functionalities; however, this feature may also be seen in some commercial samples of h-BN. The x-ray powder pattern (Figure 3) is also similar to patterns reported in the literature (29). The line at 2θ ~26°corresponds to the 002 reflection, a line at 2θ ~42-44° corresponds to overlap of the 100 and 101 reflections, and a line at 2θ ~54° is the 004 reflection. The lines are broader than typically observed from highly crystalline commercial h-BN samples, and this line broadening may result from small crystallite size or from the presence of the previously observed turbostratic structure modification (30). The measured spacing d(002) = 3.35-3.36 Å agrees very well with the reported d(002) = 3.330 Å for h-BN prepared by conventional methods. Most turbostratic modifications show a greater deviation in d(002) = 3.5-3.6 Å; therefore, it is assumed that any turbostratic disorder is small.

The macrostructure of the boron nitride obtained here is porous with pores ~2 µm in diameter. There is no evidence for microporosity and the BET surface area is ~35 $m^2\ g^{-1}$. Transmission electron micrographs (Figure 4) show regions of well developed crystallinity. The crystalling grains are ~5-10 nm on a side and 30-40 nm long. The BN (002) lattice fringes are clearly visible.

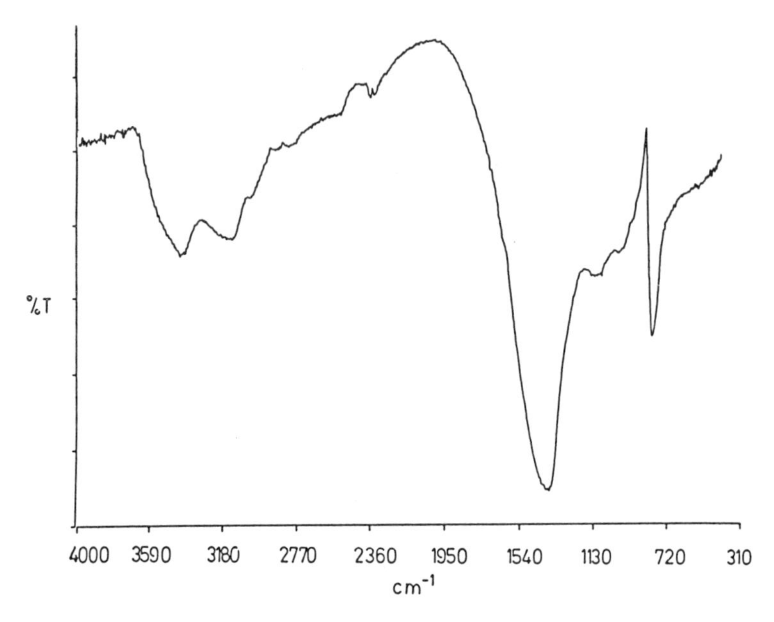

Figure 2. Infrared spectrum of BN.

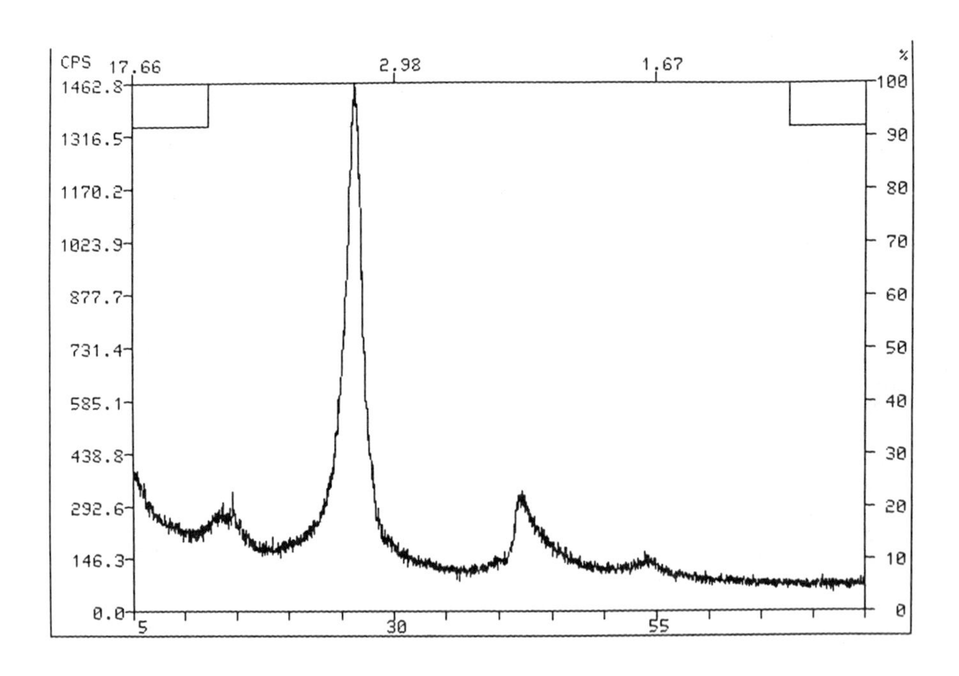

Figure 3. X-ray powder diffraction analysis of BN.

Figure 4. Transmission electron micrograph of BN.

Conclusion

The results of this study clearly reveal that crosslinking of borazene rings may be achieved with silylamines and the resulting oligomeric gels may be efficiently processed to give hexagonal boron nitride in good yields and with good crystallinity. The characteristics of the gels formed by this synthetic route lend themselves to a variety of processing modes, e.g. sol-gel and aerogel processing, not available in the synthetic procedures normally applied in boron nitride preparations. Additional studies are in progress which are designed to take advantage of these features.

Acknowledgment is made to Sandia National Laboratory for support of this research.

Literature Cited

1. Wynne, K.J.; Rice, R.W. Ann. Rev. Mater. Sci. 1984, 14, 297.
2. Hoard, J.L.; Hughes, R.E. In The Chemistry of Boron and Its Compounds; Muetteries, E.L. Ed. J. Wiley and Sons, NY 1967.
3. Meller, A. Gmelin Handbuck der Anorganische Chemie. Boron Compounds Second Supplement 1983, 1, 304.

4. Kalyoncu, C.S. Ceram. Eng. Sci. Proc. 1985, 1356.
5. Archer, N.J. Chem. Soc. (London) Spec. Pub. No. 30 1977, 167.
6. Rand, M.J.; Roberts, J.F. J. Electro. Chem. Soc. 1968, 115, 423.
7. Pierson, H.O. J. Composit. Mat. 1975, 9, 228.
8. Takahashi, T.; Itoh, H.; Takeuchi, A. J. Cryst. Growth 1979, 47, 245.
9. Sano, M.; Aoki, M. Thin Solid Films 1981, 83, 247.
10. Chopra, K.L.; Agarwal, V.; Vankar, V.D.; Deshpandey, C.V.; Bunshah, R.F. Thin Solid Films 1985, 126, 307.
11. Constant, G.; Feurer, R. J. Less Common Metals 1981, 82, 113.
12. Bender, B.A.; Rice, R.W.; Spann, J.R. Ceram. Eng. Sci. Proc. 1985, 6, 1171.
13. Paciorek, K.J.L.; Harris, D.H.; Kratzer, R.H. J. Polym. Sci. Polym. Chem. 1986, 24, 173.
14. Lappert, M.F. In Developments in Inorganic Polymer Chemistry; Lappert, M.F. Ed., Elsevier Publ. Co., New York, 1962.
15. Gaines, D.F.; Borlin, J. In Boron Hydride Chemistry; Muetterties, E.L. Ed., Academic Press, NY 1975.
16. Steinberg, H.; Brotherton, R.J. In Organoboron Chemistry; Vol. 2, J. Wiley, NY 1966.
17. Wagner, R.I. U.S. Pat. 3,288,726 Nov. 29, 1966; Chem. Abstr. 1977, 66 38349w.
18. Wagner, R.I.; Bradford, J.L. Inorg. Chem. 1962, 1, 99.
19. Meller, A.; Füllgrabe, H.J. Z. Naturforsch. 1978, 33b, 156.
20. Gerrand, W.; Mooney, E.F.: Pratt, D.E. J. Appl. Chem. 1963, 13, 127.
21. Aubrey, D.W.; Lappert, M.F. J. Chem. Soc. 1959, 2927.
22. Gerrand, W.; Hudson, H.R.; Mooney, E.F. J. Chem. Soc. 1962, 113.
23. Paciorek, K.J.L.; Kratzer, R.H.; Harris, D.H.; Smythe, M.E.; Kimble, P.F.U.S. Pat 4,581,468 Apr. 8, 1986; Chem Abstr. 1986, 105, 80546g.
24. Paciorek, K.J.L.; Kratzer, R.H.; Harris, D.H.; Smythe, M.E. Polym. Preps. 1984, 25, 15.
25. Taniguchi, I.; Koichi, H.; Takayoshi, M. Jpn. Kokai 76 53,000. Chem Abstr. 1976, 85, 96588.
26. Nöth, H. Z. Naturforschg. 1961, 16b, 618.
27. Narula, C.K.; Paine, R.T.; Schaeffer, R. manuscript in preparation.
28. Brame, E.G. Margrave, J.L.; Meloche, V.W. J. Inorg. Nucl. Chem. 1957, 5, 48.
29. Pease, R.S. Acta Cryst. 1952, 5, 356.
30. Biscoe, J.; Warren, B.E. J. Appl. Phys. 1942, 13, 364; Thomas, J.; Weston, N.E.; O'Connor, T.E. J. Am. Chem. Soc. 1963, 84, 4619; Economy, J.; Anderson, R. Inorg. Chem. 1966, 5, 989; Matsuda, T.; Uno, N.; Nakae, H.; Hirai, T. J. Mat. Sci. 1986, 21, 649.

RECEIVED September 1, 1987

Chapter 31

Boron–Nitrogen Polymer Precursors

S. Yvette Shaw, Donn A. DuBois, and Robert H. Neilson [1]

Department of Chemistry, Texas Christian University, Fort Worth, TX 76129

The synthesis of tractable, linear B-N polymers $(RBNR)_n$ is usually prevented by the ready formation of the extremely stable trimeric borazene ring system $(RBNR)_3$. In an attempt to circumvent this problem, we have prepared a series of new mono- and diborylamines, $\overline{EN(CH_2)_3N(E')B}(Ph)$ [E = H, E' = $SiMe_3$ (**2**), $B(NMe_2)_2$ (**3**), $B(Ph)NMe_2$, (**4**); E = $SiMe_3$, E' = $SiMe_3$ (**5**), $B(NMe_2)_2$ (**6**), $B(Ph)NMe_2$ (**7**); **8**: E = E' = $B(NMe_2)_2$; **9**: E = E' = $SiMe_2H$], by lithiation of the 1,3,2-diazaboracyclohexane ring system (**1**: E = E' = H) followed by quenching with electrophiles. The crystal and molecular structure of **8** has been determined by single crystal X-ray diffraction. Further chemistry of these compounds has been explored by reactions involving their pendent groups. For example, the bis(trimethylsilyl) compound **5** reacts with $PhBCl_2$ *via* Si-N bond cleavage to afford **11** [E = $SiMe_3$, E' = B(Ph)Cl]. Also, the bis(dimethylamino)boryl derivative **8** undergoes transamination reactions with 1,3-diaminopropane to give compounds containing two (**12**) or three (**13**) BN_2C_3 rings linked together. Preliminary thermolysis studies show that some of the diborylamines (e.g., **3**, **4**, and **11**) are potentially useful precursors to boron-nitrogen oligomers/polymers.

There has been considerable interest in recent years in the preparation of ceramic materials via pyrolysis chemistry of preceramic polymers. In particular, many important advances, most of which are described elsewhere in this Symposium volume, have been made in the synthesis of silicon carbide (SiC) and silicon nitride (Si_3N_4) from appropriate polymeric materials. As pointed out by Wynne and Rice [1], however, polymeric routes to boron nitride (BN), another important ceramic material, have been largely unexplored. This stems mainly from a lack of well characterized polymers containing boron and nitrogen. In the 1950s and 1960s, the syntheses of numerous amine- and aminoboranes were reported along with many unsuccessful attempts to polymerize such reagents. [2] The high thermal stability of the cyclic trimers [i.e., borazenes, $(RBNR)_3$] is generally cited as the reason for the failure of the B-N "monomers" to polymerize. In fact, to our knowledge, all of the more recent attempts to prepare BN polymer precursors involve linking these borazene rings together to form cyclomatrix or at least cyclolinear polymers. [3,4] Within this context, our main objective is to prepare and characterize tractable, high molecular weight, *linear* polymers containing the boron-nitrogen backbone. Aside from their potential as preceramic materials, the *linear* B-N polymers are expected to exhibit interesting and potentially useful properties themselves. For example, since they are the exact isoelectronic analogues of polyacetylene, these polymers should have similar electrical conductivity and possibly better physical/mechanical properties.

[1]Correspondence should be addressed to this author.

0097–6156/88/0360–0385$06.00/0

In order to overcome the problem of borazene ring formation, two different approaches, both involving diborylamines as condensation monomers, are being studied in our laboratory. The incorporation of a linear N-B-N-B unit along with other structural features should prevent these systems from condensing to the 6-membered borazene rings upon thermolysis. In the first method, acyclic diborylamines containing both the Si-N and the B-X groups (**I**) are the starting compounds. A few such species have recently been prepared and characterized. [5] In this case, the bulky t-Bu group on boron serves both to stabilize the primary aminoborane precursors to **I** and to help inhibit [6] formation of the borazene thermolysis products.

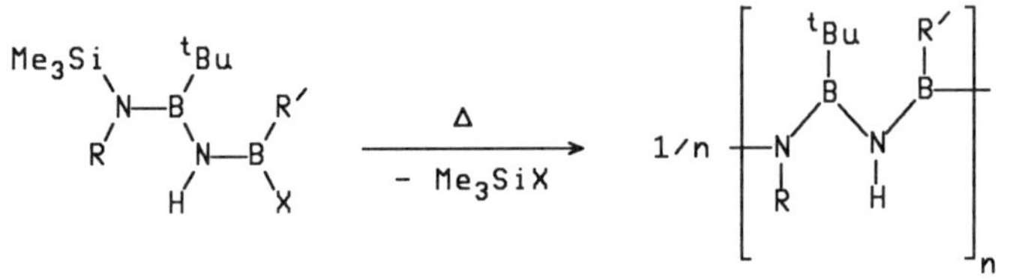

$$Me_3Si(R)N{-}B(^tBu){-}N(H){-}B(R')X \xrightarrow[-\ Me_3SiX]{\Delta} 1/n\ \left[N(R){-}B(^tBu){-}N(H){-}B(R') \right]_n \qquad (1)$$

I: R, R′, = Me, Et, ^{t}Bu, Ph, etc.

X = Cl, NMe_2, OCH_2CF_3

The second approach involves linking the nitrogen atoms of the N-B-N-B backbone through bridging alkylene units by use of the 1,3,2-diazaboracycloalkane ring systems (**II**) (eq 2). The bridges are intended to provide structural rigidity in order to prevent the boron-nitrogen backbone from condensing to the cyclic trimer. The alkylene moieties should also enhance the solubility of the polymeric products in organic solvents, thereby facilitating their characterization.

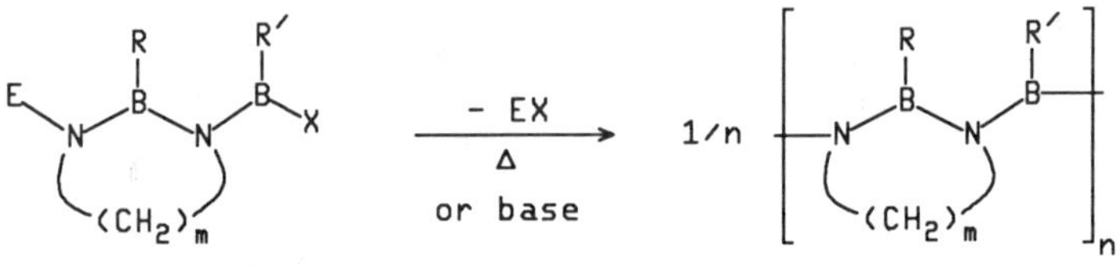

$$E{-}N{-}B(R){-}N{-}B(R')X\ [N,N\ \text{bridged by}\ (CH_2)_m] \xrightarrow[\Delta \text{ or base}]{-\ EX} 1/n\ \left[N{-}B(R){-}N{-}B(R')\ [N,N\ \text{bridged by}\ (CH_2)_m] \right]_n \qquad (2)$$

II: R, R′, X = as in eq 1

E = H, $SiMe_3$

m = 2, 3, 4

We report here some of our recent results in the second of these areas. Specifically, the synthesis, structure, reactivity, and preliminary thermolysis studies of several new mono- and diborylamines based on the 2-phenyl-1,3,2-diazaboracyclohexane ring system (**1**) are described.

Results and Discussion

Synthesis. The 2-phenyl-1,3,2-diazaboracyclohexane ring system (**1**) was selected as the starting material in this study because: (1) it is prepared easily and in high yield by a three step synthesis from BCl_3; (2) the N-H bonds are potential sites for deprotonation and substitution reactions; and (3) the trimethylene bridge enhances the rigidity of the N-B-N backbone which should prevent cyclization upon thermolysis of an appropriate precursor.

The compounds described herein were prepared by three methods. The first route involves deprotonation/substitution at the N-H sites of **1**, the second consists of a cleavage reaction of an Si-N derivative of **1** with $PhBCl_2$, and the third route is a transamination reaction between a bis(dimethylamino)boryl derivative of **1** and an aliphatic diamine. In the first approach, compound **1** is deprotonated by treatment with one equivalent of n-BuLi. Quenching of the resulting anion with various electrophiles produces the monosubstituted products **2-4** (eq 3).

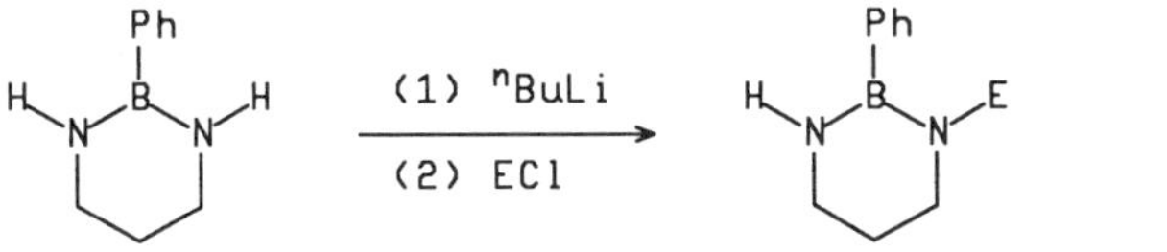

(1) ^{n}BuLi, (2) ECl (3)

1

2: E = $SiMe_3$

3: E = $B(NMe_2)_2$

4: E = $B(Ph)NMe_2$

These new derivatives were isolated in good yields (60-94%) as high boiling liquids and were fully characterized by NMR spectroscopy (^{1}H, ^{13}C, and ^{11}B) and elemental analysis. The proton NMR of the starting material **1** shows a well-resolved multiplet and quintet for the trimethylene bridge. Upon monosubstitution, however, three complex multiplets are observed, indicative of the unsymmetrical structures of these derivatives. Also, the nonequivalence of the N-C carbon atoms is clearly apparent in the ^{13}C NMR spectra of **2-4**.

Further substitution of the diazaboracyclohexane ring system may be accomplished by the same reaction sequence, starting with compounds **2-4**, to afford the disubstituted products **5-8** (eq 4). The yields of these reactions ranged from 70 to 82%. The symmetrically

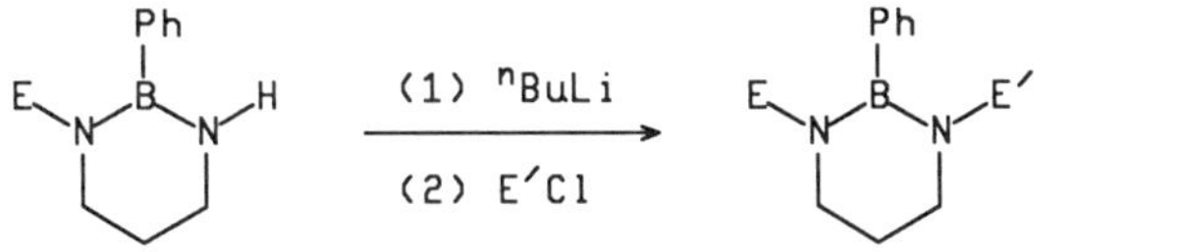

(1) ^{n}BuLi, (2) E′Cl (4)

2-4

5: E = E′ = $SiMe_3$

6: E = $SiMe_3$, E′ = $B(NMe_2)_2$

7: E = $SiMe_3$, E′ = $B(Ph)NMe_2$

8: E = E′ = $B(NMe_2)_2$

substituted compounds (**5** and **8**), like the starting material **1**, exist as white crystalline solids while the mixed silyl/boryl derivatives **6** and **7** are clear liquids. Like their mono substituted analogues (**2** and **3**), the disilyl (**5**) and diboryl (**8**) derivatives are especially useful for further derivative chemistry. Furthermore, their ^1H NMR spectra are quite simple, consisting of a triplet/quintet pattern for the trimethylene bridge. Interestingly, both of the -B(Ph)NMe_2 derivatives (**4** and **7**) exhibit nonequivalence of their N-methyl groups in the ^1H and ^{13}C NMR spectra, indicating that there are substantial barriers to rotation about the exocyclic B-N bonds in these systems. This observation is corroborated by the solid state structure of the crystalline compound **8** as described below.

As shown in equation 5, compound **1** may also be treated with two equivalents of n-BuLi and disubstituted *in one step* by addition of HMe_2SiCl. Several attempts have been made to prepare the mono substituted (with the HMe_2Si group) product but mixtures of mono and disubstituted derivatives were obtained and could not be separated by fractional distillation.

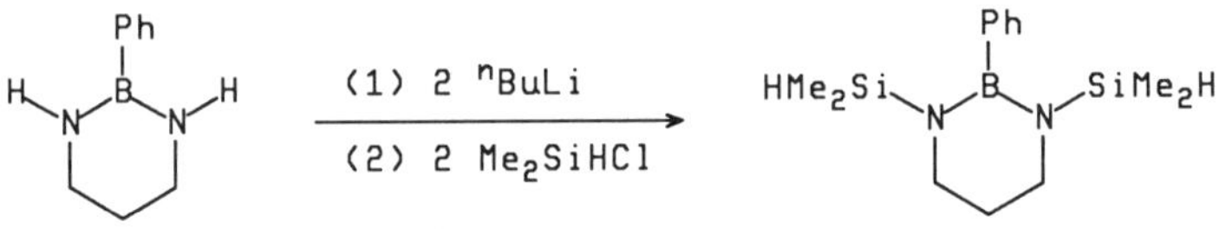

Ph
H–N–B–N–H
1

(1) 2 ^nBuLi
(2) 2 Me_2SiHCl

Ph
HMe_2Si–N–B–N–$SiMe_2H$
9

(5)

Other methods for derivatizing the diazaboracyclohexane ring system (**1**) are based on the reactivity of some of the pendent groups that have already been attached by the deprotonation/substitution method just described. For example, the mono and bis(trimethylsilyl) compounds, **2** and **5**, undergo a facile silyl cleavage reaction with $PhBCl_2$ at 0°C in CH_2Cl_2 (eq 6) to form the reactive chloroborane derivatives **10** and **11**. Unsuccessful attempts were made to isolate each of these intermediates. Upon solvent

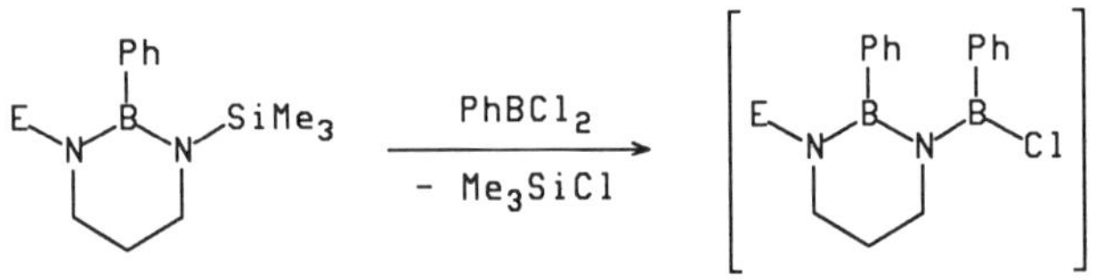

Ph
E–N–B–N–$SiMe_3$

$PhBCl_2$
- Me_3SiCl

[Ph Ph
E–N–B–N–B–Cl]

(6)

2: E = H
5: E = $SiMe_3$

10: E = H
11: E = $SiMe_3$

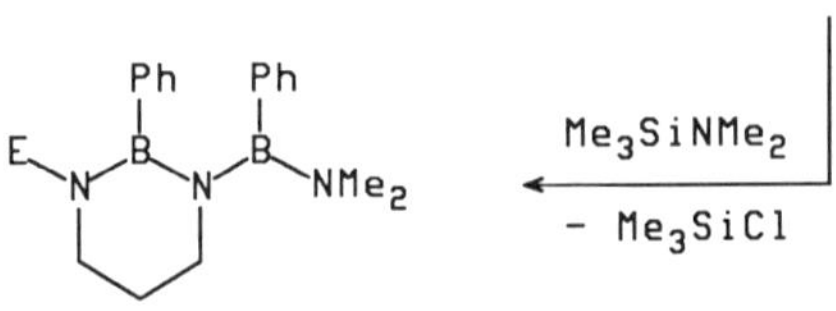

Ph Ph
E–N–B–N–B–NMe_2

Me_3SiNMe_2
- Me_3SiCl

(7)

4: E = H
7: E = $SiMe_3$

removal, **10** decomposed into an uncharacterizable solid residue, probably via HCl elimination since compounds containing both N-H and B-Cl bonds are inherently unstable. The N-silyl compound **11** was isolated in crude form as a yellow gum whose ^{1}H and ^{13}C NMR spectra are consistent with the proposed structure. The ^{11}B NMR spectrum, which is usually broad and featureless for compounds containing two or more boron atoms (e.g., **3**, **4**, **6**, and **7**), in this case consists of two distinct signals at 36.7 and 45.4 ppm. The nature of these intermediates was substantiated by another silyl cleavage reaction with Me_3SiNMe_2 to afford **4** and **7** (eq 7), which had been previously synthesized as shown above.

The third method that we have used to prepare these diazaboracyclohexane systems is the transamination reaction of 1,3-diaminopropane with the bis(dimethylamino)boryl compound **8** (eq 8). The solid products **12** and **13**, which contain, respectively, two or three of the BN_2C_3 heterocycles linked by B-N bonds, could not be distilled (or crystallized), but they were thermally stable to at least 200°C. After minor impurities were removed from **12** by vacuum distillation and from **13** by washing with hexane, the structures were confirmed by ^{1}H, ^{13}C, and ^{11}B NMR spectroscopy.

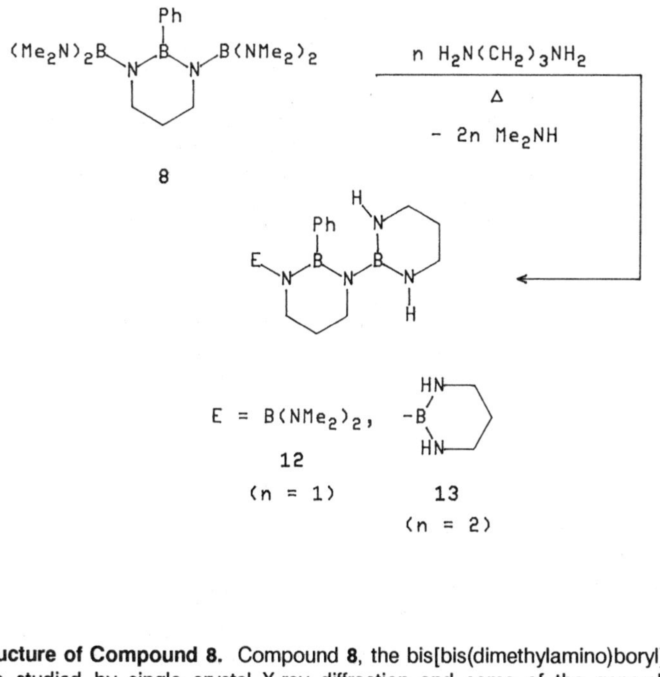

(8)

Structure of Compound 8. Compound **8**, the bis[bis(dimethylamino)boryl] derivative of **1**, was studied by single crystal X-ray diffraction and some of the general features of the molecular structure are reported here. This structure is significant since it contains a linear backbone of six B-N bonds and, as such, may serve as a close model for the structure of a linear high polymer. The diazaboracyclohexane ring is only slightly puckered with the phenyl group rotated out of the plane of the ring by 42.1°. A closer inspection of the BN skeleton (see figure below) shows that the backbone is essentially planar as expected. The atoms in the plane defined by B(10)-B(7) are coplanar with minor deviations at N(6) and B(10). The B-N bond lengths within this skeleton are characteristic of other aminoboranes and agree well with those of 1,8,10,9-triazaboradecalin. [8] However, the B-N bonds branching from the ring [i.e., N(6)-B(10) and N(2)-B(7)] are longer by 0.02-0.05 Å. Apparently, there is less dative π interaction within these bonds due to the rotation of the $B(NMe_2)_2$ moieties out of the plane of the

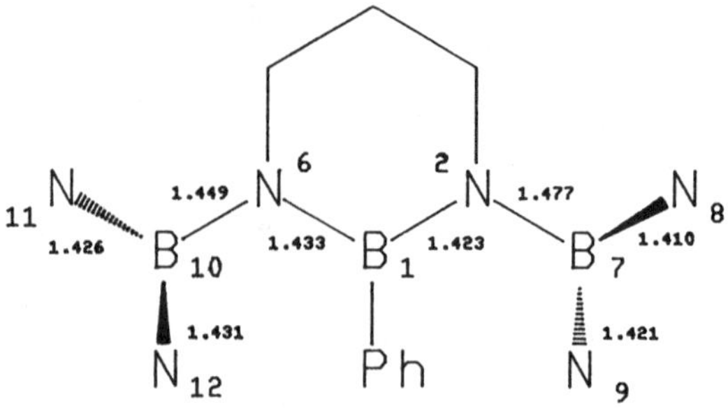

BN_2C_3 ring by approximately 57°. Furthermore, the bond length between N(2)-B(7) is 0.03 Å longer than that between N(6)-B(10) probably because of the twisted orientation of the phenyl substituent.

Preliminary Thermolysis Studies. A cursory survey of the thermal decomposition reactions of some of these potential B-N polymer precursors has been carried out. In several cases (e.g., the silylated diborylamines **6** and **7**), the compounds were thermally stable for long periods at 200°C. Others, such as the N-H substituted diborylamines **3** and **4**, show promise as condensation monomers. When heated in sealed glass ampoules at 200°C for *ca.* one week, neat samples of compound **3** decomposed *via* elimination of Me_2NH (*ca.* 50% of theory based on eq 2) to yield thick oily residues. These materials were soluble in organic solvents including THF and, by size exclusion chromatography [9], were found to have molecular weights of *ca.* 5,500, which correspond to an average of about 25 repeat units (i.e., 50 B-N bonds) in the chain. In other experiments, the Si-N/B-Cl derivative **11** decomposed readily to give an orange solid residue (molecular weight ≈ 1,000) whose NMR spectra (^1H and ^{13}C) are consistent with the type of condensation polymer shown in equation 2 (R = R' = Ph; m = 3). Thermal gravimetric analysis of this material (in argon) showed an onset of decomposition near 200°C, and beyond 550°C the residual weight was *ca.* 14% of the original value. Although this corresponds roughly to the amount of boron nitride that should be produced from this type of preceramic "polymer", the residue was a black solid of undetermined composition.

Conclusion

Thus far, this work has demonstrated three major polymer-related aspects of the chemistry of the 1,3,2-diazaboracyclohexane ring system. (1) A wide variety of mono- and disubstituted derivatives, including several novel diborylamines, can be easily prepared. (2) The structures of these small molecules, especially that of compound **8** with its rather long acyclic B-N skeleton, may be good models for the linear B-N polymers. (3) Based on some preliminary thermolysis studies, it appears that some of these derivatives will be useful as condensation monomers for the eventual synthesis of a high molecular weight B-N polymer. Our current efforts toward this objective include more detailed, systematic studies of variations in monomer structure, leaving groups, and thermolysis conditions.

Acknowledgments

The authors thank the Office of Naval Research, the Robert A. Welch Foundation, and the TCU Research Fund for financial support of this research. The X-ray diffraction study of compound **8** was performed by W. H. Watson of this department.

References and Notes

1. Wynne, K. J.; Rice, R. W. *Ann. Rev. Mater. Sci.* **1984**, *14*, 297.
2. Atkinson, I. B.; Currell, B. R. *Inorg. Macromol. Rev.* **1971**, *1*, 203.
3. Paciorek, K. J. L.; Harris, D. H.; Kratzer, R. H. *J. Polym. Sci., Polym. Chem. Ed.* **1986**, *24*, 173.
4. Narula, C. K.; Paine, R. T. ; Schaeffer, R. *Polym. Prepr., (Am. Chem. Soc. Div. Polym. Chem.)* **1987**, *28*, 454.
5. Li, B.-L.; Neilson, R. H. *Inorg. Chem.* **1986**, *25*, 361.
6. Paetzold, P.; von Plotho, C. *Chem. Ber.* **1982**, *115*, 2819 (and references cited therein).
7. Niedenzu, K.; Fritz, P.; Dawson, J. W. *Inorg. Chem.* **1964**, *8*, 1077.
8. Bullen, G. L.; Clark, N. H. *J. Chem. Soc.* **1969**, 404.
9. Molecular weights determined by SEC must be considered as only approximate values since they were measured relative to polystyrene standards. SEC conditions: μStyragel columns (500, 10^4, and 10^5 Å); THF mobile phase; 30°C; flow rate of 1.5 mL/min.

RECEIVED September 9, 1987

Chapter 32

Boron Nitride and Its Precursors

K. J. L. Paciorek [1], W. Krone-Schmidt [1], D. H. Harris [1], R. H. Kratzer [1], and Kenneth J. Wynne [2]

[1]Ultrasystems Defense and Space, Inc., Irvine, CA 92714
[2]Office of Naval Research, Arlington, VA 22217

Stepwise condensation of a series of borazines was investigated. The compounds studied were B-trichloro-N-triphenylborazine, B-triamino-N-triphenylborazine, B-trianilinoborazine, B-tris[di(trimethylsilyl)amino]-borazine, B-trichloro-N-tris(trimethylsilyl)borazine, and B-triamino-N-tris(trimethylsilyl)borazine. The formation of preceramic polymers occurs by ring opening mechanism which leads to singly and doubly joined borazine rings. This process requires the presence of a combined total of six protons on the ring and exocyclic nitrogens. Compounds where this arrangement is absent, B-trichloro-N-triphenylborazine, B-tris[di(trimethylsilyl)amino]borazine, and B-trichloro-N-tris(trimethylsilyl)borazine, failed to undergo condensation to any significant degree. Materials obtained from the phenyl-substituted borazines gave essentially nonprocessible polymers on pyrolysis; from B-triamino-N-tris(trimethylsilyl)borazine, preceramic polymers amenable to fiber manufacture were obtained. Investigation of potential linear precursor to boron-nitrogen polymers resulted in the synthesis of μ-imido-bis[bis(trimethylsilyl)-aminotrimethylsilylamino]borane (I) and μ-oxo-bis-[bis(trimethylsilyl)aminotrimethylsilylamino]borane (II). The crystal structures of the two compounds were essentially identical with the exception of the bridging atoms and their immediate environments. The same applied to mass spectral breakdown patterns wherein the major fragments differed by one amu, with the exception of the ion 332^+ which is characteristic to the oxygen-bridged compound.

Refractories such as boron nitride, silicon nitride, silicon carbide, and boron carbide are of great importance for the production or protection of systems which can be operated in very high

0097-6156/88/0360-0392$06.00/0

temperature environments. The applicability of these materials as coatings, fibers, as well as bulk items depends on the ability to produce a readily processible polymer, which can be formed into the desired shape of the final item prior to exhaustive pyrolysis and transformation into a ceramic. The technology to provide such precursors for silicon carbide, silicon nitride, and graphite, in particular the latter, has been developed and although work is proceeding on improvements, the premises are relatively well established (1). The situation in the case of boron nitride is entirely different. The extensive work performed in the sixties was directed at thermally stable polymers. The products were poorly characterized and the literature data are conflicting (2). More recently, claims have been made to the synthesis of BN preceramic polymers amenable to fiber production using the pyrolysis of B-triamino-N-triphenylborazine (3). Unfortunately, this work could not be reproduced (4,5).

Boron nitride, in view of its unique properties, namely absence of electrical conductivity, oxidation resistance, optical transparency, and high neutron capture cross-section for special applications, offers advantages over other ceramics. Thus, for the past several years the group at Ultrasystems has been actively involved in investigating routes leading to preceramic BN polymers using both cyclic and linear starting materials. Some of the findings generated will be discussed.

Results and Discussion

Borazine synthesis. The ease of borazine ring condensation would be expected to depend on the nature of the substituents. To investigate these aspects, a series of compounds were synthesized, namely B-trichloro-N-triphenylborazine, B-triamino-N-triphenylborazine, B-trianilinoborazine, B-tris[di(trimethylsilyl)amino]borazine, B-trichloro-N-tris(trimethylsilyl)borazine, and B-triamino-N-tris-(trimethylsilyl)borazine.

B-trichloro-N-triphenylborazine, mp 290-292°C, was obtained in 86% yield following the procedure of Groszos and Stafiej (6). This material was then transformed into B-triamino-N-triphenylborazine in 67% yield using the method of Toeniskoetter and Hall (7). B-trianilinoborazine and the novel B-tris[di(trimethylsilyl)-amino]borazine (mp, 131.5-132°C; characterized by GC/MS, molecular ion 558 amu, and elemental analysis) were synthesized in 76 and 71% yields, respectively, by interaction of aniline and hexamethyldisilazane with chloroborazine in the presence of triethylamine.

The synthesis of B-trichloro-N-tris(trimethylsilyl)borazine was much more complicated than that of the other borazines. The overall scheme is given below:

$$BCl_3 + Et_3N \longrightarrow BCl_3 \cdot NEt_3$$

$$BCl_3 \cdot NEt_3 \xrightarrow{HN(SiMe_3)_2 + NEt_3} Cl_2BN(SiMe_3)_2$$

$$Cl_2BN(SiMe_3)_2 \xrightarrow{\text{Pyrolysis}} [ClBN(SiMe_3)]_3$$

(cyclic borazine: ring of alternating B(Cl) and N(SiMe$_3$) units — Cl–B, Me$_3$SiN, NSiMe$_3$, ClB, BCl, N–SiMe$_3$)

Boron trichloride-triethylamine adduct, mp 89-91°C, was obtained in 80% yield following essentially the procedure of Ohashi, et al. (8). The product was washed with methanol, but not crystallized from it. It was found that crystallization from dilute ethanol reported by Gerrard, Lappert, and Pearce (9) caused extensive degradation . Bis(trimethylsilyl)aminodichloroborane was prepared following the procedure of Wells and Collins (10).

The pyrolysis of bis(trimethylsilyl)aminodichloroborane to B-trichloro-N-tris(trimethylsilyl)borazine, contrary to literature (11), did not take place in boiling xylene. Temperatures above 150°C were necessary for trimethylchlorosilane elimination. The highest yield of the relatively pure product was around 20%. The transformation of the B-trichloro-N-tris(trimethylsilyl)borazine to B-triamino-N-tris(trimethylsilyl)borazine proceeded readily using liquid ammonia.

Borazine Pyrolysis Studies. In a simplistic manner, one can visualize the formation of boron nitride via stepwise eliminations between rings, i.e.,

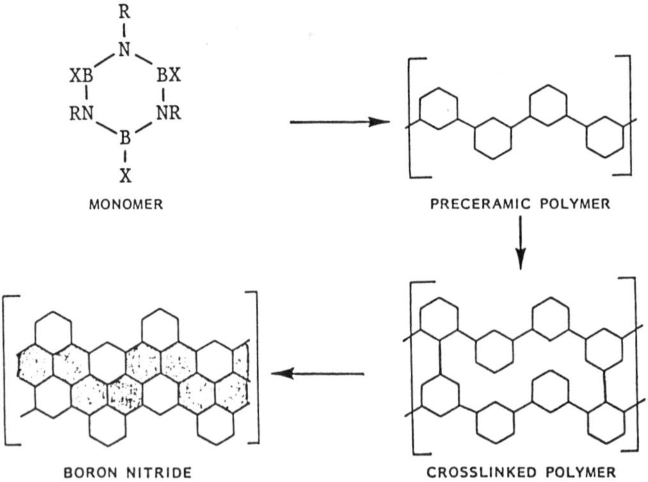

MONOMER

PRECERAMIC POLYMER

BORON NITRIDE

CROSSLINKED POLYMER

giving initially the "linear" preceramic polymer, then the lightly crosslinked structure B, followed by a partial B-N system, C, and finally pure boron nitride. The pyrolysis techniques utilized were fully described previously (4) and thus will not be reiterated here. The potential borazine candidates can be broadly divided into two classes of materials, one where the substituent on the ring boron is nitrogen-free, e.g., a halogen, and the second where the substituent is an amino moiety. With respect to the suitability of any given borazine, the important aspects are the ease of formation of the leaving molecule and its volatility. In this respect, B-trichloro-N-tris(trimethylsilyl)borazine would seem an ideal candidate. Unfortunately, pyrolysis at 260°C liberated only 6.1% of the expected trimethylchlorosilane; 75% of the starting material was recovered. Heating at 360°C for 19 h resulted in 56.0% yield of trimethylchlorosilane and a glassy residue. Further heating at

360 °C for 72 h afforded only an additional 7.3% of the expected trimethylchlorosilane. This result shows clearly that this type of borazine does not lead to a desirable preceramic polymer, inasmuch as at 260 °C the condensation process proceeds only to a limited degree, with the starting material being largely recovered and the polymeric product formed being both insoluble and infusible. B-trichloro-N-triphenylborazine would not be expected to undergo condensation readily inasmuch as the formation of chlorobenzene is not a favored process. The experimental data supported this postulation; essentially no degradation occurred at 300 °C.

B-triamino-N-triphenylborazine, and in particular its isomer, B-trianilinoborazine, the latter either alone or in co-reaction with B-triamino-N-triphenylborazine, gave fusible, <200 °C, and organic solvent soluble preceramic polymers after exposure to temperatures below 250 °C (4,5). However, in particular using B-triamino-N-triphenylborazine alone, the process was irreproducible; nonfusible products were obtained. In this system, carbon retention was invariably observed after exposure to 1000 °C, contrary to the Japanese claims (3).

The thermal degradation of B-trianilinoborazine, when conducted in ammonia even below 215 °C, produced 87.7% of the available aniline. This was brought up to 92.3% by further pyrolysis at 275-299 °C which compares with the maximum of 53% observed under the same heating regime in nitrogen atmosphere (4). Thus, it is obvious that the presence of ammonia facilitates the elimination process. However, since 31% of the aniline evolved was replaced by ammonia, based on the weight of the residue, the degree of condensation was not as high as would appear from the aniline collected. Due to the materials' insolubility in organic solvents, the molecular weights could not be determined.

From the investigations of the phenylaminoborazines it was established that the condensation does occur via ring opening (4), postulated earlier by Toeniskoetter and Hall (7), followed by elimination and bridge formation. Isomerization is inherently associated with this mechanism, as illustrated in Scheme 1. For this process to take place, it is necessary that the boron substituent is either a NH_2 or NHR moiety. Namely, the ring opening mechanism requires the presence of a combined total of six protons on the ring and exocyclic nitrogens. In agreement with the above, B-tris[di(trimethylsilyl)amino]borazine, which contains only three hydrogens, was recovered unchanged after exposure to 410 °C for 89 h. One could visualize here the intermolecular condensation to occur, in the absence of ring opening, with elimination of hexamethyldisilazane, e.g.,

$N(SiMe_3)_2$
B
HN NH
2
$(Me_3Si)_2NB$ $BN(SiMe_3)_2$
N
H

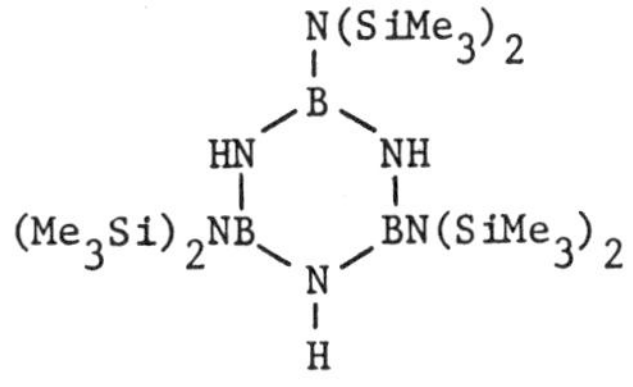

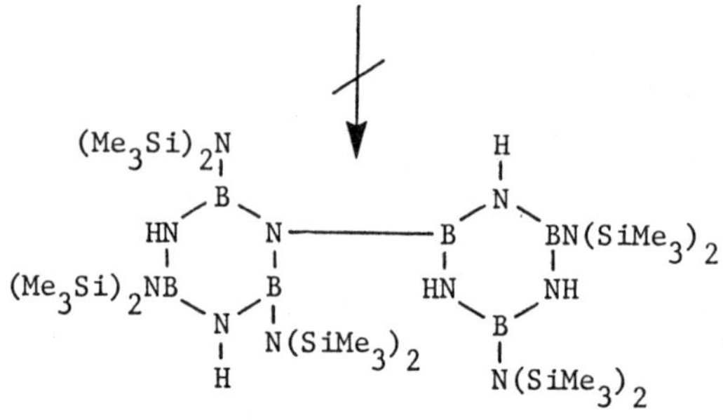

This apparently is not the case. On prolonged exposure in ammonia at 250-260°C some liberation of hexamethyldisilazane took place, but only to a very low degree.

The behavior of B-triamino-N-tris(trimethylsilyl)borazine was in good agreement with the ring opening mechanism. This compound was much more reactive than its phenyl analogues; thus, the pure monomer could not be isolated. The product obtained from the interaction of ammonia and B-trichloro-N-tris(trimethylsilyl)-borazine consisted of a 1:1 mixture of the monomer and a singly-bridged dimer, x=1.

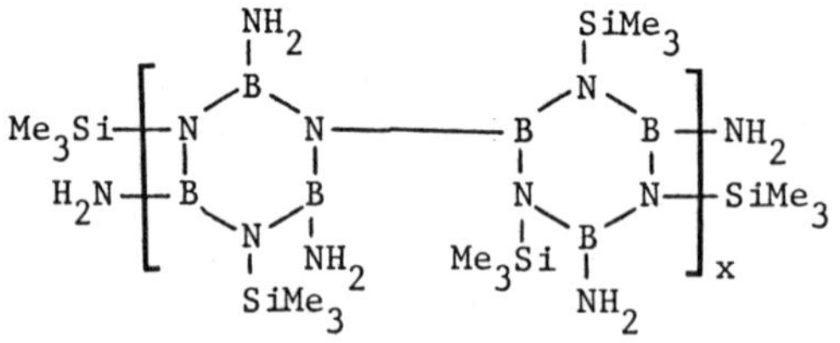

Both the monomer and dimer were composed most likely of isomers with some protons residing on ring nitrogens and with the $NHSiMe_3$ moiety replacing the amino group on some of the boron ring atoms in accordance with the Scheme 1. This assumption is supported by the liquid nature of the product and its infrared spectrum which exhibited two broad bands centered at 3430 and 3530 cm^{-1}. The infrared spectrum of pure B-triamino-N-tris(trimethylsilyl)borazine would be expected to be similar to that of B-triamino-N-tris(tri-phenyl)borazine where three sharp bands at 3430, 3505, and 3530 were observed (4). In the latter case, once the condensation process was initiated, the sharp bands disappeared (4).

To remove volatile impurities, the material was heated in vacuo at 135 ^{o}C. This treatment resulted in further condensation. Thus, the distillation residue was composed of a mixture of doubly-bridged dimers and tetramers, x = 1 and 2 as determined from the molecular weight, boron and nitrogen analyses and the volatile condensibles produced.

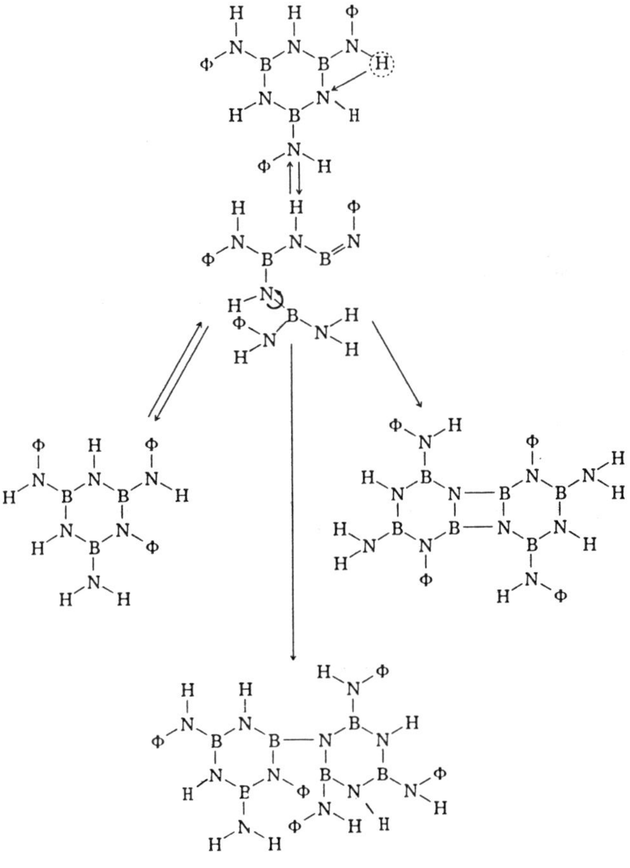

Scheme 1.

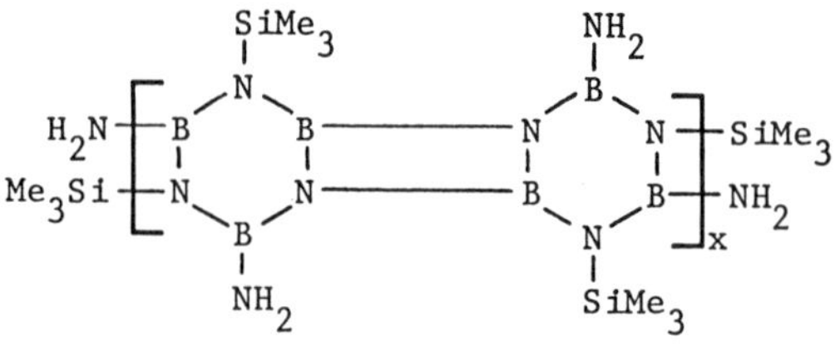

$$\left[\begin{array}{c} H_2N-B(N(SiMe_3))B(NH_2)N(SiMe_3)-N \ldots \end{array} \right]_x$$

The condensation process apparently involves elimination of trimethylsilylamine, $(CH_3)_3SiNH_2$, which, being unstable, disproportionates into hexamethyldisilazane and ammonia. In the condensible volatiles collected, these materials were present in a 1:1 ratio supporting the postulated process. Further thermolysis at 200°C (52 h), followed by 4 h at 250-260°C, resulted in a mixture of doubly-bridged tetramers and octamers, x=2 and 4. At this stage, 53% of the potential leaving groups were liberated. From the tetramer/octamer mixture, fibers with 10-20 μ diameter could be melt drawn as shown in Figures 1 and 2. Gradual heat treatment in ammonia, from 65-950°C, resulted in white, pure boron nitride fibers. The fibers exhibited no weight loss under thermogravimetric conditions, both in air and in nitrogen up to 1000°C (12).

Potential Non-Cyclic Precursors of Preceramic Polymers. Boranes such as bis(trimethylsilyl)aminotrimethylsilylaminochloroboranes can be viewed as monomers for preceramic polymer and, ultimately, boron nitride production. Intermolecular dehydrohalogenation of this borane would be thus expected to yield either the dimer or the polymeric system.

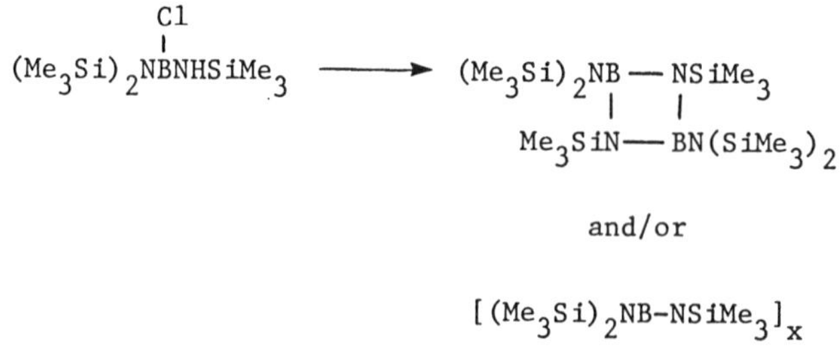

$$(Me_3Si)_2NB(Cl)NHSiMe_3 \longrightarrow \begin{array}{l} (Me_3Si)_2NB - NSiMe_3 \\ \quad\quad\quad\quad | \quad\quad | \\ \quad Me_3SiN - BN(SiMe_3)_2 \end{array}$$

and/or

$$[(Me_3Si)_2NB\text{-}NSiMe_3]_x$$

Bis(trimethylsilyl)aminotrimethylsilylaminochloroborane was obtained in a 70% yield following the procedure of Wells and Collins (10). No triethylamine hydrochloride was formed and the starting material was recovered unchanged after prolonged refluxing with a five-fold excess of specially dried triethylamine. However, when the triethylamine employed was not so rigorously dried and 10- to 20-fold excess was used, an oxygen-bridged compound,

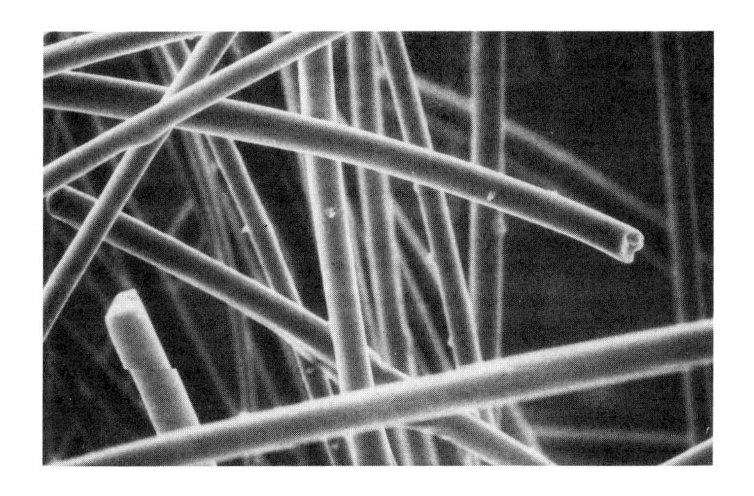

Figure 1. Melt spun BN precursor fibers, magnification 500x.

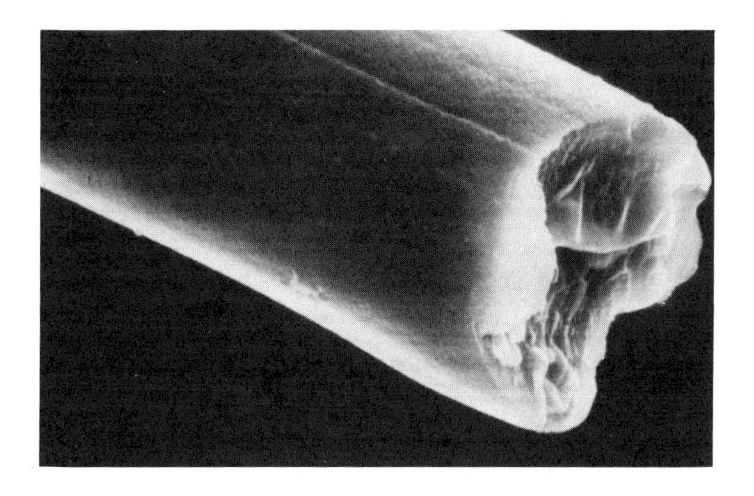

Figure 2. Melt spun BN precursor fibers, magnification 5,000x.

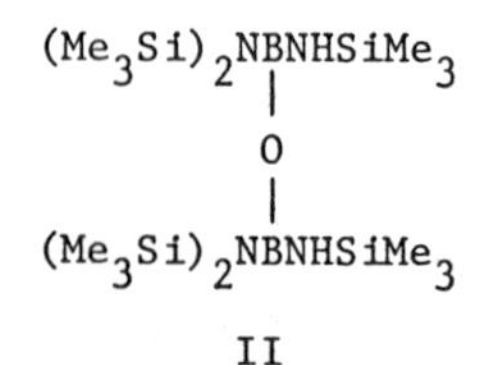

II

was obtained in a 35% yield. Conducting the reaction in benzene using the required quantity of water in the presence of triethylamine gave 55% yield of triethylamine hydrochloride, but only a 7% yield of the oxygen-bridged product. The surprising finding was the relatively high yield (∿18%) of material which is either $(Me_3SiNH)_2BN(SiMe_3)_2$ or $[(Me_3Si)_2N]_2BNH_2$ (the differentiation could not be made from the mass spectral fragmentation pattern), together with hexamethyldisilazane and aminobis(trimethylsilyl)aminotrimethylsilylaminoborane, $(Me_3Si)_2$-$NB(NH_2)NHSiMe_3$, or its isomer, $B(NHSiMe_3)_3$. These products were also observed, but in low concentrations, in the reaction where the high yield of the oxygen-bridged compound was realized. The major difference between these reactions was the relative ratio of triethylamine and the presence of benzene. Based on these results, it is clear that the process involving hydrogen chloride elimination and oxygen substitution,

$$2\ (Me_3Si)_2N\underset{|}{\overset{Cl}{B}}NHSiMe_3 + H_2O \xrightarrow{NEt_3} \begin{matrix}(Me_3Si)_2NBNHSiMe_3\\ |\\ O\\ |\\ (Me_3Si)_2NBNHSiMe_3\end{matrix} + 2\ Et_3N\cdot HCl$$

is not the only one occurring. Apparently, a more extensive hydrolysis of the chloroborane takes place concurrently, i.e.,

$$(Me_3Si)_2N\overset{Cl}{\overset{|}{B}}NHSiMe_3 \xrightarrow[NEt_3]{H_2O} (Me_3Si)_2NH + B(OH)_3 + [H_2NSiMe_3] + Et_3N\cdot HCl$$

Inasmuch as trimethylsilylamine is unstable, it will decompose into ammonia and hexamethyldisilazane. The ammonia thus formed could be visualized to form the amino compound $(Me_3Si)_2NB(NH_2)$-$NHSiMe_3$. Since $(Me_3Si)_2NB(NHSiMe_3)_2$ was prepared by Wells and Collins (13) via interaction of the chloroborane with hexamethyldisilazane, its production here could follow the same path.

The nitrogen-bridged analogue, $[(Me_3Si)_2NBNHSiMe_3]_2NH$, Compound I, was formed in a 57% yield from bis(trimethylsilyl)aminotrimethylsilylaminochloroborane and aminobis(trimethylsilyl)aminotrimethylsilylaminoborane in the presence of triethylamine by heating at 100°C over an 18 h period, i.e.

$$(Me_3Si)_2N\overset{\overset{NH_2}{|}}{B}NHSiMe_3 + (Me_3Si)_2N\overset{\overset{Cl}{|}}{B}NHSiMe_3 \xrightarrow{Et_3N} \begin{array}{c}(Me_3Si)_2NBNHSiMe_3\\ |\\ NH\\ |\\ (Me_3Si)_2NBNHSiMe_3\end{array} + Et_3N\cdot HCl$$

I

The nitrogen-bridged compound exhibited physical and spectral characteristics very similar to those of the oxygen-bridged compound. This is illustrated by the close melting points (169-171°C and 158-160°C, respectively) and essentially identical IR spectra (the spectrum of μ-imido-bis[bis(trimethylsilyl)aminotrimethylsilylamino]borane is given in Figure 3). Thermal behavior of I and II is also similar as shown by DSC traces, where in the oxygen-bridged compound two strong endotherms were observed at 70 and 75°C and in the nitrogen-bridged material at 85 and 105°C. No phase transition responsible for these endotherms could be visually observed on heating. Both materials, on remelting in air, exhibited the same melting points as those obtained under inert conditions pointing to their thermal, oxidative, and hydrolytic stabilities. Investigation of the condensation reactions of Compound I are currently in progress.

The mass spectral breakdown patterns for the two bridged compounds were very similar, with the exception that the peaks were one mass unit higher in the case of the oxygen-bridged material. Both compounds produced strong molecular ions at 533^+ (I) and 534^+ (II). The processes responsible for several of the major fragments could be identified from the metastables. In the case of μ-imido-bis[bis(trimethylsilyl)aminotrimethylsilylamino]borane, these are:

533^+ (M) ⟶ 518^+ + 15 [Me] m* 503.4

518^+ ⟶ 429^+ + 89 [H_2NSiMe_3] m* 355.3

429^+ ⟶ 341^+ + 88 [$HNSiMe_3$] m* 271.0

429^+ ⟶ 316^+ + 113 [$BNHNSiMe_3$] m* 232.8

In view of the absence of the additional proton, the equivalent of the 341^+ ion, namely the 342^+ ion, would not be expected to be formed to any significant extent in the case of the oxygen-bridged compound. The fragmentation pattern confirms it and no metastable for this process was found. In the oxygen-bridged compound, metastables were observed at m/e 504.5, 357.5, 213, and 82.5. The processes responsible have been identified as:

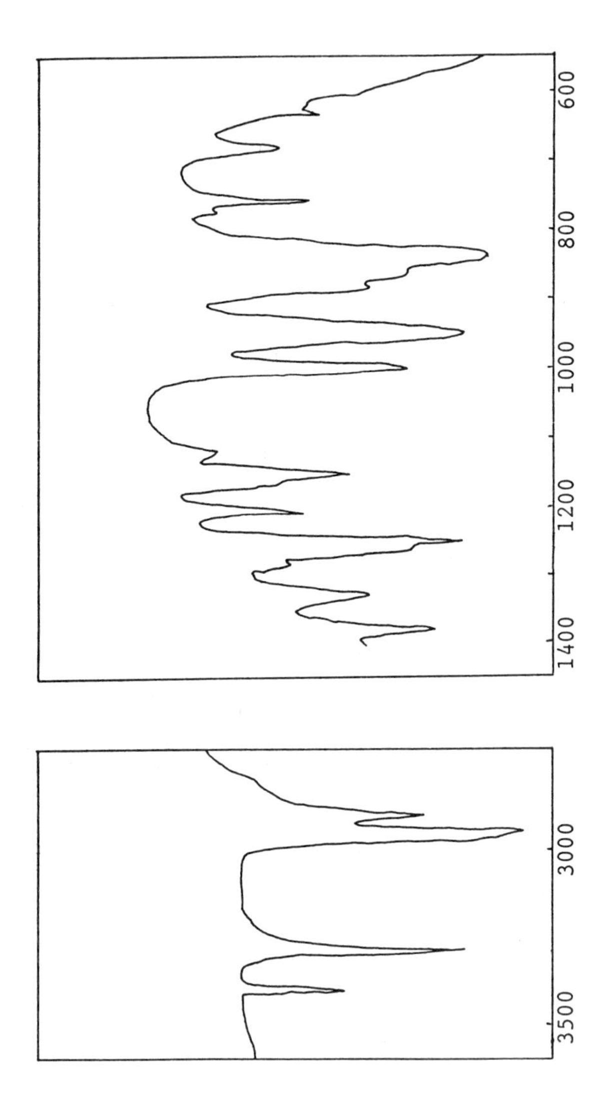

Figure 3. Infrared spectrum of μ-imido-bis[bis(trimethylsilyl)-aminotrimethylsilylamino]borane (I).

534^+ (M) ⟶ 519^+ + 15 [Me]		m* 504.4
519^+ ⟶ 430^+ + 89 [H_2NSiMe_3]		m* 356.3
519^+ ⟶ 332^+ + 187 [$B(NHSiMe_3)_2$]		m* 212.4
259^+ ⟶ 146^+ + 113		m* 82.3

The 332^+ ion is specific to the oxygen-bridged product. It is speculated that it is formed by the following path from the 519^+ ion:

519^+ ⟶ 332^+

In order to obtain insight into the nature of bonding in catenated BN species, the crystal and molecular structures of Compounds I and II were obtained. Both compounds crystallize isomorphously in the monoclinic space group $P2_1/n$. For I, $\underline{a}$ = 12.876(2) Å, $\underline{b}$ = 14.828(3) Å, $\underline{c}$ = 18.628(4) Å; β = 97.07(2)°; V = 3529(1) $Å^3$; Z = 4, and the molecular weight based on a calculated density of 1.005 g/cm^3 was 533.85, in excellent agreement with the molecular ion observed in the mass spectrum (533 m/e). Refinement led to R_f = 4.1% and R_{wf} = 4.2% for 5093 independent reflections. The structure of I, given in Figure 4, confirms the presence of a central N_2BNBN_2 framework that is essentially planar with a B(1)-N(1)-B(2) angle of 135.9(2)°. The remaining B-N-B angles are close to 120°. The B-N bond distances reflect a competition for π-electron density. The shortest B-N bonds, (B(2)-N(5) 1.405(3) Å and B((1)-N(3) 1.423(3) Å, are to nitrogens bonded to one silicon, while the longest B-N bonds B(2)-N(4) 1.485(3) Å and B(1)-N(2) 1.487(3) Å are to nitrogens bonded to two silicons. For II, $\underline{a}$ = 12.826(3) Å, $\underline{b}$ = 14.822(4) Å, and $\underline{c}$ = 18.557(4) Å; β = 95.98(2)°; V = 3509(1) $Å^3$; Z = 4, F.W. = 534.8 amu, and the calculated density is 1.012 g/cm^3. Refinement led to R_f = 4.3% and R_{wf} = 4.0% for 4458 independent reflections. Structure II, given in Figure 5, differs from I only in the existence of a central B-O-B unit wherein the boron oxygen distances [B(1)-O(1) 1.380(3) and B(2)-O(1) 1.389(3)] are somewhat shorter than the corresponding B-N distances for I [B(1)-N(1) 1.428(3) and B(2)-N(1) 1.446(3)].

Acknowledgment

Support of this research from the Strategic Defense Sciences Office through Contract N00014-85-C-0659 from the Office of Naval Research and the crystal structure analysis by C. S. Day of Crystalytics Co. are gratefully acknowledged.

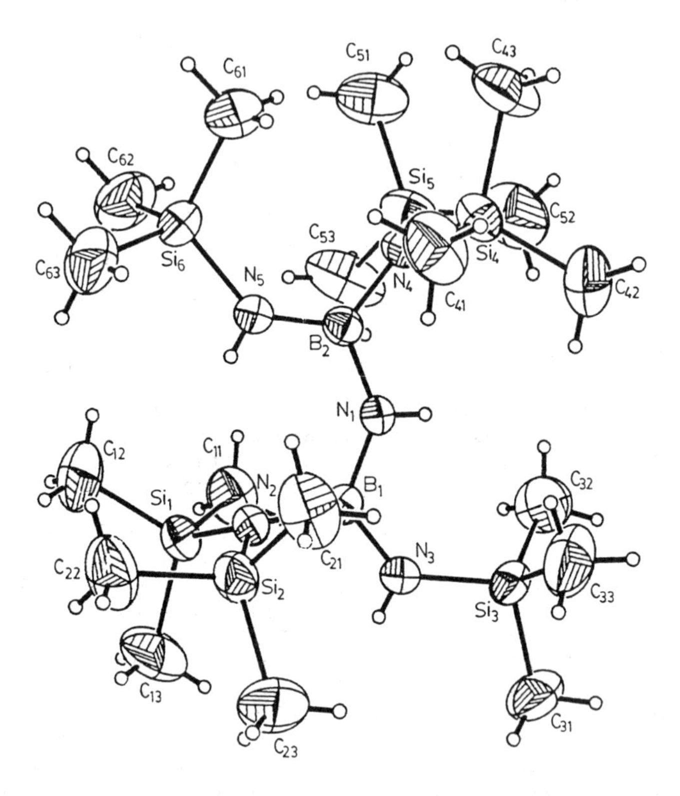

Figure 4. A perspective drawing of $C_{18}H_{57}N_5Si_6B_2$ with nonhydrogen atoms represented by thermal vibration ellipsoids drawn to encompass 50% of their electron density; hydrogen atoms are represented by arbitrarily small spheres which are in no way representative of their true thermal motion.

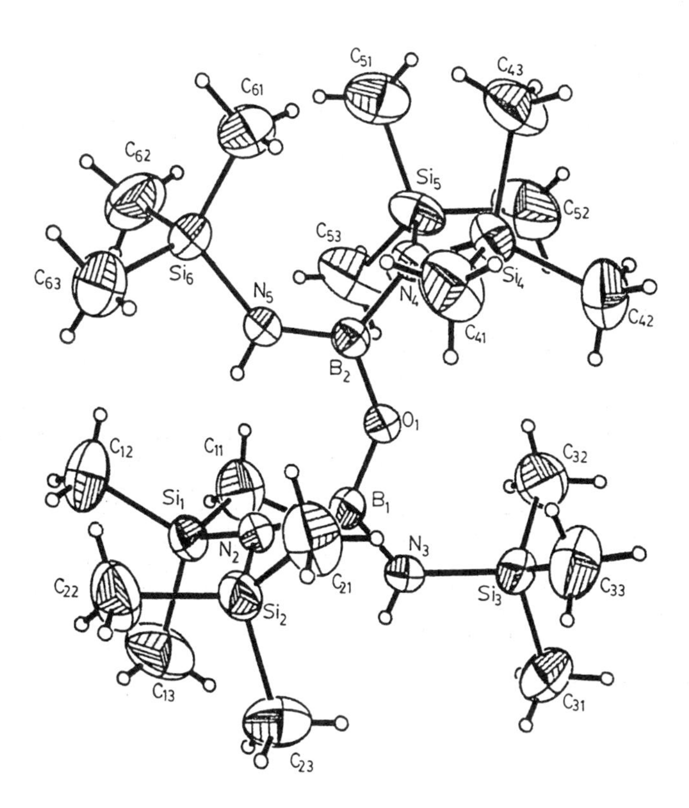

Figure 5. A perspective drawing of $C_{18}H_{56}N_4OSi_6B_2$ with nonhydrogen atoms represented by thermal vibration ellipsoids drawn to encompass 50% of their electron density; hydrogen atoms are represented by arbitrarily small spheres for purposes of clarity.

Literature Cited

1. Wynne, K. J., Rice, R. W. Ann. Rev. Mater. Sci. 1984, 14, 297.
2. Steinberg, H., Brotherton, R. J. Organoboron Chemistry, Vol. 2, pp. 175-434; J. Wiley, New York, 1966.
3. Taniguchi, I., Harada, K., Maeda, T. Jpn. Kokai 76 53,000, 11 May 1976.
4. Paciorek, K. J. L., Harris, D. H., Kratzer, R. H. J. Polym. Sci. 1986, 24, 173.
5. Paciorek, K. J. L., Kratzer, R. H., Harris, D. H., Smythe, M. E. Polymer Preprints 1984, 25, 15.
6. Groszos, S. J., Stafiej, S. F. J. Am. Chem. Soc. 1958, 80, 1357.
7. Toeniskoetter, R. H., Hall F. R. Inorg. Chem. 1963, 2, 29.
8. Ohashi, O., Kurita, Y., Totani, T., Watanabe, H., Nakagawa, T., Kubo, M. Bull. Chem. Soc. Japan 1962, 35, 1317.
9. Gerrard, W., Lappert, M. F., Pearce, C. A. J. Chem. Soc. 1957, 381.
10. Wells, R. L., Collins, A. L. Inorg. Chem. 1966, 5, 1327.
11. Geymayer, P., Rochow, E. G. Monatsh. 1966, 97, 429.
12. Without additional heat treatment above 1000°C (annealing), the fibers are amorphous based on transmission electron microscopy (TEM) with islands of microfibrils or microcrystallites. Selected area electron diffraction (SAED) of the microcrystallites indicates that they could be BN, B_2O_3, or graphite. Auger electron spectroscopy showed the fibers to be essentially free of carbon which is supported by their colorless appearance. X-ray diffraction produces a weak signal corresponding to hexagonal BN. The fact that the fibers do not change in shape when heated at 1000°C rules out the presence of significant amounts of B_2O_3, which melts at 460°C. Further characterization of the fibers in regard to composition, structure, susceptibility to "graphitization", as well as strength and density is in progress.
13. Wells, R. L., Collins, A. L. Inorg. Chem. 1968, 7, 419.

RECEIVED September 1, 1987

OTHER METAL- AND METALLOID-CONTAINING POLYMERS

Chapter 33

Electron Transport in and Electrocatalysis with Polymeric Films of Metallotetraphenylporphyrins

B. A. White[1], S. A. Raybuck[2], A. Bettelheim[3], K. Pressprich, and Royce W. Murray[4]

Kenan Laboratories of Chemistry, University of North Carolina, Chapel Hill, NC 27514

Amino-, hydroxy-, and pyrrole-substituted tetraphenylporphyrins undergo electro-oxidative polymerizations analogous to those of aniline, phenol, and pyrrole, to form electroactive poly-porphyrin films on electrode surfaces. The voltammetry of differently metallated forms of these poly-porphyrins has been studied as has the rate of electron-hopping diffusion within the polymers. The films are demonstrably permeated by axial ligands and are remarkably pinhole free. The rate constants for electron hopping in the polymers have values parallel to those of heterogeneous electron transfer rate constants for corresponding metalloporphyrin monomers at solid electrode surfaces, consistent with lack of dominance of ionic or polymer chain motions on the barrier to electron transfers within the poly-porphyrin films. The cobalt-metallated polymer is an electrocatalyst for reduction of dioxygen.

This research rests upon oxidative electrochemical polymerization (ECP) of solutions of the functionalized metallotetraphenylporphyrin monomers shown in Fig. 1. These oxidations lead to formation of thin, cross-linked polymeric films of the metallotetraphenylporphyrins on the electrode surface. The films contain from *ca.* 4 to 500 monolayer-equivalents of porphyrin sites, which are in high concentration (*ca.* 1*M*) since the polymer backbone consists solely of the porphyrins themselves as the backbone units. The polymeric films adhere to the electrode and

[1]Current address: The Upjohn Company, Kalamazoo, MI 49001
[2]Current address: Department of Chemistry, Harvard University, Cambridge, MA 02138
[3]Current address: Nuclear Research Center—Negev, Atomic Energy Commission, POB 9001 Beer Sheva, Israel
[4]Correspondence should be addressed to this author.

0097-6156/88/0360-0408$06.00/0

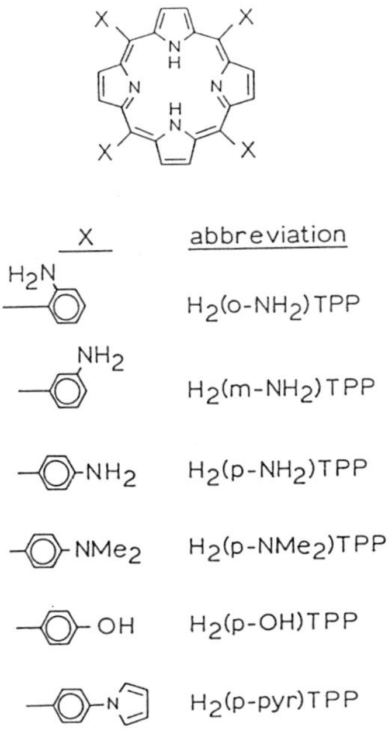

X	abbreviation
H_2N	H_2(o-NH_2)TPP
NH_2	H_2(m-NH_2)TPP
NH_2	H_2(p-NH_2)TPP
NMe_2	H_2(p-NMe_2)TPP
OH	H_2(p-OH)TPP
N	H_2(p-pyr)TPP

Fig. 1. Structures of oxidatively electropolymerizable tetraphenylporphyrins. The porphyrins can be polymerized as metallated forms, or the metal can be inserted into the polymer in some cases.

are soluble in most solvents, which allows investigation of their electrochemical reactivity as surface films (1,2), the electrocatalytic reactivity of the cobalt porphyrin as a dioxygen reduction catalyst (2,4), and the permeability of the films as membranes (5). Additionally, since the porphyrin sites are in close contact electron hopping occurs, often freely, between sites which are adjacent members of a redox couple pair. The currents due to these electron hopping (or self exchange) reactions have been investigated for a series of porphyrin redox couples (6) in order to probe the molecular factors controlling their rate.

Electrochemical Polymerization of Metallotetraphenylporphyrins

Our strategy for electrochemical polymerization (ECP) of metal complexes was based on an analogy to the ECP of small functionalized monomers (7-12) like vinyl-substituted aromatics, pyrrole, aniline, and phenol. The idea is to attach such activating groups onto the periphery of the desired metal complex (or organic moiety), and then electrochemically oxidize or reduce this functionalized metal complex to form it into a polymer. Forming a film on the electrode from this polymer depends on the electrochemically oxidized or reduced functionalized metal complex either grafting directly onto some functional group present on the electrode surface, or undergoing rapid coupling, oligomerization, nucleation, and precipation as a film before mass transport carries the activated monomer (or oligomer) away into the solution. We have had considerable previous success with this strategy for metal polypyridine complexes (13-15), for a large number of functionalized metal complexes, and it works for the compounds shown in Fig. 1, as free bases and as variously metallated forms (1,2).

Tetra(o-aminophenyl)porphyrin, $H_2(o\text{-}NH_2)TPP$, can for the purpose of electrochemical polymerization be simplistically viewed as four aniline molecules with a common porphyrin substituent, and one expects that their oxidation should form a "poly(aniline)" matrix with embedded porphyrin sites. The pattern of cyclic voltammetric oxidative ECP (1) of this functionalized metal complex is shown in Fig. 2A. The growing current-potential envelope represents accumulation of a polymer film that is electroactive and conducts electrons at the potentials needed to continuously oxidize fresh monomer that diffuses in from the bulk solution. If the film were not fully electroactive at this potential, since the film is a dense membrane barrier that prevents monomer from reaching the electrode, film growth would soon cease and the electrode would become passified. This was the case for the phenolically substituted porphyrin in Fig. 1.

Characterization of the Electropolymerized Films

Each metallotetraphenylporphyrin has its own particular current-potential growth pattern; those for other complexes besides that in Fig. 2A are displayed elsewhere (16). We have not attempted to unravel the details of the oxidative redox states involved in the film growth. That of Fig. 2A obviously is not simply

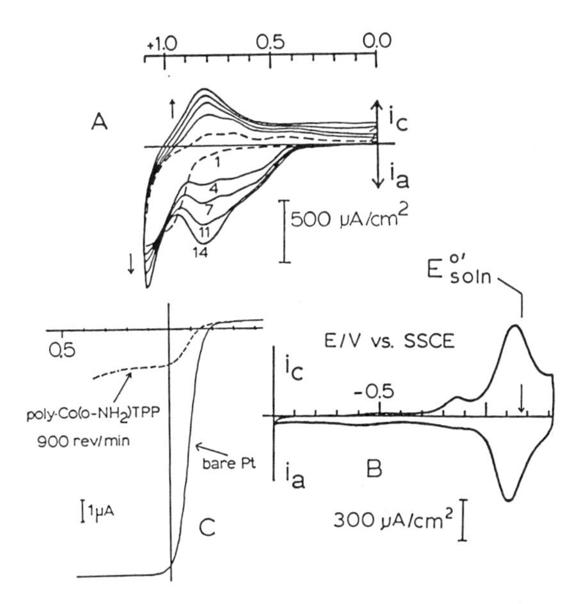

Fig. 2. Curve A: Electropolymerization of 1m$\underline{H}$ H_2(o-NH_2)TPP in 0.1$\underline{M}$ Et_4NClO_4/CH_3CN by sweeping potential at 200mV/s on Pt electrode. Numbers represent scan number. Curve B: Cyclic voltammogram of an electropolymerized film of poly-[H_2(o-NH_2)TPP] on a Pt electrode, in 0.1$\underline{M}$ Et_4NClO_4/CH_3CN at 200 mV/s. Integration of the charge under the wave shows that coverage is $3.5X10^{-9}$ mol/cm^2 of the porphyrin sites. Curve C: Rotated disk electrode voltammetry of the Os(III,II) reaction for 0.2 m$\underline{M}$ [Os(bpy)$_2$Cl$_2$] in 0.1$\underline{M}$ Et_4NClO_4/CH_3CN at a naked (———) and at a polymer coated (-----) Pt disk electrode. The polymer is poly-[Co(o-NH_2)TPP] at $2.8X10^{-9}$ mol/cm^2.

poly(aniline)-like nor in fact would we expect it to be given the overlap of the aniline and porphyrin ring oxidation potentials and the interruption of growth of long linear poly(aniline) chains expected given from the steric bulkiness of the porphyrin substituent. However, we do believe the coupling chemistry of the oxidized porphyrin involves mainly (perhaps not exclusively) short (dimer, trimer?) head-to-tail coupled aniline units. Such coupling has been supported by Raman studies of the amino-porphyrin films by Spiro and associates (who independently discovered their ECP as part of their own interest in this topic) (17). Many uncertainties on the other hand, remain about the details of the radical cation coupling reactions, and their products, that act to bind the central metallotetraphenylporphyrin units together. The problem is complicated in an analytical sense, since the extremely small quantities of polymer that are formed by ECP in a film limit the number of techniques usefully brought to bear on the interconnecting polymer structure.

The electrochemical reduction reactions of the central metallotetraphenylporphyrin moieties are, fortunately, much more straightforwardly analyzed (1,2). With few exceptions, when transferred to a fresh supporting electrolyte solution, films formed from ECP reactions like Fig. 2A exhibit electrochemical reduction waves at or very near the potentials observed for reductions of the corresponding monomers dissolved in solutions. For example, a film formed oxidatively as in Fig. 2A gives in fresh electrolyte the reductive cyclic voltammogram of Fig. 2B. The indicated formal potential $E^{o\prime}_{soln}$ of the corresponding monomer (-1.17V) in solution is very near that of the surface film (-1.13V *vs*. SSCE). That formal potentials of surface films on chemically modified electrodes are near those of their corresponding dissolved monomers (13,18) is actually a common, and quite useful, observation. In the present case, it demonstrates that the electronic structures of the porphyrin rings embedded in the polymer film are not seriously perturbed from that of the monomer.

The reduction electrochemistry of ECP porphyrin films furthermore responds to added axial ligands in the expected ways. We have tested this (2,6) for the ECP form of the iron complex of tetra(o-amino)phenyl)porphyrin by adding chloride and various nitrogeneous bases to the contacting solutions, observing the Fe(III/II) wave shift to expected potentials based on the monomer behavior in solution. This is additional evidence that the essential porphyrin structure is preserved during the oxidation of the monomer and its incorporation into a polymeric film.

Another point of importance about the film structure is the degree to which it can be permeated by various ions and molecules. It is of course essential that supporting electrolyte ions be able to penetrate the film, else the electrical double layer at the electrode/polymer interface could not be charged to potentials that drive electron transfers between the polymer and the electrode. The electroneutrality requirements of porphyrin sites as their electrical charges are changed by oxidation or reduction also could not be satisfied without electrolyte permeation. With the possible exception of the phenolic structure in Fig. 1, this level of permeability seems to be met by the ECP porphyrins.

Another level of permeability is that of axial bases, such as pyridine. As noted above, we have seen the porphyrin electrochemistry to respond to addition of axial bases to the contacting solutions; this implicitly demonstrates that bases such as pyridine are able to penetrate the films to the axial porphyrin sites in its interior.

If the permeating molecule is sufficiently bulky, of course, the degree to which it can partition into the polymer (dissolve in, with partition coefficient $P = C_{out}/C_{in}$), and its mobility within the polymer (diffusion coefficient within, D_{pol}), may become small. This is in fact observed, as illustrated by the rotated disk electrode voltammogram of a solution of the bulky permeant $[Os(bpy)_2Cl_2]$ at a ECP film of cobalt tetra(o-aminophenyl)porphyrin in Fig. 2C. In order to be oxidized, this complex must permeate through the film to the electrode/polymer interface. The difference between the limiting currents observed for the osmium(II/III) oxidation at the naked (---) as opposed to the polymer-coated (- - -) Pt disk represents the transport impedance of the polymer film as a membrane barrier to the bulky complex. Analysis of how the limiting currents depend on the electrode rotation rate (5) gives a permeability PD_{pol} = $4X10^{-9} cm^2/s$. This is over 1,000 X smaller than the diffusion coefficient for this osmium complex diffusing freely in the acetonitrile solvent (obtained from the limiting current at the naked Pt electrode), and the observed PD_{pol} corresponds to a very low permeability of the polymer film to such bulky permeants.

Given our ability to prepare sandwich electrode microstructures from these polymeric porphyrins in the face of the known (19) sensitivity of these microstructures to failure via pinhole defects in the film, we believe that the above film permeation (Fig. 2C) occurs mainly by a membrane-dissolving mechanism rather than by transport of the osmium complex through an assortment of holes and defects in the polymer film. That is, we believe the "holes" in the films that permeants pass through in order to reach the electrode/polymer interface, are approximately molecular sized and mainly represent the interstices between randomly coupled porphyrin units, rather than holes that are molecularly large spaces as might be caused by intervention by particulate matter or by mechanical or thermally-induced stress tears in the film. Better evidence for this assertion could come from a comparison of PD_{pol} among permeants of a graded molecular size, as done previously for a different ECP polymer film (20). The results of current experiments (5) are promising in this regard.

Electron Transport in ECP Metallotetraphenylporphyrin Films

The ability, in a voltammetric experiment like Fig. 2B, to electrochemically charge the equivalent of many monolayers of porphyrin sites even though the individual sites are relatively immobile within the polymer, implies the existance of an efficient site-site electron transport mechanism. The electron transport mechanism for films like the polymeric porphyrins was in fact identified some time ago, as a hopping of electrons (21) between the localized electronic sites that the redox monomer sites in the polymer represent. The process can be represented, for a polymeric iron tetra(o-aminophenyl)porphyrin chloride, of first a poly-[Fe(III)TPP(X)] site next to the electrode/polymer interface becoming reduced by an electron transfer from the electrode:

$$\text{Electrode}(e^-) + \text{poly-[Fe(III)TPP(X)]} \longrightarrow \text{Electrode} + \text{poly-[Fe(II)TPP]} + X^-$$

The reduced poly-[Fe(II)TPP] porphyrin site now finds itself next to a fresh poly-[Fe(III)TPP(X)] site one polymer lattice unit further into the polymer. An electron hopping - or self exchange - reaction can then ensue, repeatedly, in successive layers and sites:

$$\text{poly-[Fe(II)TPP]} + \text{poly-[Fe(III)TPP(Cl)]} \longrightarrow \text{poly-[Fe(III)TPP(Cl)]} + \text{poly-[Fe(II)TPP]}$$

The reaction amounts to a vectorically directed current in the sense of occurring down a concentration gradient of reduced poly-[Fe(II)TPP] sites emanating from the reducing electrode/polymer interface. The magnitude of the current clearly conveys information about the rate of the poly-[Fe(III)TPP(X)] - poly-[Fe(II)TPP] self exchange reaction.

The above mechanistic aspect of electron transport in electroactive polymer films has been an active and chemically rich research topic (13-18) in polymer coated electrodes. We have called (19) the process "redox conduction", since it is a non-ohmic form of electrical conductivity that is intrinsically different from that in metals or semiconductors. Some of the special characteristics of redox conductivity are non-linear current-voltage relations and a narrow band of conductivity centered around electrode potentials that yield the necessary mixture of oxidized and reduced states of the redox sites in the polymer (mixed valent form). Electron hopping in redox conductivity is obviously also peculiar to polymers whose sites comprise spatially localized electronic states.

In the present case, the electron hopping chemistry in the polymeric porphyrins is an especially rich topic because we can manipulate the axial coordination of the porphyrin, to learn how electron self exchange rates respond to axial coordination, and because we can compare the self exchange rates of the different redox couples of a given metallotetraphenylporphyrin polymer. To measure these chemical effects, and avoid potentially competing kinetic phenomena associated with mobilities of the electro-neutrality-required counterions in the polymers, we chose a steady state measurement technique based on the sandwich electrode microstructure (19).

Figure 3 illustrates the construction of a sandwich electrode, which is a delicately assembled yet powerful experimental tool. The porosity of the Au electrode evaporated onto the thin polymer film (thickness d) allows inspection of the film's voltammetry, and permits the ion and solvent access for adjustment of its average oxidation state to a mixed valent form.

Figure 4 shows the application (6) of potentials to the Pt and Au electrodes of the sandwich (vs. a reference electrode elsewhere in the contacting electrolyte solution) so that they span the $E^{o\prime}$ of the poly-[Co(II/I)TPP] couple (Fig. 4B). There is a consequent redistribution of the concentrations of the sites in the two oxidation states to achieve the steady state linear gradients shown in the inset. Figure 4C represents surface profilometry of a different film sample in order to determine the film thickness from that the actual porphyrin site concentration (0.85M). The flow of self exchange-supported current is experimentally parameterized by applying Fick's first law to the concentration-distance diagram in Fig. 4B:

$$i_{lim} = nFAD_e C_{Co}/d$$

where C_{Co}/d represents the gradient of oxidized or reduced porphyrin sites and the mobility of the electron is given as the "electron diffusion coefficient" D_e (13,18). D_e is proportional to the apparent self exchange rate as formulated by Saveant (22)

$$D_e = 10^3 k_{ex}^{app} C_{Co} (\Lambda)^2 \quad (2)$$

where C_{Co} is the total concentration (mol/cm^3) of porphyrin sites separated by the average distance Λ. Application of these two equations to limiting currents like that of Fig. 4B, for films polymeric porphyrins containing a variety of metals, mixed valent states, and added ligands gives results plotted in the important Fig. 5.

We can say several things about the apparent self exchange rate constants. First, they vary by nearly 10^3X, even though the electrolyte ions and the mode of ECP coupling are the same throughout. (Tetra(o-aminophenyl)porphyrin, in differently metallated forms, was used in each case.) This is commanding evidence that the primary response of k_{ex}^{app} is to the particular chemistry of the porphyrin that pertains to the porphyrin's metal, its oxidation state, and axial ligand. The alternative result would have been a k_{ex}^{app} that varied little with porphyrin metal because the electron hopping was energetically controlled by the rate of movements of electrolyte ion or polymeric linkages within the polymers.

Secondly, Fig. 5 shows that the polymeric rate constants parallel values of heterogeneous rate constants that have been observed for the electrochemical reactions of solutions of the corresponding dissolved porphyrin monomers. (The slope of the line is 0.5). This re-emphasizes what was said above, that measurements of electron hopping in polymers can give rate constants that are meaningful in the context of the metalloporphyrin's intrinsic electron transfer chemistry.

Thirdly, one experiment to measure D_e was carried out with the sandwich of electrodes and polymeric porphyrin film exposed to a gas phase rather than a fluid liquid bathing environment. This

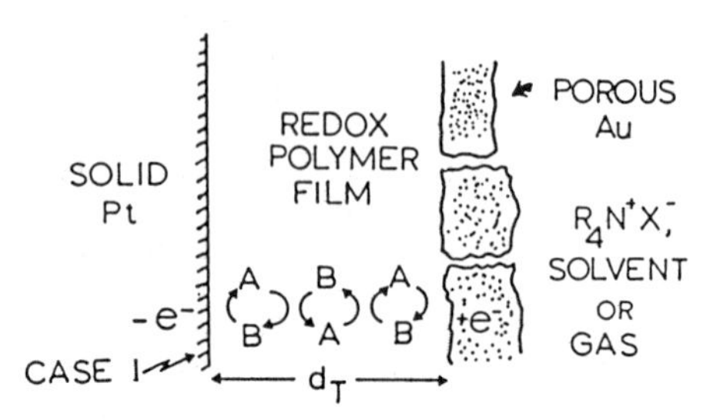

Fig. 3. Diagram (cross-section) of a sandwich electrode.

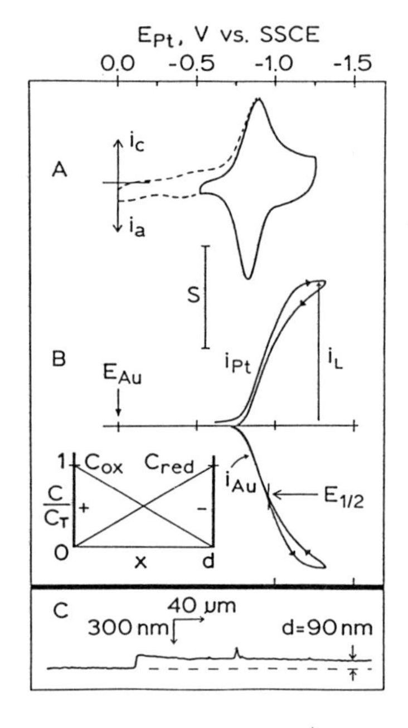

Fig. 4. Voltammograms in 0.1M Et_4NClO_4/CH_3CN; $\Gamma_T = 1.2X10^{-8}$ mol/cm^2. Curve A: Cyclic voltammetry of Pt/poly-Co(o-NH_2)TPP at 20 mV/s; S = 200μA/cm^2. Curve B: Four-electrode voltammetry of Pt/poly-Co(o-NH_4)TPP/Au sandwich electrode with E_{Au} = 0.0 V; E_{Pt} scanned negatively at 5 mV/s; S = 400μA/cm^2. Curve C: Surface profilometry of a poly-Co(o-NH_2)TPP film on SnO_2/glass; Γ_T = $7.6X10^{-9}$ mol/cm^2. (Reproduced from Ref. 6. Copyright 1987 American Chemical Society.)

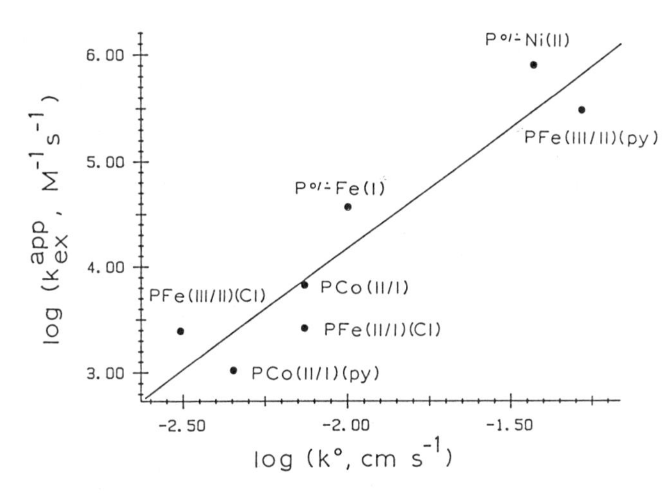

Fig. 5. Plot of apparent electron self exchange rate constants k_{ex}^{app}, derived from polymer D_e values for films containing the indicated metals, mixed valent states, and ligands, all in acetonitrile, using Equation 2, vs. literature heterogeneous electron transfer rate constants k^0 for the corresponding monomers in nitrile solvents. See Ref. 6 for details. (Reproduced from Ref. 6. Copyright 1987 American Chemical Society.)

novel result was obtained by electrolytically converting a porphyrin film to the 1:1 Fe(III/II) mixed valent form, then removing the sandwich from the electrolyte solution, drying it, exposing it to a pyridine/acetonitrile vapor bath, and scanning the voltage applied to the sandwich (7). The rate constant for electron self exchange, as measured by D_e, is by this experiment shown to be sensitive to a change of polymer solvation state, decreasing in the bathing gas by 3X as compared to immersing the film in pyridine/acetonitrile liquid. Hopefully, continuing experiments will give us insights into the molecular effects that are important to solvent-deprived electron self-exchanges.

Electrocatalysis of Dioxygen Reduction by Cobalt Porphyrins.

Dioxygen reduction electrocatalysis by metal macrocycles adsorbed on or bound to electrodes has been an important area of investigation (23) and has achieved a substantial molecular sophistication in terms of structured design of the macrocyclic catalysts (24). Since there have been few other electrochemical studies of polymeric porphyrin films, we elected to inspect the dioxygen electrocatalytic efficacy of films of electropolymerized cobalt tetraphenylporphyrins. All the films exhibited some activity, to differing extents, with films of the cobalt tetra(o-aminophenylporphyrin) being the most active (2-4). Curiously, this compound, both as a monomer in solution and as an electropolymerized film, also exhibited two electrochemical waves in the range of potentials expected for Co(III/II) reactivity. The situation is complex because the relative prominence of these two features depends on the electrolyte and in the case of the polymer, the thickness of the polymer film (3). Both waves show catalytic activity, and rotated ring disk measurements reveal that electrolysis at the potential of the first (more positive) Co(III/II) wave produces a larger portion of the two electron H_2O_2 reduction product than does electrolysis at the second wave potential. The proportions furthermore depend on the polymer film thickness; being 26 and 10% for 3.4×10^{-10} (about four porphyrin monolayers) and 1.6×10^{-9} mol/cm^2 (about 16 layers), respectively, at the potential of the first wave. Generally similar results were obtained in 1M NaOH, except that thicker polymer films in that case rather stably effect reduction of dioxygen with no detectable H_2O_2 product, i.e., a four electron pathway.

Interpreting these results on a detailed molecular basis is difficult because we have at present no direct structural data proving the nature of the split Co(III/II) voltammetry (which seems critical to the electrocatalytic efficacy). Experiments on the dissolved monomeric porphyrin, in CH_2Cl_2 solvent, reveal a strong tendency for association, especially for the tetra(o-aminophenyl)porphyrin. From this observation, we have speculated (3) that the split Co(III/II) wave may represent reactivity of non-associated (dimer?) and associated forms of the cobalt tetra(o-aminophenyl)porphyrins, and that these states play different roles in the dioxygen reduction chemistry. That dimeric cobalt porphyrins in particular can yield more efficient four electron dioxygen reduction pathways is well known (24). Our results suggest that efforts to incorporate more structurally well defined dimeric porphyrins into polymer films may be a worthwhile line of future research.

Acknowledgments

This research was supported by grants from the National Science Foundation and the Office of Naval Research. Helpful conversations with Professor J. P. Collman of Stanford University are also acknowledged.

Literature Cited

1. White, B. A.; Murray, R. W. J. Electroanal. Chem. 1985, 189, 345.
2. Bettelheim, A.; White, B. A.; Raybuck, S. A.; Murray, R. W. Inorg. Chem. 1987, 26, 1009.
3. Bettelheim, A.; White, B. A.; Murray, R. W. J. Electroanal. Chem. 1987, 217, 271.
4. Bettelheim, A.; Reed, R. A.; Hendricks, N. H.; Collman, J. P.; Murray, R. W. J. Electroanal. Chem., in press.
5. Pressprich, K., University of North Carolina, unpublished results, 1987.
6. White, B. A.; Murray R. W. J. Am. Chem. Soc. 1987, 109, 2576.
7. Faulkner, L. R.; Shaw, B. R.; Haight, G. P., Jr. J. Electroanal. Chem. 1982, 140, 147.

8. Kanazawa, K. K.; Diaz, A. F.; Geiss, R. H.; Gill, W. D., Kwak, J. F.; Logan, J. A.; Rabolt, J. F., Street, G. B. J. Chem. Soc. Chem. Commun. 1979, 854.
9. Frommer, J. E.; Chance, R. R. Encyclop. Science and Technol., 2nd Ed., Vol. 5, John Wiley and Sons, 1986, p. 462.
10. Diaz, A. F.; Logan, J. A. J. Electroanal. Chem. 1980, 111, 111.
11. Bazier, M.; Lund, H., Eds., Organic Electrochemistry, M. Dekker, N. Y., 1983.
12. Adams, R. N.; Electrochemistry at Solid Electrodes, M. Dekker, Inc., N. Y., 1969, p. 329, 363.
13. Murray, R. W.; Ann. Rev. Mats. Sci. 1984, 14, 145.
14. Calvert, J. M.; Schmehl, R. H.; Sullivan, B. P.; Facci, J. S.; Meyer, T. J.; Murray, R. W. Inorg. Chem. 1983, 22, 2151.
15. Leidner, C. R.; Sullivan, B. P.; Reed, R. A.; White, B. A.; Crimmins, M. T.; Murray, R. W.; Meyer, T. J. Inorg. Chem. 1987, 26, 882.
16. White, B. A.; Ph.D. thesis, University of North Carolina, Chapel Hill, NC, 1986.
17. Su, Y. O.; Macor, K. A.; Miller, L. A., Spiro, T. G. 191 ACS National Meeting, N. Y. City, April 1986, Div. Inorg. Chem. Abstr. 50.
18. Murray, R. W. Electroanalytical Chemistry, Vol. 13, A. J. Bard, Ed., M. Dekker, 1984.
19. Pickup, P.; Kutner, W.; Leidner, C. R.; Murray, R. W. J. Am. Chem. Soc. 1984, 106, 1991.
20. Ikeda, T.; Schmehl, R.; Denisevich, P.; Willman, K.; Murray, R. W. J. Am. Chem. Soc. 1982, 104, 2683.
21. Kaufman, F. B.; Schroeder, A. H.; Engler, E. M.; Kramer, S. R.; Chambers, J. Q. J. Am. Chem. Soc. 1980, 102, 483.
22. Andrieux, C. P.; Saveant, J. M. J. Electroanal. Chem. 1981, 123, 171.
23. Yeager, E. J. Molec. Cat. 1986, 38, 5.
24. Collman, J. P.; Denisevich, P.; Konai, Y.; Marrocco, M.; Koval, C.; Anson, F. C. J. Am. Chem. Soc. 1980, 103, 6027.

RECEIVED September 1, 1987

Chapter 34

Electronically Conducting Films of Poly(trisbipyridine)–Metal Complexes

C. Michael Elliott, J. G. Redepenning, S. J. Schmittle, and E. M. Balk

Department of Chemistry, Colorado State University, Fort Collins, CO 80523

Polymers films formed from trisbipyridineruthenium complexes and coated on electrode surfaces have been found to have interesting electrochromic and conductivity properties.

Many examples of electrodes modified with polymers containing metal trisbipyridine complexes have been reported in the last decade. Most of these materials fall into a sub-category of ionic conductors often referred to as "redox" conductors (1-2). The driving force responsible for charge transport across a redox conducting film coated on an electrode and having spatially fixed electroactive sites is the establishment of a concentration gradient within the film. This gradient is generated by the oxidation or reduction of electrochemically active sites which are in direct contact with the electrode. The charge is then propagated through the film by site-site electron exchange or so-called "electron hopping." In contrast, electronically conducting polymers, such as polypyrrole, are ohmic in nature and charge transport is in response to an electrostatic potential gradient rather than a concentration gradient (3).

We report here studies on a polymer film which is formed by the thermal polymerization of a monomeric complex tris(5,5'-bis[(3-acrylyl-1-propoxy)carbonyl]-2,2'-bipyridine)ruthenium(II) as its tosylate salt, **I** (4). Polymer films formed from **I** (poly-**I**) are insoluble in all solvents tested and possess extremely good chemical and electrochemical stability. Depending on the formal oxidation state of the ruthenium sites in poly-**I** the material can either act as a redox conductor or as an electronic (ohmic) conductor having a specific conductivity which is semiconductor-like in magnitude.

0097-6156/88/0360-0420$06.00/0

Results and Discussion

A cyclic voltammogram of a poly-I film coated on a tin oxide electrode is shown in Figure 1. One of the more interesting properties of I and polymers formed from it, as evidenced by the voltammogram in Figure 1, is the extensive array of stable oxidation states that are accessible by electrochemical reduction. As the electrode potential is varied between -0.8 and -2.2 V vs. Ag/Ag^+ (-0.303 V vs. ferrocene/ferricenium) the ruthenium sites are converted, by sequential one electron steps, from the 2+ oxidation state all the way to the formal 4- state. In the course of these reductions the ruthenium almost certainly remains as a d^6 ruthenium(II) throughout and the reducing equivalents enter the low-lying π^* orbitals of the very electron deficient, ester-substituted bipyridine ligands (5).

It was noted early in our studies of monomeric analogs of I that significant changes occurred in the visible spectrum upon conversion between oxidation states; changes which translate into vivid, visual differences in color. These differences in color are also very apparent in poly-I and thus make it of interest as an electrochromic material (6).

In addition to its unusual optical properties, poly-I is rather unique in several other respects. In the formal 2+ oxidation state the fixed sites in poly-I are positively charged. At -2.2 V, however, in the 4- formal oxidation state, the fixed sites are negatively charged. Thus by simply changing the applied potential the polymer may be converted from an anion exchanger, to a neutral polymer and finally to a cation exchanger. It is also possible to sterically block oxidation states of a particular charge-type from forming in the polymer by using a solution electrolyte having one charge-type ion that is too large to enter the polymer (7-8). Poly(1,1-dimethyl-3,5-dimethylenepiperdinium hexafluorophosphate),II, is just such an electrolyte.

Figure 2 is a cyclic voltammogram of a platinum electrode coated with a film of poly-I in an acetonitrile solution containing a normal electrolyte, $TBA^+PF_6^-$ (tetra-N-butylammonium hexafluorophosphate) and the same electrode in an acetonitrile solution of II. As can be seen the first two reductions, 2+/1+ and 1+/0, (which overlap one another) are present in both voltammograms at about the same potentials. In both of these cases charge balance within the polymer is maintained by transport of PF_6^- ions in and out of the film. The third and subsequent reductions, however, require the transport of cations into the film in order to maintain charge balance. In the case of the $TBA^+PF_6^-$ electrolyte the cations can enter the film and the reductions take place, but for the polymeric cation case the reductions are absent because the cations are to large to penetrate the poly-I film (8).

Figure 3 shows the cyclic voltammogram of an acetonitrile solution containing 9-fluorenone on a bare Pt electrode and on a Poly-I coated Pt electrode using II as the electrolyte. The only significant difference between the voltammograms is the presence of the poly-I reduction and oxidation waves for the coated electrode. We have previously determined by other means that poly-I films should not be permeable to molecules as large as 9-

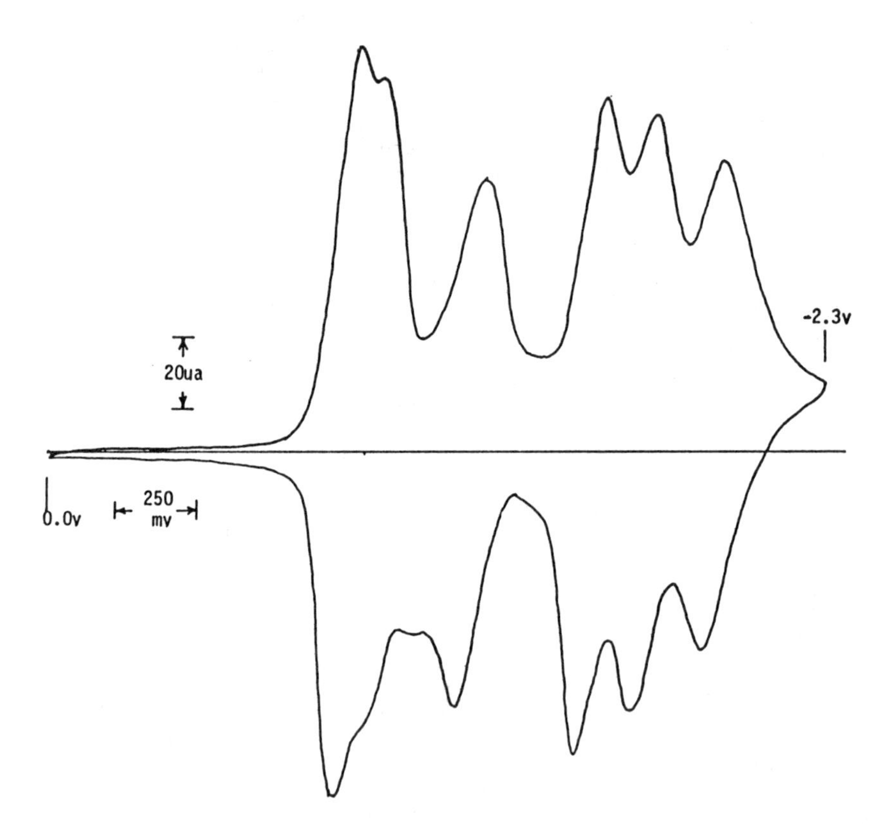

Figure 1. Cyclic voltammogram on a tin oxide electrode modified with a thin film of poly-I. A sweep rate of 50mV/s was employed in CH_3CN containing 0.1 M $TBAPF_6$. E vs. Ag^+(0.1 M $AgNO_3$ in DMSO)/Ag.

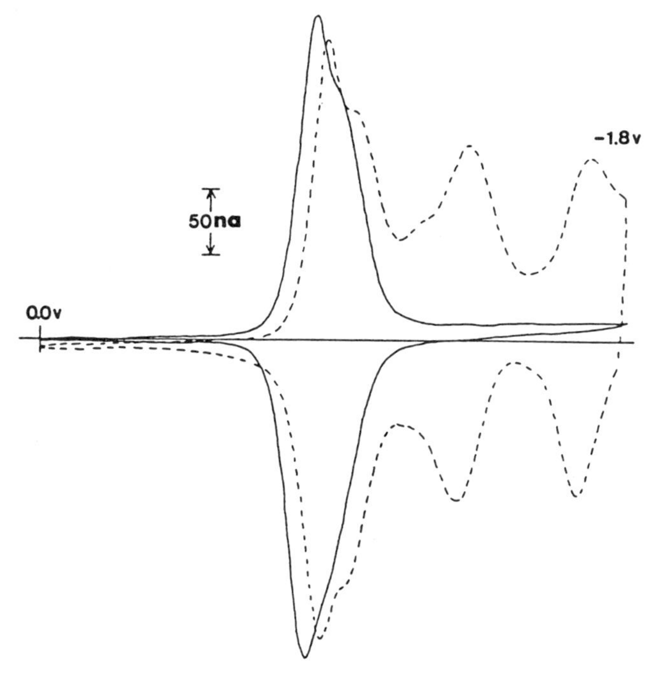

Figure 2. Cyclic voltammograms on a Pt electrode coated with poly-I: (-----), CH_3CN, 0.1 M $TBAPF_6$; (———), CH_3CN, 50 mM polycationic electrolyte, II. Scan rate = 50 mV/s. E vs. Ag^+/Ag. (Reproduced from Ref. 7. Copyright 1985 American Chemical Society.)

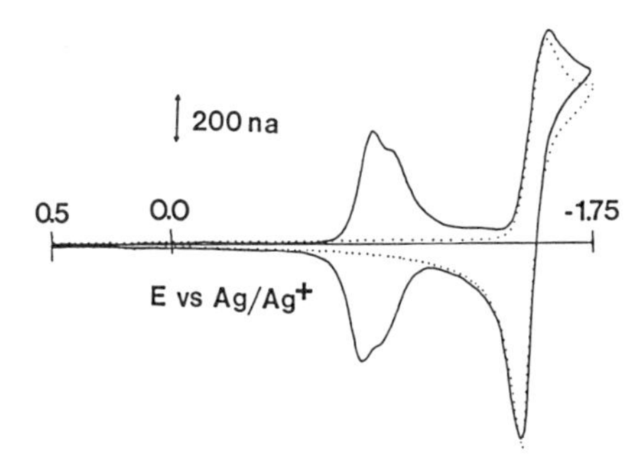

Figure 3. Cyclic voltammograms showing 9-fluorenone redox activity at a bare Pt electrode (-----) and at a Pt electrode coated with poly-I (———). Both voltammograms were run in the same CH_3CN, 50 mM II electrolyte solution and referencedto At^+/Ag. (Reproduced with permission from Ref. 8. Copyright 1986 Elsevier.)

fluorenone; therefore, Figure 3 suggests that the fluorenone must be being reduced at the polymer/solution interface. In order to produce identical voltammograms for 9-flourenone on the bare and coated electrodes under these conditions requires that the film in the formal zero-valent oxidation state have significant electronic conductivity (7). This interpretation is borne out by the results shown in Figure 4 below.

Murray, et. al. have described the construction of so-called "sandwich electrodes" in which a polymer film is sandwiched between two metallic electrodes (9). If one of the electrodes is porous, ions from solution will be able to enter and leave the polymer, thus the oxidation state of the polymer can be changed . Also, the two electrodes are electrically separated by the polymer film and thus each can be controlled at a different potential. A sandwich electrode of poly-**I** was constructed and examined as follows: In a solution of **II**, using a bipotentiostat, the two electrodes were controlled such that a small potential difference (ca. 20 mV) was maintained between them. The potential of both electrodes was scanned in the negative direction through, and beyond, the poly-**I** 2+/1+ and 1+/0 couple while maintaining the small, constant potential difference between them. Figure 4 shows the current flowing between the two electrodes as a function of potential. At potentials positive of the 2+/1+ couple no measurable current flows between the electrodes as a result of the large resistance of the film ($>1 \times 10^{12}$ Ω-cm) when in the 2+ oxidation state. However, at potentials sufficiently negative to fully reduce the poly-**I** film to the formal zero-valent oxidation state, the film passes considerable current. Based on estimates of the film thickness the calculated specific resistance of the reduced film is about 1×10^{3} Ω-cm which is at least 1×10^{9} more conductive than it is in the 2+ form. Performing the same experiment as above using a conventional electrolyte allows one to access the negative formal oxidation states of poly-**I**. These experiments confirm that the zero oxidation state of poly-**I** is by far the most conductive form.

Once a poly-**I** film in a sandwich electrode configuration is reduced to the zero-valent form it can be removed from the electrochemical cell and examined dry, e.g. after the acetonite that swells the film evaporates, (the film in this oxidation state must, however, be protected from O_2). In the dry form zero-valent poly-**I** is about a factor of two more conductive than when swollen with acetonitrile. Generally, the conductivity of redox conducting polymers decreases when the solvent is removed, ostensibly because of more restricted local motion of the counterions. The zero-valent poly-**I** film contains no counterions. When immersed in toluene the film gives the same resistance as when in the dry form. Figure 5 is a plot of the logarithm of the film resistance (measured with the electrode in a toluene bath) vs. T^{-1}. The curve is fairly linear over the entire range from room temperature down to 77K, suggesting that the conduction mechanism is a thermally activated process. However, before considering this dry film conductivity data further, a model must be developed to explain why zero valent poly-**I** films should conduct at all.

Zero-valent poly-**I** differs in a number of ways from other

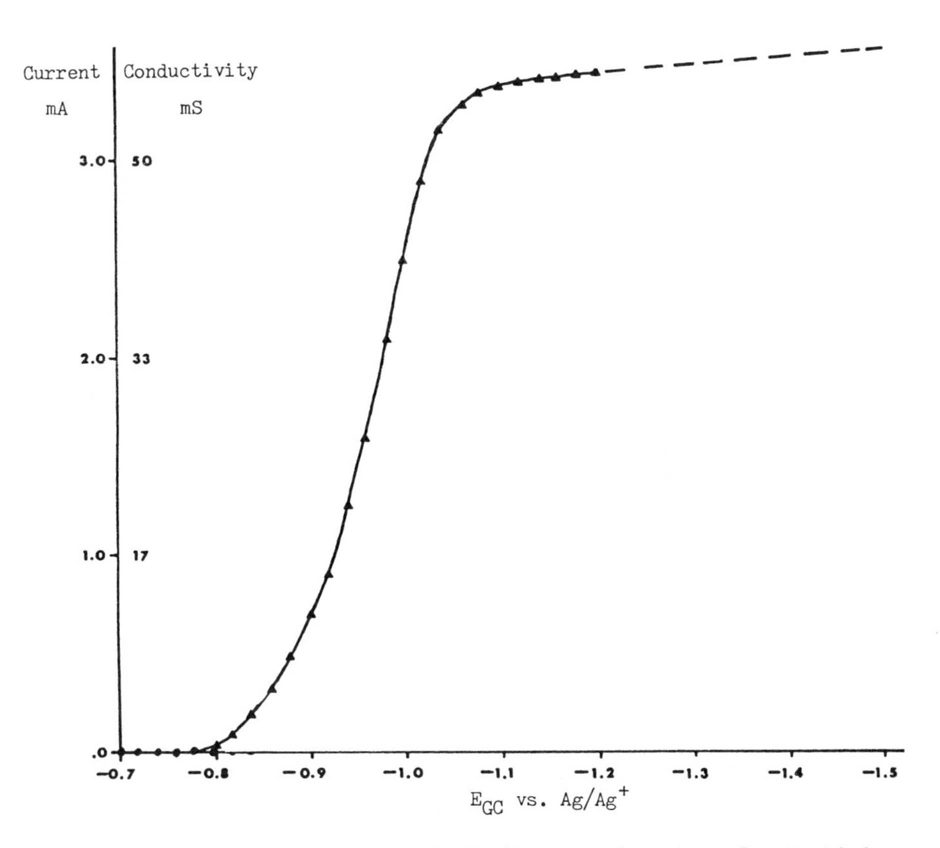

Figure 4. Plot of poly-I conductivity as a function of potential. A series of potential step of 20mV were employed on a sandwich electrode. Each potential was held until Faradaic current ceased, where upon a DC conductivity measurement, ΔE = 60MV, was taken, before proceding with the next potential. The results are for 0.05 M II electrolyte in acetonitrile vs. Ag^+/Ag.

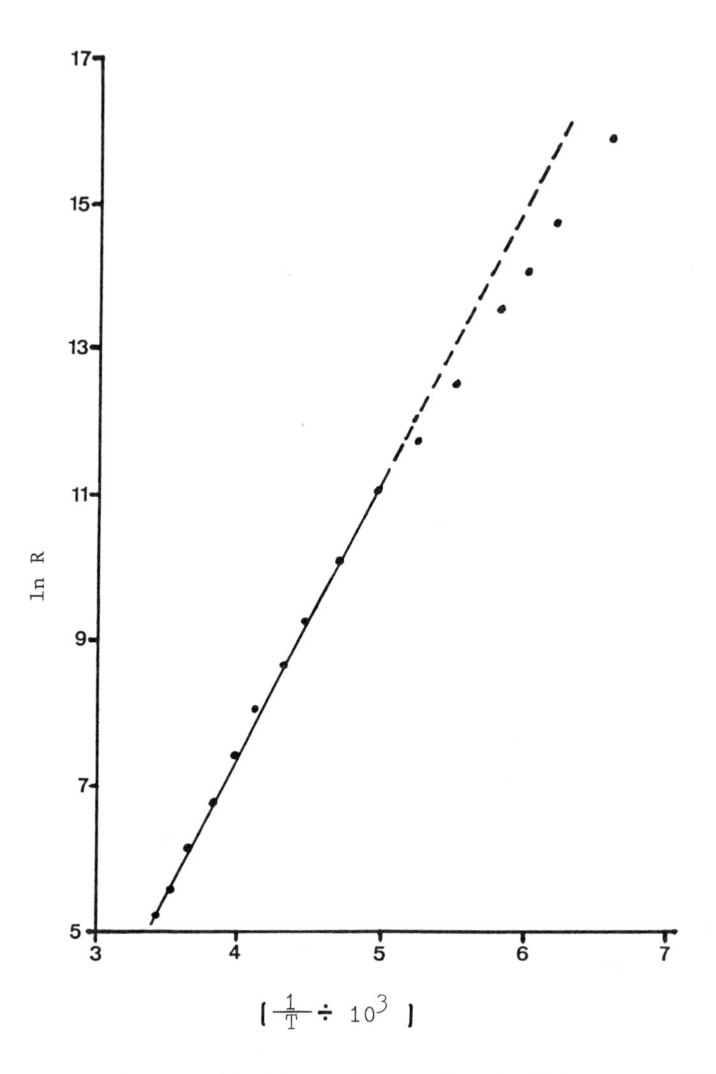

Figure 5. Conductivity of a poly-I film vs. 1/T(K). Experimental procedure: A poly-**I** sandwich electrode was potentiostated in its Ru^0 state using polycationic electrolyte, **II**, then removed from the electrochemical cell and placed in a toluene bath. The toluene was quickly frozen to 77K, where resistance of the Ru^0 poly-**I** was determined using a DC bias of 20mV. The temperature was slowly returned to 298K, with periodic resistance measurements. The break from linearity occurs at the toluene melting point.

electronically conducting polymers. Firstly, it is amorphous and probably highly crosslinked. Secondly, it contains no extended conjugation. Lastly, at least in a formal sense, it is not mixed valent. A close examination of the electrochemistry of both the poly-I film and the monomer in solution, suggests that this latter point, while true in the macroscopic sence, probably is not true on the microscopic level. The redox potentials for the 1+/0 and 0/1- couples in solution are respectively -0.850 and -1.006 V vs. SCE. The exact potentials for the analogous couples in the polymer are more difficult to measure but, whatever their absolute values are, their separation is likely to be small. Consideration of the Nernst equation shows that if the same 156 mV separation in potentials exists between the 1+/0 and 0/1- couples in the polymer as exists in solution, then the zero valent state should be disproportionated to about 5% into the 1+ and 1- forms at room temperature. In reality then, zero-valent poly-I _is_ a mixed valent material. Two unique features of this material should be noted at this point. First, unlike other trisbipyridineruthenium polymers, there are no counterions present in the polymer other than the counterion pair generated by the disproportionation. This means that both the cations and anions can move by electron exchange under force of a voltage gradient. Second, unlike other mixed valence polymers the degree of mixed valence character is not determined by the dopant concentration. In a sense the population of 1+/1- state is thermally induced, and thus the concentration of charge carriers is temperature dependent in a Boltzmann fashion. To an extent then, reduced poly-I has some similarities to an intrinsic semiconductor: the number of positive and negative charge carriers is equal and their absolute concentration is exponentially dependent on the absolute temperature. Extending this analogy one step further, the separation between the 1+/0 and 0/1- potentials is analogous to a band gap energy. There is, however, no evidence, direct or otherwise, to suggest that reduced poly-I has any long range electron delocalization analogous to a band structure. In fact, the visible spectrum of the reduced film is qualitatively similar to the solution spectrum of soluble analogs in the same oxidation state; a fact which would argue that the redox states are probably fairly well localized within the polymer.

The data in Figure 5 can now be considered in light of the conduction model developed above. As stated previously, conduction in reduced poly-I behaves like an activated process. There are two sources that potentially could be responsible for this behavior. The first is the Boltzmann type concentration dependence of the 1+ and 1- states discussed above. The number of charge carriers is expected to decrease approximately exponentially with T^{-1}. The second is the activation barrier to self-exchange between 1+ and 0 sites and 0 and 1- sites. For low concentration of charge carriers both processes are expected to contribute to the measured resistance.

Another factor which appears to be of importance in the conductivity behavior of zero-valent poly-I is the absence of non-electroactive ions in the polymer. The difference in potential between the 2+/1+ and 1+/0 couples in solution is significantly smaller than the difference in potential between the 1+/0 and 0/1-

couples. The degree of disproportionation and thus the numbers of potential charge carriers, should thus be significantly larger in the 1+ form. Nonetheless, the zero valent form is at least five times more conductive. One is drawn to the conclusion that either the self-exchange rates between the oxidation states involved (ie. 2+, 1+ and 0) are innately lower for the 1+ state or that the presence of ions in the polymer adversely affects charge transport. Based on the cyclic voltammetry of the monomer in solution the heterogeneous charge transfer for the 2+/1+ couple is at least as fast as the charge transfer for the 1+/0 couple. This would argue that the differences in conductivity between the 1+ and 0 forms of poly-I lies in the presence and absence, respectively, of non-electroactive ions within the polymer.

One possible way that the presence of non-electroactive ions in the polymer could adversely affect the conductivity of the films is by deepening the potential well for charge localization. Consider, for example, PF_6^- ions residing in the polymer matrix. The cationic sites within the polymer would be expected to localize around the PF_6^- anions. In order for the cationic sites to move either the anion must also physically move or considerable energy must be expended to overcome the electrostatic forces holding the ion pairs together. In the zero-valent polymer the cations and anions are both capable of "moving" by site-site exchange. Additionally, to the extent that the charges of both the 1+ and 1- ion sites are delocalized, the potential wells for each will necessarily be shallower than for a totally localized ion case (ie. PF_6^-). It is thus possible to rationalize in several ways why, despite the smaller degree of disproportionation, the neutral, "ion-free" zero valent form is a much better conductor than the 1+ form.

Summary

Polymer films of **I** have a wide range of reversible reduction electrochemistry. The fixed sites are stable in oxidation states that range form the formal 2+ to 4- forms. Over this range of oxidation states the polymer changes through an array of colors. Additionally, because the fixed sites change formal charge, the film can be either a cation exchanger, a neutral polymer or an anion exchanger depending upon the applied voltage.

By employing a soluble cationic polymer as the solution electrolyte polymer films can be sterically blocked from reducing beyond the formal zero valent form. In the zero-valent form the polymer is an ohmic conductor both in solution and dry. A model has been proposed which describes the conductivity of the polymer and in part accounts for its ohmic nature and semiconductor-like temperature dependence.

Acknowledgment

Support of this work under a grant from the U. S. Department of Energy, Office of Basic Energy Sciences, DE-FG02-87ER13666 is gratefully acknowledged.

References

1. R. W. Murray, "Chemically Modified Electrodes", in Electroanalytical Chemistry, Vol. 13, (A. J. Bard, ed.), M. Dekker, NY, 191 (1984).
2. a) H. D. Abruna, P. Denisevich, M. Umana, T. J. Meyer, and R. W. Murray, J. Am. Chem. Soc., 103, 1 (1981); b) P. Denisevich, K. W. Willman, and R. W. Murray, J. Am. Chem. Soc., 103, 4727 (1981); c) P. Denisevich, H. D. Abruna, C. R. Leidner, T. J. Meyer, and R. W. Murray, Inorg. Chem., 21, 2153 (1981); d) P. K. Ghosh and T. G. Spiro, J. Electrochem. Soc., 128, 1281 (1981); e) J. M. Calvert, B. P. Sullivan, and T. J. Meyer, Adv. Chem. Ser., 192, 159 (1982); f) J. S. Facci, R. H. Schmehl, and R. W. Murray, J. Am. Chem. Soc., 104, 4959 (1982); g) J. M. Calvert, R. H. Schmehl, B. P. Sullivan, J. S. Facci, T. J. Meyer, and R. W. Murray, Inorg. Chem., 22, 1983, 2151.
3. J. L. Bredas and G. B. Street, Acc. Chem. Res., 18, 309 (1985).
4. C. M. Elliott and J. G. Redepenning, J. Electroanal. Chem., 197, 219 (1986).
5. C. M. Elliott and E. J. Hershenhart, J. Am. Chem. Soc., 104, 7519 (1982).
6. Chem. and Eng. News, 61, 24, 21 (1983).
7. C. M. Elliott, J. G. Redepenning, and E. M. Balk, J. Am. Chem. Soc., 107, 8302 (1985).
8. C. M. Elliott, J. G. Redepenning, and E. M. Balk, J. Electroanal. Chem., 213, 203 (1986).
9. P. Pickup, W. Kutner, C. R. Leidner, and R. W. Murray, J. Am. Chem. Soc., 106, 1991 (1984).

RECEIVED September 1, 1987

Chapter 35

Copper Chloride Complexes with Poly(2-vinylpyridine)

A. M. Lyons [1], E. M. Pearce [2], M. J. Vasile [1], A. M. Mujsce [1], and J. V. Waszczak [1]

[1]AT&T Bell Laboratories, Murray Hill, NJ 07974
[2]Polytechnic University, Brooklyn, NY 11201

Metal containing polymers were prepared by complexing copper(II) chloride to poly(2-vinylpyridine)(P2VPy) in aqueous methanol. Through control of the steric environment of the ligand and selection of the solvent system, soluble, linear complexes were formed, where one pyridine moiety is coordinated to each copper atom. The thermal decomposition of the complexes was studied by thermogravimetric analysis, pyrolysis mass spectroscopy, optical, infrared and x-ray photoelectron spectroscopy, and magnetic susceptibility measurements. The complexes decompose in several steps including loss of water, reduction of Cu(II), evolution of halogen, and fragmentation and condensation of the polymer structure. Thermal decomposition of the complex results in a composite of metallic copper and carbon. Increasing the copper chloride concentration in the complex significantly increases the polymer char yield.

The synthesis of inorganic compounds via the thermal decomposition of metal containing polymer precursors offers greater flexibility than conventional processing methods. The object of this study is to extend the processing advantages inherent to polymeric materials, such as film forming, fiber drawing, and molding of complex shapes, to inorganic compounds. In order to exploit these processing techniques, high molecular weight *linear* systems are required since crosslinked networks are not easily processable.

The study of inorganic compound formation from polymer precursors has been limited to refractory materials. Ceramics such as silicon carbide[1], silicon nitride[2], and silicon carbide-titanium carbide composites[3] have been prepared by pyrolyzing silicon based polymers. Electrically conductive films, fibers, and molded articles of carbon, prepared from polymeric precursors such as poly(acrylonitrile) and phenol-formaldehyde, are well known[4]. Recently, conductive materials prepared from poly(phosphazenes) have been reported[5]. The conductivity of these materials is limited to $\approx 10^{-3}$ Ω-cm without subsequent doping.

In this study, we extend the range of inorganic materials produced from polymeric precursors to include copper composites. Soluble complexes between poly(2-vinylpyridine) (P2VPy) and cupric chloride were prepared in a mixed solvent of 95% methanol 5% water. Pyrolysis of the isolated complexes results in the formation of carbonaceous composites of copper. The decomposition mechanism of the complexes was studied by optical, infrared, x-ray photoelectron and pyrolysis mass spectroscopy as well as thermogravimetric analysis and magnetic susceptibility measurements.

0097–6156/88/0360–0430$06.00/0

Experimental

Poly(2-vinylpyridine), with $\overline{M}_v$=36,000, was purchased from Aldrich Chemical Co. and purified by two precipitations from ethanol into deionized water. $CuCl_2{\cdot}2H_2O$ (ACS purity) and poly(4-vinylpyridine) from Mallinckrodt and Polysciences Inc. respectively, were used as received. Complexes with different ratios of Cu to pyridine moiety were formed in solution (95% methanol, 5%H_2O). Due to decreasing solubility of the complex with increasing copper concentration, the concentration of reagents was varied as shown in Table I.

Table I. Reactant specifications for the preparation of $CuCl_2$:P2VPy complexes

Complex (Cu:Py Ratio)	$CuCl_2{\cdot}2H_2O$ (moles)	P2VPy (moles)	Solvent (mls)
1:4	1.25×10^{-3}	5.00×10^{-3}	22
1:2	2.50×10^{-3}	5.00×10^{-3}	84
1:1	5.00×10^{-3}	5.00×10^{-3}	225

The complexes were isolated by freezing the solutions in liquid nitrogen and removing the solvent under a vacuum of $\approx$ 100μm of Hg as the solution thawed. Total solution concentration of 10^{-2} M was employed for the continuous variation analysis. A series of $CuCl_2$ solutions was used to generate a curve from which the background correction factors were determined, as $CuCl_2$ does not follow Beer's Law in this region.

Infrared spectra were recorded on a Perkin Elmer model 680 spectrophotometer as mulls in nujol or fluorolube. The magnetic susceptibility of the copper complexes was measured from 4.2 to 300 K by the Faraday method[6]. X-ray photoelectron spectroscopy (XPS) was performed with a Perkin Elmer hemispherical spectrometer.

TGA analysis was performed on a Perkin Elmer TGS-2 thermobalance in purified N_2 at a flow rate of 30cc/min, heating rate of 10°C/min, and sample weight of 30mg. The materials were dried by heating in N_2 for 2 hours at 30°C before beginning each experiment. The material formed in the TGA was analyzed by reheating to 600°C in O_2. The carbonaceous char volatilized, leaving a residue of CuO (as confirmed by x-ray diffraction).

Pyrolysis mass spectroscopy was conducted with a Hewlett-Packard model 5985B gas chromatograph/quadrupole mass spectrometer, operated at $\approx 10^{-6}$ Torr and 70eV electron-impact ionization energy. Samples were introduced into the mass spectrometer via a glass lined direct insertion probe (DIP). The samples were decomposed in the DIP to a nominal temperature of 300°C at a heating rate of 30°C/min.

Results and Discussion

Complex Preparation. Three complexes were isolated with nominal copper to pyridine ratios of 1:4, 1:2, and 1:1. The green solutions yielded either green (1:4 and 1:2) or yellow brown (1:1) hygroscopic powders. The complexes could be redissolved after isolation. Films prepared by evaporating the solutions were transparent, and of the same color as the powders.

Upon complexation, shifts in the UV-visible spectra of cupric chloride are manifested as a shoulder at approximately 370 nm, and a shift in the visible absorption from 865 to 850 nm. The method of continuous variation[7] (Job's Method) was employed using the new, 370 nm, absorption. The results indicate one monomer residue (pyridine moiety) is coordinated to each copper cation. When a sterically unhindered amine, such as poly(4-vinylpyridine), is used as a ligand, two monomer residues complex to one copper atom. This results in the formation of insoluble, blue, crosslinked gels.

Coordination of copper to pyridine is also observed in the IR spectrum of the complexes. Shifts relative to the uncomplexed material are observed which are in agreement with model compounds[8] and dilute complexes of cupric chloride and P2VPy[9]. In addition, new bands were observed at 1622 and 1540 cm^{-1} which could be attributed to charge transfer from the pyridine ring to $Cu^{(II)}$. Partial reduction of $Cu^{(II)}$ to $Cu^{(I)}$ is suggested by the magnetic susceptibility results. Curie-Weiss behavior was observed from 20-300°C for the three P2VPy-$CuCl_2$ complexes, and moments of 1.5 - 1.6 B.M. were calculated. These values are low compared to the spin-only value of 1.73 B.M., or the experimentally observed moments of 1.8-2.0 B.M. for Cu(II) complexes[10]. These results indicate that 22% of the copper is in the $Cu^{(I)}$ state. XPS results also show that copper is predominately in the $Cu^{(II)}$ state with some $Cu^{(I)}$ also present.

Decomposition Process. Thermograms for P2VPy and the 1:4, 1:2, and 1:1 complexes are shown in Figure 1. The first step in the decomposition sequence occurs at 170°C with the evolution of water and reduction of $Cu^{(II)}$ to $Cu^{(I)}$. The change in weight measured by TGA from 35 to 220°C is in good agreement with the loss of one water molecule per copper atom. Loss of water was confirmed by MS results (Figure 2). The reduction of Cu(II) to Cu(I) was demonstrated by the XPS results. The bands for $Cu^{(II)}$ disappear with the concomitant growth of the $Cu^{(I)}$ bands. In addition, the magnetic moments decrease from 1.5 to 0.2 B.M., and significant shifts are observed in the visible absorption spectrum. All complexed pyridine rings acquire a net positive charge during this transition, accounting for the perseverance of the 1622 and 1540 cm^{-1} vibrations.

The second major event observed in the TGA results occurs with an extrapolated onset temperature of 320°C and is due to the decomposition of the polymer. This value is 70°C lower than that for uncomplexed P2VPy. Polymer decomposition was studied by MS and a stepwise evolution of gaseous products was observed, as shown for selected ions evolved from the 1:4 complex in Figure 2. This decomposition pattern is similar for all complexes, regardless of copper concentration. Initially, HCl is generated and two maxima are observed in the mass spectra at nominal temperatures of 153°C and at 170°C. The second maximum of HCl is associated with the decomposition of the polymer backbone. The polymeric ligand forms a mixture of compounds upon decomposition. Initially, a conjugated dimer (207 a.m.u.) and trimer (311 a.m.u.) with relative ratios of 100:24 are evolved (the concentration of an ion is not directly proportional to the number of counts, however changes in ion counts are related to changes in concentration). The structure of the ions is postulated in Figure 3. Both species exhibit maxima at $\approx$170°C, but the dimer continues to be the most abundant species throughout the polymer decomposition. Lower molecular weight molecules begin to appear at 165°C, but increase continuously with higher temperatures. These include a conjugated dimer (194 a.m.u., Figure 3), monomers (106, 93 and 79 a.m.u.), and benzene (78 a.m.u.). The relative ratios of these species are compared to the 207 ion at nominal temperatures of 170, 220, and 270°C in Table II.

Examination of the mass spectrum of P2VPY taken during the maximum decomposition rate reveals the major decomposition products as methylpyridine (93 a.m.u.), protonated vinyl pyridine (106 a.m.u.), and protonated dimer (211 a.m.u.) with ion ratios 74:100:59 respectively. Trimeric and tetrameric protonated species (316 and 421 a.m.u.) are also observed but in relatively small amounts. Protonated ions, rather than the simple monomers and dimers observed for the decomposition of poly(styrene) by MS[11], may be created by a mechanism similar to that reported for the decomposition of 2-(4-heptyl)pyridine[12] in the mass spectrometer.

The unsaturated nature of the volatiles generated during decomposition of the complexes, compared to the monomeric species released by the virgin polymer, probably

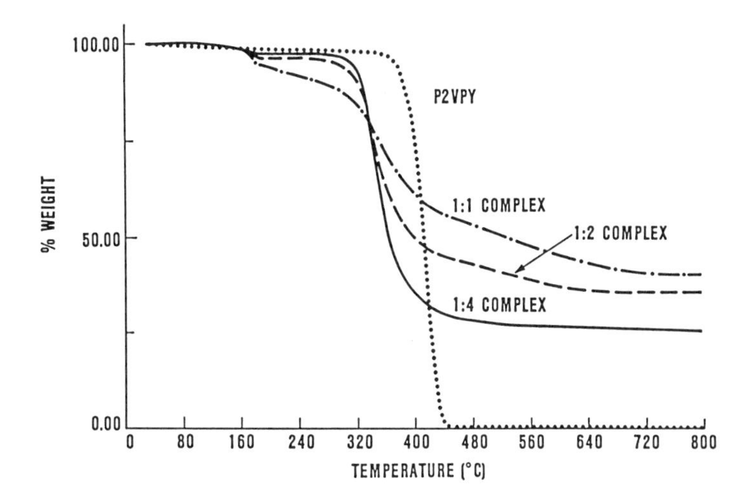

Figure 1. Thermograms of P2VPy and complexes of $CuCl_2$ with P2VPy with ratios of 1:4, 1:2, and 1:1.

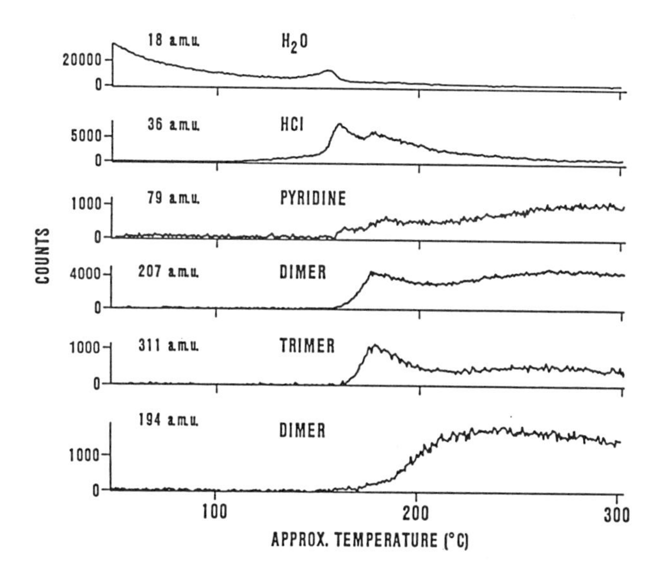

Figure 2. Evolution of selected ions from pyrolysis of the 1:4 complex in the mass spectrometer as a function of temperature.

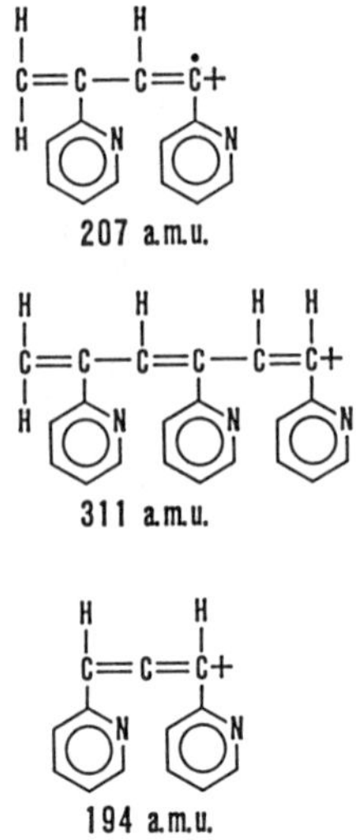

207 a.m.u.

311 a.m.u.

194 a.m.u.

Figure 3. Proposed structures of selected ions evolved during the pyrolysis of $CuCl_2$:P2VPy complexes.

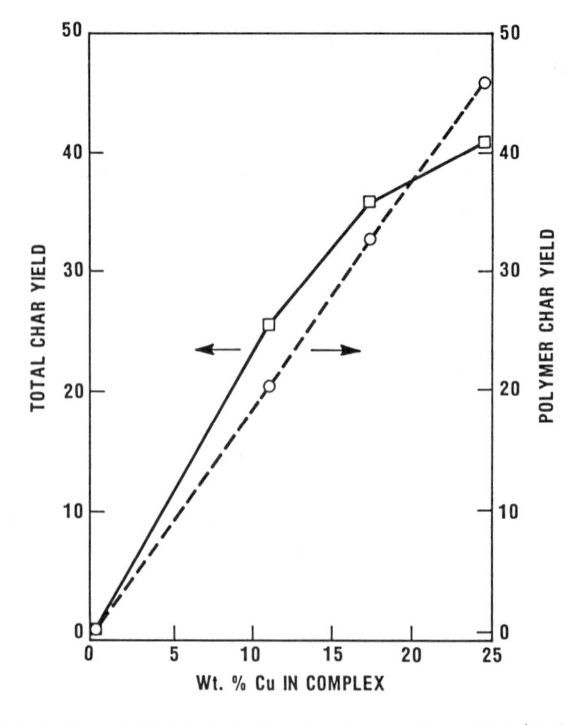

Figure 4. Total char yield (solid line) and polymer char yield (dashed line) of complexes pyrolyzed to 800°C vs weight % copper in the complex.

Table II. Relative mass intensities evolved from the 1:4 chloride complex at three temperatures during thermal decomposition in the mass spectrometer. Nominal temperature values are given.

a.m.u.	Nominal Temperature		
	170°C	220°C	270°C
207	100	100	100
311	24	11	10
194	7	47	35
193	7	53	35
130	7	18	25
106	8	22	35
93	8	19	35
79	11	19	22
78	22	28	29
36	144	67	20

results from dehydrogenation of the polymer during HCl evolution. Double bonds may be created along the chain backbone which would react to form crosslinks or cyclized structures. Both mechanisms are known to increase char yields in organic polymer systems[13]. The increases in both total char yield ($\frac{\text{wt. of char}}{\text{wt. of precursor}}$) and polymeric char yield ($\frac{\text{wt. of carbonaceous products}}{\text{wt. of polymer in precursor}}$) are shown in Figure 4. Copper$^{(I)}$ is reduced to copper metal during the decomposition of the polymer.

Conclusions

P2VPy and $CuCl_2$ react in solution to form a soluble complex in which one pyridine moiety is coordinated to each copper atom. Approximately 22% of the copper atoms are in the Cu(I) state.

The decomposition of the polymer complexes occurs via a stepwise process. A copper composite results where the where the polymer char yield increases with increased copper complexation. P2VPy was initially thought to be an ideal ligand for the pyrolytic formation of pure copper as it thermally decomposes to form no char. The presence of $CuCl_2$ significantly changes the decomposition mechanism of the polymeric ligand and results in high polymer char yields upon pyrolysis. Dehydrogenation of the backbone may occur when HCl is generated. This would result in the formation of a conjugated polymer intermediate, which would subsequently decompose to form high char yields.

Acknowledgments

The authors would like to thank Pat Gallagher, Frank DiSalvo, William Reents Jr., and Costas Tzinis for helpful discussions.

Literature Cited

[1] Yajima, S.; Hayashi J.; Omori, M.; Okamura, K. Nature 1976, **261,** 685.

[2] Penn, B. G.; Ledbetter, F. E. III; Clemons, J. M.; Daniels, J. G. J. Appl. Polym. Sci. 1982, **27,** 3751.

[3] Yajima, S. *et. al.*, J. Materials Sci. 1981, **16,** 1349.

[4] Lyons, A. M. J. Non-Crystalline Solids 1985, **70,** 99. Knop, A.; Scheib, W. Chemistry and Applications of Phenolic Resisns; Springer: Berlin, 1979.
[5] Allcock, H. R. International Conference on Ultrastructure in Organic and Inorganic Polymers,
[6] DiSalvo, F. J.; Safran, S. A.; Haddon, R. C.; Waszczak, J. V. Phys. Rev. B 1979, **20,** 4883. DiSalvo, F. J.; Waszczak, J. V. Phys. Rev. B 1981, **23,** 457.
[7] Skoog, D. A.; West, D. M. Fundamentals of Analytical Chemistry; Saunders College Pub.: New York, 1982; 4th ed., p 552.
[8] Goldstein, M.; Mooney, E. F.; Anderson, A.; Gebbie, H. A. Spectrochimica Acta 1965, **21,** 105.
[9] Inagaki, N.; Suganuma, R.; Katsuura, K. Europ. Polym. J. 1978, **14,** 151.
[10] Hathaway, B. J.; Billing, D. E. Coordin. Chem. Rev. 1970, **5,** 143.
[11] Luderwald, I.; Vogl, O. Makromol. Chem. 1979, **180,** 2295.
[12] Biemann, K. Mass Spectrometry; McGraw-Hill: New York, 1962; p 130.
[13] Quinn, C. B. J. Polym. Sci., Polym. Chem. Ed. 1977, **15,** 2587.

RECEIVED September 1, 1987

Chapter 36

Cationic and Condensation Polymerization of Organometallic Monomers

Kenneth E. Gonsalves [1] and Marvin D. Rausch [2]

[1]Department of Chemistry and Chemical Engineering, Stevens Institute of Technology, Hoboken, NJ 07030
[2]Department of Chemistry, University of Massachusetts, Amherst, MA 01003

1,1'-Bis(β-aminoethyl)ferrocene was synthesized via a 6-step process starting with ferrocene. This monomer was then copolymerized with various aromatic and aliphatic diacid chlorides as well as with diisocyanates, leading to ferrocene-containing polyamides and polyureas having moderately high to low viscosities. The above monomer and 1,1'-bis(β-hydroxyethyl)ferrocene were also utilized as chain extenders. Three types of isopropenylmetallocene monomers were synthesized and subjected to polymerization and copolymerization by cationic initiators: (1) isopropenylferrocene, (2) (η^5-isopropenylcyclopentadienyl)dicarbonylnitrosylmolybdenum; and (3) 1,1'-diisopropenylcyclopentadienylstannocene, and related derivatives of each.

There is currently considerable interest in organometallic polymers, since polymers containing metals might be expected to possess properties different from those of conventional organic polymers.[1-4] Two major approaches to the formation of materials of this type have involved the derivatization of preformed organic polymers with organometallic functions[5] and the synthesis and polymerization of organometallic monomers that contain vinyl substituents.[6,7] For the transition metals, condensation polymerizations have also been investigated. However, the reactions have generally been conducted at elevated temperatures, and the resulting products have often not been well characterized.[4,8]

Ferrocene-Containing Polyamides and Polyureas

We now report a convenient method for the interfacial polycondensation of 1,1'-bis(β-aminoethyl)ferrocene (1) with a variety of diacid chlorides and diisocyanates, leading to ferrocene-containing polyamides and polyureas.[9] In some instances, we have been able to observe film formation at the interface. Moreover, the polymerization reactions can be conveniently conducted at ambient temperatures in contrast to earlier high-temperature organometallic condensation

0097-6156/88/0360-0437$07.25/0

polymerizations, which frequently led to undesirable side reactions.[8] We also find that the related monomer, 1,1'-bis(β-hydroxyethyl)ferrocene (2) reacts with diacid chlorides and diisocyanates to form ferrocene-containing polyesters and polyurethanes, respectively. Monomers (1) and (2) are shown in Scheme I.

Monomers 1 and 2 have been synthesized starting from ferrocene, utilizing modifications of procedures outlined previously by Sonoda and Moritani[11] and by Ratajczak et al.[12] The intermediate diacid, 1,1'ferrocenedicarboxylic acid, was synthesized according to the more convenient procedure of Knobloch and Rauscher.[10] Monomer 1 was vacuum distilled prior to use (bp 120°C, 1 mm Hg). Details of the synthetic routes are given in Scheme 1. It should be emphasized that in contrast to previous ferrocene-containing monomers,[13] 1 and 2 position the reactive amino and hydroxyl groups two methylene units removed from the ferrocene nucleus. This feature minimizes steric effects and also enables 1 and 2 to undergo the Schotten-Baumann reaction readily without the classical α-metallocenylcarbonium ion effect providing any constraints.[14,15] Polyamide formation is vigorous, exothermic, and instantaneous.

Interfacial or solution polycondensation, with or without stirring, was the general procedure utilized for the preparation of the polyamides and polyureas.[16a] Details are given in Table I. An important point to be noted is that, in the unstirred interfacial condensation polymerization of 1 with sebacoyl chloride or terephthaloyl chloride in the organic phase and triethylamine as the proton acceptor, immediate film formation took place at the interface. The polyamide films were removed after 1 h, dried, and utilized for taking electron micrographs.

Attempts to obtain molecular weights of these new iron-containing polyamides in m-cresol solution have not been successful, due to the very limited solubilities of the materials in organic solvents. Similar difficulties have previously been encountered in the molecular weight determination of nylon 66 (polyhexamethyleneadipamide).[17] However, the intrinsic viscosity values greater than 1.0 for the polyamides obtained from 1 and terephthaloyl chloride or sebacoyl chloride are comparable to intrinsic viscosities of nylons having number average molecular weights between 10,000 and 18,000.[16b] The low [η] values obtained for the polyurethanes can be attributed to premature precipitation from solution and, in the case of polymers obtained from 1 and TDl, to decreased reactivity imposed by steric effects.[18]

The polyamides and polyureas[19] exhibited broad, intense N-H stretches around 3300 cm^{-1}. A very strong carbonyl stretching vibration was present at 1630 cm^{-1}. The amide II band was evident near 1540 cm^{-1}. In addition, sp^2 C-H stretches occurred around 3100 cm^{-1} and asymmetric and symmetric sp^3 C-H stretches at 2950 and 2860 cm^{-1}, respectively. The polyurethane showed the carbonyl absorption near 1700 cm^{-1} and C-O stretches in the vicinity of

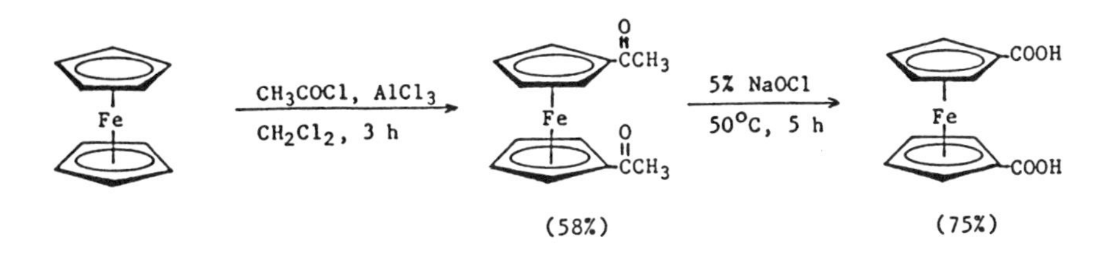

CH_3OH, H_2SO_4
Δ, 24 h
CO_2CH_3
Fe
CO_2CH_3
(90%)
$LiAlH_4$, Et_2O
Δ, 3 h
CH_2OH
Fe
CH_2OH
(98%)

PCl_3, THF
25^oC, 2 h (N_2)
KCN, H_2O
1 h
CH_2CN
Fe
CH_2CN
(85%)
$LiAlH_4$, $AlCl_3$
Et_2O, 3 h
$CH_2CH_2NH_2$
Fe
$CH_2CH_2NH_2$
1 (73%)

NaOH
EtOH, Δ
CH_2COOH
Fe
CH_2COOH
(70%)
$LiAlH_4$, THF
Δ, 16 h
CH_2CH_2OH
Fe
CH_2CH_2OH
2 (98%)

Scheme I.

Table I. Polycondensation Reactions between 1,1'-Bis(β-aminoethyl)ferrocene (1) and (β-hydroxyethyl)ferrocene (2) with Diacid Chlorides and Diisocyanates

monomer (M_1)	monomer (M_2)	process (base used)	% yield	η dL/g[e]
1[a]	terephthaloyl chloride(CH_2Cl_2)	UI[b](Et_3N)	72	1.50
1[a]	sebacoyl chloride (CCl_4)	UI (Et_3N)	85	0.37
1[a]	sebacoyl chloride (CCl_4)	UI (NaOH)	39	0.59
1[a]	sebacoyl chloride (CCl_4)	I[b] (Et_3N)	51	1.09
1[a]	adipoyl chloride (CCl_4)	UI (Et_3N)	47	0.53
1	terephthaloyl chloride(CH_2Cl_2)	S[b] (Et_3N)	45	0.80
2	terephthaloyl chloride (m-xylene, reflux)	S (pyridine)	51	0.16
2	TDI[c] (Me_2SO, 115 °C)	S	46	0.20
1[a]	TDI ($CHCl_3$)	UI	58	0.16
1	TDI ($CHCl_3$)	S	53	0.10
1	MDI[d]	S	67	f

[a]Monomer in aqueous phase

[b]UI, unstirred interfacial
S, solution
I, Stirred interfacial

[c]TDI: tolyene 2,4-diisocyanate (80%) + 2,6, isomer (20%)

[d]MDI: methylenebis(4-phenylisocyanate)

[e] Intrinsic viscosity determined in m-cresol at 3?°C

[f]insoluble in m-cresol

1220 and 1280 cm^{-1}. Similar absorptions were present in the polyester. The polyamides and polyureas are thus assessed to have structures outlined in Scheme II.

Model Compounds. Further elucidation of these polymer structures was done by synthesizing model analogs.

It has been demonstrated by Hauser and coworkers[20] that β-aminoethylferrocene (3) undergoes reactions typical of the amino functional group.

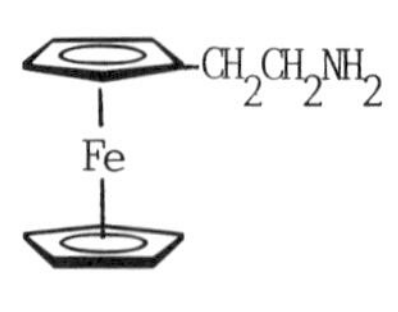

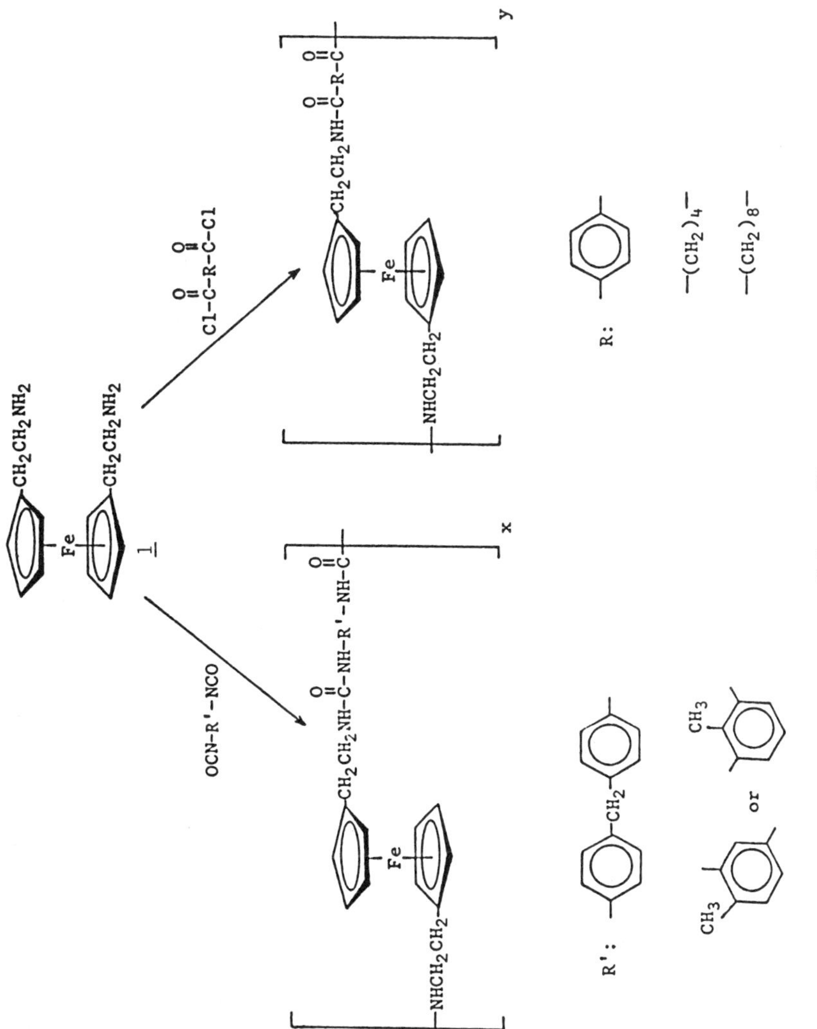
CH2CH2NH2
Fe
CH2CH2NH2
1
Cl-C-R-C-Cl
OCN-R'-NCO
NHCH2CH2
CH2CH2NH-C-R-C
y
CH2CH2NH-C-NH-R'-NH-C
x
R:
—(CH2)4—
—(CH2)8—
R':
CH2
CH3
or

Scheme II.

They showed that 3 afforded a picrate on being treated with saturated alchoholic picric acid and formed N,N,N-trimethyl-β-ferrocenylethylammonium iodide on being treated with methyl iodide. β-Aminoethylferrocene (3) was synthesized by a modification of the literature method[20] and was characterized by elemental analysis, IR and ^1H NMR spectroscopy. Following the procedure of Pittman et al.,[21] the methiodidide of N,N-dimethylaminomethylferrocene was converted to ferrocenylacetonitrile by refluxing with sodium cyanide in deoxygenated water. The latter compound was then reduced to the amine 3 with $LiAlH_4$. In order to isolate pure 3, instead of passing hydrogen chloride gas into an ether solution of 3,[20] 6N H_2SO_4 was used. The precipitated ferrocenylammonium sulfate was filtered under nitrogen and treated with aqueous sodium hydroxide to obtain the free amine 3 in the organic layer. Pure β-aminoethylferrocene (3) was obtained in 70% yield by distillation under vacuum (b.p. 120°C/1 torr).

A suspension of β-aminoethylferrocene (3) in deoxygenated water, containing an excess of sodium hydroxide, was found to form a yellow precipitate immediately on being shaken vigorously with one equivalent of pure terephthaloyl chloride in dry benzene. Elemental analysis and an IR spectrum indicated the yellow precipitate to have the structure 4 (yield 100%).

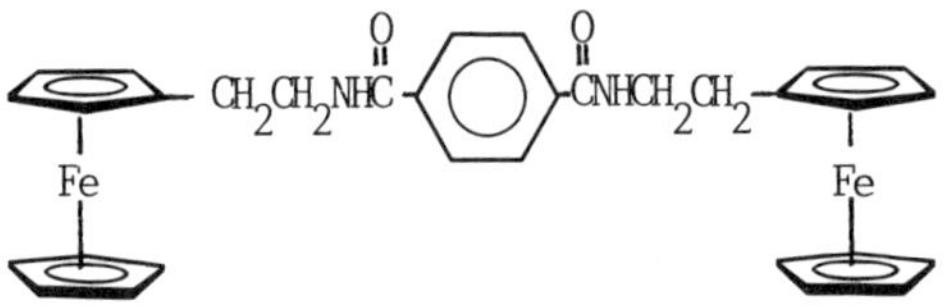

$CH_2CH_2NHC(=O)$ — C_6H_4 — $C(=O)NHCH_2CH_2$ (Fe, Fe)

4

In 3, the amino functional group is two methylene units removed from the ferrocene nucleus. It appears from the instantaneous and quantitative formation of 4 from 3 that this feature minimizes steric effects and also enables 3 to undergo the Schotten-Baumann reaction readily without the classical α-metallocenylcarbenium ion effects providing any constraints.[14] The IR spectrum of 4 showed the characteristic N-H stretch at 3320 cm^{-1}(s), the amide 1 (carbonyl) stretch at 1625 cm^{-1}(s), the amide II (N-H) stretch at 1540 cm^{-1}(s), and the amide III band at 1310 cm^{-1}(m). In addition, characteristic absorptions of the ferrocenyl group were evident at 1100 and 1000 cm^{-1} (indicating an unsubstituted cyclopentadienyl ring) and at 800 cm^{-1}.

In addition to the above reaction, 1,1'-bis(β-aminoethyl)-ferrocene (3) was reacted with two equivalents of benzoyl chloride in the presence of triethylamine. A product (5) whose infra-red spectrum closely resembled that of 4 was obtained.

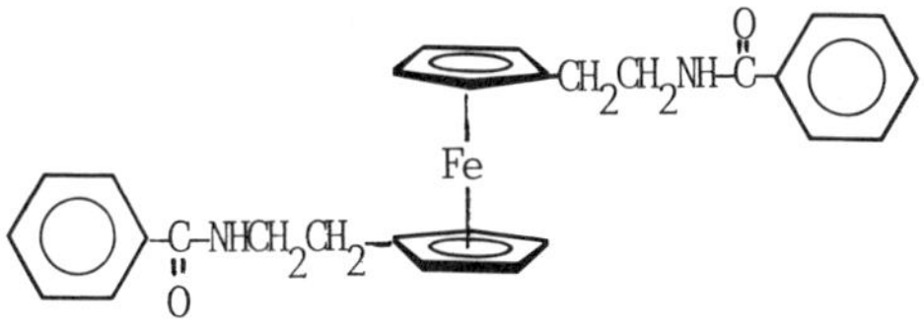

5

As before, an N-H stretch was observed at 3325 cm^{-1}(s), the amide I at 1650 cm^{-1}(s), amide II at 1550 cm^{-1}(s), and amide III at 1325 cm^{-1}(m).

It is well established that primary amino-functional groups, particularly aliphatic ones, react instantaneously with isocyanates to form ureas at ambient temperature. Indeed this was observed when β-aminoethylferrocene (3) and freshly distilled phenylisocyanate were shaken vigorously. A yellow precipitate separated out immediately. Elemental analysis and an IR spectrum of the product indicate this compound to have the structure 6.

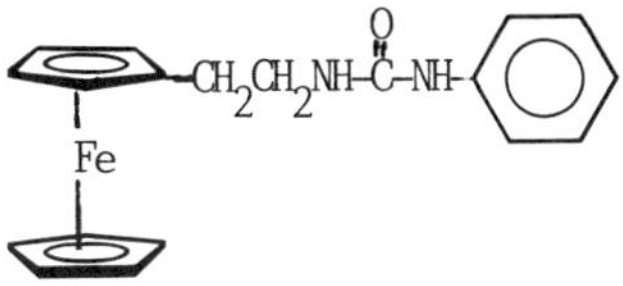

6

The characteristic absorptions of the urea group were evident in the IR spectrum: -NH stretch 3320 cm^{-1}(s); amide I 1625 cm^{-1}(m); amide II 1560 cm^{-1}(s); amide III 1240 cm^{-1}(m). The yield of 6 was again quantitative. When the diamine 1 was similarly reacted with phenylisocyanate, a yellow precipitate was observed immediately. The IR spectrum of this product (7) was similar to that of 6. The -NH stretch occurred at 3300 cm^{-1}(s) and the carbonyl absorption at 1630 cm^{-1}(s). The amide II band occurred at 1550 cm^{-1}(s,br) and the amide III at 1260 cm^{-1}(m).

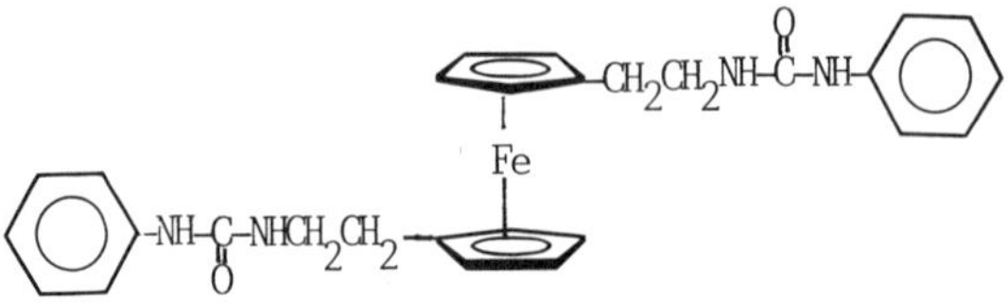

7

Segmented Poly(ether urethane) Films Containing Ferrocene Units in the Hard Segments

Segmented poly(ether urethanes) were synthesized from polypropylene glycol (PPG) and 4,4'methylene-bis(phenyl-isocyanate) (MDI), using 1,1'-bis(β-aminoethyl)ferrocene (1) and 1,1'-bis(β-hydroxyethyl)-ferrocene (2) as chain extenders.

Synthesis of Polyurethanes. In the "prepolymer method" employed in this study, MDI (2 equivalents) and PPG (1 equivalent) were reacted at 60°C in the presence of 1% dibutyltin dilaureate as catalyst, in the melt. The course of the polymerizations was followed spectroscopically by observing the intensity of the —NCO peak in the IR (Scheme III).

MDI itself exhibits a strong NCO peak at 2260 cm^{-1}. After the above reaction had proceeded for 30 min, a small aliquot was withdrawn from the reaction vessel and its IR spectrum recorded. It showed a strong absorption at 2260 cm^{-1} characteristic of the —NCO group. Similarly, after 60 min the absorption for —NCO was still strong. The prepolymer thus obtained had reactive —NCO groups. When this prepolymer was cured in a vacuum oven for 12 h at 60°C, the resulting material also exhibited a strong broad absorption at 2245 cm^{-1}. Likewise, a similar prepolymer allowed to cure at ca. 20°C in air showed the above absorption. Thus in all cases, the prepolymer possessed reactive isocyanate end groups.

After the MDI and PPG had reacted to form the prepolymer, 1 (1 equivalent) in dry DMF was added as the chain extender. The reaction was then continued at ambient temperature, ca. 20°C, for 3 h followed by curing at 20°C for 24 h in vacuum. No isocyanate group absorption was observed in the IR spectrum of the block poly(urea urethane) polymer (BPU1a). Alternatively, the curing was also carried out at 60°C in a vacuum oven for 24 h. No isocyanate group absorption was observed in the IR spectrum of this polymer (BPU1). However, the difference in curing procedures produced dark brown films having different solubilities. The former was soluble and the latter insoluble in DMF.

The complete disappearance of the —NCO peak in the IR spectra of these dark-brown translucent films is a good indication that complete chain extension or cure had occurred. In the latter case,

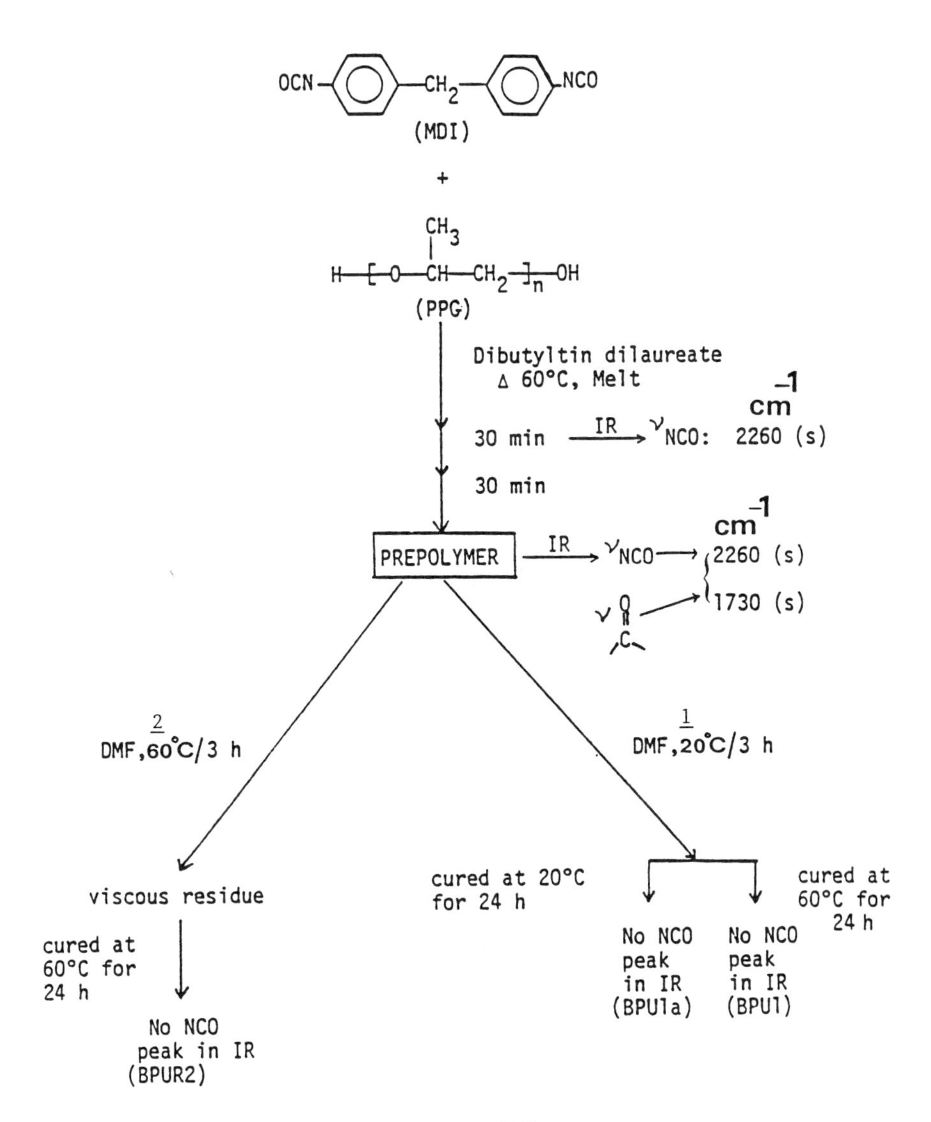
OCN
CH_2
NCO
(MDI)
+
CH_3
H—[—O—CH—CH_2—]$_n$—OH
(PPG)
Dibutyltin dilaureate
Δ 60°C, Melt
30 min
IR
ν_{NCO}: 2260 (s)
cm^{-1}
30 min
PREPOLYMER
IR
ν_{NCO}
2260 (s)
1730 (s)
cm^{-1}
2
DMF,60°C/3 h
1
DMF,20°C/3 h
viscous residue
cured at 60°C for 24 h
No NCO peak in IR (BPUR2)
cured at 20°C for 24 h
cured at 60°C for 24 h
No NCO peak in IR (BPU1a)
No NCO peak in IR (BPU1)

Scheme III.

however, the insolubility of the film in DMF also indicates that biuret branching and crosslinking could have occurred on curing at 60°C for 24 h, as might be anticipated.

The above prepolymer on treatment with 2 as the chain extender in dry DMF did not proceed at ambient temperature. The mixture had to be heated to 60°C for 3 h before the reaction was complete. After curing at 60°C for 24 h, the yellow, translucent block polyurethane film (BPUR2) again showed the absence of the —NCO peak in the IR spectrum indicating that curing had been complete. The fact that a higher temperature had to be used in the case of 2 as the chain extender compared to 1 is in keeping with the lower order of reactivity of diols with diisocyanates as compared to diamines with diisocyanates.

On the basis of their infrared spectra the polymers BPU1a and BPUR2 were assessed to have the structures 8 and 9 respectively, as given in Figure 1.

Polymer BPU1a showed the characteristic NH stretch at 3350 cm^{-1} along with the corresponding amide I and II absorptions at 1735(s) and 1530 (br) cm^{-1} respectively. A carbonyl absorption at 1650 cm^{-1} was ascribed to the urea group in this polymer. A strong C-O-C (br) peak at 1100 cm^{-1} and CH stretches at 3050, 2940, and 2860 cm^{-1} were also observed. The polymer BPUR2 also showed the characteristic —NH stretch at 3300 cm^{-1} and the amide I and II absorptions at 1720 (s) and 1530 (s) cm^{-1}, respectively. Again the C-O-C stretch was observed at 1100 cm^{-1} and the CH stretches at 3050, 2940, and 2860 cm^{-1}. These correlations compare well with the IR spectra of similar segmented polyurethanes.[22,23]

The inclusion of 1 as an integral part of the polyurethane system was confirmed in the synthesis of BPU1a. After the reaction had been completed, the mixture was poured into excess diethyl ether, resulting in a yellow gelatinous precipitate. The yellow viscous material was separated and allowed to cure at ambient temperature for 24 h. The resulting translucent yellow film was soluble in DMF. The amount of iron determined by elemental analysis in this sample was 1.1% (4% ferrocene). Thus the IR spectra, elemental analyses, as well as precipitation procedure for BPU1a all point towards the inclusion of 1 in the polyurethanes via chemical linkage, and not by just mere physical compounding. The percentage of iron in the other polyurethanes, BPU1 and BPUR2, were determined to be approximately 1.9 and 1.2%, respectively, corresponding to 6.7% and 5% ferrocene units in the copolymers.

The molecular weight distribution (MWD) of the linear polyurethanes were determined by GPC. The solvent used was THF and the instrument calibrated by narrow MWD polystyrenes. Polymer BPU1a had an $\bar{M}_w$ of 56,000 ($\bar{M}_n$:12,500); and BPUR2 $\bar{M}_w$ of 97,000 ($\bar{M}_n$:9,100).

Mass Spectrometry. Mass spectrometric (MS) analysis has been utilized for polymer and copolymer structural identification[24]. Recently Dussel et al.[25] utilized pyrolysis-MS to characterize

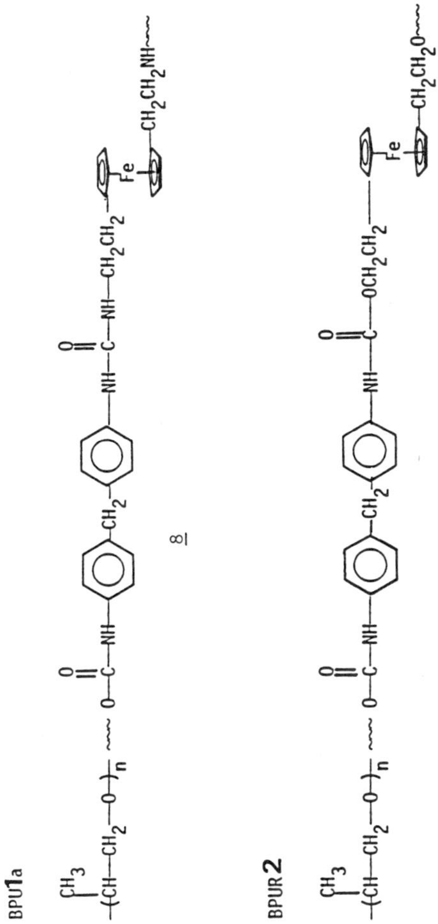

BPU1a

8

BPUR 2

9

Figure 1. Average structure of segmented polyurethane containing ferrocene units. (Reproduced with permission from Ref. 53. Copyright 1986 John Wiley.)

segmented copoly(ether urethane-urea) (PEUU), by identifying key fragments and relating them to the building blocks of PEUU — polyethers, diisocynates, and the diamine. Richards et al.[26] have also utilized this technique for the structural analysis of a poly(ether urethane-urea) (Biomer). Generally, the decomposition of the macromolecular chains is initiated by scission of the urea bonds followed by splitting of the urethane bonds This is followed by the evaporation of the polyether building blocks.[25]

Based on the above studies, the identity of the diisocyanate component in BPU1a was deduced by the presence of fragments at m/z 250 [methylene-bis(4)phenyl-isocyanate] (MDI), 224 (4-amino-4'-isocyanato-diphenylmethane), and 198(diamino-diphenylmethane).[27] The presence of the diamine, as the chain extender was evident from the presence of the fragments m/z 324, assignable to 1,1'-bis(β-isocyanatoethyl)ferrocene, and 298, assignable to the structure 10.[26]

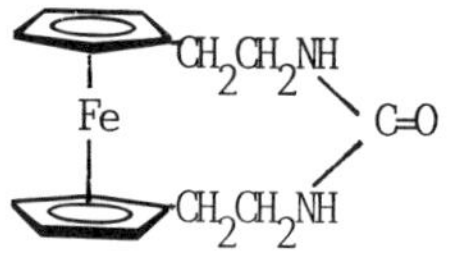

CH_2CH_2NH
Fe C=O
CH_2CH_2NH

10

The intense signals belonging to the ion series m/z 59, 117, 175, 133, 291, 349, etc. are 58 mass units apart, corresponding to the mass of the repeat units of PPG.[26] Thus the composition of BPU1a can be regarded as PPG, end-capped with MDI and chain extended by 1. The MS of BPUR2 was similar.

^{13}C-NMR. The ^{13}C-NMR spectrum of BPUR2 in acetone-d_6 also provided information regarding the incorporation of BHF into the polymer via chain extension. The δ shifts at 68.8, 69.7, and 85.4 are within the range expected for cyclopentadienyl carbons.[28] The methylene carbons adjacent to the cyclopentadienyl rings exhibit the resonances at δ 72.3 (C_β) and 65.7 (C_α) ppm respectively. The absorbances at 75.9 and 73.8 ppm can be assigned to the methine and methylene carbons and the δ shift at 17.8 ppm to the methyl carbons in the PPG segments. The aromatic carbons were assigned the δ shifts 119.2, 129.7, 136.4 and 138.2 ppm and the methylene carbon, bridging the benzene rings, 41 ppm.[29] The absorbance at δ 154.3 ppm belongs to the carbonyl carbon of the urethane group. The intense signals at 30.2 and 205.9 are of the solvent acetone. The ^{13}C spectrum of BPU1a was similar but of lesser resolution owing to its limited solubility in acetone. The above spectra of BPUR2 is given in Figure 2.

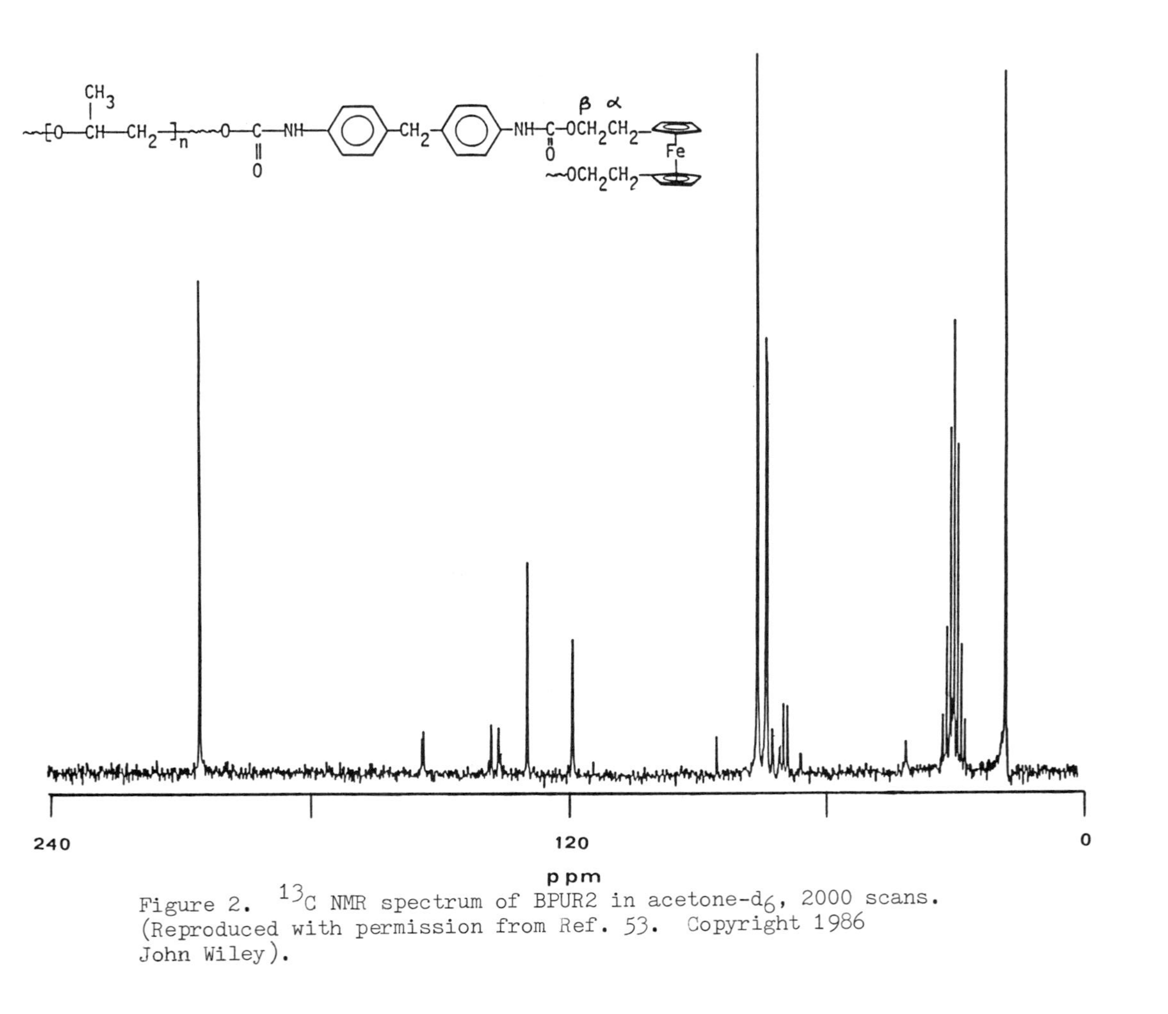

Figure 2. ^{13}C NMR spectrum of BPUR2 in acetone-d_6, 2000 scans. (Reproduced with permission from Ref. 53. Copyright 1986 John Wiley).

Cationic Polymerization and Copolymerization of Isopropenylmetallocene Monomers

Only a few examples of the cationic polymerization reactions of organometallic monomers have been reported in the literature. Kunitake and coworkers[30] first reported the cationic polymerization of vinylferrocene, and Korshak and co-workers[31] polymerized 1,1'-diisopropenylferrocene with cationic initiators. Recently, Jablonski and Chisti further investigated the cationic polymerization of 1,1'-diisopropenylferrocene.[31]

The Q-e values of vinylferrocene determined in the free radical copolymerization studies[32] have shown that the metallocene group is strongly electron contributing, and it is also known that the isopropenyl group is strongly electron contributing. Furthermore, the isopropenyl group has substantial advantage over the vinyl group in cationic polymerization reactions, as seen in a comparison of the relative polymerizabilities of α-methylstyrene and styrene. For these reasons, and since we have been able to synthesize a wide range of η^5-cyclopentadienyl-metal monomers which contain isopropenyl units,[33] we felt that these new organometallic monomers should be potentially attractive candidates for polymers prepared by cationic initiation conditions. As an initial step in our investigations in this uncharted area, we studied in some detail the cationic polymerization reactions of isopropenylferrocene (IF) itself. We were particularly interested in determining the effect of the interaction of the propagating carbenium ion with the ferrocenyl group on the reactivity and polymerizability of this monomer and of the related isopropenylmetallocene monomers, whose structures are shown below, including α-trifluorovinylferrocene (TVF), (η^5-isopropenylcyclopentadienyl)dicarbonylnitrosylmolybdenum (IDM), and its tricarbonylmethyl analog (ITMM)

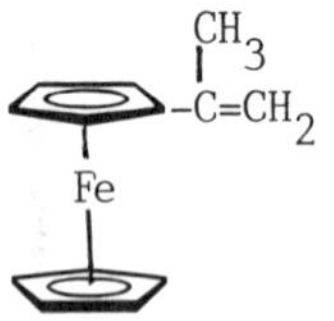

(IF) 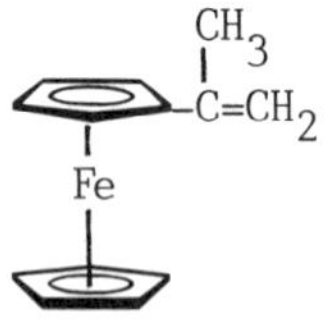 (TVF) 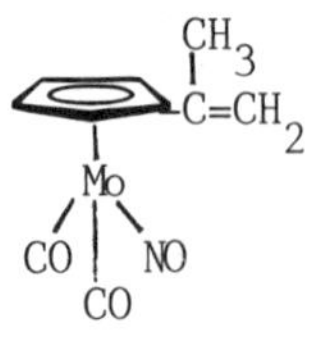 (IDM)

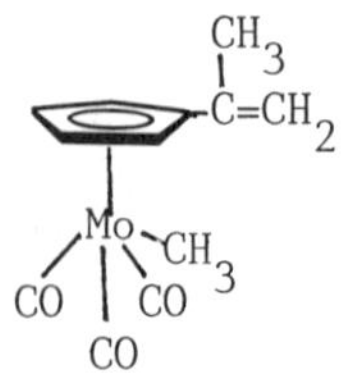

(ITMM) 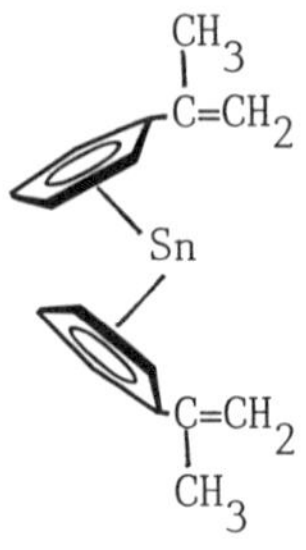 (DIS)

Ferrocene Monomers. Although the isopropenylferrocene monomer has been reported previously by several groups of investigators,[34-37] we were able to develop a more facile, dependable, high yield synthesis for this compound, utilizing the three-step procedure shown below:

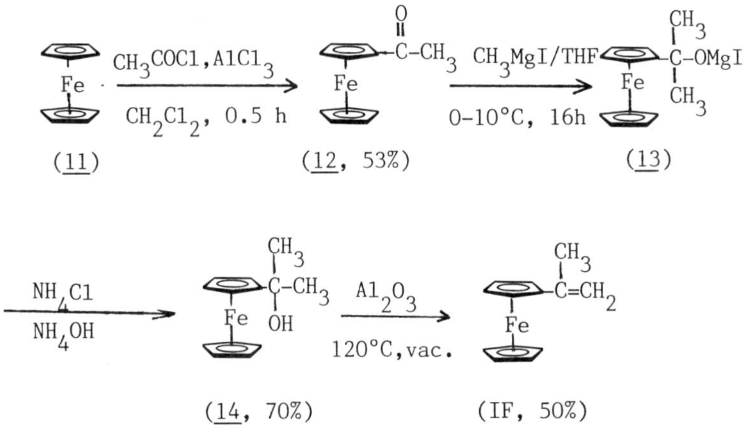

$$\text{Ferrocene } (\underline{11}) \xrightarrow[CH_2Cl_2,\ 0.5\ h]{CH_3COCl,\ AlCl_3} FcC(O)CH_3\ (\underline{12},\ 53\%) \xrightarrow[0\text{-}10°C,\ 16h]{CH_3MgI/THF} FcC(CH_3)_2OMgI\ (\underline{13})$$

$$\xrightarrow[NH_4OH]{NH_4Cl} FcC(CH_3)_2OH\ (\underline{14},\ 70\%) \xrightarrow[120°C,\ vac.]{Al_2O_3} FcC(CH_3){=}CH_2\ (IF,\ 50\%)$$

Ferrocene (11) was converted to acetylferrocene (12) by a Friedel-Crafts acylation involving a modification of literature methods.[38,39] 12 was reacted with methylmagnesium iodide in tetrahydrofuran (THF) and the reaction mixture containing 13 was hydrolyzed to 2-ferrocenyl-2-propanol (14). This was subjected to sublimation pyrolysis on neutral alumina under reduced pressure in a vacuum sublimer at about 120°C, following the procedure developed by Rausch and Siegel for the dehydration of 1-ferrocenyl-ethanol.[40] This dehydration procedure was highly advantageous, because isopropenylferrocene was obtained in both higher yields and purities than previously possible. Elemental analyses and spectroscopic data (IR, ^{1}H-NMR, ^{13}C-NMR) were used to fully characterize the monomer.

Homopolymerization reactions of isopropenylferrocene were carried out using four different cationic initiators: $BF_3{\cdot}OEt_2$, $SnCl_4$, $AlCl_3$, and $Ph_3C^+SbCl_6^-$, by methods similar to those used by Kunitake and co-workers for vinylferrocene.[30] Polymerization temperatures ranged from —78 to 20°C, and flame-dried Schlenk tubes purged with argon were utilized for this purpose. The amount of monomer for each experiment was 2-4 g dissolved in 20 mL of dried, deoxygenated solvent (either CH_2Cl_2, THF, or benzene). The amount of initiator used varied from 2 to 5% of the weight of monomer. The results so obtained under various conditions are summarized in Table II.

Table II. Homopolmerization of Isopropenylferrocene

Polymer No.	Initiator		Solvent[a]	Temp. (°C)	Time (h)	Polymer yield (%)	Mol. wt. $\bar{M}_n$[b]
	Type	Wt. %					
1	$BF_3 \cdot OEt_2$	2	THF	-78	24	Oligomer[d]	
2	$Ph_3C^+ SbCl_6^-$	c	CH_2Cl_2	-78	24	Oligomer[d]	
3	$BF_3 \cdot OEt_2$	2	CH_2Cl_2	-65	24	Oligomer[d]	
4	$BF_3 \cdot OEt_2$	2	CH_2Cl_2	-40	24	Oligomer[d]	
5	$BF_3 \cdot OEt_2$	2	CH_2Cl_2	0→20	24	11.1	1300
6	$Ph_3C^+ SbCl_6^-$	c	CH_2Cl_2	0	24	3.0	
7	$BF_3 \cdot OEt_2$	2	CH_2Cl_2	20	24	52.0	3900
8	$SnCl_4$	5	CH_2Cl_2	30	24	51.0	2500
9	$BF_3 \cdot OEt_2$	2	C_6H_6	20	24	Oligomer[d]	
10	$BF_3 \cdot OEt_2$	2	CH_2Cl_2	40	24	17.0	3000
11	$BF_3 \cdot OEt_2$	2	C_6H_6	80	24	26.6	1300

[a]The charge in each experiment was 1.50-2.00 g of monomer in 20 mL of solvent

[b]Determined by VPO method in toluene solution

[c]11×10^{-5} M.

[d]The oily viscous product obtained was believed to be oligomers.

Source: Reproduced with permission from Reference 28. Copyright 1985 John Wiley.

In contrast to the results from previous studies with related monomers, at low temperatures, from —78 to —40°C, no polymerization reaction apparently occurred. However, if the polymerization reactions initiated with either $BF_3 \cdot OEt_2$ or $SnCl_4$ were carried out at 0°C and the system was allowed to attain ambient temperature (20°C) over a period of 24 h, or if initiation was done directly at ambient temperature and stirring was continued for 24 h, good yields of low molecular weight polymers, which were insoluble in methanol, were obtained. The latter procedure was found to be the most effective, but at 0°C only viscous residues resulted. However, for shorter polymerization periods, even at 20°C, no products insoluble in methanol were obtained, and the monomer was recovered virtually unreacted.

With $AlCl_3$ and $Ph_3C^+ SbCl_6^-$ as inititors, unsatisfactory results were obtained. With $AlCl_3$, the system turned black immediately on transferring the initiator into the polymerization tube, and a black precipitate was observed. After 24 h, on pouring the contents of the tube into methanol, no insoluble fraction was observed except the black residue observed earlier.

The number average molecular weights, $\bar{M}_n$, of the polymers obtained in Table II ranged from 1300 to 3900 as determined by vapor pressure osmometry (VPO), and they were further characterized by ^{1}H- and ^{13}C-NMR spectrometry at ambient temperature and at 90°C, as well as by IR spectroscopy. The homopolymers of isopropenylferrocene were found to have the expected structure, P, shown below, obtained by polymerization through the isopropenyl units, as indicated by spectroscopic characterization:

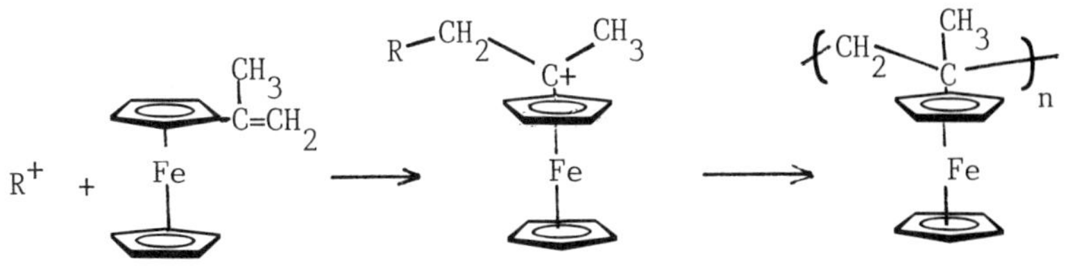

R^+ + (IF) ⟶ (P^+) ⟶ (P)

The IR spectra of the polymer (P) contained two sharp absorptions near 1000 and 1100 cm^{-1}, indicative of the presence of unsubstituted cyclopentadienyl rings in the products.[38] The 250-MHz ^{1}H-NMR spectrum, shown in Figure 3, contained the expected peaks for the methyl, methylene, and cyclopentadienyl protons, respectively, at δ1.52, 1.57 and 4.04 ppm. No olefinic proton resonances were present, and all of the samples of the polymer in Table II exhibited the same ^{1}H-NMR spectrum.

In the cationic polymerization reactions of isopropenylferrocene, as for α-methylsytrene, an expected side reaction would be for the intermediate ferrocenyl carbenium ion (P^+) to react with its monomer by an electrophilic substitution on the ring, as well as by isopropenyl group propagation. However, the propagation pathway shown above results in the preferential formation of the more thermodynamically stable carbenium ion[43] than that from the pathway involving electrophilic aromatic substitution, so normal propagation predominates in most cases. The formation of stable α-ferrocenyl carbenium ions, such as those involved in isopropenyl group propagation, have been reported to be very facile.[41-43] Nevertheless, ring alkylation probably still occurs to some extent under all reaction conditions, and this reaction would be an effective chain transfer or termination step. A particularly facile chain termination step of this type would be that in which the propagating carbenium ion undergoes electrophilic ring substitution on the unsubstituted cyclopentadienyl ring of the penultimate group, to form a bridge system of the type shown below:

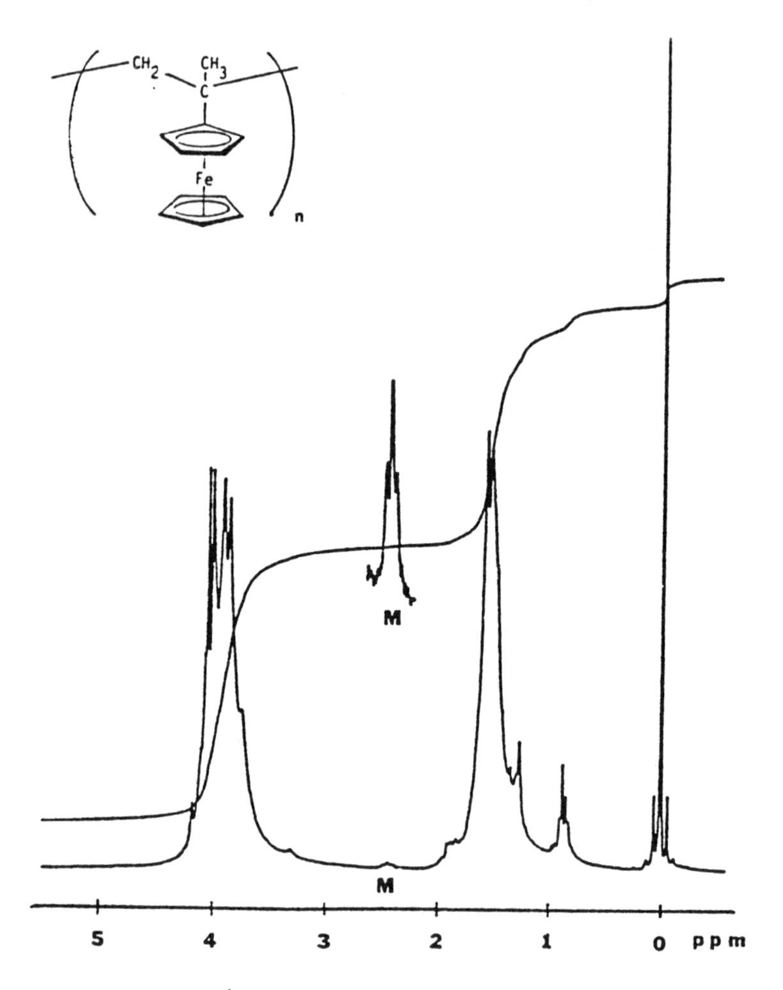

Figure 3. 250-MHz ^{1}H-NMR spectrum of polyisopropenylferrocene (polymer 7, Table II) in $CDCl_3$ solution. Internal standard TMS. (Reproduced with permission from Ref. 28. Copyright 1985 John Wiley.)

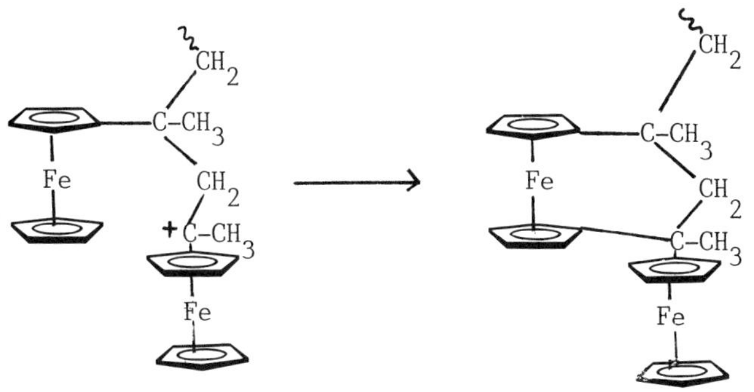

The ^{1}H-NMR in Figure 3 establishes this possibility. A multiplet (M, enlarged in scale in the figure) is indicative of a bridge methylene end group, as the chemical shift of the methylene group is in agreement with such bridged systems.[44]

The unusually high stability of isopropenylferrocenyl carbenium ion (P^+) can account for the inability of isopropenylferrocene to undergo cationic-initiated polymerization at low temperatures, because the unreactive carbenium ions so formed would not undergo rapid propagation. This type of unreactivity has been shown by Chisti and Jablonski[31] in the polymerization of 1,1'-diisopropenylferrocene. They demonstrated the inertness of the ferrocenyl carbenium ion at −78°C and even at 0°C, while at ambient temperature their intermediate carbenium ions were found to be more reactive. Similarly, in our investigations, at − 78° and 40°C no polymerization occurred. At 0°C only oily viscous oligomers resulted, but at ambient temperature the carbenium ions were apparently not so inert and polymerization occurred. In the case of vinylferrocene,[30] polymerization also occurred only at 0°C, resulting in the formation of polymers having $\bar{M}_n$ in the same range as shown in Table II. In the present case, the isopropenylferrocene carbenium ion is further stabilized by the α-methyl group.

In the copolymerization of isopropenylferrocene with α-methylstyrene at 0°C, using varying molar ratios of isopropenylferrocene and α-methylstyrene, traces of polymer formation were obtained only at a 30/70 ratio of the two monomers, as shown in the data in Table III. Because α-methylstyrene has a much lower ceiling temperature than styrene, we also decided to use styrene as a comonomer under conditions similar to those employed with α-methylstyrene. The reaction temperature for the copolymerization with α-methylstyrene was 20°C.

Table III. Cationic Copolymerization of Isopropenylferrocene (M_1)

Polymer No.	Monomer M_2	M_1 (in feed) (mol %)	Initiator (mol %)	Solvent	Time (h)	Temp. (^oC)	Result
12	α-Methylstyrene	10	$SnCl_4$(1.7)	CH_2Cl_2	1	0	
13	α-Methylstyrene	30	$SnCl_4$(1.6)	CH_2Cl_2	1	0	Trace polymer formation
14	α-Methylstyrene	50	$SnCl_4$(1.7)	CH_2Cl_2	1	0	
15	α-Methylstyrene	10	$SnCl_4$(1.6)	CH_2Cl_2	1	0	
16	Styrene	23	$BF_3 \cdot OEt_2$(18)	CH_2Cl_2	20	20	White-orange ppt, $\bar{M}_n$-2900
17	Styrene	23	$BF_3 \cdot OEt_2$(10)	CH_2Cl_2	1	20	
18	Styrene	23	$BF_3 \cdot OEt_2$(10)	CH_2Cl_2	24	20	Trace polymer formation
19	p-Methoxy-α-methylstyrene	50	$BF_3 \cdot OEt_2$(2)	CH_2Cl_2	36	0	White solid, $\bar{M}_n$-1600

Source: Reproduced with permission from Ref. 28. Copyright 1985 John Wiley.

Copolymerization of isopropenylferrocene with styrene was accomplished in two ways. In one method (polymer 16 of Table III) styrene and isopropenylferrocene were mixed together in CH_2Cl_2 at 20°C at a mole ratio of 23/77 of isopropenylferrocene to styrene, and polymerization was initiated with $BF_3 \cdot OEt_2$. From the 250-MHz NMR spectrum of the product, 27% styrene and 73% isopropenylferrocene units were found to be present in the copolymer, which had an $\bar{M}_n$ of 2900. This ratio was also confirmed by elemental analysis. In the second method (polymer 18) isopropenylferrocene and CH_2Cl_2 were placed in the polymerization tube at 0°C and stirred for approximately 10 min. then styrene was added. This reaction was continued for 20 h at 20°C, but only a trace of polymer formation was visible in this case.

Copolymerization reactions with p-methoxy-α-methylstyrene were also attempted at 0°C, and a methanol-insoluble product having an $\bar{M}_n$ of 1600 was obtained over a period of 36 h. From the 250-MHz ^1H-NMR spectrum, it was found that p-methoxy-α-methylstyrene was present in the polymer to the extent of 14%.

In the copolymerization of isopropenylferrocene with α-methylstyrene at 0°C, it appears that the stable carbenium ion of isopropenylferrocene acted as an inhibitor for the polymerization. Even the copolymerizations with styrene at 20°C were very slow and gave only low molecular weight copolymers in extremely low yields. In these styrene copolymers with isopropenylferrocene, the low inclusion of styrene units could be attributed to the greater reactivity of isopropenylferrocene, and vinylferrocene has been assumed to have an *r* value greater than styrene by Aso and Kunitake.[30]

All of these observations on the copolymerizations of isopropenylferrocene can again be attributed to the extreme stability of the intermediate carbenium ions. In order to obtain more conclusive evidence of this unreactivity, an attempt was made to destabilize the vinylferrocene by introducing a deactivating electron-withdrawing trifluoromethyl substituent at the α-position. For this purpose, the new monomer α-trifluoromethylvinylferrocene (TVF) was synthesized by the route shown below:

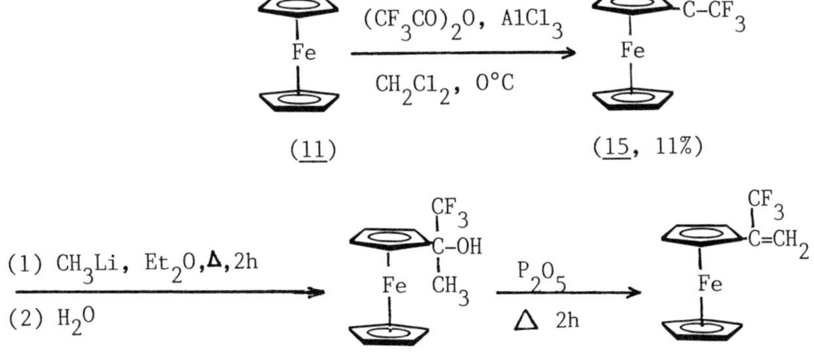

Fe (11) $\xrightarrow[CH_2Cl_2,\ 0°C]{(CF_3CO)_2O,\ AlCl_3}$ Fe–C(=O)–CF_3 (15, 11%)

$\xrightarrow[(2)\ H_2O]{(1)\ CH_3Li,\ Et_2O,\ \Delta,\ 2h}$ Fe–C(CF_3)(OH)–CH_3 (16, 76%) $\xrightarrow[\Delta\ 2h]{P_2O_5}$ Fe–C(CF_3)=CH_2 (TVF, 80%)

The deactivating influence of the CF_3 group was evident even in the monomer preparation because on subjecting the intermediate carbinol (16) to sublimation-pyrolysis, no dehydration occurred, in contrast to the behavior of ferrocenylcarbinol (14). Furthermore, no homopolymerization of the monomer occurred with cationic initiators such as $SnCl_4$ and $BF_3 \cdot OEt_2$ at 20°C. However, in a copolymerization reaction with styrene, at 0°C and using a mole ratio of TVF to styrene mole ratio of 1:1 over a period of 20 h using $BF_3 \cdot OEt_2$ as the initiator at 0°C, a small yield of polymer containing largely styrene units was obtained according to the ^{1}H-NMR spectrum, which showed a predominance of atactic polystyrene peaks. On repeating the above reaction for a period of 36 h at 0°C, a light-yellow solid insoluble in methanol was obtained in 52% yield, $\bar{M}_n$-4200. From the ^{1}H-NMR spectrum and elemental analysis, the copolymer was found to contain 45% TVF. These results are consistent with a lower order of stability for the carbenium ion derived from TVF compared to that derived from isopropenylferrocene. This result is of significance since α-trifluoromethylstyrene has not been homo or copolymerized by cationic initiation owing to the drastic deactivating influence of the system by the α-trifluoromethyl group.[45]

To further demonstrate the stability and unreactivity of the isopropenylferrocene carbenium ion, a stable derivative of this ion was prepared and evaluated as an initiator for the cationic polymerization of styrene. The derivative prepared for this purpose was the tetrafluoroborate salt of the isopropenylferrocene carbenium ion.[46] Even at 20°C a 26% yield ($\bar{M}_n$ = 11,315, $\bar{M}_w$ = 18,815) of polystyrene was obtained in 3 h of reaction) in CH_2Cl_2, and at 0°C only a 10% yield ($\bar{M}_n$ = 8444, $\bar{M}_w$ =12,314) of polystyrene resulted.

Significant evidence for the extreme stability of the carbenium ions derived from isopropenylferrocene was also provided by its ^{1}H-NMR spectrum in trifluoroacetic acid at 20°C which was similar to that reported earlier.[47] However, repeated scans of this sample at 1 h intervals showed no changes until a period of 10 h had elapsed. Thus it seems that the formation of the carbenium ion is very facile and once formed it is extremely stable in a non-nucleophilic environment. On pouring the contents of the NMR tube into methanol, no precipitate was seen.

New Metallocene Monomers

The attempted cationic polymerizations of the IDM and ITMM monomers with $SnCl_4$,$BF_3 \cdot OEt_2$ at 0° and − 78°C were not successful, and the monomers were recovered largely unreacted as indicated by their IR and ^{1}H-NMR spectra. The attempted cationic polymerization reaction at 20°C with $SnCl_4$ gave methanol-insoluble products which showed negligible carbonyl peaks in their IR spectra at 1950 and 2015 cm^{-1}, but a broad, strong nitrosyl stretching band was observed at 1675 cm^{-1}. This result may be attributed to side reactions which can occur with IDM and ITMM under cationic conditions. Although it has been reported that $SnCl_4$ itself does not complex with (η^5-cyclopentadienyl)dicarbonylnitrosylmolybdenum, yet it can complex with stronger Lewis acids.[48] Therefore, on initiation by the relatively weak Lewis acid $SnCl_4$, the IDM and ITMM monomers could form carbenium ions which could complex with the basic sites in the monomer, i.e., with the carbonyl and particularly with the nitrosyl groups. This type of interaction would prevent propagation through the isopropenyl group, and it would give products of complex structures, as was evidenced by their IR spectra. The results of attempted copolymerization reactions of these two monomers with styrene indicated that the same type of complications occurred. In the precipitated product an extremely strong, broad nitrosyl stretching band was observed at 1650 cm^{-1} as well as three weak carbonyl bands at 1950, 2020, and 2075 cm^{-1}.

Isopropenylferrocene does not homopolymerize under free radical conditions using AIBN as an initiator, but it does copolymerize with styrene.[49] Preliminary results indicate that the IDM monomer also copolymerizes with styrene using AIBN. In benzene solvent at 50°C in 24 h, a 10.6% yield of copolymer (IR $\nu_{C=O}$ 2020, 1945 cm^{-1}, $\nu_{N=O}$ 1675 cm^{-1}, $\nu_{C=C}$ 1601 cm^{-1}, ν_{CH} 3020 cm^{-1}) resulted having an $\bar{M}_n$ of 5700 and containing 6.8% of IDM as determined by elemental analysis. The initial monomer mixture contained 17% of IDM.

The DIS monomer, unlike its iron analogue, did not homopolymerize with $SnCl_4$ initiator even on heating. A plausible reason for this result is that this monomer contains a lone pair of electrons available for donation to Lewis acids.[50] Thus side reactions similar to those of the previous two monomers would prevent propagation. However, the DIS monomer also underwent a free radical copolymerization reaction with styrene and AIBN initiation.

From the above studies, it can be concluded that structural and electronic effects are not conducive to the cationic polymerization of isopropenylmetallocene monomers in general, in contrast to the high reactivities of their organic analogues, α-methylstyrene (isopropenyl benzene), for which polymerization through the isopropenyl group occurs very rapidly even at low temperatures. However, free radical polymerization using repeated initiations with AIBN over prolonged reaction periods can be utilized to prepare high molecular weight copolymers[33] for the last three monomers. Significant destabilization of the unreactive carbenium ion in isopropenylferrocene can also be achieved by introducing structural modifications, such as the α-trifluoromethyl group, and high molecular weight polymers should be preparable under cationic conditions.

Conclusions

We have demonstrated, by carefully manipulating the design of organometallic monomers 1 and 2, that their condensation polymerizations can be very facile.[51,52,53] Simultaneously, the introduction of flexible methylene groups can introduce elements of processability and mechanical integrity in such polymers.[53]

From our cationic polymerization studies of isopropenyl organometallic monomers, it can be concluded that structural and electronic effects are not conducive to this type of initiation and propagation[54,55] in general. This is in contrast to the high reactivities of their organic analogs.

Ferrocene containing condensation polymers have been utilized by us to modify the surfaces of electrodes.[56] Materials of this type that incorporate organo-iron compounds into a polymer matrix, either through chemical bonding or by formation of blends, have the potential of being thermally processed to yield iron oxides. If γ-Fe_2O_3 could be selectively synthesized, this could be a method of obtaining magnetic coatings.[57]

Recently we have also extended these low temperature polycondensation synthetic techniques to the preparation of borazine containing polyureas as precursors for BN/G_4C ceramics.[58]

Acknowledgments

Acknowledgment is made to the Donors of the Petroleum Research Fund, administered by the American Chemical Society, and to the Materials Research Laboratory, University of Massachusetts (Amherst), for grants in support of this research. The National Science Foundation, through the Expedited Award for Novel Research at Stevens Institute, has enabled K.E.G. to develop the processing of such transition metal containing organometallic polymers.[58]

References

1. Culbertson, B.M.; Pittman, C.U., Jr. "New Monomers and Polymers" Polymers"; Plenum Press: New York, 1984.
2. Carraher, C.E., Jr.; Sheats, J.E.; Pittman, C.U., Jr. "Advances in Organometallic and Inorganic Polymer Science"; Marcel Dekker: New York, 1982.
3. Carraher, C.E., Jr.; Sheats, J.E.; Pittman, C.U., Jr. "Organometallic Polymers", Academic Press: New York, 1978.
4. Neuse, E.W.; Rosenberg, H. "Metallocene Polymers"; Marcel Dekker: New York, 1970.
5. Pittman, C.U., Jr. In "Polymer Supported Reactions in Organic Synthesis"; Hodge, P.; Sherrington, C.D., Eds.; Wiley: New York, 1980.
6. Pittman, C.U., Jr. Organomet. React. Synth. 1977, 6 1.
7. Macomber, D.W.; Hart, W.P.; Rausch, M.D.; Priester, R.D.; Jr.; Pittman, C.U., Jr. J. Am. Chem. Soc. 1982, 104, 884.
8. Carraher, C.E., Jr. In "interfacial Synthesis. Volume II Polymer Applications and Technology"; Millich, F.; Carraher, C.E.; Jr., Eds.; Marcel Dekker: New York, 1988; p. 251.
9. One previous synthesis of ferrocene-containing condensation polymers via interfacial methods at room temperature has been reported by Knobloch and Rauscher,[10] who formed low molecular weight polyamides and polyesters by reacting 1,1'-bis(chloroformyl)ferrocene with various diamines and diols. Further, Carraher and co-workers[8] have utilized interfacial techniques in the formation of other types of organometallic polymers.

10. Knobloch, F.W.; Rauscher, W.H. J. Poly. Sci. 1961, 54, 651.
11. Sonoda, A.; Moritani, I. J. Organomet. Chem. 1971, 26, 133.
12. Ratajcak, A.; Czech, B.; Drobek, L. Synth. React. Inorg. Met. Org. Chem. 1982, 12, 557.
13. Pittman, C.U., Jr. J. Poly. Sci., Polym. Chem. Ed. 1968, 6, 1687.
14. Trifan, D.W.; Backskai, R. Tetrahedron Lett. 1960, 1.
15. Watts, W.E. In "Comprehensive Organometallic Chemistry"; Wilkinson; G., Stone, F.G.A.; Abel, E., Eds.: Pergamon Press: New York vol. 8, p. 1052.
16. (a) Morgan, P.W. "Condensation Polymers by Interfacial and Solution Methods"; Intersci.: NY, 1965; (b) Chapter 10, p. 446.
17. Waltz, J.E., Taylor, G.B., Anal. Chem. 1947, 19, 448.
18. Saunders, J.H.; Frisch, K.C. "Polyurethanes. Chemistry and Technology. Part I. Chemistry"; Interscience: New York, 1962; p. 174.
19. (a) Snider, O.E.; Richardson, R.J. In "Encyclopedia of Polymer Science and Technology"; Bikales, N.M., Ed.; Wiley-Interscience: New York, 1969, Vol. 10, p. 391 (b) Hummel, D.O. "Polymer Spectroscopy"; Verlag Chemie: Wienheim, 1974; Chapter 2.2.
20. Lednicer, D.; Lindsay, J.K.; Hauser, C.R. J. Org. Chem. 1958, 23(5), 653.
21. Pittman, C.U., Jr.; Voges, R.L.; Jones, W. Macromolecules 1971, 4(3), 296.
22. Hepburn, C. "Polyurethane Elastomers", Applied Science, London, 1982; Saunders, J.H., Frisch, K.C. "Polyurethane Chemistry and Technology, Part I, Chemistry", Interscience, New York, 1962, p. 174.
23. Bailey, F.E., Jr., Critchfield, F.E. "Urethane Chemistry and Applications", ACS Symposium Series, 1981, Chapter 11, p. 134.
24. Bark, L.S., Allen, N.S. "Analysis of Polymer Systems", Applied Science, London, 1982, p. 103.
25. Dussel, H.J., Wenzel, N., Hummel, D.O. Angew. Makromol. Chem., 1985, 129, 121.
26. Dussel, H.J., Wenzel, N., Hummel, D.O. Angew. Makromol. Chem. 1982, 106, 107.
27. Richards, J.M., McClennen, W.H., Menzelaar, H.L.C., Gregonis, D.E., Reichert, W.M., Helle, M.A., Macromolecules, 1985, 18, 496.
28. Gonsalves, K., Lin, Z.R., Lenz, R.W., Rausch, M.D. J. Polym. Sci. Polym. Chem. Ed. 1985, 23, 1707.
29. Silverstein, R.M., Bassler, C.G., Morrill, T.C. "Spectrometric Identification of Organic Compounds", fourth ed., Wiley, New York, 1981, Chapter 5, p. 249.
30. Aso, C., Kunitake, T , Nakashima, T. Makromol. Chem. 1969, 124 232.
31. Sosin, S.L., Korshak, V.V. Dokl. Akad. Nauk SSR, 1968, 179, 1124; Chisti, A.S., Jablonski, C.R. Makromol. Chem. 1983, 184, 1837.
32. Pittman, C.U., Jr. "Organometallic Polymers", Carraher, C.E., Jr. Sheats, J.E., Pittman, C.U., Jr., Eds., Academic, New York, 1977, p. 5; Rausch, M.D., Macomber, D.W., Fang, F.G., Pittman, C.U., Jr., Jayaraman, T.V., Priester, R.D. "New Monomers and Polymers", Culbertson, B.M., Pittman, C.U., Jr. Eds., Plenum, New York, 1984, p. 243.

33. (a) Macomber, D.W.; Hart, W.P., Rausch, M.D.; Priester, R.A., Jr.; Pittman, C.U., Jr. J. Am. Chem. Soc. 1982, 104, 882; (b) Rausch, M.D.; Macomber, D.W., Gonsalves, K., Fang; F.G., Lin, Z.R.; Pittman, C.U., Jr. Polym. Maters.: Sci. Eng. Prepr. 1983, 49, 358.
34. Rinehart, K.L.; Kittle, R.A., Ellis; A.F. J. Am. Chem. Soc. 1960, 82 2083.
35. Ellis, A.F. Dissertation Abstr. 1963, 24, 510.
36. Fitzgerald, W.P., Jr. Dissertation Abstr. 1964, 24, 2687.
37. Pittman, C.U., Jr. Organomet. React. & Synth. 1977, 6, 1.
38. Rosenblum, M.; Woodward. R.B. J. Am. Chem. Soc. 1958, 80, 5443.
39. Vogel, M.; Rausch, M.D., Rosenberg, H. J. Org. Chem. 1957, 22, 1016.
40. Rausch, M.D.; Siegel, A. J. Organomet. Chem. 1968, 11, 317.
41. Cais, M.; Eisenstadt, A. J. Org. Chem. 1965, 30, 1148.
42. Lupan, S.; Kapon, M.; Cais, M.; Herbstein, F.H. Angew. Chem. Int. Ed. 1972, 11, 1025.
43. Watts, W.E. Organomet. Chem. Rev. 1979, 7, 399.
44. Rosenblum, M.; Banerjee, A.K.; Danieli, N.; Fish, R.W.; Schlatter, V. J. Am. Chem. Soc. 1963, 85, 316; Watts, W.E.; Lentzner, H.L. Tetrahedron 1971, 27, 4343.
45. Antonucci, J.M., "High Polymers XXV: Fluoropolymers", Wall, L., Ed., Interscience, New York, 1972, p. 70.
46. Allenmark, S. Tetrahedron Lett. 1974, 4, 371; Allenmark, S.; Kalen, K.; Sandblom, A. Chemica Scripta. 1975, 7, 97.
47. Turbitt, T.D.; Watts, W.E. J. Chem. Soc., Perkin Trans. 2 1974, 189.
48. Lokshin, B.V.; Rusach, B.B.; Kolobova, N.E.; Makamov, Y.V.; Ustynyuk, N.A.; Zdanovich, V.I.; Zh. Zhakaeva, A.; Setkina, V.N. J. Organomet. Chem. 1976, 108, 353.
49. Howard, M.; Reed, S.F., Jr. J. Polym. Sci. A-1 1971, 9, 2085.
50. Harrison, P.G.; Richards, J.A. J. Organomet. Chem. 1976, 108, 35; Holliday, A.K.; Makin, P.H.; Puddephat, R.H. J. Chem. Soc., Dalton Trans. 1976, 435.
51. Gonsalves, K.; Lenz, R.W.; Rausch, M.D. Appl. Organomet. Chem. 1987, 1, 81.
52. Gonsalves, K.; Lin, Z.; Rausch, M.D. J. Am. Chem. Soc. 1984, 106, 3862.
53. Gonsalves, K.; Rausch, M.D. J. Polym. Sci: Part A: Polym. Chem. 1986, 24, 1599.
54. Gonsalves, K.; Lin, Z.; Lenz, R.W.; Rausch, M.D. J. Polym. Sci: Polym. Chem. Ed. 1985, 23, 1707.
55. Lin, Z.; Gonsalves, K.; Lenz, R.W.; Rausch, M.D. J. Polym. Sci: Polym. Chem. Ed. 1986, 24, 347.
56. Gonsalves, K.; Rausch, M.D.; Bard, A.J.; Kepley, L. unpublished studies.
57. Gonsalves, K.; Rausch, M.D. J. Polymer Sci., Chem. Ed., (in press).
58. Gonsalves, K.; Agarwal, R. presentation at "High Temperature Structural Composites: Synthesis, Characterization & Properties" Symposium May 87, sponsored by MRS, NJ.

RECEIVED September 24, 1987

Chapter 37

Soluble Metal Chelate Polymers of Coordination Numbers 6, 7, and 8

Ronald D. Archer, Bing Wang, Valentino J. Tramontano, Annabel Y. Lee, and Ven O. Ochaya

Department of Chemistry, University of Massachusetts, Amherst, MA 01003

Metal coordination polymer syntheses are often plagued by insolubility, which prevents oligomers growing into polymers; however, four 6-coordinate cobalt(III) chelate polymers with acetylacetonato and leucinato ligands, a sizable number of 7-coordinate dioxouranium-(VI) dicarboxylate polymers, and two 8-coordinate zirconium(IV) polymers with Schiff-base ligands can be prepared with degrees of polymerization which are only limited by reagent stiochiometry. Chain growth of the inert cobalt(III) center is provided through ligand-centered condensation reactions of the acetylacetonato ligands with a sulfur or thionyl chloride. The problem of low ligand solubility of aromatic Schiff-bases has been side stepped by ligand synthesis during the polymerization reaction. Characterization has been accomplished primarily through infrared spectroscopy, nuclear magnetic resonance spectroscopy (both solution and solid state) and gel permeation chromatography.

Intractability (insolubility) has long plagued the development of linear transition metal coordination polymers; cf., Bailar (1). Classically the problem appears to be related to the concern for thermally stable species; thus, flat conjugated ligands were often joined with divalent metal ions which possess square planar coordination spheres. Intermolecular "stacking" forces between such planar arrays typically results in the precipitation of metal oligomers before adequate chain growth occurs. We have attempted to overcome this stacking problem with 1) 8-coordinate nonrigid metal-ion cores, 2) cis octahedral coordination, 3) oxo-metal ions, 4) strongly coordinating solvents, and 5) bulky side groups. Since another recent article (2) discusses the effects of side groups on phthalocyanine solubility [C_6 or greater side chains on eight sites of the monomer provide good solubility for the otherwise insoluble phthalocyanines] examples in which 1) to 4) are potentially able to solve the stacking problem are noted herein.

0097-6156/88/0360-0463$06.00/0

The synthesis of new metal coordination polymers also requires a careful consideration of reaction kinetics. That is, inert metal centers do not undergo ligand substitution reactions fast enough to obtain clean polymerizations with bridging ligands, except when a complete substitution of all ligands can be tolerated through elevated temperature reactions, when a photo-assisted or photo-catalyzed reaction is possible to accomplish the same results, or when one type of ligand can be removed by a factor of over 50:1 relative to other ligands on the metal. That is, a 98% reaction yield with 99% (or better) pure ligands at exactly a 1:1 mole ratio of reactants is required in order to obtain even a reasonable average chain length (degree of polymerization, DP=50) for random step-growth polymerizations of the type needed for linear coordination polymers (3,4). Bridging axial ligands on macrocycles are one class of species in which the 50:1 reaction ratio is vastly exceeded, and uranyl chelate polymers (5) provide a second type because the oxo ligands are inert relative to other ligands in non-aqueous solvents.

On the other hand, labile metal ions, for which 98% plus reactions should be easily obtained, often undergo side reactions and depolymerizations (the reverse reactions) are favored from entropy considerations.

One method of solving the kinetics dilemma is well known in coordination chemistry; that is, start with a labile metal ion and render it inert during the course of the synthetic reaction. We have accomplished this in the case of zirconium(IV) by starting with tetrakis(salicylaldehydo)zirconium(IV), which is quite labile, and polymerization with 1,2,4,5-tetraaminobenzene in a Schiff-base condensation reaction in situ (6). The polymeric product contains a "double-headed" quadridentate ligand, which is much more inert to substitution. However, 1,2,4,5-tetraaminobenzene has become very expensive. Therefore, the synthesis of a zirconium polymer with 3,3',4,4'-tetraaminobiphenyl (commercially 3,3'-diaminobenzidine) with zirconium salicylaldehyde, $Zr(sal)_4$ (7) has been undertaken as shown below:

$$n\ Zr(sal)_4 + n\ (NH_2)_2(C_6H_3)(C_6H_3)(NH_2)_2 \Longrightarrow [Zr(tsdb)]_n + 4n\ H_2O \quad (1)$$

where tsdb is the tetraanion of the product of the Schiff-base condensation reaction between four salicylaldehyde molecules and 3,3'-diaminobenzidine (Archer, R. D.; Wang, B., paper in preparation). See structure 1. The reaction is conducted in dimethyl sulfoxide (DMSO), because DMSO is a very good solvent for the growing oligomers and its strong interaction with water prevents hydrolysis of the zirconium(IV) and favors the forward Schiff-base condensation reaction.

A second method of solving the kinetics dilemna consists of driving a polymerization reaction to completion by removing a volatile coproduct as in the case of uranyl dicarboxylate polymers from uranyl acetate and the dicarboxylic acid (5):

$$n\ UO_2(O_2CCH_3)_2 + n\ HO_2CRCO_2H \Longrightarrow [UO_2(O_2CRCO_2)(DMSO)_m]_n + 2n\ HO_2CCH_3 \quad (2)$$

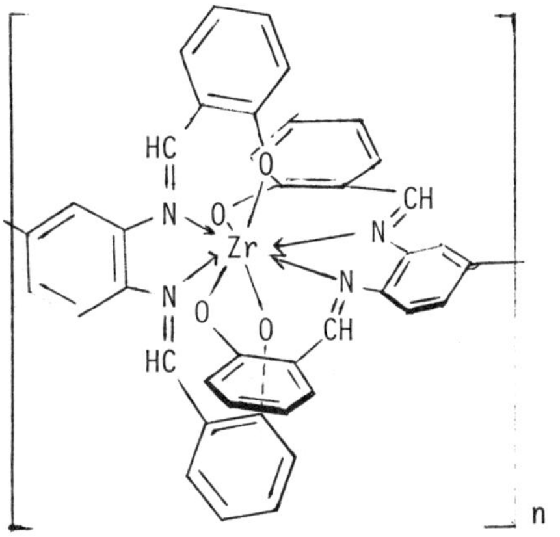

1 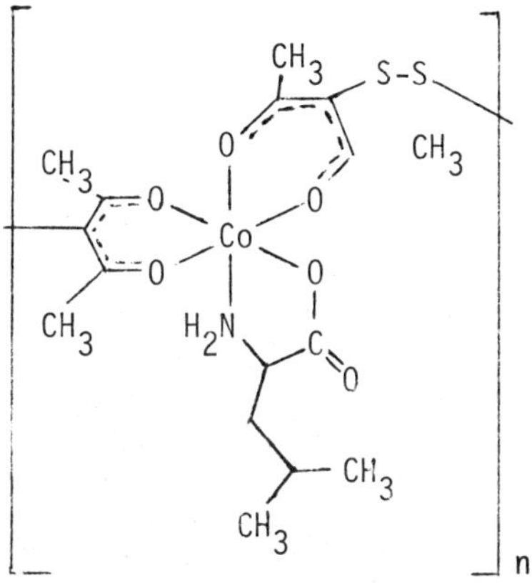 2

where R is an alkyl group, an alkene, an alkyne, a phthalic group, or a dialkylthio group. The acetic acid is removed at about 50°C *in vacuo*. More recently we have synthesized polymers in which R is an analogous dithiodialkyl and methylenedithiodialkyl group using the same reaction. [The extreme radiation sensitivity of these species is shown in Table I and will be detailed elsewhere (Archer, R.D.; Hardiman, C.J.; Lee, A.Y. Proc. 7th Intl. Symp. Photochem. Photophys., Springer-Verlag, in press; full papers in preparation.) The G values can be compared with the widely used poly(methyl methacrylate), which has a G_s value of about 1.3.]

A third method of solving the kinetics dilemma is to use organic reactions with functionalized inert coordination compunds. An example of a cobalt(III) polymer which has been prepared in this manner is shown below.

$$n\ Co(acac)_2(leu) + n\ S_2Cl_2 \Longrightarrow [Co(acacSSacac)(leu)]_n + 2n\ HCl \qquad (3)$$

where acac is the anion of acetylacetone (2,4-pentanedione enolate) and leu is the anion of leucine. The polymer has been prepared in dimethylacetamide with sodium carbonate as a base to eliminate reverse reactions. The salts were removed with water prior to fractionation with DMAC/acetone. Best yields have been obtained when the reaction was conducted at -10°. (Archer, R. D.; Tramontano, V. J., paper in preparation). See structure 2. Similarly, we have prepared the analogous thio-bridged species by the reaction of SCl_2 with the same $Co(acac)_2(leu)$ chelate in acetonitrile at room temperature or methylene chloride at -10°, the analogous sulfoxo bridged polymer by the reaction of thionyl chloride with the same cobalt(III) chelate in acetonitrile at room temperature or methylene chloride at -10°, and the analogous sulfone bridged polymer by oxidation of the sulfoxo bridged polymer with hydrogen peroxide. Preliminary data indicates a high radiation sensitivity for these polymers as well. (G_s values based on polystyrene calibrated molecular weights are calcu-

Table I. Gamma-ray Sensitivity of Uranyl Polymers

Empirical Formula Unit[a] (bridging carboxylate)	$\bar{M}_n$[b]	$(G_s - G_x)$[c]
$UO_2[O_2CC(CH_3)_2CH_2CO_2](C_2H_6SO)$ (2,2-dimethylsuccinate)	9,000[d]	- 12[d]
$UO_2[O_2CC(CH_3)_2CH_2CH_2CO_2](C_2H_6SO)$ (2,2-dimethylglutarate)	48,000[e]	- 2.7[e]
$UO_2[O_2CCH_2C(CH_3)_2CH_2CO_2](C_2H_6SO)$ (3,3-dimethylglutarate)	12,000[d]	- 6.9[d]
$UO_2[O_2CC(CH_3)_2(CH_2)_3C(CH_3)_2CO_2](C_2H_6SO)$ (2,2,6,6-tetramethylpimelate)	12,000[d]	- 37[d]
$UO_2(Z\text{-}O_2CH{=}CHCO_2)(C_2H_6SO)_{1.75}$ (maleate)	9,300	- 2.9
$UO_2(o\text{-}O_2CC_6H_4CO_2)(C_2H_6SO)_2$ (phthalate)	17,800[d]	- 3.3[d]
$UO_2(O_2CCH_2SCH_2CO_2)(C_2H_6SO)_2$ (thiodiglycolate)	14,000[d]	+ 50[d]
$UO_2(O_2CCH_2SSCH_2CO_2)(C_2H_6SO)_{1.5}$ (dithiodiglycolate)	12,000[d-g]	+280[d-g]
$UO_2(O_2CCH_2SCH_2SCH_2CO_2)(C_2H_6SO)_2$ [methylenebis(thioglycolate)]	15,000[d,f]	+ 93[d,f]
$UO_2(E\text{-}O_2CCH{=}CHCO_2)(C_2H_6SO)_2$ (fumarate)	26,000[f,g]	+ 55[f,g]

[a]Repeating unit of polymer chain including solvation
[b]$\bar{M}_n$ of samples not irradiated; based on GPC in NMP with polystyrene standards, or by NMR end-group analysis if so indicated
[c]Measured net G value after cesium-137 irradiation corrected relative to Fricke dosimeter
[d]$\bar{M}_n$ values for this polymer--end-group calibrated by NMR
[e]Polymer has poor molecular weight distribution and poor agreement with end-group determination
[f]Polymer undergoes both scission and crosslinking
[g]Polymer very sensitive to radiation--nonirradiated samples change molecular weight distribution significantly during irradiation of other samples

lated as 8.2, 13.2, and 16.8 for the thio, sulfoxo, and dithio bridged polymers, respectively, for cesium-137 gamma irradiation.)

Other examples of this synthetic strategy are known; for example, a recent zirconium polymer by Illingsworth and Burke (8), who joined amine side groups of a zirconium bis(quadridentate Schiff-base) with an acid dianhydride to give amide linkages. Once again, caution is necesary, as Jones and Power (9) learned when they attempted to link metal bis(β-diketonates) with sulfur halides; that is, they obtained insoluble metal sulfides because the β-diketone complexes which they used were fairly labile and the insolubility drove the reactions to completion in the wrong direction.

From gel permeation chromatography, nuclear magnetic resonance spectroscopy end-group analysis, and viscosity measurements in N-methyl-2-pyrrolidone (NMP), d_6-DMSO, and NMP, respectively, the number-average molecular weights are estimated at about 10,000 for the new zirconium Schiff-base polymer, 6,000 to 24,000 for the uranyl polymers, and from 20,000 to greater than 40,000 for fractionated samples of the cobalt polymers. {Gel permeation chromatography calibration with polystyrene can lead to erroneous molecular weights for the heavy metal species as shown by NMR analysis of the acetate end groups of the uranyl polymers, which have the average formula $H[O_2C\text{-}RCO_2UO_2]_nO_2CCH_3$, neglecting the coordinated solvent. The gamma proton of an end-group acetylacetonate provides a similar NMR handle for the cobalt(III) complexes.} A higher molecular weight zirconium polymer could undoubtedly be obtained (cf. ref. 6), but our current interest in this polymer has to do with strong surface interactions with silica; and in fact, we have been deliberately using an excess of the zirconium salicylaldehyde component. We have spectral evidence for the strong adhesion of the zirconium end groups to silica.

NMR measurements also provide information on the coordination of the ligands in the uranyl polymers. Solid-state ^{13}C-NMR confirms the coordination modes of the carboxylate ligands to the uranyl ion; that is, both monodentate and bidentate carboxylate coordination modes are evident. The uranyl dicarboxylate polymers which possess two moles of coordinated DMSO exhibit two carbon-13 carbonyl resonances, one at about 175 ppm downfield from tetramethylsilane (TMS) and one at about 185 ppm. The polymers which possess only one mole of coordinated DMSO exhibit only the carbonyl peak near 185 ppm. Based on other known coordination compounds, the 175 ppm peak can be assigned to monodentate carboxylate and the 185 ppm peak to bidentate carboxylate. Thus, 7-coordination predominates in the polymers with either one or two moles of solvent coordinated to the uranyl ion, which is consistent with the infrared results reported elsewhere (5).

The proton NMR of the cobalt monomer and polymers also confirms the presence of diastereomers due to the S-leucine moiety. Peaks due to the diastereomers is most evident in the gamma proton of the β-diketones in the monomer, for which four resonances are observed in pairs of different relative intensities based on the method of preparation.

Infrared and thermal studies also add to our knowledge of these polymers. For example, the aromatic C-O stretch of the Schiff-base ligands is shifted above 1300 cm^{-1} when coordinated, and coordinated carboxylate shows no CO stretch above 1600 cm^{-1} when bidentate. The thermal stability of the zirconium chelate polymers is outstanding, with temperatures of 600° required for decomposition.

Whereas the preceding provides only three examples of overcoming the kinetics delimma associated with the synthesis of coordination polymers, the messages are clear:

1. Labile metal ions can be converted into inert polymers if a sufficient increase in the dentate number occurs during the reaction. Double-headed terdentate ligands with metals of modest inertness [e.g., nickel(II)] should work similarly. Furthermore, cobalt(II) to cobalt(III) and chromium(II) to chromium(III) should provide further possibilities. Ingenuity is required for the last two cases since the odd charge makes a simple step-growth reaction more difficult, though a charge per unit could be an advantage in the preparation of water-soluble ionomers (similar to proteins in nature and polyacrylamides in industry).

2. The possibilities of linking inert metal coordination centers together appears almost limitless. The use of macrocycles allows the use of metal ions which would normally be labile. Manganese(II) or (III) with funcionalized phthalocyanines or porphyrins could be linked to form strongly interacting sheets of potentially conducting materials. The size is normally limited by solubility; however, large alkyl groups on the rings can overcome the intractability (2).

The preceding is merely the "tip of the iceberg" to show the potential which exists in this field.

Acknowledgments

We acknowledge with gratitude the financial support of the Office of Naval Research for the radiation sensitive polymer research and the University of Massachusetts Institute for Interface Science and IBM for the surface active polymer research.

Literature Cited

1. Bailar, J.C., Jr. In Organometallic Polymers, Carraher,C.E.,Jr.; Sheats, J.E.; Pittman, C.U., Jr., Eds.; Academic Press, New York, 1978.
2. Orthmann, E.; Wegner, G. Makromol. Chem., Rapid Commun. 1986, 7, 243.
3. Ward, T.C. J. Chem. Educ., 1981, 58, 867.
4. Billmeyer, F.W., Jr. Textbook of Polymer Science 2nd ed., Wiley-Interscience, New York, 1971.
5. Hardiman, C.J.; Archer, R.D. Macromolecules, 1987, 20,
6. Archer, R.D.; Illingsworth, M. L.; Rau, D.N.; Hardiman, C.J. Macromolecules, 1985, 18, 1371.
7. Archer, R.D.; Illingsworth, M.L. Inorg. Nucl. Chem. Lett., 1977, 13, 661.
8. Illingsworth, M.L.; Burke, S. Abstr. Am. Chem. Soc. Mtg., Apr. 1986, INOR 148.
9. Jones, R.D.G.; Power, L.F. Australian J. Chem., 1971, 24, 735.

RECEIVED September 1, 1987

Chapter 38

A New Class of Oligomeric Organotin Compounds

Robert R. Holmes, Roberta O. Day, V. Chandrasekhar, Charles G. Schmid, K. C. Kumara Swamy, and Joan M. Holmes

Department of Chemistry, University of Massachusetts, Amherst, MA 01003

Oligomeric organotin oxycarboxylates based on the compositions, $[R'Sn(O)O_2CR]_6$ and $[(R'Sn(O)O_2CR)_2R'Sn(O_2CR)_3]_2$ have "drum" and "ladder" structures, respectively. They may be prepared by condensation of a stannoic acid with the respective carboxylic acid. Reaction of $RSnCl_3$ with the silver salt of the carboxylic acid also gives oxycarboxylate compositions. With the use of diphenylphosphinic acid instead of a carboxylic acid, an oxygen-capped cluster results. In addition, cubic and butterfly arrangements are obtained with dicyclohexylphosphinic acid. All of these represent new structural forms for organotin compounds. X-ray analysis shows the presence of four- and six-membered rings as a primary geometrical feature. The cluster can be viewed as a hydrolysis product of the drum just as the drum is viewed as a hydrolysis product of the ladder. Interconversion of the latter two forms has been followed by ^{119}Sn NMR and is shown to be reversible. Consideration of their structural relations suggests additional forms yet to be discovered.

In 1922 Lambourne (1) reported that reactions of methyl stannoic acid with carboxylic acids yielded compounds whose formulation corresponded to $[MeSn(O)O_2CR']_3$. In a subsequent paper, Lambourne (2) reported additional oxycarboxylates of methyl stannoic acid, $[MeSn(O)O_2CR]_6$ where R = Et, n-Pr, i-Pr having hexameric compositions established from cryoscopic molecular weight measurements in benzene. Both the trimer and hexamer derivatives were assigned cyclic structures containing tetracoordinated tin atoms. From our work, we know now that these were erroneous assignments. The hexamer derivatives of this composition that we have isolated are "drum" shaped molecules containing hexacoordinated tin atoms. Figure 1 shows a representative member, $[PhSn(O)O_2CC_6H_{11}]_6$, whose geometry was determined by a single crystal X-ray diffraction study (3).

0097-6156/88/0360-0469$06.00/0

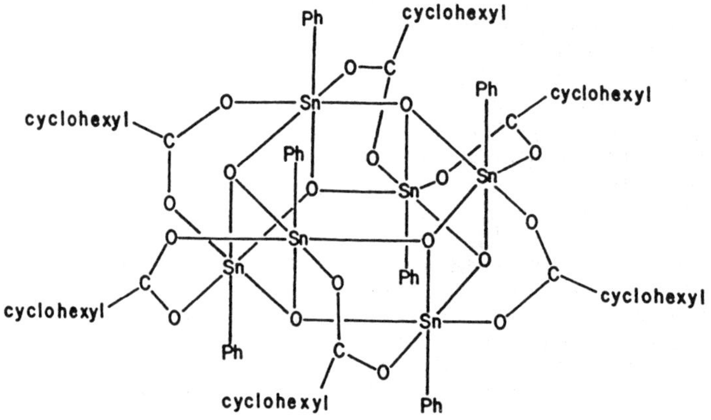

Figure 1. Schematic representation of the drum structure of $[PhSn(O)O_2CC_6H_{11}]_6$. (Reproduced from Ref. 3. Copyright 1985 American Chemical Society.)

As outlined elsewhere (4), we have employed a variation of the original reaction by Lambourne (3) in exploring condensation products leading to the drum composition as well as to a mixed oxycarboxylate-tricarboxylate formulation, $[(R'Sn(O)O_2CR)_2R'Sn(O_2CR)_3]_2$. The latter was identified as having an unfolded drum or ladder structure (4). The reaction consists of a condensation of an organostannoic acid with a carboxylic acid.

The application of this reaction leads to the formation of a soluble drum compound, $[n\text{-}BuSn(O)O_2CC_5H_9]_6 \cdot C_6H_6$, 1, Figure 2, containing cyclopentane units, and the formation of an unusual ladder compound, $[(n\text{-}BuSn(O)O_2CPh)_2\ n\text{-}BuSn(Cl)(O_2CPh)_2]_2$, 2, Figure 3 (4). In addition the ladder composition, $[(n\text{-}BuSn(O)O_2CC_6H_{11})_2\ n\text{-}BuSn(O_2CC_6H_{11})_3]_2$, 3, Figure 4, was prepared from the reaction of n-$BuSnCl_3$ with silver cyclohexanoate (5). The latter reaction was employed by Anderson (6) in synthesizing organotin tricarboxylates. Both the drum 1 and ladder 3 undergo structural changes in solution in interconversion processes as indicated by ^{119}Sn NMR. A mechanism associated with this process is proposed (5).

Discussion

Synthesis. Both the cyclopentanoate drum composition, 1, and the cyclohexanoate drum, $[n\text{-}BuSn(O)O_2CC_6H_{11}]_6$, 4, were prepared by a condensation reaction of n-butyl stannoic acid with the corresponding carboxylic acid, Equation 1.

$$6\ n\text{-}BuSn(O)OH + 6\ RCO_2H \longrightarrow [n\text{-}BuSn(O)O_2CR]_6 + 6\ H_2O \qquad (1)$$

In a subsequent reaction, n-$BuSnCl_3$ was reacted with the silver salt of cyclohexane carboxylic acid in the presence of wet solvent. This reaction gave the ladder formulation, 3, identified above, Equation 2.

$$6\ n\text{-}BuSnCl_3 + 10\ Ag^+C_6H_{11}CO_2^- + 4\ H_2O \longrightarrow$$

$$[(n\text{-}BuSn(O)O_2CC_6H_{11})_2\ n\text{-}BuSn(O_2CC_6H_{11})_3]_2 + 10\ AgCl + 8\ HCl \qquad (2)$$

Other than the chloro derivative, 2, the drum and ladder compounds are prepared in high yield, > 70%. All are soluble in organic solvents and show characteristic infrared spectra. The drum compounds 1 and 4 exhibit a symmetrical doublet for the carboxylate stretching frequency, ν_{COO}, centered near 1550 cm^{-1} and a single Sn-O stretch, ν_{Sn-O}, near 600 cm^{-1}. In contrast, the open-drum structures, 2 and 3, show an unsymmetrical ν_{COO} doublet in the same region as that for the drums and the presence of two Sn-O stretches near 600 cm^{-1}.

The drum compounds are thermally quite stable. For example, the cyclopentane carboxylic acid drum, 1, was heated at 300°C for 3 h in vacuum. The material obtained was soluble in $CDCl_3$, and ^{119}Sn NMR showed a single line at -491.4 ppm, compared to the starting material -485.8. The shift is presumably due to loss of solvent molecules present in the crystal in the starting material.

^{119}Sn NMR Data. Hydrolytically, the drum formulations are more stable than the open-drum forms. The hydrolysis reaction in Equation 3 illustrates this.

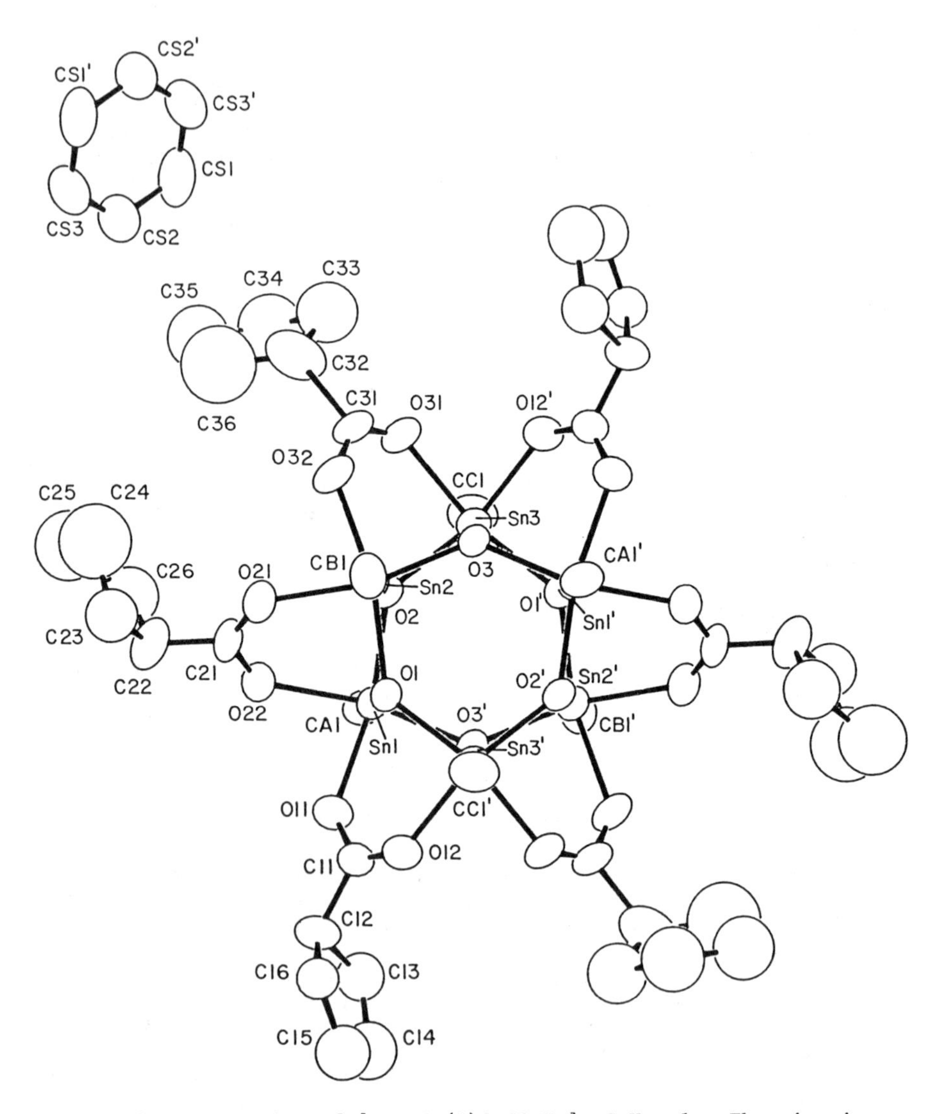

Figure 2. ORTEP plot of [n-BuSn(O)$O_2CC_5H_9]_6 \cdot C_6H_6$, 1. The view is down the pseudo S_6 axis. The terminal carbon atoms of the n-Bu groups are omitted for purposes of clarity. (Reproduced from Ref. 5. Copyright 1987 American Chemical Society.)

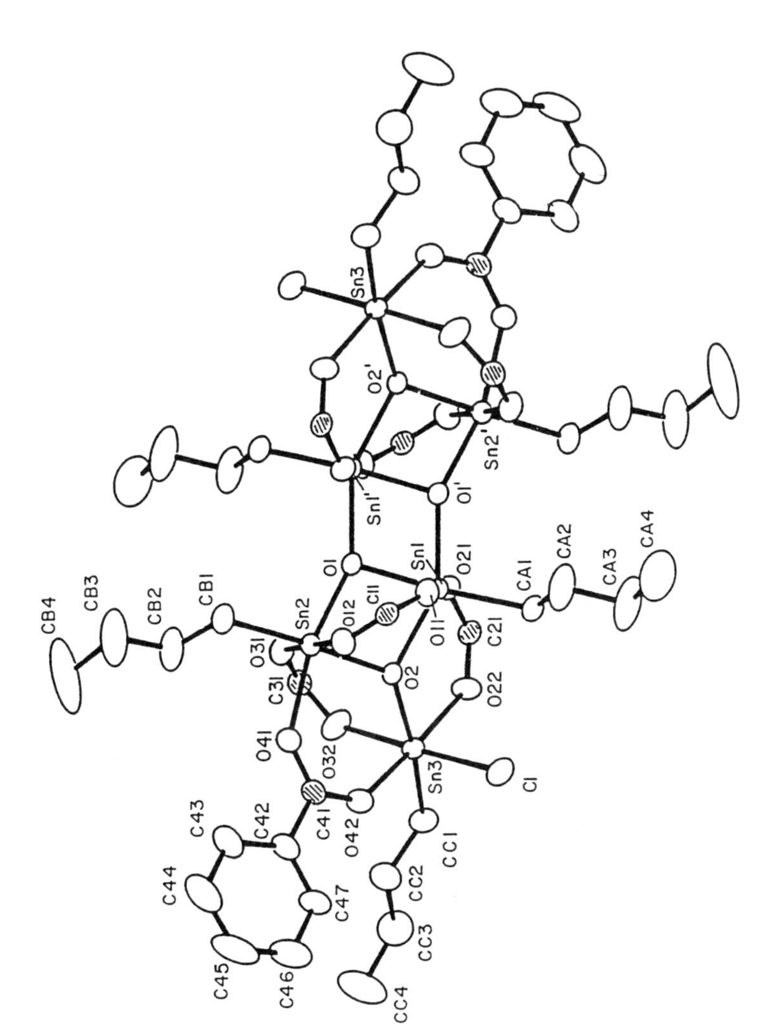

Figure 3. ORTEP plot of $[(\underline{n}\text{-BuSn(O)O}_2\text{CPh})_2\ \underline{n}\text{-BuSn(Cl)(O}_2\text{CPh})_2]_2$, 2. Six of the eight phenyl groups and all hydrogen atoms have been omitted for purposes of clarity. (Reproduced from Ref. 5. Copyright 1987 American Chemical Society.)

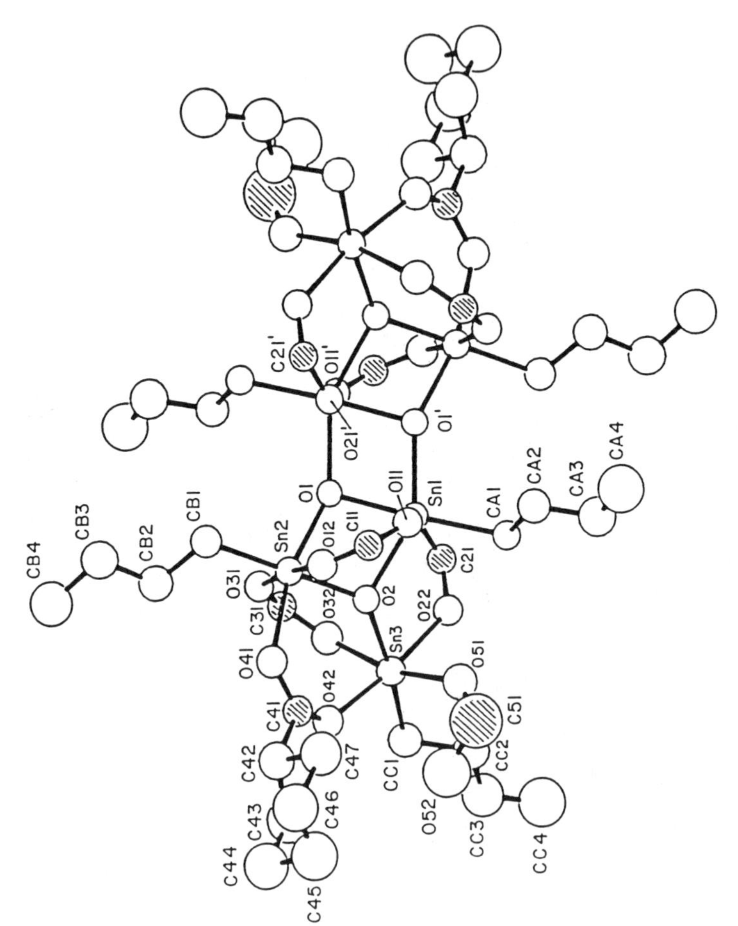

Figure 4. ORTEP plot $[(\underline{n}\text{-BuSn(O)O}_2\text{CC}_6\text{H}_{11})_2\ \underline{n}\text{-BuSn(O}_2\text{CC}_6\text{H}_{11})_3]_2$, 3. Carbon atoms of only two of the cyclohexyl groups are shown. Reproduced from Ref. 5. Copyright 1987 American Chemical Society.)

$$[(R'Sn(O)O_2CR)_2R'Sn(O_2CR)_3]_2 + 2\ H_2O \longrightarrow [R'Sn(O)O_2CR]_6 + 4\ RCO_2H \quad (3)$$

Our investigation using ^{119}Sn NMR establishes the retention of the drum and ladder structures in solution and shows their interconversion according to Equation 3.

Figure 5 displays ^{119}Sn NMR spectra of a sample of the ladder compound, 3, in slightly moist $CDCl_3$. Hydrolysis is evident as the drum peak at -486 ppm grows in intensity relative to signals assigned to the ladder. As Figure 5 emphasizes, the ^{119}Sn peak at -532 ppm for 3 appears as a major intermediate. Signals in this region are present in the spectra of all samples except that for the chloro derivative, 2, which also lacks a high field signal in the 600-630 ppm region.

The action of excess cyclopentanecarboxylic acid on a $CDCl_3$ solution of the cyclopentane drum, 1, causes its signal at -485.8 ppm to disappear and gives rise to four other peaks which correspond to formation of the ladder and an intermediate. These NMR experiments have shown that the hydrolysis process given in Equation 3 is reversible, i.e., a drum forms from a ladder composition and, in the presence of excess acid, the drum can be opened up to yield the ladder formulation.

Structural Details. For 1, the hexameric "drum" has idealized S_6 molecular symmetry. The geometry of the stannoxane framework of the molecule consists of six-membered rings in a chair conformation. Each Sn atom is bonded to three framework oxygen atoms, where the Sn-O bonds are all of comparable strength and have lengths ranging from 2.075(7)Å to 2.093(7)Å. The oxygen atoms of the framework are tricoordinate. The sum of the three Sn-O-Sn angles about these oxygen atoms range from 331.8° to 333.9°. The Sn atoms, which are all chemically equivalent, are hexacoordinated, with the coordination sphere being completed by an n-butyl group and two oxygen atoms from different carboxylate groups. Each of the four-membered rings of the core is spanned by a carboxylate group that forms a symmetrical bridge between two Sn atoms. The Sn-O bonds to the bridging carboxylate atoms are longer than the core bonds and range from 2.155(8)Å to 2.173(8)Å.

The "unfolded-drum" or "ladder" compound 2 has crystallographic C_i symmetry. This corresponds to the idealized molecular symmetry and, therefore, there are three chemically inequivalent types of Sn atoms in the molecule, although all are hexacoordinated. The oxygen atoms in the open form can be subdivided into two types, as in the case of the drum molecule: tricoordinate framework oxygen atoms and the dicoordinate oxygen atoms of the bridging carboxylate ligands.

The geometry about the oxygen atoms of the framework of the open form tends toward planarity. In this case the sums of the angles about the tricoordinate oxygen atoms is 357.4° for O1 and 355.3° for O2. The geometry about the Sn atoms is best described as distorted octahedral.

The general structural features of the "ladder", $[(n\text{-}BuSn(O)O_2CC_6H_{11})_2\ n\text{-}BuSn(O_2CC_6H_{11})_3]_2$, 3, shown in Figure 4, are very similar to that found for 2, the principal difference being a

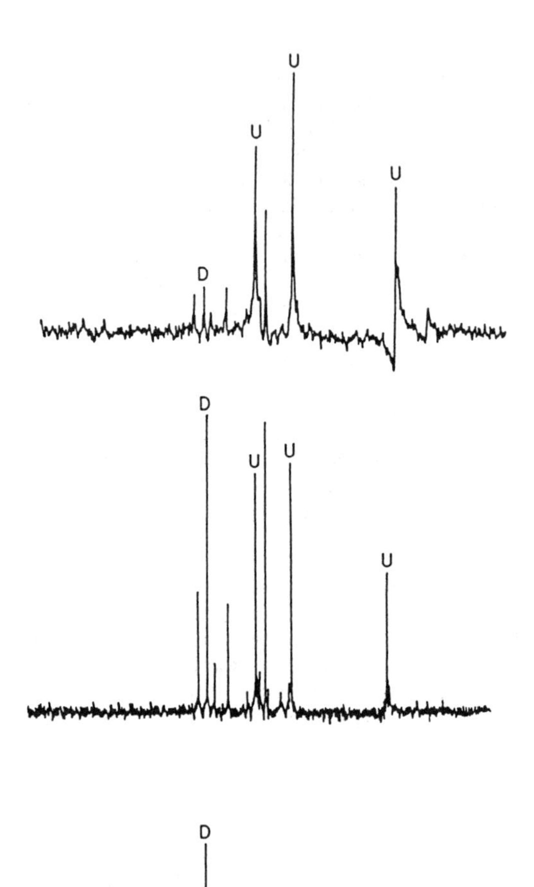

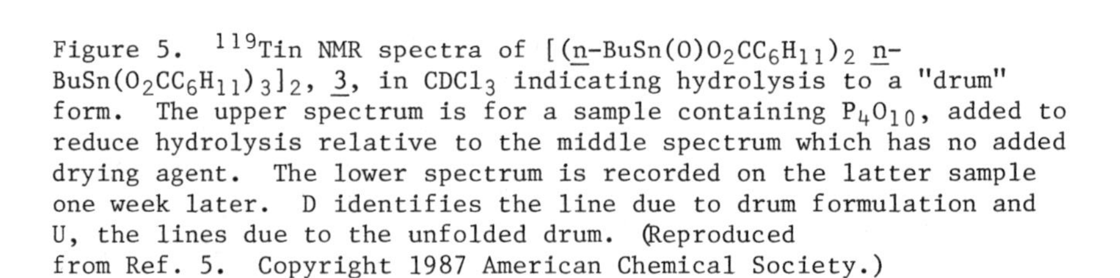

Figure 5. ^{119}Tin NMR spectra of $[(\underline{n}\text{-BuSn(O)O}_2\text{CC}_6\text{H}_{11})_2\ \underline{n}\text{-BuSn(O}_2\text{CC}_6\text{H}_{11})_3]_2$, $\underline{3}$, in $CDCl_3$ indicating hydrolysis to a "drum" form. The upper spectrum is for a sample containing P_4O_{10}, added to reduce hydrolysis relative to the middle spectrum which has no added drying agent. The lower spectrum is recorded on the latter sample one week later. D identifies the line due to drum formulation and U, the lines due to the unfolded drum. (Reproduced from Ref. 5. Copyright 1987 American Chemical Society.)

replacement of the Cl atom in 2 by a pendant cyclohexanoate group. Like 2, the molecule, 3, has crystallographic C_i symmetry. The Sn-O bond lengths of the bridging carboxylates are in general longer (averaging 2.15(3)Å) than the Sn-O framework bonds (averaging 2.06(2)Å).

It is interesting to note that the bond angles at the "pyramidal" framework oxygen atoms in the drum, 1, average 100.4(3)° in the distannoxane ring and are at 133.2(3)° in the tristannoxane ring. This compares with an average bond angle at oxygen within the distannoxane ring of 102.9(1)° in the unfolded drum, 2, which has nearly planar oxygen atoms and slightly shorter Sn-O framework bonds, averaging 2.063(3)Å for 2 compared to 2.086(7)Å for 1. Although the number of bonds broken and formed is conserved in the hydrolysis of an unfolded drum of the type 3 to a drum structure, Equation 3, the entropy change is favorable.

Ladder to Drum Conversion. It is interesting to speculate how the drum structure is formed from the ladder. The implication is that the energy difference between the two basic structures is relatively small. By way of illustration, we apply the hydrolysis process to the ladder formulation, 3. Mechanistically, for this example, a conservation is expected in the number of bonds to be broken and formed in executing the hydrolysis of Equation 3. A minimum of nine Sn-carboxylate oxygen bonds are required to be cleaved. During this cleavage the Sn-carboxylate oxygen bonds must assume *cis* positions at the tin atoms. These chelating groups are in *trans* positions in the open-drum structure. The bonds formed in completing the conversion to the drum structure with this constraint consist of three Sn-carboxylate oxygen bonds and six Sn-O bonds. Figure 6 shows a possible initial intermediate and Figure 7 illustrates a proposed form just prior to closure to yield the drum.

Phosphinic Acid Reactions. Reaction of *n*-butylstannoic acid with diphenylphosphate instead of a carboxylic acid also results in the formation of a drum composition $[n\text{-BuSn(O)}O_2P(OPh)_2]_6$ (Chandrasekhar, V.; Holmes, J. M.; Day, R. O.; Holmes, R. R., unpublished work). However, when diphenylphosphinic acid is reacted with *n*-butylstannoic acid under reflux in toluene, a new structural form of tin is obtained (7). The reaction proceeds according to Equation 4 giving the stable oxide composition in 90% yield, mp 198-208°C dec.

$$3\ n\text{-BuSn(O)OH} + 4\ Ph_2PO_2H \longrightarrow [(n\text{-BuSn(OH)}O_2PPh_2)_3O][Ph_2PO_2] + 2\ H_2O \qquad (4)$$

X-ray analysis (Figure 8) shows tin(IV) present in an oxygen-capped cluster molecule. The basic framework consists of a tristannoxane ring in a cyclohexane chair arrangement. Hydroxyl groups comprise the oxygen components of the ring system. A tricoordinated oxygen atom caps one side of this framework while three additional diphenylphosphinate groups bridge adjacent hexacoordinated tin atoms. It is noted that three Sn_2O_2 ring units form as a consequence of the presence of the unique capping oxygen atom. These three four-membered rings contain the latter atom and form a portion of a cube.

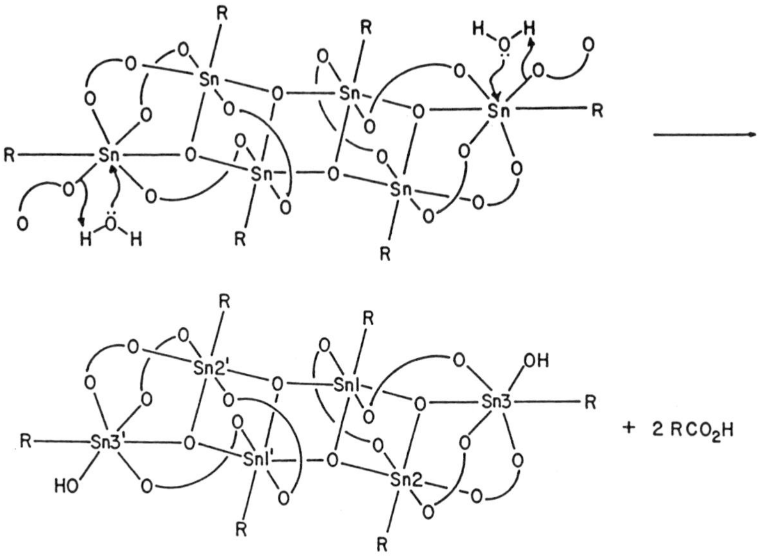

$$+ \; 2\,RCO_2H$$

Figure 6. A suggested initial intermediate resulting in the hydrolysis of a ladder structure as represented by Equation 3. (Reproduced from Ref. 5. Copyright 1987 American Chemical Society.)

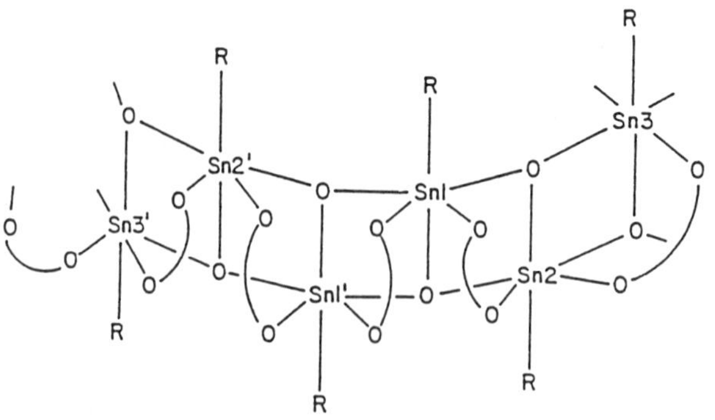

Figure 7. Suggested structural representation just prior to closure to yield the drum. (Reproduced from Ref. 5. Copyright 1987 American Chemical Society.)

The oxygen-capped cluster can be viewed as a hydrolysis product of the drum just as the drum is viewed as a hydrolysis product of the ladder, cf. Equation 5 (R = Ph_2P).

$$[R'Sn(O)O_2R]_6 + 2\ RO_2H + 2\ H_2O \longrightarrow 2[(R'Sn(OH)O_2R)_3O][RO_2] \quad (5)$$

drum cluster

The schematic (Figure 9) for a drum indicates how it is related to two cluster molecules. Formally, two bridging phosphinates rearrange, two oxygen atoms are added, four Sn-O bonds are cleaved, the six Sn-O-Sn linkages become Sn(OH)Sn units, and two phosphinates are added to hydrogen bond to the hydroxyls.

The ^{119}Sn NMR spectrum for the oxygen-capped cluster exhibits a single resonance with triplet character centered at -498.5 ppm ($|^2J(^{119}Sn\text{-}O\text{-}^{31}P)|$ = 132.0 Hz). This observation is consistent with the presence of three equivalent tin atoms provided by a cluster unit which has the hydrogen-bonded anionic phosphinate undergoing fast exchange among the three hydroxyl groups of the tristannoxane ring.

A cubic arrangement, $[\underline{n}\text{-}BuSn(O)O_2P(C_6H_{11})_2]_4$, is prepared (Kumara Swamy, K. C.; Day, R. O.; Holmes, R. R., accepted by JACS for publication) from a condensation of n-butylstannoic acid with dicyclohexyl phosphinic acid in toluene solution; mp 263-265°C, ^{119}Sn NMR, -462.8 ppm; $^2J(\underline{Sn}\text{-}O\text{-}\underline{P})$ = 116 Hz. As shown in Figure 10, the core of the molecule is defined by tin atoms and trivalent oxygen atoms which occupy the corners of a distorted cube, each face of which is defined by a four-membered Sn_2O_2 ring. The top and bottom faces of the cube are open, while each of its four sides is spanned diagonally by a phosphinate bridge between two tin atoms. The phosphinate bridges are required by symmetry to be symmetrical.

In addition, a butterfly structure (Figure 11) which contains two tin atoms in a dimeric representation $[\underline{n}\text{-}BuSn(OH)(O_2P(C_6H_{11})_2)_2]_2$ arises in the same reaction as the cube (Kumara Swamy, K. C.; Day, R. O.; Holmes, R. R., unpublished work). The cube may be viewed as a condensation product of the dimeric cluster according to the expression in Equation 6. Hence, the dimeric cluster may serve as a precursor on the way to the cubic form.

$$2[\underline{n}\text{-}BuSn(OH)(O_2P(C_6H_{11})_2)_2]_2 \longrightarrow [\underline{n}\text{-}BuSn(O)O_2P(C_6H_{11})_2]_4 + 4(C_6H_{11})_2PO_2H \quad (6)$$

The common structural unit in all of the oligomeric forms discovered so far is the four-membered dimeric distannoxane ring, Sn_2O_2. The number of tin atoms found in these structures ranges from two to six, excluding five. Further, condensation and coupling reactions are expected to produce additional oligomeric forms, revealing other interesting features of this new structural class of organotin compounds.

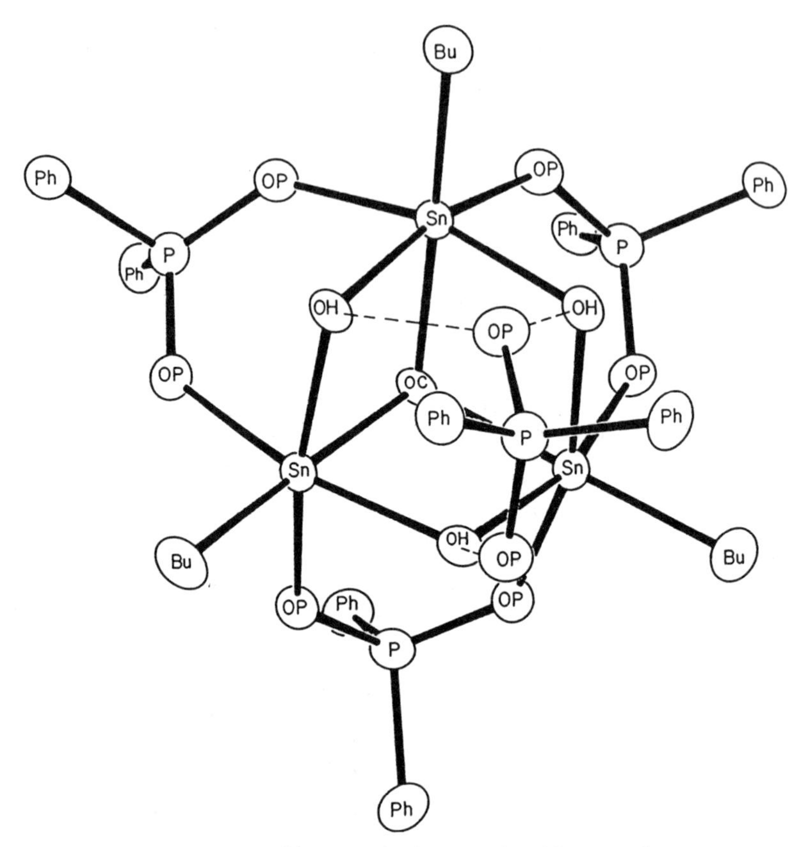

Figure 8. ORTEP plot of $[(\underline{n}\text{-BuSn(OH)}O_2PPh_2)_3O][Ph_2PO_2]$. Pendant atoms of the three n-Bu groups and of the eight Ph groups are omitted for purposes of clarity. Hydrogen-bonding interactions are shown as dashed lines. (Reproduced from Ref. 7. Copyright 1987 American Chemical Society.)

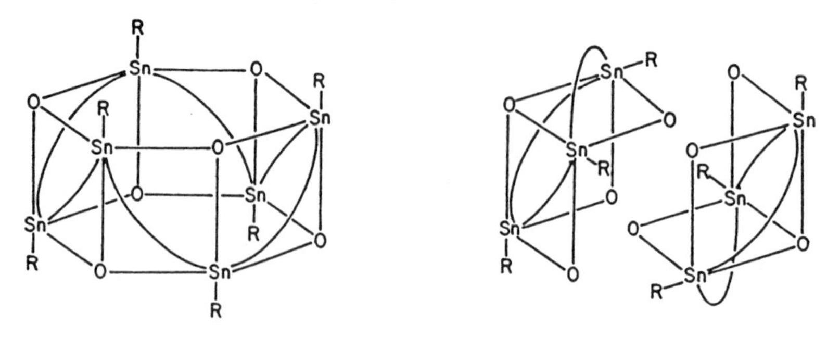

Figure 9. Schematic of a drum and two oxygen-capped clusters. (Reproduced from Ref. 7. Copyright 1987 American Chemical Society.)

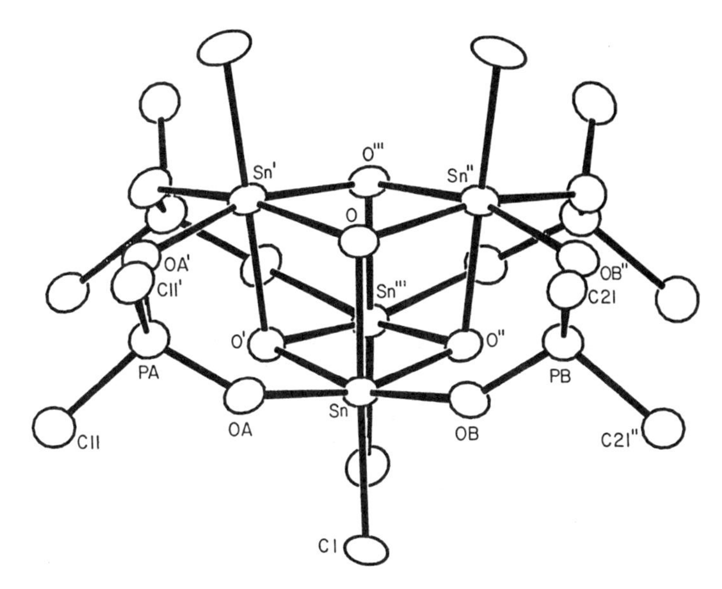

Figure 10. ORTEP plot of $[\underline{n}\text{-BuSn(O)O}_2\text{P(C}_6\text{H}_{11})_2]_4$.

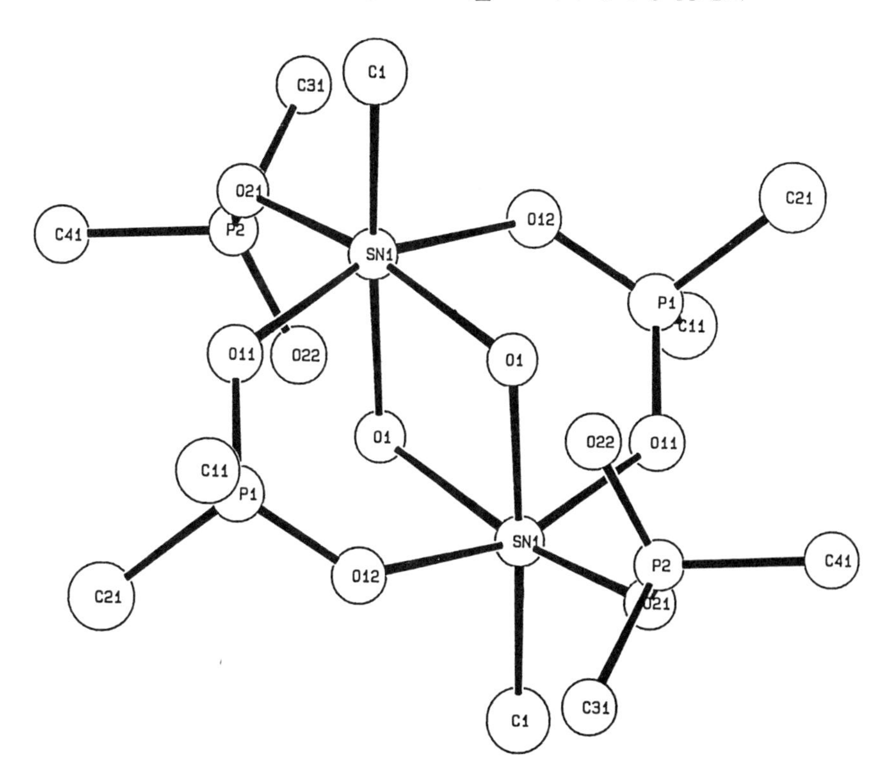

Figure 11. ORTEP plot of $[\underline{n}\text{-BuSn(OH)(O}_2\text{P(C}_6\text{H}_{11})_2)_2]_2$.

Acknowledgments

Support of this research by the donors of the Petroleum Research Fund, administered by the American Chemical Society and by the National Science Foundation (CHE 8504737) is gratefully acknowledged. We thank the University of Massachusetts Computing Center for generous allocation of computing time.

Literature Cited

1. Lambourne, H. J. Chem. Soc. 1922, 121(2), 2533.
2. Lambourne, H. J. Chem. Soc. 1924, 125, 2013.
3. Chandrasekhar, V.; Day, R. O.; Holmes, R. R. Inorg. Chem. 1985, 24, 1970.
4. Holmes, R. R.; Schmid, C. G.; Chandrasekhar, V.; Day, R. O.; Holmes, J. M. J. Am. Chem. Soc. 1987, 109, 1408.
5. Chandrasekhar, V.; Schmid, C. G.; Burton, S. D.; Holmes, J. M.; Day, R. O.; Holmes, R. R. Inorg. Chem. 1987, 26, 1050.
6. Anderson, H. H. Inorg. Chem. 1964, 3, 912.
7. Day, R. O.; Holmes, J. M.; Chandrasekhar, V.; Holmes, R. R. J. Am. Chem. Soc. 1987, 109, 940.

RECEIVED September 1, 1987

Chapter 39

NMR Characterization of the Compositional and Configurational Sequencing of Tri-*n*-butyltin Polymers

Jon M. Bellama and William F. Manders

Department of Chemistry and Biochemistry, University of Maryland, College Park, MD 20742

The chemical shift and structure of the ^{119}Sn NMR spectrum of the copolymer poly(tri-n-butyltin methacrylate/methyl methacrylate) exhibit strong solvent functionality. The values of $^{n}J(^{119}Sn,^{13}C)$ are also solvent dependent, increasing as the donor ability of the solvent increases. The ^{119}Sn chemical shift and structure are linearly related to $^{3}J(^{119}Sn,^{13}C)$ in oxygen- and nitrogen-containing solvents.

The tri-n-butyltin groups were removed from the tin-containing polymers to give poly(MAA/MMA). The tin-stripped polymers were better resolved and free from the intense butyl-carbon signals. The compositional and configurational sequencing of the original polymer was investigated. The α-methyl and carbonyl regions of poly(MAA/MMA) contained well-resolved configurational triads. The carbonyl region, however, is also split by compositional effects. The α-methyl region is insensitive to compositional sequencing; integration of α-methyl triad gave quantitative information concerning tacticity.

Compositional sequences were assigned in the carbonyl region of the ^{13}C spectra of poly(MAA/MMA) and in the ^{119}Sn spectra of poly (TBTM/MMA). Both syndiotactic and heterotactic splittings were observed in the carbonyl spectrum. In the ^{119}Sn spectrum only the syndiotactic compositional triad was observed.

Tributyltin and triphenyltin compounds are used as wood preservative fungicides, agrochemical fungicides, and as miticides and biocides in marine paints.

A copolymer of tri-n-butyltin methacrylate (TBTM) and methyl methacrylate (MMA), which is the active agent in antifouling

0097-6156/88/0360-0483$06.00/0

paints, can be synthesized either by the free radical copolymerization of the respective vinyl monomers with a peroxide initiator (1,2) or by esterification of free acid groups in a copolymer of methacrylic acid (MAA) and methyl methacrylate by bis(tri-n-butyltin) oxide (TBTO) (3).

The objective of this study is to characterize tin-containing polymers on a molecular level by means of high field, high resolution, multinuclear Fourier Transform Nuclear Magnetic Resonance (FT-NMR) (4). This study is generally an applied approach dealing with composition and configuration of specific formulations of the copolymer.

Publications concerning pendant triorganotin polymers are limited to synthetic methods and applications (for a recent review, see reference 5). There are also several pertinent reports on the chemistry of monomeric triorganotin carboxylates, alkoxides, and halides (6-8).

EXPERIMENTAL

Syntheses. Isotactic poly(methyl methacrylate) was synthesized by the method of Tsuruta et al. (9). Under a nitrogen atmosphere, a quantity of 6 mL (0.056 mole) of methyl methacrylate (MMA) dried over 4A molecular sieve was dissolved in 24 mL of similarly dried toluene. To the glass vial containing the reaction was added 0.65 mL of 1.6 M n-butyllithium, and the reaction was kept at -78°C in a dry ice/isopropanol bath. The polymerization was halted 24 hr later with the addition of hydrochloric acid and methanol (methanol/water = 4.1 by volume). The polymer was dried in vacuo at 50°C, redissolved in methylene chloride, precipitated by being poured into water-containing methanol, and dried in vacuo at 50°C. Tacticity and composition were verified with ^{1}H NMR. Yield: 47%.

Isotactic poly(methyl methacrylate/methacrylic acid), a copolymer of methyl methacrylate and methacrylic acid, was synthesized by the partial hydrolysis of isotactic poly(MMA) according to the method of Klesper et al. (10-13). A hydrolyzing mixture of 8 mL dioxane and 4 mL methanolic KOH (10% by weight KOH) was mixed with 250 mg of polymer in closed vials at 85°C for 48 hr. Saponified polymer separated from the solution and adhered to the walls of the vial. The precipitated polymer was dissolved in water and then precipitated again with a few drops of HCl. The solution was warmed and the coagulated polymer removed, washed with water, and dried in vacuo at 50°C. The ^{13}C NMR spectrum indicated approximately 25% hydrolysis. Yield: 73%.

Atactic poly(methyl methacrylate/methacrylic acid), the copolymer of methyl methacrylate (MMA) and methacrylic acid (MAA), was synthesized "directly" as a prepolymer to be esterified with bis(tri-n-butyltin) oxide (TBTO). Two formulations of poly (MMA/MAA) were synthesized, a 1:1 and a 2:1 MMA and MAA copolymer whose syntheses differ only in the proportion of monomer reacted.

The preparation was carried out in a 3 L 4-neck round-bottom flask fitted with thermometer, stirrer, addition funnels, and reflux condenser. The initial components were mixed and heated to a moderate reflux (66-67°C) for 10 min. The added charges were then fed simultaneously and evenly over a 3 hr period while maintaining moderate reflux. The final charges were then fed evenly over a one hr and a 5 min period while maintaining the same temperature and agitating for at least one hr to insure that all the polymer is dissolved from the sides of the vessel. The solvent is removed in vacuo at 60°C.

In a second method, a 1 g sample of poly(tri-n-butyltin methacrylate/methyl methacrylate) was dissolved in 4 mL of chloroform. One mL of concentrated HCl was added dropwise with shaking until no more precipitate appeared. The precipitate was removed and then shaken again with a clean batch of chloroform and HCl. The precipitate was dried in vacuo at 60°C. The yield of poly(MMA/MAA) was 70%.

Tri-n-butyltin methacrylate (TBTM) was synthesized by chemists at the David Taylor Naval Ship Research and Development Center (DTNSRDC), Annapolis, Maryland. The methods of Dyckman et al. (1) and Montermoso et al. (2) were used, in which 30 g methacrylic acid was slowly added to 103.8 g bis(tri-n-butyltin) oxide in 300 mL benzene in a flask equipped with a stirrer and reflux condenser. Cooling kept the temperature below 25°C. After all acid had been added, the solution was heated gradually and maintained at 30°C while the water of reaction was removed in vacuo and benzene was added to replace the benzene lost during this period. When the solution became clear, the benzene was removed and the resulting pale yellow viscous liquid was diluted with 100 mL ether and cooled to -20°C. The product separated as long, thick, transparent needles. The yield was close to theoretical and the m.p. was 18°C.

A free-radical synthesis of poly(tri-n-butyltin methacrylate/methyl methacrylate) was also carried out by the DTNSRDC chemists.

Two different formulations of the tin-containing copolymer were prepared. The first contained a 1:1 ratio of TBTM to MMA units and the second contained a 1:2 ratio of TBTM to MMA units. The synthesis for each differs only in the relative proportion of monomer loaded into the reaction vessel. Commercial benzoyl peroxide (1 percent by weight based on the sum of TBTM and MMA) was employed as the initiator. 0.8 mol of TBTM freshly recrystallized from ether and 0.8 mol MMA were added to 860 g of benzene in a flask equipped with a thermometer and reflux condenser. To this mixture was added 4.3 g (0.018 mol) of benzoyl peroxide. The flask and contents were warmed at the reflux temperature of benzene (80.1°C) with stirring. After 24 hours reaction time the benzene was removed in vacuo at 50°C.

Nuclear Magnetic Resonance. All spectra used in this study were collected on either a Bruker Instruments Inc. CXP-200 or WM-400 superconducting spectrometer operating at field strengths of 4.7 T and 9.4 T respectively. For a field strength of 4.7 T the NMR frequencies for ^{1}H, ^{13}C, and ^{119}Sn nuclei are 200.00, 50.31, and 74.60 MHz, respectively; at 9.4 T: 400.00, 100.62, and 149.20 MHz, respectively. The ^{119}Sn spectra were collected at 4.7 and 9.4 T, while most ^{1}H and ^{13}C spectra were collected at 9.4 T. Conditions for spectral acquisition always included quadrature phase detection and broad-band proton decoupling for ^{13}C and ^{119}Sn spectra. Decoupling for ^{119}Sn spectra was gated on during acquisition to minimize a negative NOE, while decoupling for ^{13}C spectra employed a two-level scheme with low power on during the relaxation delay to induce a favorable NOE. Normal levels were used during acquisition. Normally a $\pi/2$ pulse was used with a relaxation delay of $3(T_1)$. The value of T_1 for ^{119}Sn nuclei in tin-containing polymers was measured as approximately 1 s. For ^{13}C spectra a relaxation delay of 3 s was used. Primary chemical shift references used were internal TMS for ^{1}H spectra, external TMS for ^{13}C and internal tetramethyltin for ^{119}Sn. Most spectra were indirectly referenced to the above standards using the internal deuterium lock frequency as a secondary reference. All spectra were acquired with a deuterium lock. Where the approriate deuterated solvent was not available, a concentric tube containing a deuterated solvent was employed.

RESULTS AND DISCUSSION

Solvent Effects in the ^{119}Sn Spectra of Poly(TBTM/MMA). Samples of poly(MMA/TBTM) synthesized by the free-radical copolymerization of the appropriate monomers were solutions in benzene with approximately 33% solids (weight to volume). The particular formulation chosen as representative of the class contained a 1:1 ratio of pendant methyl to tri-n-butyltin groups. In preparing the dry polymer, the benzene was removed in vacuo with nominally 5% by weight residual solvent.

A phenomenological study was performed to determine the effect of solvent on ^{119}Sn NMR spectra of these organometallic polymers. Samples were dissolved in chloroform, benzene, n-hexane, acetone, tetrahydrofuran, methanol, and pyridine. The ^{119}Sn NMR spectra in these solvents are given in Figure 1. The appearance and location of the ^{119}Sn resonance changes drastically over the range of selected solvents. The chemical shift moves upfield in the order chloroform, benzene, n-hexane, acetone, tetrahydrofuran, pyridine, and methanol. The amount of structural information and, conversely, the broadening of the resonance increases in the same order with methanol and pyridine reversed.

Several phenomena may contribute to the appearance and location of the ^{119}Sn resonance. The situation is further complicated by the fact that the tri-n-butyltin group is attached to a high molecular weight polymer whose solvent interactions may also bear on the ^{119}Sn resonance.

It is well known that organotin compounds with at least one electronegative substituent can readily expand their coordination number to 5 or 6, and that donor solvents may promote such expansion. In the absence of coordinating solvents, it appears likely that tin will expand its coordination number by self-association with ester groups in other parts of the polymer. The addition of one solvent molecule will rehybridize tin from sp^3 to sp^3d with planar R groups. It is generally observed that addition of electron density at the tin center in R_3SnOR monomers will move the ^{119}Sn resonance upfield. It has been shown that $^{1}J(^{119}Sn,^{13}C)$ is sensitive to changes in bond orientation in Me_3SnCl brought about by coordinating solvents. Coupling between active nuclei is more efficient through bonds with a higher percent s character, and the

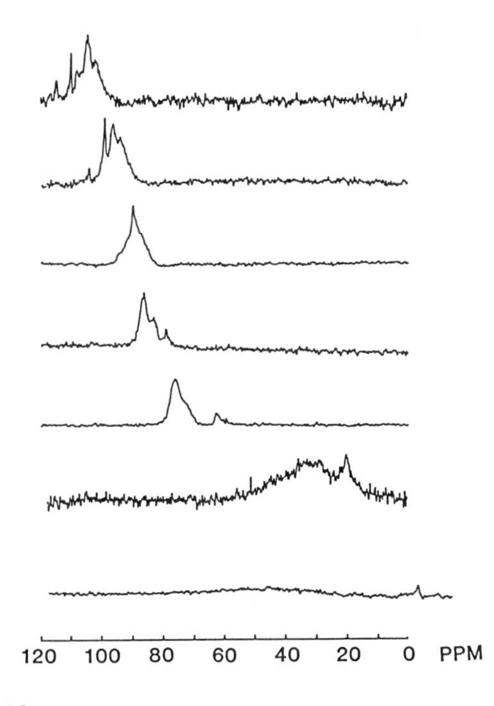

Figure 1. ^{119}Sn NMR spectra of 10% w/v poly(TBTM/MMA) in several solvents. From the top: chloroform, benzene, n-hexane, acetone, tetrahydrofuran, methanol, and pyridine.

increased coupling is manifested by an increase in the coupling constant. On rehybridization of tin from sp^3 to sp^3d the Sn-C bonds in the equatorial positions rehybridize from sp^3 to sp^2 (25% s to 33% s character).
The $^{119}Sn-^{13}C$ coupling constants observed in ^{13}C spectra in various solvents are given in Table I. A representative spectrum

Table I. Observed $^{119}Sn-^{13}C$ Coupling Constants in Poly(Tri-n-butyltin Methacrylate/Methyl Methacrylate)in Various Solvents [a]

Solvent	$^{1}J(^{119}Sn,^{13}C)^{b}$ (Hz)	$^{2}J(^{119}Sn,^{13}C)^{b}$ (Hz)	$^{3}J(^{119}Sn,^{13}C)^{b}$ (Hz)
Benzene	360		63.0
Hezane	350	18.0	64.3
Chloroform	360		65.2
i-Propanol		19.3	66.0
Dioxane	370	20.4	66.6
Acetone	370	19.1	67.9
Tetrahydrofuran	380	21.3	68.3
n-Butanol		21.5	69.6
Ethanol		23.2	70.7
Pyridine	410	23.6	72.4
Methanol	420	24.9	74.4

[a] Observed in ^{13}C spectra. Where no values are given the coupling constant was either too small to be resolved (^{2}J) or simply not recorded (^{1}J).

[b] The values listed for $^{n}J(^{119}Sn,^{13}C)$ are actually an average of $^{n}J(^{119}Sn,^{13}C)$ and $^{n}J(^{117}Sn,^{13}C)$, which were not resolved.

electron density at the tin center in R_3SnOR monomers will move the ^{119}Sn resonance upfield. It has been shown that $^{1}J(^{119}Sn,^{13}C)$ is sensitive to changes in bond orientation in Me_3SnCl brought about by coordinating solvents. Coupling between active nuclei is more efficient through bonds with a higher percent s character, and the increased coupling is manifested by an increase in the coupling constant. On rehybridization of tin from sp^3 to sp^3d the Sn-C bonds in the equatorial positions rehybridize from sp^3 to sp^2 (25% s to 33% s character).

The $^{119}Sn-^{13}C$ coupling constants observed in ^{13}C spectra in various solvents are given in Table I. A representative spectrum is given in Figure 2. The coupling is not observed in ^{119}Sn spectra due to the lack of resolution. The values of ^{1}J, ^{2}J and ^{3}J are larger for recognized donor solvents such as methanol and pyridine, and smaller for non-coordinating solvents such as benzene, hexane, and chloroform. The steric demands of the coordinating phenomenon are evident in the series MeOH, EtOH, i-PrOH, and n-BuOH. Methanol (with ^{3}J = 74.4 Hz) is most efficient in donating to a tin flanked by three bulky n-butyl groups. The coupling constants for EtOH and n-BuOH are 70.7 and 69.6 Hz, respectively, while that for the secondary alcohol, i-PrOH, is 66.0 Hz, comparable to chloroform. It appears that substitution at the α-carbon greatly hinders donation, while the size of the unbranched

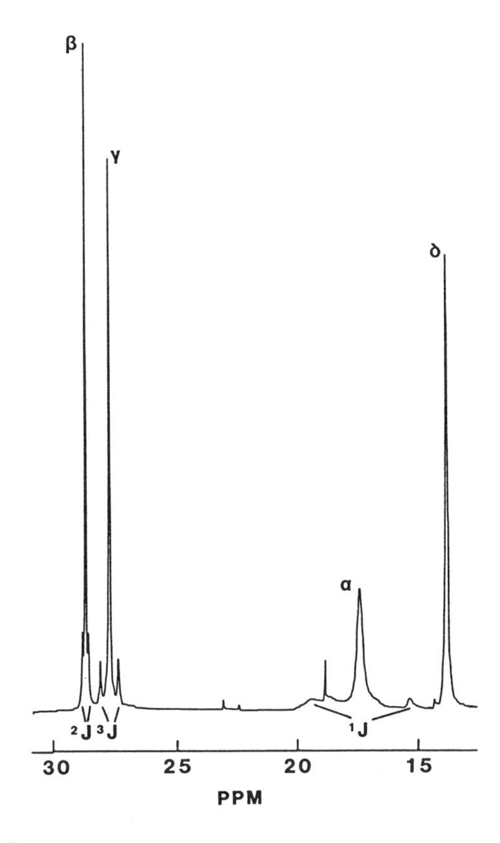

Figure 2. ^{13}C NMR spectrum of n-butyl region of 25% w/v poly(TBTM/MMA) in methanol at 300 K with $^{n}J(^{119}Sn,^{13}C)$ indicated.

primary alcohol is less critical. The same reasoning may also explain the order MeOH > pyridine for the coupling constants. Pyridine is regarded as the better coordinating solvent, but its bulk may hinder its ability to donate.

The ^{119}Sn spectra of the polymer in methanol, n-butanol, and i-propanol shows broadening of the ^{119}Sn signal, and the $^{3}J(^{119}Sn,^{13}C)$ values both increase in the order i-propanol < n-butanol < methanol. Both the appearance and the position of the ^{119}Sn signal in donating solvents can be attributed to the ability of the solvent to coordinate with tin.

The move upfield of the ^{119}Sn NMR signal as the donation to tin increases can be interpreted as an increase in shielding of the tin center. The broadening of the signal, however, cannot be explained as easily. Averaging of the chemical shift between a coordinated and a non-coordinated tin would broaden the signal, although monomeric triorganotin carboxylates do not show a marked broadening. The rate of solvent exchange on the tin attached to the polymer may be slowed relative to the monomeric form, and the line-broadening may not appear in the latter case. Variable temperature studies were performed on the solution in pyridine with any appreciable change in the appearance of the spectrum. An increase in temperature would be expected to increase the rate of solvent exchange and narrow the signal. The broadening of the ^{119}Sn signal may also result from the many processes influencing the transverse relaxation time, T_2, which determines the linewidth when averaging of chemical environments is not significant.

In donating solvents the subtle effects determining the chemical shift in chloroform, benzene, and hexane are apparently masked. In hexane, which is considered a poor solvent, self-association is possible and would explain the appearance of the ^{119}Sn spectrum. Chloroform and benzene are excellent solvents for organometallic polymers, and the structure and downfield position support a well-solvated, unassociated environment.

Determination of Configurational Sequences. During the polymerization of methyl methacrylate, configurational sequences are generated as a result of the inequivalence of the methyl and methyl formate substituents on the carbon backbone (14). This isomerism, termed tacticity, results in three different arrangments: isotactic, with all methyl groups on one side of the extended chain; syndiotatic, with methyl groups alternating regularly from one side to the other of the backbone; and atactic, with the methyl groups distributed randomly. Configurational sequences can be further characterized as catenated monomer dyads of either meso (m) with both methyl groups on the same side of the backbone, or racemic (r) with methyl groups on opposing sides of the backbone. At the triad level three different sequences are possible: isotactic, mm; syndiotactic, rr; and heterotactic, mr.

It is well known that the mechanical and physical properties of vinyl polymers are dependent upon their stereochemical configuration. It is critical, therefore, that the stereoregularity of poly(TBTM/MMA) be determined accurately and conveniently if the field performance of the material is to be predicted with any certainty. The effectiveness of organometallic polymers as an antifouling agent is a result of the gradual hydrolysis of the organotin moiety and its subsequent solubilization (1). It has been shown that the rate of hydrolysis of poly(methyl methacrylate) is strongly dependent on the tacticity of the polymer. The rate of hydrolysis of the isotactic polymer is at least an order of magnitude greater than either the heterotactic or syndiotactic placement. The relevance of tacticity to field performance is obvious.

Nuclear magnetic resonance has proved to be a valuable tool in determination of configurational sequences in poly(MMA) (14). In Figure 3 is shown the ^1H NMR of poly(MMA) synthesized with an anionic polymerization catalyst known to produce predominantly isotactic sequences. In these polymers, the ^1H NMR spectrum of the methylene units in the polymer backbone gives an unequivocal determination of tacticity. The methylene signal, occurring about 1.8 ppm, is diagnostic for dyad sequences (r or m). In the syndiotactic groups (r), the protons are magnetically equivalent and appear as a single peak at the chemical shift position. In isotactic groupings (m), however, the protons are magnetically inequivalent and hence appear as a doublet approximately equally spaced (by about 0.35 ppm) on either side of the syndiotactic peak. Because of the inequivalence, each line of the doublet is split into two signals by the geminal ^1H-^1H coupling of about 15 Hz so that a total of 4 lines are seen. Integrals of the singlet and the quartet allow the tacticity at the dyad level to be determined quantitatively. It is, of course, more desirable to increase the length of the sequence analyzed. The number of distinguishable n is generally limited by the resolution of the spectrometer. The tacticity of the polymer in Figure 4 is predominantly isotactic at the dyad level. The three α-methyl resonances at 0.8, 1.0, and 1.2 ppm comprise the α-methyl triad (mm, mr, rr, respectively). This identification can only be made with the knowledge provided by the methylene dyad analysis. Integration of the α-methyl resonance shows the polymer to be 86.4% isotactic, 8.4% heterotactic, and 3.1% syndiotactic.

The ^1H spectra of copolymers of MMA and MAA are poorly resolved, and although the β-methylene and α-methyl signals show structure, very little quantitative information is available. The butyl resonances at 0.8, 1.2, 1.3, and 1.6 ppm overwhelm the useful proton signals at the β-methylene and α-methyl psoitions, i.e., the resolution is poor because of compositional sequencing.

The ^{13}C spectrum of several samples of poly(MMA/MAA) are given in Figures 4-5. In Figure 4 the six unique carbons are matched with their respective signals. The downfield carbonyl resonance is the free acid. The resolution has improved relative to the ^1H spectra, and quantitation should be possible. In Figure 5 the ^{13}C spectrum of a partially saponified sample of isotactic poly(MMA) is given. It is likely the poly(MMA/MAA) produced in the hydrolysis will also be isotactic. Based on this assumption the isotactic carbon resonances can be assigned: α-methyl, 21.0 ppm; β-methylene, 51.0 ppm; $-CO_2H$, 180.0 ppm; and $-CO_2CH_3$, 177.6 ppm. The carbonyl resonances are given as the center frequency of the compositional triad to be discussed later. The quaternary and methoxy carbons do not contain configurational information. These polymers were synthesized with free-radical polymerization catalysts and are

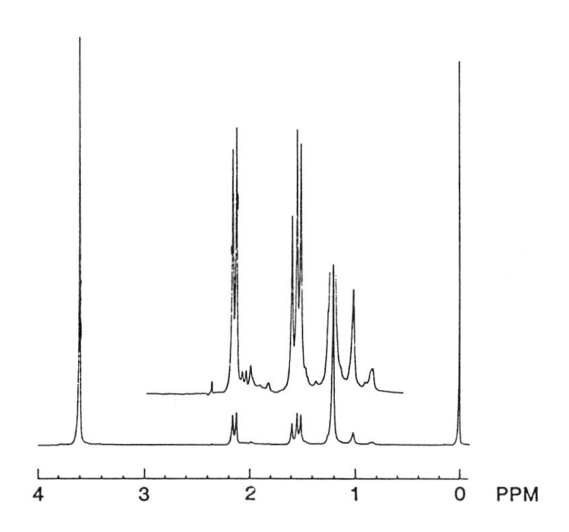

Figure 3. ^{1}H NMR of isotactic poly(MMA) in $CDCl_3$ at 300 K with 1% TMS. Upper trace is x8 vertical expansion.

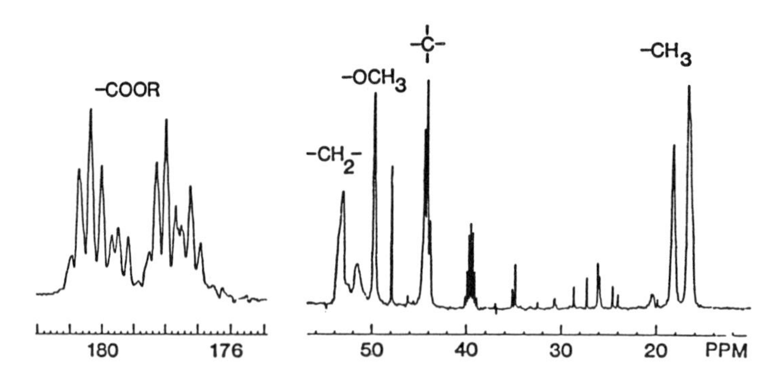

Figure 4. ^{13}C NMR spectrum of 25% w/v poly(MMA/MAA) prepolymer in pryidine at 360 K.

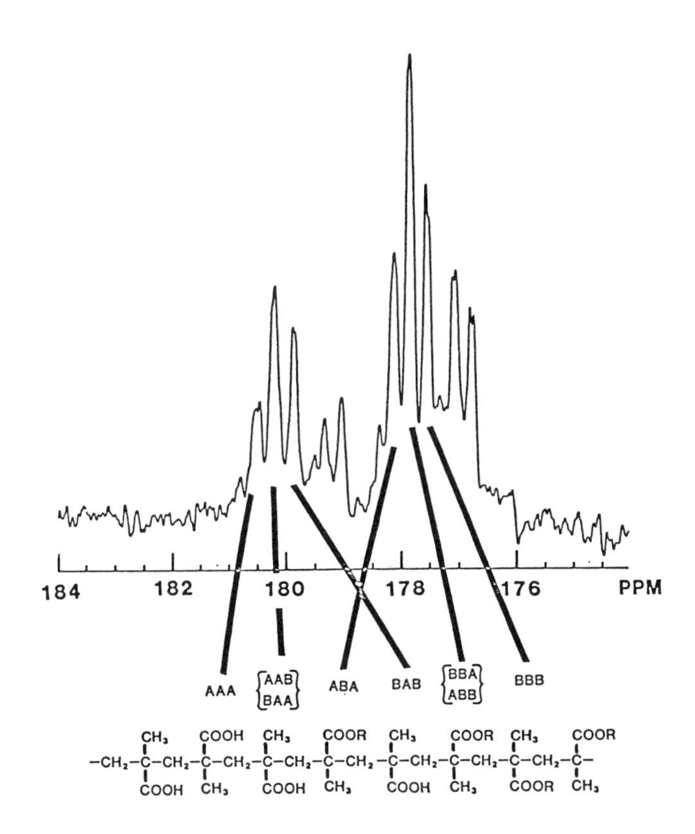

Figure 5. ^{13}C NMR of carbonyl region of 25% w/v poly(MMA/MAA) prepolymer in pyridine at 360 K. The syndiotactic peaks are labelled according to composition. The weak higher field peaks are caused by heterotactic polymer; there is too little isotactic material to be seen.

expected to be primarily syndiotactic and heterotactic. The α-methyl regions of the ^{13}C spectrum of the tin containing polymer are essentially identical, which is indicative of a configurational sequence unaffected by the change in composition. The isotactic α-methyl peak occurs at 21 ppm, while the larger peaks at 19 and 18 ppm are assigned as heterotactic and syndiotactic, respectively, as shown in Figure 6. The β-methylene region provides an analysis at the dyad level; the peak at 53 ppm is racemic, while the peak at 51 ppm is meso. The carbonyl region includes 4 sets of triads; the downfield pair is attributed to the free acid, while the upfield pair is the methyl ester. The identity was determined by comparing the polymer containing an equal number of free acid and methyl ester groups with the polymer containing twice as many methyl ester units as free acid units. The triad centered at 180.5 ppm is the syndiotactic sequence.

The ^{13}C spectrum of the α-methyl region is well resolved with respect to tacticity, but, unlike the carbonyl region, is insensitive to compositional sequences. Results of the integration of the α-methyl region are given in Table II.

Table II. Composition and Tacticity Ratios Taken from ^{13}C NMR Spectra of Poly(MMA/MAA) [a]

Copolymer Designation	$COOCH_3$ / $COOH$	Isotactic (%)	Heterotactic (%)	Syndiotactic (%)
A[b]	---	3.0	32	65
B[b]	1.0	4.0	37	59
C[b]	2.0	5.0	37	58
D[c]	0.9	2.0	35	63
E[c]	2.4	2.0	40	58

[a] Tacticity percents Are ±0.5%
[b] Tin-stripped polymers
[c] Prepolymers

Determination of Compositional Sequences. In the copolymerization of MMA and MAA or TBTM, compositional dyads and triads are generated. These sequences are determined by the relative concentration of monomers as well as by their relative reactivity. These compositional sequences characterize the material and allow predictions of activity based on structure by comparison with field tested polymers.

The relative reactivities (in free-radical copolymerizations) of TBTM and MMA are 0.79 and 1.00 respectively (15). With equal concentrations of monomer, an excess of MMA in the polymer would be expected. In the following discussion A will represent the MAA or TBTM unit and B will represent the MMA unit. For A-centered triads four different arrangements are possible: AAA, AAB, BAA, and BAB. Analogous sequences apply to the B-centered triads. For a random compositon, Bernoullian statistics should apply (14). With P_A (the proportion of TBTM) equal to 0.5, the probabilities of each of the A-centered triads is P_A^2 or 0.25. The AAB and BAA triads are indistinguishable and appear as a single resonance.

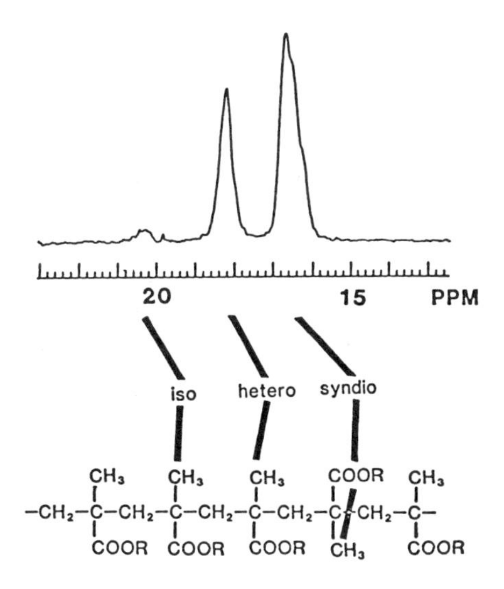

Figure 6. ^{13}C NMR spectrum of α-methyl region of 25% w/v poly(MMA/MAA) prepolymer in pyridine at 360 K. Labelled according to tacticity.

As with the configurational analysis, the tri-n-butyltin groups are stripped from the polymers for compositional analysis. The ^{13}C NMR spectra of the resulting poly(MMA/MAA) and prepolymers show that the carbonyl spectra of the prepolymer and the tin-stripped polymer are nearly identical. The syndiotactic compositonal triads are assigned in Figure 5 with A representing the free acid centered triad and B representing the methyl ester centered triad. The ratio of the intensities of the A-centered triad are approximately correct for a random composition.

The structures of the ^{119}Sn resonance of signals of the several polymers are similar if the impurities are ignored (the sharp peak on the downfield shoulder is TBTM). The observed triad is comparable to the compositional triads identified in the ^{13}C carbonyl spectrum. The ^{119}Sn triad signals appear as shoulders rather than as well-resolved peaks (as in the ^{13}C spectrum). The ^{119}Sn spectra contain no information regarding tacticity, although it is assumed the observed triad is syndiotactic. By comparing the various polymers, compositional sequences can be identified. Beginning with the downfield peak, the assignment is BAB, AAB/BAA, and AAA where A is the TBTM unit and B is the MMA unit.

Literature Cited

1. Dyckman, E. J.; Montemarano, J. A. In Antifouling Organometallic Polymers: Environmentally Compatible Materials, Report No. 4186, Naval Ship R & D Center, Bethesda, MD, 1974.
2. Montermoso, J. C.; Andrews T. M.; Marinelli, L. P. J. Polym. Sci., 1958, 32, 523.
3. Okawara, R.; O'Hara, M. In Organotin Compounds; Sawyer, A. K., Ed.; Marcel Dekker, Inc., New York, 1971; Vol. 2, Chapter 5.
4. Farrar, T. C.; Becker, E. D. Pulse and Fourier Transform NMR, Academic: New York, 1971.
5. Davies, A. G.; Smith, P. J. Comprehensive Organometallic Chemistry, Wilkinson, G., Ed., Pergammon Press: New York, 1982, Chapter 11.
6. Somasekharan, K. N.; Subramanian, R. V. ACS Symposium Series No. 121, American Chemical Society; Washington, D.C., 1980; pp 165-181.
7. Neumann, W. P. The Organic Chemistry of Tin, Interscience Publishers: New York, 1970; Chapter 2.
8. Neumann, W. P. ibid., Chapter 7.
9. Tsuruta, T.; Makisomoto, T.; Kanai, H. J. Macromol. Sci., 1966, 1, 31.
10. Klesper, E.; Johnsen, A.; Gronski, W.; Wehrli, F. Makromol. Chem., 1975, 176, 1071.
11. Klesper, E. J. Polym. Sci., 1968, B 6, 313.
12. Klesper, E. J. Polym. Sci., 1968, B 6, 633.
13. Klesper, E.; Gronski, W. J. Polym. Sci., 1969, B 7, 727.
14. Bovey, F. A. High Resolution NMR of Macromolecules, Academic: New York, 1972, Chapter 3.
15. Ghanem, N. A.; Messiha, N. N.; Ikaldesus, N. E.; Shaaban, A. F. Eur. Poly. J., 1979, 15, 823.

RECEIVED October 23, 1987

INDEXES

Author Index

Allcock, Harry R., 250
Allen, Christopher W., 290
Archer, Ronald D., 463
Arnold, C. A., 180
Balk, E. M., 420
Barendt, Joseph M., 303
Bellama, Jon M., 483
Bent, Elizabeth G., 303
Bettelheim, A., 408
Blum, Yigal D., 124
Bowmer, T. N., 296
Brandt, P. J. A., 180
Brinker, C. J., 314
Bunker, B. C., 314
Carr, Thomas M., 156
Chandrasekhar, V., 469
Chen, C. L., 78
Chichester, S. V., 296
Day, R. O., 469
Dejak, B., 166
Downing, J. W., 61
DuBois, Donn A., 385
Elliott, C. Michael, 420
Elsbernd, C. S., 180
Farmer, B. L., 43
Fickes, G. N., 43
Fife, Wilmer K., 199
Fleming, W., 43
Ford, R. R., 283
Gaudiello, John G., 224
Glaser, Raymond H., 354
Glaser, Robert, 124
Gonsalves, Kenneth E., 437
Goodwin, George B., 238
Haddon, R. C., 296
Haltiwanger, R. Curtis, 303
Hani, R., 283
Harris, D. H., 392
Harrod, John F., 89
Holms, Joan M., 469
Holms, Robert R., 469
Huang, Hao-Hsin, 354
Huang, Horng-Yih, 112
Ishikawa, Mitsuo, 209
Kamiya, K., 345
Karatsu, T., 61
Kellogg, Glen E., 224
Kenney, Malcolm E., 238
Kilic, S., 180
Kim, H. K., 78
Kirkpatrick, R. J., 314
Klingensmith, 61
Kratzer, R. H., 392
Krone-Schmidt, W., 392
Kulpinski, J., 166
Kuzmany, H., 43
Laine, Robert M., 124
Lasniak, E., 166
Lasocki, Z., 166
Lee, Annabel Y., 463
Lipowitz, Jonathan, 156
Lyons, A. M., 430
Manders, William F., 483
Marks, Tobin J., 224
Matyjaszewski, K., 78
Maxka, Jim, 6
McGrath, J. E., 180
Michl, J., 61
Miller, R. D., 43,61
Mujsce, A. M., 430
Murray, Royce W., 408
Narula, C. K., 378
Nate, Kazuo, 209
Neilson, Robert H., 283,385
Norman, Arlan D., 303
Ochaya, Ven O. 463
Paciorek, K. J. L., 392
Paine, P. T., 378
Pearce, E. M., 430
Penton, H. R., 277
Piechucki, S., 166
Poutasse, Charles A., 143
Pressprich, K., 408
Rabe, James A., 156
Rabolt, J. F., 43
Rausch, Marvin D., 437
Raybuck, S. A., 408
Redepenning, J. G., 420
Riffle, J. S., 180
Roy, A. K., 283
Sakka, S., 345
Sawan, Samuel P., 112
Schaeffer, R., 378
Scheide, G. M., 283
Schmid, Charles G., 469
Schmidt, H. K., 333
Schmittle, S. J., 420
Schwark, Joanne M., 143

Sennet, Michael S., 268
Seyferth, Dietmar, 21,143
Shaw, S. Yvette, 385
Singler, Robert E., 268
Sooriyakumaran, R., 43
Sormani, P. M., 180
Summers, J. D., 180
Swamy, K. C. Kumara, 469
Tallant, D. R., 314
Tetrick, Stephen M., 224
Tian, Cheng-xiang, 199
Tramontano, Valentino J., 463
Tsai, Yi-Guan, 112
Tse, Doris, 124
Vasile, M. J., 430
Wallraff, G. M., 61
Wang, Bing, 463
Ward, K. J., 314
Waszczak, J. V., 430
West, Robert, 6
Wettermark, U. G., 283
White, B. A., 408
Wilkes, Garth L., 354
Willingham, Reginald A., 268
Wiseman, Gary H., 143
Wisian-Neilson, P., 283
Witekowa, M., 166
Worsfold, D. J., 101
Wynne, Kenneth J., 1,392
Xu, Jian-min, 199
Yilgor, I., 180
Yoko, Y., 345
Yu, Yuan-Fu, 143
Zeldin, Martel, 199

Affiliation Index

Army Materials Technology Laboratory, 268
AT&T Bell Laboratories, 296,430
Bell Communications Research, Inc., 296
Carnegie Mellon University, 78
Case Western Reserve University, 238
Colorado State University, 420
Dow Corning Corporation, 156
Ethyl Corporation, 277
Fraunhofer-Institut für Silicatforschung, 333
Hiroshima University, 209
Hitachi Ltd., 209
IBM Almaden Research Center, 43
Indiana University–Purdue University at Indianapolis, 199
Kyoto University, 345
Massachusetts Institute of Technology, 21,143
McGill University, 89
Mie University, 345
National Research Council of Canada, 101
Northwestern University, 224
Office of Naval Research, 1,392
The Pennsylvania State University, 250
Polytechnic University, 430
SRI International, 124
Sandia National Laboratories, 314
Southern Methodist University, 283
Stevens Institute of Technology, 437
Technical University, Lodz, Poland, 166
Texas Christian University, 283,385
Ultrasystems Defense and Space, Inc., 392
University of Colorado, 303
University of Illinois, 314
University of Lowell, 112
University of Maryland, 483
University of Massachusetts, 437,463,469
University of Nevada, 43
University of New Mexico, 378
University of North Carolina, 408
University of Texas at Austin, 61
University of Vermont, 290
University of Vienna, 43
University of Wisconsin, 6
Virginia Polytechnic Institute and State University, 180,354
Washington State University, 43

Subject Index

A

α form of boron nitride
  preparation, 378–379
  structure, 378

Absorption spectra, poly(di-*n*-alkylsilanes), 64,65*f*

N-Acetyltetramethylcyclodisilazoxane, formation, 167

Acyclic diborylamines, prevention of borazene ring formation, 386
Acyclic silane oligomers, intermediates in irradiation, 54
Adhesive, synthesis via organic network formers, 337,338*f*,339
Aerogels, description, 317
Aging effects of tetraethoxysilane–poly(tetramethylene oxide) systems
occurrence, 372–373
on mechanical properties, 373t
SAX profile, 374,375*f*
temperature, 373,374*t*
Alkenyl polysilanes, photochemical cross-linking, 10
Alkenylphosphazenes
electronic structure of olefinic center, 292–293
properties, 291–292
structure, 292
synthesis of ceramic solids, 293
Alkoxy-substituted polysilanes, formation, 86
Alkyl silicates
effect of alcohols on synthesis, 244–245
structures, 245,246*f*
synthesis
by elemental-silicon process, 240,241*f*
by silicate-based substitution approach, 240–247
synthesis step 1, 244–245
synthesis step 2, 245,247
synthesis step 3, 247
Alkylated polysilanes, thermochromic behavior, 64
Alkylpolysilanes
photodegradation, 9–10
^{29}Si-NMR spectra, 14,15*f*
Amine–phosphorus(III) halide condensation
products, 305–307
reaction, 305
Amino-terminated polysiloxane oligomers, preparation, 182–183
Aminoboranes, formation via silylamines and haloboranes, 379
(β-Aminoethyl)ferrocene, reactions, 440,442
Aminolysis of dihalosilanes, reaction, 127
Aminopropyl-terminated poly(dimethylsiloxane) oligomers, characteristics, 184
Ammonolysis
analysis of products, 149
preparation of silicon-containing ceramics, 149
products, 148
Ammonolysis of dihalosilanes
modified reaction, 127
reaction, 126
Amorphous silicon carbide, formation from polysilanes, 16
Angular scattering variable, definition, 361
Anionic polymerization, 1,3-disilacyclobutanes, 27
Aromatic polyester copolymers
dynamic mechanical behavior, 186,188*f*
preparation, 186
stress–strain behavior, 186,187*f*
Aryl-substituted polysilanes, photodegradation, 9–10
N-Arylcyclosilazoxanes
polymerization, 170–172
preparation, 170
resistance to hydrolysis, 173
Arylpolysilanes, ^{29}Si-NMR spectra, 14,15*f*
Atactic, definition, 490
Atactic poly(methyl methacrylate/methacrylic acid), synthesis, 484–485

B

Bimodal molecular-weight distributions, polysilanes, 8–9,12*f*
Biocompatible–bioactive polyphosphazenes
biologically inert water-insoluble polymers, 259
water-insoluble but bioerodible polymers, 259,261
water-insoluble polymers that bear biologically active surface groups, 259
water-soluble polymers that bear bioactive side groups, 261
water-swellable polymeric gels–amphiphilic polymers that function as membranes, 259–260
Bis(aminopropyl)poly(dimethylsiloxane), synthesis via an equilibration polymerization process, 183–184
1,1′-Bis(β-aminoethyl)ferrocene
polycondensation, 437–438,440*t*
reactions
with benzoyl chloride, 442–443
with phenylisocyanate, 443
structure, 438–439
synthesis, 438
1,1′-Bis(β-hydroxyethyl)ferrocene
polycondensation, 438,440*t*
structure, 438–439
Bis(borazinyl)amines, characterization and preparation, 380
Bis(trimethylsiloxy)phenylsilyl radical, role in photo-cross-linking, 211–213
1,1-Bis(trimethylsiloxy)-1-phenyl(trimethyl)disilane
irradiation, 211–212
photolysis, 211
synthesis, 211

Bis(trimethylsilyl)aminotrimethylsilylaminochloroborane
formation of nitrogen-bridged analogue, 400–401
intermolecular dehydrohalogenation, 398,400
mass spectral breakdown, 398
preparation, 398
1,2-Bis-(trifluoromethanesulfonyloxy)tetramethyldisilane, reactions, 84
Boranes, intermolecular dehydrohalogenation, 398
Borate systems
FTIR spectra, 322,323*f*
stability of B–O–B bonds toward hydrolysis, 322,324*f*,325
Borazene gels, processing, 380,381*f*
Borazene ring formation, prevention, 386
Borazine
pyrolysis studies, 394–399
synthesis, 393–394
Boron nitride
crystalline modifications, 378
formation, 394
infrared spectrum, 381*f*,383
transmission electron micrograph, 382*f*,383
X-ray powder diffraction analysis, 382*f*,383
Boron nitride precursors
crystal, 398
mass spectra, 401,403
melt spin, 398,399*f*
Boron–nitrogen polymer precursors
preliminary thermolysis studies, 390
preparation, 385–390
problems in cyclic formation, 385
Branched poly(carbosilanes), structures, 29,31*f*
Bulk-thermal imidization
description, 193
solubilities, 195*t*
[*n*-BuSn(OH)($O_2P(C_6H_{11})_2)_2]_2$, ORTEP plot, 479,481*f*
[(*n*-BuSn(OH)$O_2PPh_2)_3O][Ph_2PO_2]$, ORTEP plot, 477,480*f*
[(*n*-BuSn(O)$O_2CC_6H_{11})_2$-*n*-BuSn-($O_2CC_6H_{11})_3]_2$, ORTEP plot, 471,474*f*
[*n*-BuSn(O)$O_2CC_5H_9]_6 \cdot C_6H_6$, ORTEP plot, 471,472*f*
[(*n*-BuSn(O)$O_2CPh)_2$-*n*-BuSn(Cl)-($O_2CPh)_2]_2$, ORTEP plot, 471,473*f*
[*n*-BuSn(O)$O_2P(C_6H_{11})_2]_4$, ORTEP plot, 479,481*f*

C

Capacitance device sensor, scheme, 341,342*f*
Catalysis in transacylation reactions
conversion of benzoyl chloride to benzoic anhydride, 205
hydrolysis of diphenyl phosphorochloridate, 206,207*t*
importance of hydrophobic binding interactions, 207
Catalytic dehydrocoupling, reaction, 131–132
Catalytic elimination of H_2, production of polysilylenes, 90
Catalytically active polymers
development, 199
examples, 199–200
Cationic polymerization, isopropenylferrocene, 453–460
Ceramic fiber with Si–C–N–O composition, preparation, 157
Ceramic science, relationship to inorganic polymers, 250,251*f*,252
Ceramics, advantages of preparation from polymers, 156–157
Chemical vapor deposition techniques, preparation of α form of boron nitride, 379
$C_6H_4(NH)_2[P(NEt_2)]$, molecular structure, 309,311*f*
$C_6H_4N_2P(NEt_2)_2]_2$, structure, 307,308*f*
$C_6H_3N_3[P(S)NMe_2]_2[P(S)(NMe_2)_2]_2$, structure, 307,308*f*
Chromophoric chain segments of polysilanes
electronic effects of substituents, 68,69*f*
lowest energy excitation, 68,69*f*,70
Si–Si molecular orbitals, 70,71*f*,72
^{13}C-NMR spectra
of poly(TBTM/MMA), 488,489*f*,490
segmented poly(ether urethanes), 448,449*f*
Coammonolysis, pyrolysis of products, 149
Cobalt(III) polymer, synthesis via organic reactions, 465
Cohydrogenation of olefins
advantages, 93
effect on rate of polymerization, 93–94
Compositional sequences of tri-*n*-butyltin polymers, ^{13}C-NMR spectra, 496
relative reactivities, 493
1,2-Condensation dimer of $C_6H_4(NH)[NP(NEt_2)_2](PNEt_2)$, molecular structure, 310,311*f*
Condensation polymerization of polyphosphazenes
advantages, 284
characterization, 285–287
derivatization reaction, 287–288
reaction, 284
synthesis, 284–285
Conduction model, for poly(trisbipyridine)–metal complexes, 420–428

Configurational sequences of tri-*n*-butyltin polymers
^{13}C-NMR spectra, 491–495
composition and tacticity ratios, 493*t*
^{1}H-NMR spectra, 491,492
Contact lenses
improvement of mechanical strength of materials, 339,340*f*
structure of polymer, 339,340*f*
synthesis of materials, 339
Contrast enhancement lithography, description, 57
Coordination polymers, synthesis, 463–468
Copoly(diorganosilanes)
change in UV absorbance vs. UV exposure time, 120,121*f*
effect
of molecular weight on UV spectra, 114,117,118*f*
of substituents on UV spectra, 114,116*f*
Fourier transform–IR spectrum, 117,119*f*
properties, 114,115*t*
proton-coupled ^{13}C-NMR spectra, 117,118–119*f*
UV spectra, 114,116*f*
UV spectra vs. UV exposure time, 117,120*f*
Copolymerization, reaction rate of phenylmethyldichlorosilane vs. that of dimethyldichlorosilane, 110,111*f*
Copper chloride complexes with poly(2-vinylpyridine)
complex preparation, 431–432
infrared spectra, 431
isolation, 431
loss of water, 432,433*f*
pyrolysis mass spectroscopy, 431
reactant specifications for preparation, 431*t*
relative mass intensities, 432,435*t*
structure of evolved ions, 432,434*f*
thermograms, 432,433*f*
thermogravimetric analysis, 431
total char yield and polymer char yield, 434*f*,435
Coupling of germanes
model for polymerization reaction, 94–95
to insoluble gels, 94
Covalently cross-linked systems, comparison with microcrystallite-based ultrastructure, 262
Cyclic bis(silylamides)
effects of substituents on stability, 167–169
properties, 167
structure, 167
structure determination by ^{29}Si-NMR spectroscopy, 168
synthesis from silazane-containing ring compounds, 166–167
Cyclic trimer, preparation of block polymers, 192
Cyclic trimer substitution–polymerization route for polyphosphazene synthesis
cyclic trimers with halogens and organic side groups, 257,260
illustration, 257–258
Cyclohexanoate drum compound
^{119}Sn-NMR data, 471,475
synthesis, 471
Cyclohexasilicate ion, structure, 238,239*f*
Cyclopentanoate drum compound
^{119}Sn-NMR data, 471,475
synthesis, 471
Cyclopentanoate ladder compound, ^{119}Sn-NMR data, 475,476*f*
Cyclosilazoxanes, anionic mechanism of polymerization, 170–172
Cyclotrisilicate ion, structure, 238,239*f*

D

Deamination–condensation polymerization reactions, reaction, 130–132
Decoupling
pyrolysis studies, 137,138*t*,139
size exclusion chromatography, 137,138*f*
synthesis of preceramic polysilazanes, 137–139
Dehydrocyclization reaction, mechanism, 133
Dehydrogenative coupling
absence of low-molecular-weight oligomers, 91–92
constraints on catalytic activity, 91
description, 91
Deprotonation–substitution of 2-phenyl-1,3,2-diazaboracyclohexane
reactions, 387
yields, 387–388
Derivatization reactions for polyphosphazenes
cothermolysis, 287–288
deprotonation, 287
Peterson olefination reaction, 287
Dialkoxydisilanes, preparation, 84
1,3,2-Diazaboracycloalkane ring systems, prevention of borazene ring formation, 386
1,3,2,4-Diazadiphosphetidines, structure, 304
Dichlorodialkylsilanes, products of reductive coupling, 101
1,1′-Diisopropenylferrocene
cationic polymerization, 450
unreactivity in polymerization, 455
Dimethyltitanocene, coupling of germanes, 94
Dioxygen reduction electrocatalysis by metal macrocycles, efficacy, 417–418

Diphenylgermane, polymerization, 94
1,3-Disilacyclobutanes
anionic polymerization, 27
formation of high-molecular-weight product, 23–26
preparation from (chloromethyl)chlorosilane, 23
synthesis, 22–29
p-Disilanylenephenylene polymers, synthesis, 215
Disilicate ion, structure, 238,239*f*
1,1-Disubstituted silacyclopentanes, ring-opening polymerization, 33,38
Drawability of $Si(OC_2H_5)_4$ solution, influencing factors, 348
Drum compounds
formation from phosphinic acid, 477,480*f*
molecular symmetry, 475
schematic, 479,480*f*

E

Electrocatalysis of dioxygen reduction by cobalt porphyrins, efficacy, 417–418
Electrochemical polymerization of metallotetraphenylporphyrins, film formation, 409
Electron hopping, description, 420
Electron transport in metallotetraphenylporphyrin films
apparent self-exchange rate constants, 415,417*f*
electron hopping, 414
mechanism, 413–414
sandwich electrode, 414,416*f*
Electropolymerized films
characterization, 409,412–413
electron transport, 413–417
permeation, 412–413
Elemental-silicon process
disadvantage, 240
variation in oxidation number of silicon and number of oxygens bonded, 240,241*f*
Emission spectra, poly(di-*n*-alkylsilanes), 62,63*f*
$(EtO)_4Si$, synthesis, 242
$(EtO)_{10}Si_6O_7$ isomers, synthesis from $Na_4Ca_4Si_6O_{18}$, 243–244
$(EtO)_3SiOSi(OEt)_3$, synthesis from $Ca_2ZnSi_2O_7$, 242–243

F

Ferrocene monomers, formation of isopropenylferrocene monomer, 451
Field effect transistor sensor, scheme, 341,342*f*
Fluorescence polarization, polysilanes, 64,66,67*f*
Functional polysiloxane oligomers
equilibration polymerization processes, 181–182
general structure, 181
overall synthesis, 181–182

G

Gel formation, availability of monomer for silicate systems, 318–319
Gel structure during consolidation
difference between evolution during consolidation and that in solution, 325
effect of exposure to water vapor, 325,326*f*
formation of three-membered rings, 328
MASS and CPMASS NMR spectra, 325,327*f*,328
Raman spectra of silicates, 325,326*f*
stability of siloxane bonds, 328–329
Gelation, description, 317
Gels
applications for dried gels, 317
densification, 317–318
description, 318
drying, 317
factors influencing structure, 318
^{29}Si-NMR spectra, 322
Germanes, coupling, 94
Glasslike inorganic network, modification by organic groupings, 334,336*f*
Group 14 hydrides, polymerization, 89–99

H

Haloboranes, formation of amino-boranes, 379
(Halomethyl)halosilanes, synthesis of high-molecular-weight poly(silmethylenes), 22
Hexachlorocyclotriphosphazene, polymerization, 270
Hexakis(pyrrolyl)cyclotriphosphazene
structure, 297,299*f*
synthesis, 297–298
High-molecular-weight polymers, molecular-weight distribution, 114,115*t*
High-molecular-weight polysilanes, ultraviolet spectra, 11,12*f*
High-molecular-weight poly(silmethylenes), synthesis by (halomethyl)halosilanes, 22
Highest occupied molecular orbital, for polysilanes, 70,71*f*,72

Hybrid inorganic–organic polymers
synthesis
via alkenylphosphazenes, 291–293
via (vinyloxy)phosphazenes, 293–294
inorganic substituents, 291–294
organic substituents, 290–291
Hydrosilylation, reaction, 38
Hydroxy-terminated poly(dimethylsiloxane)
disadvantages of use in glass networks, 355–356
incorporation into tetraethoxysilane glass network, 355
Hydroxy-terminated poly(methyl-3-pyridinylsiloxane)
^{1}H-NMR spectrum, 202,203*f*
IR spectrum, 202
synthesis, 201–202
thermal analysis, 202,204*f*,205

I

Imidization, techniques, 193
μ-Imidobis[bis(trimethylsilyl)amino-trimethylsilylamino]borane
infrared spectrum, 401,402*f*
mass spectral breakdown, 401,403,404–405*f*
INDO/S method, nature of chromophoric chain segments, 66,68–71
Infrared spectrum, of boron nitride, 381*f*,383
Inorganic and organometallic macromolecules
applications, 3
with network structures, 3
Inorganic compound formation from polymer precursors, examples, 430
Inorganic networks
addition of organic groupings by polymerization, 334,336*f*
addition of organic groupings by substitution, 334,336*f*
characterization, 334
typical structure, 334,336*f*
Inorganic polymer chemistry, relationship to organic polymers, ceramic science, and metals, 250,251*f*,252
Inorganic polymer research, main purposes, 252
Inorganic polymers, definition, 1
Inorganic polymers based on alternating main group element–nitrogen skeletons, structures, 303
Intractability, problem for linear transition metal coordination polymers, 463
Inverse addition, definition, 23
Isopropenylferrocene, cationic polymerization, 453,455
Isopropenylferrocene carbenium ion, stability, 458
Isopropenylferrocene copolymerization with α-methylstyrene, 455,456*t*,457
Isopropenylferrocene homopolymers
^{1}H-NMR spectrum, 453,454*f*
IR spectra, 453
structure, 453
Isopropenylferrocene monomer
homopolymerization reactions, 451,452*t*
synthesis via ferrocene monomers, 451
Isopropenylferrocenyl carbenium ion, stability, 455
Isopropenylmetallocene monomers
cationic polymerization, 450
structures, 450
Isotactic, definition, 490
Isotactic poly(methyl methacrylate), synthesis, 484
Isotactic poly(methyl methacrylate/methacrylic acid), synthesis, 484

L

Ladder compounds
conversion to drum form, 477,478*f*
general structural features, 475,477
molecular symmetry, 475
Linear boron–nitrogen polymers
preparation, 385–390
properties, 385
Linear chain macromolecules
behavior, 3
preparation, 2
Linear phosph(III)azanes, formation, 303–304
Linear polysilane, preparation, 6
Linear poly(silmethylenes), preparative procedures, 22
Linear polysilylenes, production, 90
Linear transition metal coordination polymers
problems with intractability, 463
reaction kinetics, 464
Low-molecular-weight product of reductive coupling, isolation, 103–104
Low-molecular-weight silicon catenates, electronic spectra, 46
Low-temperature coupling in the presence of ultrasound
chain-growth mechanism, 80
degradation of polysilanes, 81,84
dependence of yield, polymerization degree, and polydispersity, 81
effect on molecular weight, 81,83*f*
gel permeation chromatographic traces, 81,82*f*
limitation of chain growth, 80
monomodality of polysilanes, 80

Low-temperature reductive coupling in presence of ultrasound, preparation of polysilanes, 79
Lowest unoccupied molecular orbital, for polysilanes, 70,71*f*,72

M

Macromolecular substitution route for polyphosphazene synthesis
advantages, 254
effects of changes in side group structure, 254,256–257
illustration, 254–255
preparation of glucosylphosphazene polymer, 254,256–257
Macromolecules
critical properties, 252
definition, 1
development, 1
Mass spectrometry, segmented poly(ether urethanes), 446,448
$(Me_2Si)_{16}$, X-ray crystal structure, 11,13*f*
$Me_{10}Si_6O_7$ isomers, synthesis from $(EtO)_{10}Si_6O_7$, 244
Metallocene monomers, synthesis, 458–459
Metallophosphazenes, description, 261
Metallotetraphenylporphyrins
electrochemical polymerization, 409
structure, 408, 410*f*
Metals, relationship to inorganic polymers, 250,251*f*,252
Methyl-terminated poly(dimethylsiloxane) oligomers, preparation, 181–182
3-(Methyldichlorosilyl)pyridines
hydrolysis, 201–202
infrared spectrum, 201
mass spectrum, 201
synthesis, 201
Methylpoly(silylmethylenes), preparation, 22
Microcrystallite-based ultrastructure, comparison with covalently cross-linked systems, 262
Microlithography
problems, 57
single and multilayer process, 55,56*f*
Microscale electrochemical experiments, apparatus, 225,226*f*,227
Modification of polysilanes
^{1}H-NMR spectra, 84,85*f*
model reactions, 84
Molecular metals, structure, 225
Molecular-weight distributions, determination, 102–103

N

Network formers, definition, 334
Nicalon fibers, formation, 32
Nonoxide ceramic materials
classical preparation, 378
preparation via polymeric ceramic precursor, 378
Nonoxide ceramics
expense in preparation, 124–125
properties, 124
Nonpolymerizability, causes, 86
Nuclear magnetic resonance, collection of spectra, 485–486
Number-average molecular weights, determination, 184–185

O

Octaphenylcyclotetrasilane, polymerization, 86–88
Olefins, cohydrogenation, 93–94
Oligomeric organotin oxycarboxylates
ladder to drum conversion, 477,478*f*
phosphinic acid reactions, 477,479,480*f*
structural symmetry, 475,477
synthesis, 471–475
Oligosilazanes, structure from formation via modified ammonolysis, 127–128
Organic network formers
applications, 337–342
basic principles, 335,337
effect of reaction conditions on reaction sequence, 337,338*f*
synthesis of adhesive material, 337,338*f*,339
Organic polymers, relationship to inorganic polymers, 250,251*f*,252
Organofunctional cyclophosphazenes, precursors of inorganic–organic polymers, 291
Organometallic polymers
definition, 1
formation, 437
Organometallics, effect on reductive coupling of dichlorosilanes, 104,106,107*f*
Organosilicon polymers, lithographic applications, 221,222*f*
Organosilicon polymers bearing phenyldisilanyl units, photochemical behavior, 209
Organosiloxane copolymers, synthesis, 185
Organosiloxanes, synthesis by elemental-silicon process, 240,241*f*
Ormocer, definition, 334
Ormosils, definition, 334
Orthosilicate ion, structure, 238,239*f*
Oxidatively electropolymerizable tetraphenylporphyrins, structures, 409,410
Oxygen-capped cluster, schematic, 479,480*f*

P

Pentamethylphenyldisilane, products of irradiation, 210
Perfectly alternating siloxane poly(arylene ether sulfone) copolymers, preparation, 186,191
2-Phenyl-1,3,2-diazaboracyclohexane ring system
advantages to study, 387
cleavage reaction, 387–389
deprotonation–substitution, 387–388
transamination reaction, 387,389
2-Phenyl-1,3,2-di[bis(dimethylamino)-boryl]azaboracyclohexane, molecular structure, 389–390
Phenylaminoborazines, condensation mechanism, 395–396
N-Phenylsilazoxane linear polymers, thermostability, 173
Phosph(III)azane oligomers/polymers
properties, 305,307
structure, 305
Phosphinic acid reactions, formation of a drum composition, 477
Phosphorus–nitrogen skeletons with stabilizing units, structure, 304
Photochemistry of polysilanes
primary process, 73
secondary processes, 73–75
Photoconductors, polysilanes, 18
Photoelectrochemical properties of titanium dioxide films
apparatus, 350*f*
current–bias potential curves, 350*f*,351
photocurrent vs. heating time, 351*f*
porosity of film vs. heating temperature, 351,352*f*
preparation of gel films, 348,350
scanning electron micrograph, 351,352*f*
Photoinitiation by polysilanes
efficiency, 17
photopolymerization, 17*t*
$[(PhP)(C_6H_4N_2PPh)]_2$, structure, 305,306*f*
$[PhSn(O)O_2CC_6H_{11}]_6$, schematic of drum structure, 469,470*f*
Phthalocyanine molecular metals
apparatus for electrochemical experiments, 225,226*f*,227
effect of counterions on electrochemistry, 231,232*f*,233,234*f*
electrochemical structural relationships, 228,230*f*
microscale electrochemical voltage spectroscopy experiments, 227–228,229*f*
reductive doping, 233,235*f*
slurry electrochemical voltage spectroscopic experiments, 228,230*f*,231,232*f*
slurry-scale electrochemical doping, 227,229*f*
Phthalocyanine molecular metals—*Continued*
structural transformations accompanying electrochemical doping, 227,229*f*
Platinum-catalyzed polymerization, poly(dimethylsilylene), 28
P(III)–N oligomer–polymer interconversion, occurrence, 304
Poly(amic acid), conversion to fully imidized polyimide, 193
Polyamides, structures, 440–441
Poly(aryloxyphosphazene) elastomers
applications, 279–280
properties, 280–281*t*
Polybis(pyrrolyl)phosphazene
cross-linking, 300–301
electroxidation, 297
future prospects, 300,302
pentacoordinate transition state, 298–299
^{31}P-NMR spectrum of preparation reaction, 298,300*f*
properties, 298,300
thermogravimetric analysis, 300,301*f*
Polycarbonate copolymers
dynamic mechanical behavior, 186,188*f*
preparation, 186
stress–strain behavior, 186,187*f*
Poly(carbosilane)
definition, 21
preparation, 26,29,32*f*,145–146
structure, 21
variations, 32
Polycyclic carbosilanes, examples, 33,34–37*f*
Poly(cyclohexylmethyl-*co*-isopropylmethylsilane), 117,118*f*
Poly(cyclohexylmethyl-*co*-*n*-propylmethylsilane) copolymers, ^{13}C-NMR spectra, 117,118*f*
Poly(diarylsilane) homopolymer, preparation, 49,52
Poly(dichlorophosphazene)
characteristics, 277
effect of organic groups on end product, 278
polymerization process, 270,271*f*
preparation routes, 278,279*f*
solution polymerization with BCl_3, 270,271*f*
Poly[*p*-(diethyldimethyldisilanylene)-phenylene]
formation of C=Si intermediates, 216,218
^{1}H-NMR spectrum, 216
molecular weights of products vs. irradiation time, 216,217*f*
photochemistry of thin films, 220–221
preparation, 215
properties, 215
Poly(di-*n*-hexylsilane)
absorption maximum, 49,50*f*

Poly(di-*n*-hexylsilane)—*Continued*
absorption spectra, 46,48*f*,64,65*f*
CP/MAS ^{29}Si-NMR spectra, 46,49,51*f*
emission spectra, 62,63*f*
experimental procedures for photochemistry, 62
film diffraction pattern, 49,51*f*
IR and Raman spectroscopic studies, 46
phases, 47,48*t*
Raman spectra, 49,50*f*
thermochromic behavior, 47
Poly[*p*-(1,2-dimethyldiphenyldisilanylene)-phenylene]
formation of silaethene intermediates and hydrosilanes, 220
^{1}H-NMR spectrum, 216
lithographic applications of a double-layer resist system, 221,222*f*
molecular weights of products vs. irradiation time, 218,219*f*
photochemistry of thin films, 220–221
preparation, 215
properties, 215
Poly(dimethylsilane), properties, 112
Poly(dimethylsiloxane)
bulk microphase separation characteristics, 192
structure, 21
Poly(dimethylsiloxane) chain, structure, 180–181
Poly(dimethylsilylene)
applications, 90
platinum-catalyzed polymerization, 28
properties, 90
synthesis, 6–7
Poly(dimethylsilylmethylene), structure, 21
Poly(diorganophosphazenes), synthesis, 1
Poly(diorganosilane) copolymers
molecular-weight determination, 113
synthesis, 113
Polydiorganosilanes, synthesis, 2
Poly(di-*n*-pentylsilane)
absorption maximum, 49,50*f*
CP/MAS ^{29}Si-NMR spectra, 49,51*f*
film diffraction pattern, 49,51*f*
helical conformation, 49,53*f*
Raman spectra, 49,50*f*
Poly(*n*-dipentylsilane), ultraviolet spectrum vs. temperature, 11,13*f*
Poly(diphenylsilylene), preparation, 6
Poly[*p*-(disilanylene)phenylenes]
lithographic applications, 221,222*f*
synthesis, 39
Poly(fluoroalkoxyphosphazene) elastomers
applications, 279
preparation, 278
properties, 279,280*t*
Poly(imide–siloxane) copolymers, structure, 193,194*f*
Polymer–silylamide
applications, 153
properties, 152–153
pyrolysis products, 153
reactions, 153
synthesis, 152
Polymeric boron–nitrogen compounds, formation, 379–383
Polymeric films of metallotetraphenylporphyrins
characterization, 409,412–413
electrocatalysis of dioxygen reduction, 417–418
electron transport, 413–417
structure, 408
Polymerization by aminolysis, reaction, 127
Polymerization by ammonolysis
modified reaction, 127
reaction, 126
sensitivity to steric factors, 126
Polymerization mechanisms, elucidation, 2
Polymerization of *N*-arylcyclosilazoxanes
mechanism, 171–172
reaction, 170–171
Polymerization of primary organosilanes
degrees, 93
mechanism, 92
properties of polymers, 92
rates, 92
Polymerization of primary silanes
mechanisms, 95,98
products, 95,96*f*
pseudoequilibrium, 95
rate law, 95
routes for the propagation step, 95,97
termination reaction, 98–99
Polymerization of silanes, cation complexation, 44
Polymers, ceramic formation, 156–157
Polymers with triflate groups, properties, 86
Poly(methyl methacrylate) (MMA), ^{1}H-NMR spectra, 491,492*f*
Poly(methyl methacrylate–methyl methacrylate) (MMA–MMA)
^{13}C-NMR spectra, 491,492*f*,493,494–495*f*
composition and tacticity ratios, 493*t*
Poly(methyldisilylazane)
^{13}C-NMR spectrum, 158,159*f*
experimental procedures, 157–158
^{15}N-NMR chemical shifts, 163*t*,164
reaction scheme, 160,162–163*f*
^{29}Si-NMR shifts of low-molecular-weight species, 160,161*t*
^{29}Si-NMR spectrum, 158,159*f*
Poly(organophosphazenes)
classes, 257–265
effect of substituents, 272,273*f*,273*t*,274*f*
morphology, 272,273*t*,274
synthesis, 278,279*f*

Poly(organosilylenes), thermochromic behavior, 90
Poly(phenylmethyl-*co*-*n*-propylmethylsilane) copolymers, change in UV absorbance, 120,121*f*
Polyphosphazene chemistry, historical development, 252,253*f*,254
Polyphosphazenes
- applications, 278–281
- biocompatible–bioactive compounds, 259–261
- characteristics, 250
- characterization, 285–287
- commercial development, 277–278
- definition, 268
- derivatization reactions, 287–288
- diffraction patterns, 287
- electronic structure, 296
- historical perspective, 264*f*,265
- intrinsic viscosities, 285
- melt transition, 285
- NMR spectra, 285
- polymerization process, 270,271*f*,288
- synthesis, 254
- thermogravimetric analysis, 286–287
- two-step synthesis process, 268,269*f*
- various end products from single precursor, 278
- weight-average molecular weights, 285

Poly(phosphazenes)
- condensation polymerization, 284
- synthesis by ring-opening substitution method, 283

Polyphosphazenes with rigid, stackable side groups
- formation, 262
- liquid-crystalline polymers, 264–265
- polyphosphazene–phthalocyanine structures, 262,265
- tetracyanoquinodimethane–polyphosphazene systems, 262–264

Polyphosphazenes with transition metals in side groups, methods for linking metals, 261–263
Poly(*n*-propylmethyl-*co*-isopropylmethylsilane) copolymers, 114,116*f*,117,120*f*
Poly(*n*-propylmethylsilane)
- ^{13}C-NMR spectra, 117,119*f*
- FT–IR spectrum, 117,119*f*

Poly(silaacetylides), preparation, 38
Poly(silaarylene–siloxanes), synthesis, 38
Polysilane high polymers, photodegradation mechanism, 54–55
Polysilanes
- applications, 78
- bimodal molecular-weight distribution, 8–9,12*f*
- bleaching upon irradiation, 52,54,56*f*
- catalytic polymerization routes, 2

Polysilanes—*Continued*
- cross-linking, 10
- definition, 6,43
- discovery, 43–44
- electronic spectra, 11,12–13*f*,14
- high-resolution patterns, 57,58*f*
- microlithography, 55,56*f*,57
- modification, 84
- nature of chromophoric chain segments, 66,68–71
- photochemical quantum yields, 54
- photoconductors, 18
- photodegradation reactions, 9–10
- photoinitiators, 17*t*
- polarization of the emission, 64,66,67*f*
- precursors to silicon carbide, 14,16–17
- preparation, 79–88
- primary photochemical process, 73
- properties, 8*t*,61–62
- reactions, 9
- secondary photochemical processes, 73–75
- sensitivity toward ionizing radiation, 54
- ^{29}Si-NMR spectra, 14,15*f*
- submicron images, 57
- substituent effects, 46,48*f*
- synthesis, 7–8,44,45*t*
- technological applications, 14–18
- unique properties, 112–113
- variable-temperature studies, 46
- Wurtz condensation reaction, 78–79

Polysilanes with alkoxy groups, properties, 86
Polysilastyrene
- formation of silicon carbide, 17
- synthesis, 7

Polysilazanes
- acid-catalyzed ring-opening polymerization, 129
- catalytic polymerization routes, 2
- curing of fibers, 148
- general synthetic approaches, 125–126
- melt spinning, 147–148
- properties, 147,148
- structure, 129
- use in applications of preceramic polymers, 147

Polysilazanes with aromatic spacing groups
- structure, 175–176
- synthesis by polycondensation, 175–177

Polysilazoxanes, synthesis, 2
Poly(silmethylenes)
- formation, 26,28–29
- properties, 28–29
- structures, 29,30*f*

Polysiloxanes
- copolymerization reaction, 185–186
- structures, 185
- use as hydrophobic catalysts, 200

Polysiloxanes containing phenyldisilanyl units
irradiation, 211,213,215
photochemical behavior, 213,215
preparation, 210
UV spectra of thin liquid film, 213,214*f*
UV spectrum, 210
Poly[*p*-(tetramethyldisilanylene)phenylene], preparation, 215
Poly(tetramethylene oxide)-modified sol–gel glasses
effects
of aging, 372–375
of molecular weight, 359–364
of tetraethoxysilane content, 365–368
of titanium addition, 368–372
example procedures, 357–358
materials, 356
reaction scheme, 356–357
sample nomenclature, 358*t*
Poly(tri-*n*-butyltin methacrylate–methyl methacrylate) (TBTM–MMA)
^{13}C-NMR spectrum, 488,489*f*,490
determination
of compositional sequences, 493,496
of configurational sequences, 490–495
^{1}H-NMR spectra, 491,492*f*
^{119}Sn–^{13}C coupling constants, 488*t*
solvent effects in the ^{119}Sn-NMR spectra, 486,487*f*,490
Poly(trisbipyridine)–metal complexes
absence of nonelectroactive ions, 427–428
conduction as activated process, 427
conductivity vs. potential, 421,424,425*f*
cyclic voltammogram, 421,422–423*f*
film resistance vs. reciprocal temperature, 424,426*f*
optical properties, 421
properties vs. those of other electronically conducting polymers, 424,427
sandwich electrode configuration, 424
Polyureas, structures, 440–441
Polyurethanes
average structure, 446,447*f*
molecular-weight distribution, 446
synthesis, 444–445
Poly(4-vinylpyridine 1-oxide), use as nucleophilic catalysts, 200
Preceramic organosilicon polymers, synthesis, 143–154
Preceramic polymers
applications, 143–144
design, 144–145
effect of pyrolysis, 145
factors determining usefulness, 144
preparation of silicon nitride mixtures, 148
preparation using chlorosilane, 146
Preceramic polymers—*Continued*
problems with composition of ceramic product, 145
products, 146–147
synthesis by decoupling, 137–139
Precursors to silicon carbide, polysilanes, 14,16–17
Preparation of silica glass fibers
composition and properties of tetraethoxysilane solutions, 346*t*
viscosity vs. time, 346,347*f*
Primary organosilanes
mass spectra, 92–93,96*t*
polymerization, 92
Primary silanes, polymerization mechanism, 95–99
Properties, polysilanes, 8*t*
$(n\text{-PrO})_3\text{SiO}(n\text{-PrO})_2\text{SiOSi}(\text{O}n\text{-Pr})_3$, synthesis from $\text{Ca}_3\text{Si}_3\text{O}_9$, 243
Pyridine 1-oxide, use as nucleophilic catalysts, 200
Pyrolysis of monosilacyclobutanes, synthesis of monosilacyclobutanes, 26
Pyrrolylphosphazenes, preparation, 297

R

Radical disproportionation reaction, documentation, 74–75
Reaction kinetics of reductive coupling of dichlorosilanes, effect of sodium surface area, 106,108*f*,109*t*
Red shifts, origin for soluble poly(diarylsilanes), 52
Redistribution reactions, sequence, 132
Redox conduction, description, 414
Redox conductors, charge transport, 420
Reductive coupling of dichlorosilanes with sodium
addition of organometallics, 101
chain nature of reaction, 104*t*,105*f*
change in molecular weight of intermediate product, 103*t*
experimental procedures, 102–103
gel permeation chromatographic traces of complete products, 103,105*f*
low-molecular-weight product, 103–104
reaction kinetics, 106,108*f*,109*t*
reaction termination reagent, effect, 109–110
sodium surface area, effect, 106,108*f*
Refractories, applications, 392–393
Restructuring polymers
effect after gelation, 320
effects at short-length scales, 320,322
process, 319–320

Ring-opening polymerization
discussion, 86
1,1-disubstituted silacyclopentanes, 33,38
preparation of polysilanes, 79
Ring-opening polymerization of silazanes
catalysis by transition metals, 129–130
reaction, 128–129

S

Sandwich electrode
application of potentials, 415,416*f*
description, 414–415,416*f*,424
Science of solids, definition, 250,252
Scratch-resistant coating,
properties, 339,341*t*
Segmented poly(ether urethanes)
average structure, 446,447*f*
^{13}C-NMR spectrometry, 448,449*f*
mass spectrometry, 446,448
molecular-weight distribution, 446
synthesis, 444–446
Silazane systems
effect of substituents, 169
equilibrium, 169
thermodynamic stability, 169
Silazane-containing ring compounds,
synthesis of useful polymers, 170–177
Silazanes
catalytic dehydrocoupling, 131–132
deamination–condensation polymerization
reactions, 130–132
formation
via aminolysis, 125–128
via deamination–condensation
reaction, 126
via dehydrocoupling reaction, 125–126
via redistribution reaction, 126
problems in synthesis, 139–140
ring-opening polymerization, 128–129
transition metal catalyzed
dehydrocoupling, 134–137
Silazoxane linear polymers and copolymers
formation, 170–173
properties, 173
Silica glass fibers, preparation, 345–348
Silicate gels, Raman spectra, 320,321*f*
Silicate systems
availability of monomer, 318–319
hydrolysis, 318
restructuring, 318
Silicate-based substitution process
examples, 242–244
reactions, 240,242
utility, 247
Silicon carbide, conventional method of
preparation, 143
Silicon carbide fiber precursor,
preparation, 145
Silicon carbide fibers, process, 32–33
Silicon carbide precursor preparation
silylamide-catalyzed reactions, 151–152
starting materials, 150
yield, 150–151
Silicon catenates, dependence of electronic
properties on backbone conformation, 47
Silicon nitride
conventional method of preparation, 143
preparation, 149–150
Silicon-containing ceramics,
description, 143
Silicon–nitrogen bond, polarity, 169
Siloxane-modified polyimide copolymers
solubilities, 195*t*
upper glass transitions, 195,196*t*
Siloxane-modified polyimides
microphase separation, 193
thermal–mechanical properties, 193
Siloxanes, beneficial effects on polymeric
systems, 192
Siloxane–cyclodisilazane block copolymers
polycondensation reaction, 174
synthesis, 173–174
thermostability, 174–175
Silyl cleavage reaction of 2-phenyl-
1,3,2-diazaboracyclohexane derivatives
reaction, 388
yields, 388–389
N-Silyl-*P*-(trifluoroethoxy)phosphoranimine
reagents
formation of polyphosphazenes, 284–285
preparation, 284
Silylamines, formation of aminoboranes, 379
SiO_2 glasses
applications, 335
introduction of organic network
modifiers, 335,338*f*
$Si(OC_2H_5)_4$ solution
factors influencing drawability, 348
spinnability, 346,347*f*,348
Slurry-scale electrochemical experiments,
apparatus, 225,226*f*,227
^{119}Sn-NMR spectra, ^{119}Sn–^{13}C coupling
constants, 488*t*
solvent effect of
poly(TBTM–MMA), 486,487*f*,490
Sodium surface area, effect on reductive
coupling of dichlorosilanes, 106,108*f*,109*t*
Sol–gel process
advantages, 333
formation of pure inorganic
materials, 333–342
limitation, 355
silica glass fiber preparation, 346–348
Sol–gel process for making glasses and
ceramics
schematic, 314,315*f*
shrinkage vs. temperature, 314,316*f*
uses, 314,317

Sol–gel reactions
production of low-temperature glasses, 354
two-step network-forming polymerization process, 355
Sol–gel techniques
inorganic networks, 334–336
interpenetrating networks, 341
organic network formers, 335–341
Sol–gel-derived inorganic polymers, structural evolution, 318–329
Soluble high-molecular-weight polysilanes, preparation, 101
Soluble poly(diarylsilanes)
absorption spectra, 52,53*t*
origin of large spectral red shifts, 52
Soluble polysilanes
discovery, 7
polymerization process, 44,45*t*
yield of high polymer, 44,45*t*
Solution imidization
description, 193
solubilities, 195*t*
structural representation, 193,194*f*
Spinnability of $Si(OC_2H_5)_4$ solution, log $\overline{M}_n$ versus log [η] plots, 346,347*f*,348
Stöber process, description, 319–320
Styrylphosphazene, polymerization, 293
Syndiotactic, definition, 490

T

Tacticity, definition, 490
Tensile strength of silica glass fibers, vs. cross-sectional area, 348,349*f*
Tetra(*o*-aminophenyl)porphyrin, electrochemical polymerization, 409,411*f*
Tetraethoxysilane–poly(dimethylsilane) systems
calculated electron densities of components, 364*t*
dynamic mechanical spectra, 359,361,362*f*
effect of tetraethoxysilane content, 365–368
model, 361,363
Tetraethoxysilane–poly(tetramethylene oxide) systems
aging effects, 372–375
calculated electron densities of components, 364*t*
characterization methods, 358
dynamical mechanical spectra, 359,360*f*,361
example procedures, 357–358
materials, 356
mechanical properties, 359*t*
mechanical properties of titanium-containing systems, 368,369*t*
reaction scheme, 356,357
sample nomenclature, 358*t*
Tetraethoxysilane–poly(tetramethylene oxide) systems—*Continued*
SAX profiles, 361,362*f*,363
schematic model for structure, 364,366*f*
stress–strain behavior, 359,360*f*
tetraethoxysilane contents, effect, 365
titanium addition, effect, 368–372
1,1,3,3-Tetramethyl-1,3-disilacyclobutane, preparation, 23
Thermolytic polymerization of 1,3-disilacyclobutanes, reaction, 26
Thermoplastic elastomers, bulk microphase separation characteristics, 192
Titanium dioxide coating films
color, 348,349*f*
photoelectrochemical properties, 348,350–352
preparation, 348–352
Titanium-catalyzed polymerization of primary silanes, mechanisms, 98
Titanium-containing tetraethoxysilane–poly(tetramethylene oxide) systems
dynamic mechanical behavior, 370,371*f*
mechanical properties, 368–369*f*
SAX profiles, 370,371*f*
Transacylation
catalysis, 205–206,207*t*
phase-transfer process, 200
Transamination condensations
intermediates, 307,309
monomer–dimer equilibrium, 310
products, 309
reactions, 307
Transamination reaction of 1,3-diaminopropane, reaction, 389
Transition metal catalysts, initiation of ring-opening polymerization of 1,3-disilacyclobutanes, 27
Transition metal catalyzed dehydrocoupling polymerization reactions
mechanism, 134–136
reaction, 134
steric constraints, 136
Transmission electron micrograph, of boron nitride, 382*f*,383
β-Triamino-*N*-triphenylborazine, pyrolysis, 395
β-Triamino-*N*-tris(trimethylsilyl)borazine
composition of monomer and dimer, 396–398
ring-opening mechanism, 396
β-Trianilinoborazine, pyrolysis, 395
Tributyltin compounds, applications, 483
Tri-*n*-butyltin methacrylate, synthesis, 485
Tri-*n*-butyltin polymers, synthesis, 483–484
β-Trichloro-*N*-triphenylborazine, synthesis, 393
β-Trichloro-*N*-tris(trimethylsilyl)borazine
pyrolysis, 394–395
synthesis, 393–394

Trimethylsiloxyl end-capped *N*-oxidized poly(methyl-3-pyridinylsiloxane)
  catalyst for transacylation reactions, 205–206,207*t*
  ^{1}H-NMR spectrum, 202,203*f*
  IR spectrum, 202
  synthesis, 202
  thermal analysis, 202,204*f*,205
Trimethylsiloxyl end-capped poly(methyl-3-pyridinylsiloxane)
  ^{1}H-NMR spectrum, 202, 203*f*
  IR spectrum, 202
  synthesis, 202
  thermal analysis, 202,204*f*,205
Trimethylsilyl radicals, disproportionation, 74
Triphenyltin compounds, applications, 483
Tris{5,5′-bis[(3-acrylyl-1-propoxy)carbonyl]-2,2′-bipyridine}ruthenium(II), properties of polymer films, 420

U

Unsymmetrically substituted, atactic polysilanes, spectroscopic studies, 46
Uranyl dicarboxylate polymers, polymerization, 464
Uranyl polymers
  coordination of ligands, 467
  γ-ray sensitivity, 465,466*f*
  infrared and thermal studies, 467–468
Urea-linked copolymers
  chemistry, 186,191
  dynamic mechanical behavior, 186,190*f*
  properties, 186,191–192
  stress–strain behavior, 186,189*f*

V

(Vinyloxy)phosphazenes, synthesis of inorganic–organic polymers, 293–294
Vinylferrocene, destabilization, 457–458

W

Wurtz condensation reaction, synthesis of polysilanes, 78–79

X

X-ray powder diffraction analysis, of boron nitride, 382*f*,383
Xerogels
  description, 317
  formation of dense glasses and ceramics, 317

Z

Zirconium polymer, synthesis, 464–465
Zirconium Schiff base polymer, number-average molecular weights, 467

Production by Meg Marshall
Indexing by Deborah H. Steiner
Jacket design by Carla L. Clemens

Elements typeset by Hot Type Ltd., Washington, DC
Printed and bound by Maple Press, York, PA

Recent Books

Personal Computers for Scientists: A Byte at a Time
By Glenn I. Ouchi
276 pp; clothbound; ISBN 0-8412-1000-4

The ACS Style Guide: A Manual for Authors and Editors
Edited by Janet S. Dodd
264 pp; clothbound; ISBN 0-8412-0917-0

Silent Spring Revisited
Edited by Gino J. Marco, Robert M. Hollingworth, and William Durham
214 pp; clothbound; ISBN 0-8412-0980-4

Chemical Demonstrations: A Sourcebook for Teachers
By Lee R. Summerlin and James L. Ealy, Jr.
192 pp; spiral bound; ISBN 0-8412-0923-5

Phosphorus Chemistry in Everyday Living, Second Edition
By Arthur D. F. Toy and Edward N. Walsh
362 pp; clothbound; ISBN 0-8412-1002-0

Pharmacokinetics: Processes and Mathematics
By Peter G. Welling
ACS Monograph 185; 290 pp; ISBN 0-8412-0967-7

Synthesis and Chemistry of Agrochemicals
Edited by Don R. Baker, Joseph G. Fenyes, William K. Moberg, and Barrington Cross
ACS Symposium Series 355; 474 pp; 0-8412-1434-4

Nutritional Bioavailability of Manganese
Edited by Constance Kies
ACS Symposium Series 354; 155 pp; 0-8412-1433-6

Supercomputer Research in Chemistry and Chemical Engineering
Edited by Klavs F. Jensen and Donald G. Truhlar
ACS Symposium Series 353; 436 pp; 0-8412-1430-1

Sources and Fates of Aquatic Pollutants
Edited by Ronald A. Hites and S. J. Eisenreich
Advances in Chemistry Series 216; 558 pp; ISBN 0-8412-0983-9

Nucleophilicity
Edited by J. Milton Harris and Samuel P. McManus
Advances in Chemistry Series 215; 494 pp; ISBN 0-8412-0952-9